# PHYSICS
# and
# TECHNOLOGY
# of
# NUCLEAR MATERIALS

# PHYSICS
## and
# TECHNOLOGY
## of
# NUCLEAR MATERIALS

by

IOAN URSU

*Professor at the University of Bucharest, Romania*

PERGAMON PRESS

OXFORD · NEW YORK · TORONTO · SYDNEY · FRANKFURT

U.K. — Pergamon Press Ltd., Headington Hill Hall, Oxford OX3 0BW, England

U.S.A. — Pergamon Press Inc., Maxwell House, Fairview Park, Elmsford, New York 10523, U.S.A.

CANADA — Pergamon Press Canada Ltd., Suite 104, 150 Consumers Road, Willowdale, Ontario M2J 1P9, Canada

AUSTRALIA — Pergamon Press (Aust.) Pty. Ltd., P.O. Box 544, Potts Point, N.S.W. 2011, Australia

FEDERAL REPUBLIC OF GERMANY — Pergamon Press GmbH, Hammerweg 6, D-6242 Kronberg-Taunus, Federal Republic of Germany

---



First edition 1985

**Library of Congress Cataloging in Publication Data**

Ursu, Ioan.
Physics and technology of nuclear materials.
Includes bibliographies and index.
1. Nuclear reactors—Materials. I. Title.
TK9185.U772 1985 621.48′33 85–9570

**British Library Cataloguing in Publication Data**

Ursu, Ioan
Physics and technology of nuclear materials.
1. Nuclear engineering
I. Title. II. Fizica şi tehnologica materialelor nucleare. *English*
621.48 TR9145
ISBN 0-08-032601-3

An updated translation of *Fizica Şi Tehnologica Materialelor Nucleare,* published in 1982 by Editura Academiei Republicii Socialiste România

*Printed in Great Britain by A. Wheaton & Co. Ltd., Exeter*

# Preface

About thirteen percent of the world's total electric generating capacity is now provided for by nuclear means. This is much less than expected in the fifties and sixties when the enthusiast working in the field considered nuclear energy the ultimate solution for the world's energy needs. There are several reasons why the development anticipated did not take place: there were technical problems, the demand for electricity failed to rise to the levels predicted due to underlying economic difficulties in many countries, and the public at large - as well as the mass-media - in many cases came to view nuclear power generation as a dangerous practice.

The pendulum has swung again, however. Among other developments of potential importance for the future of nuclear power, the momentum now achieved by the environmentalist movement, supported by people equally as enthusiastic as those in the fifties, has recently forced the general public to consider the consequences of using conventional fossil energy sources. The implication is that a far more benevolent view of nuclear power generation as a means of eliminating one probable reason for the devastation of forests, namely the burning of coal and oil, is warranted. At the same time, it is also obvious more energy than is generated at present will be needed to improve living conditions for the developing countries and to meet the needs arising from the increase of the world's population from the present 4500 million people to an expected 6000 million people at the turn of the century.

It is, therefore, unfortunate that such an important issue as national energy policy has often been clouded by less than objective, or less than relevant, considerations.

I share with the author of "Physics and Technology of Nuclear Materials" the opinion that nuclear energy is capable of making a much larger contribution towards satisfying our future energy needs, and Professor Ursu should be recognized for having himself contributed substantially to the basis for the future development of this technology. Through years of research, university teaching and management of nuclear programmes in his country, as well as through his interaction with the international scientific community via co-operative research projects, congresses, conferences and symposia, Ioan Ursu has developed an extensive expertise in the physics and technology of nuclear materials. In the scientific world the author of this book is known as a specialist in the field of interaction of radiation with matter, and the principles, methods and tools he has applied have served to provide deeper insight into the properties of nuclear materials, particularly of nuclear fuels and moderators and of their behaviour under irradiation. Professor Ursu has also done extensive research and development work on isotopic separation and related industrial applications. By condensing his knowledge in the present volume, he is making available valuable information for the specialist and the advanced student alike. The materials presented should also be of importance in a wide range of training activities.

Professor Ursu is now head of the Department of Atomic and Nuclear Physics at the University of Bucharest. He is a senior researcher with the Romanian National Centre of Physics, and is involved in the organization and management of various research and development programmes in his country. He served for many years as Director General of the Institute of Atomic Physics in Bucharest and as President of the Romanian State Committee for Nuclear Energy. In this book he has succeeded in combining in a natural way his approach to topics as a scientist with his perception of issues as a senior manager of nuclear projects in a country that is committed to use nuclear power. Moreover, readers from those countries with market economies may find it interesting to look at several facets of nuclear technology from the intrinsically comparative perspective provided by an author whose presentation is duly reflective of the stand taken in respect of a centrally planned economy. On the other hand, owing to his participation in international scientific events and projects, as well as his years as member of the International Atomic Energy Agency's Board of Governors and as a member of the Agency's Scientific Advisory Committee, Professor Ursu has wide experience of nuclear science and technology in the western countries, which he has constantly used to facilitate international collaboration in this field. He has contributed as well to the activities of the European Physical Society since the inception of this organization, and has served as President of the Society.

The author has often emphasized that present-day technology in many cases is able to create materials with pre-determined properties. The importance of this is obvious, and I believe that Professor Ursu's book opens many doors to guide the reader to fascinating new areas of importance for the development of nuclear energy for peaceful purposes. I recommend this work to the English-reading public, and wish it the wide readership it certainly deserves.

*Sigvard Eklund*

*Director General Emeritus*
*International ATomic Energy Agency*

# Contents

# Introduction

The current perception on science and technology, that highlights the effectiveness of solutions, the ingenuity of models, the subtleness of mechanisms, may underestimate, on occasions, the capital part which materials play in either boosting or slowing down the progress in this field. Far from being only a trivial raw substrate, easy to find and harness, materials are the object of a wide scientific and technological effort, aiming at ever more cost-effective production, convenient processing, reliable testing, continued improvement and efficient utilization. Such concerns have set the premises for a wealth of specialist technologies in the modern industrial enterprise, and also for a genuine *science of materials* that is currently subject to a very fast development. It is worth noting that, from the dawn of civilization, ages are labelled according to the dominant material extracted and exploited by man: the *Stone* Age, the *Bronze* Age, the *Iron* Age.

The study of materials and of technologies for their production and processing is of great interest nowadays, since the economic and social progress depends more and more on advances in science and technology. Betting on a strategy of intensive rather than extensive growth, aiming at a maximum performance and productivity, high quality and economic efficiency, present-day industry demands that materials display increasingly sophisticated, and, more often than not, contradictory merits, and pushes its orders beyond any conventional limit; as a result, super-hard, super-refractory, chemically inert materials, resistant to high mechanical stress and fatigue, structurally stable in radiation fields, or exhibiting special electronic structures — super-conductors, semiconductors, or super-insulators, laser or photovoltaic active media with predetermined absorption and emission capacity, etc. are searched for, actually developed, and going commercial.

Present-day technology, well grown-up after a first industrial revolution that was essentially tributary to natural, or minimally and empirically processed materials, started *creating* materials, pre-intending them for highly specialized uses and, therefore, *designing* their composition, structure and properties from the very beginning. "Materials with designed, predetermined properties" are undoubtedly one of the most fascinating promises of the contemporary technical and scientific revolution. The interest in new materials covers both those employed in the "classical" sectors of industry and in the new ones,such as the one in which the nuclear sector is outstanding.

Production and utilization of materials requires a deep *basic knowledge of their properties*, implementation of *appropriate procedures of testing and overall behaviour investigation*, development of *techniques to analyse and evaluate consequences of possible material defects* which might impair the systems they belong to. The phenomenological knowledge of material properties is to be paralleled by a *deep understanding of their microscopic structures*, which brings about the need to develop and improve *sound theoretical models* of material structure and behaviour.

A point well taken is that physics, chemistry, the technical sciences *describe* substances, but *create* materials. Material is a substance *intended for a practical purpose*. From this standpoint, interweaving theory and practice, scientific creativity and experimental evidence much more strictly than in many other fields, the science and technology of materials is expected to provide ways and means of capital importance for a timely, successful solution to several major problems of our times, such as the security of raw material supplies, more labour productivity through wise automation, improved communications and informational environment, a better natural environment and, not the least, *energy*.

The diversity of materials with special properties which are to be designed, manufactured and marketed in order to meet the requirements of the new technologies for energy generation, conversion, storage, distribution and utilization is growing steadily. The generation of heat, electricity and synthetic fuels from low-density primary sources of energy, such as solar radiation, is met with numerous constraints on materials. At the other end, the harnessing of nuclear energy, a primary source of utterly high density, comes up with requirements that lean heavily upon materials, as far as the economic efficiency and societal impact of this technology are concerned — a technology whose responsibilities are paralleled in magnitude only by its immense potential.

The design, construction and operation of a nuclear power station, as well as of other facilities in the nuclear industry, has been and is still raising problems of an extreme complexity. Apart from a compound of severe conventional requirements, such as those related to thermal and mechanical stress, corrosion and so on, new and unusual non-conventional requirements occur, such as those related to the resistance to the effects of nuclear radiation, the interaction with neutrons, the generation of fission products, etc. Therefore, nuclear power has included in its technical and economic realm preoccupations for a large number of novel, or less common, materials, also evolving new standpoints in evaluating and utilizing already known materials. In this respect, one can speak about a new and important chapter in the science of materials — that of *nuclear materials*.

This book is devoted to the physics and technology of nuclear materials. Its twelve chapters encompass basic information concerning the structure, properties, processing methods and response to irradiation of the principal materials which fission and fusion nuclear reactors have to rely upon.

*Chapter One* introduces selectively several fundamentals of nuclear physics, particularly the nuclear fission on which the nuclear reactor operation is based. The general description of the design and operation of a nuclear reactor and the main types of reactors define the utilization range of nuclear materials and outline the functions and working conditions they have to cope with; accordingly, one speaks of such categories as fuels, structural-, moderating-, cooling-, shielding materials, etc.

*Chapter Two* presents to the reader the material's science in general, introducing the basic concepts and phenomena in connection with the microscopic structure, types of defects, relation of macroscopic properties to the microscopic ones. The effects on materials of irradiation is emphasized as a specific phenomenon and limitative factor of consequence in the case of nuclear materials.

In *Chapter Three*, devoted to nuclear fuel, the discussion is geared towards the operational categories of nuclear materials. The properties, production and processing methods, the behaviour under irradiation etc., of uranium,plutonium, thorium and of their compounds are presented. For uranium, a brief description of the technology of enrichment in the fissionable isotope $^{235}U$ — of a particular importance for certain types of reactors — is given.

The materials that, though in the reactor, do not intervene directly in the progress and control of the fission process — also called structural materials — are treated in *Chapter Four*. Relevant data are presented, regarding structure and properties, processing methods and behaviour under operational reactor conditions of the main structural materials, e.g. aluminium, zirconium (and alloys), stainless steel, etc.

*Chapter Five* describes the moderator materials employed to "slow down" fission neutrons and includes reference data for the main materials in use to this end: graphite, light water, heavy water, beryllium, metal hydrides.

Special attention is devoted to the preparation, by isotope separation, of heavy water — a material of chief interest for those types of reactors using natural uranium. This chapter is reflective of the R and D commitment in Romania, from basic research to the industrial level, to foster the knowledge and utilization of heavy water.

*Chapter Six* is devoted to neutron highly-absorbent materials that serve in reactor's power control. This chapter contains data of interest in applications for the major materials employed in the nuclear reactor control: boron carbide, hafnium, rare earths, etc.

Heat, as generated in the fission process,is evacuated by cooling agents — fluids carrying the energy to its final stage of conversion into electric power. These materials are discussed in *Chapter Seven*. Again, this deals with light and heavy water, equally used as moderators, but also with other gaseous and liquid materials, such as carbon dioxide, helium, organic liquids, liquid metals, molten salts, etc.

*Chapter Eight* deals with thermal and biological shielding materials employed as screens to attenuate radiation, to the effect of containing it satisfactorily within the reactor borders.

In *Chapter Nine* some outstanding reactor components, the fuel elements — delicate contraptions of fuel and structural materials — are presented, the properties of both merging in order to face the very demanding operational conditions of nuclear power generation. Design criteria, classification by constructive types and manufacturing processes in current use in different countries are reviewed; several of these are now fully developed in Romania.

After the end of combustion in the reactor, a variety of elements are cradled in the spent fuel, some of them fissile, others highly radioactive, of a shorter or longer life. Irradiated fuel reprocessing, with a view to recycling the fissile elements, is the subject matter of *Chapter Ten*. Reprocessing technology aims at closing the nuclear fuel cycle, promising major potential savings and an increased efficiency in the utilization of nuclear fuel resources.

*Chapter Eleven* deals with nuclear material quality inspection by destructive and non-destructive methods, meant to provide, as part of a strict technological discipline, for assuring and controlling so-called "nuclear quality". Another issue in this chapter is Safeguards, aiming at preventing diversion of the fissile materials from their peaceful, original destinations.

Finally, *Chapter Twelve* deals with specific materials envisaged for use in future thermonuclear reactors. Although such reactors are still in an early development stage, the research was carried far enough to enable an at least preliminary outline of the material requirements for such types of reactors. In this chapter, following an introduction of the fusion phenomenon and of the proposed developing principles for reactors, the fuel and structural materials, coolants and other materials are presented. Here again, owing to unprecedentedly harsh operating conditions, a great many problems arise, which result in bringing the material's science to the verge of a new development stage.

A number of appendices are included, for a better understanding of certain aspects discussed in the book.

The author thought it necessary to combine in this monograph basic data and phenomena with some insights into more strictly specialist issues. To the same end, each chapter is supported with references which, without claiming to be exhaustive, cover the main topical books, papers in journals and reports of the research institutes concerned with this subject.

*

This work is, to a large extent, based on lectures delivered at the Department of Physics of the University of Bucharest; it was further substantiated in seminars held at different Romanian research institutions, as well as in the framework of training programs in industry, and on the occasion of lectures held abroad as contributions to scientific gatherings in the nuclear field. To the final accomplishment I enjoyed the collaboration of my co-workers in the physics research and academic institutions. Acknowledging it, I take this way to thank them all.

To update the information, I used as pertinent sources the works of the Scientific Advisory Committee of the International Atomic Energy Agency in Vienna. The good advice of prominent officials of the Agency and fellow-members of the Committee, with whom I collaborated in evolving the Agency's programmes in the field of material science and technology for the current decade, is gratefully recognized.

The kind co-operation of many publishing houses and Romanian and foreign authors, who authorized me to enhance the documentation with appropriate numerical and graphical material, deserves a special mention. In the editorial process I enjoyed the untiring assistance of the Publishing House of the Romanian Academy. My appreciation is equally directed to Pergamon, for affording me the hospitality of its press in a spirit of ever improved communication worldwide.

The present monograph is mainly intended for the wide range of specialists — physicists, chemists, energy experts, engineers, students and doctoral candidates wishing to gear their research and development, educational, and other, activities towards the field of nuclear power and nuclear technology in general. I would be happy if this book could contribute its share in the major endeavour of better preparing professionals, as well as interested segments of the public at large, in view of fostering sound, peaceful, applications of nuclear energy.

*Ioan Ursu*

*Bucharest,*
*February 1985*

CHAPTER I

# Elements of Nuclear Reactor Physics

*Nuclear reactors* are systems that, in a controlled manner, transform nuclear into thermal energy. The present meaning of the term implies that nuclear energy in a reactor is released as a result of the process of *nuclear fission*. At the complete "burning" of one gram of $^{235}U$, transformed in the fission process, about 8.4 x $10^{10}$ J are released, i.e. the equivalent of 3 tons of conventional coal. There are now sufficient reasons to believe that in the first decades of the next century one may, also, talk about "nuclear fusion reactors". Drawing upon the nuclear properties of the elements at the beginning of the periodic table — over-abundant in the environment — these reactors might offer the ultimate solution to the world energy problem.

## 1.1. STRUCTURE OF ATOMIC NUCLEI. BINDING ENERGY

The energy released in a nuclear reactor originates in the transformation of atomic nuclei. The mechanism of nuclear energy release is connected, in a direct yet subtle way, to the structure of the nucleus.

The atomic nucleus is made of *nucleons* — protons and neutrons. These "fundamental"[1] particles may be studied as such, in the free state. Some of their properties are presented in Table 1.1 (by comparison with the similar properties of the electron).

In this table $u$ stands for the atomic mass unit (a.m.u.) and $e$ is the elementary electric charge (i.e. 1.609 x $10^{-19}$ C); the units of magnetic moment are $\mu_N$ – the nuclear magneton and $\mu_B$ – the Bohr's magneton.

Data in this table provide interesting suggestions that are actually developed in the framework of today's physics of subnuclear particles. Thus, the existence of a magnetic moment of the neutron should imply an internal structure of positive and negative charges, equal to each other in absolute value, but with uneven spatial distributions. The neutron mass, somewhat

---

[1]Although still taken as fundamental particles, it is currently considered that protons and neutrons have an internal structure implying subnucleonic particles.

TABLE 1.1 Essential Nucleon Characteristics, as Compared to the Electron's

| *Particle* | *Mass* | | *Charge* | *Spin* | *Type of statistics* | *Magnetic moment* | | *Decay mode* |
|---|---|---|---|---|---|---|---|---|
| | *kg* | *u* | | | | $\mu_N$ | $\mu_B$ | |
| Nucleons: | | | | | | | | |
| proton (p) | $1.67252 \times 10^{-27}$ | 1.007276 | +1 | 1/2 | Fermi-Dirac | 2.79276 | | stable* |
| neutron (n) | $1.67482 \times 10^{-27}$ | 1.008665 | 0 | 1/2 | Fermi-Dirac | -1.9131 | | $n \to p + e^- + \bar{\nu}$ $T_{1/2}$=12 min |
| Electron (e) | $9.1091 \times 10^{-31}$ | $5.48597 \times 10^{-4}$ | -1 | 1/2 | Fermi-Dirac | | 1.00115 | stable |

*Although the question of the proton stability is now under keen investigation, from the point of view of this work the proton is considered stable.

larger than the proton's, should be connected with its instability and its transformation through β-decay into a proton. The nucleus is indeed entitled to be considered the massive "core" of the atom, for the electron mass is 1836 times smaller than the proton's. Having a half-integer spin, all these particles are fermions, therefore obeying the Pauli exclusion principle.

As is well known, an atomic nucleus is characterized by the number of protons and neutrons in its structure. For a neutral atom, the number $Z$ of protons, also called *the atomic number*, equals the number of electrons. If $N$ is the number of neutrons, *the mass number A* is $N+Z$, and the actual mass of the nucleus, in a.m.u., is a decimal number very close to $A$. A nuclear species having $Z$ protons and $N$ neutrons is called a *nuclide*, and is written $^A_Z X$, where $X$ stands for the symbol of the chemical element.

The stability of the nucleus is a result of the equilibrium between the attractive and repulsive interactions among its constituents. Attraction between nuclear particles evolves from *strong interactions*. Among the four types of fundamental interactions known to exist in nature — strong, electro-magnetic, weak and gravitational — the strong interactions are characterized by the largest coupling constant of particles generically called *hadrons*. The attractive interactions counterbalance the repulsive electromagnetic forces between protons (the small dimensions of the nuclei $10^{-14}$ – $10^{-15}$ m, make these repulsive forces large) and provide for the particularly great cohesion of the nucleus. From this, one may infer that the intensity of these forces has to be very large. An equivalent way of describing strong interactions among nucleons is by using the concept of *nuclear forces*. The intensity of the nuclear forces is by two orders of magnitude larger than that of the electromagnetic forces.

Other features of the nuclear forces are: a short range of action (about $10^{-15}$ m); the property of saturation; charge independence and spin dependence, etc. With respect to the nuclear forces, protons and neutrons are indiscriminable; for that reason both are called — after Heisenberg — *nucleons*. To discriminate among nucleons, strong interaction physics has introduced a new quantum number — the *isospin*. For the nucleon family, the isospin is 1/2, the difference between the proton and the neutron being the isospin projection (+1/2 and -1/2, respectively).

Owing to the nuclear attractive forces, the breaking of a nucleus into its constituents requires an amount of energy equal to the work performed against the nuclear forces to bring the nucleons at an infinite distance to each other (free nucleons), starting from their nuclear distances. The energy, which is released in the process of building a nucleus from free nucleons, or, alternatively, is absorbed in its breaking into free constituents, is called *binding energy*. It is correlated with the difference $\Delta M$ between the total mass of free nucleons and the mass of the resulting nuclide. Experimentally, it is verified that $\Delta M$ is always positive, because an established nucleus is more stable, and hence it has a lower energy than the sum of its constituents. The quantity $\Delta M$, known as *mass defect*, is given by

$$\Delta M = ZM_p + (A-Z)M_n - M(Z,A) \tag{1.1}$$

and corresponds, on the grounds of Einstein's relation between mass and energy, to the *binding energy*

$$W(A,Z) = c^2\Delta M = c^2[ZM_p + (A-Z)M_n - M(Z,A)], \tag{1.2}$$

where $M_p$ and $M_n$ are the masses of the proton and the neutron, respectively.

Atomic masses may replace in (1.2) the nuclear masses. This substitution entails neglecting the mass defect associated to the binding energy of the electrons. Such an omission is justified by its very small effect (for $^{235}U$, $W$ = 1783.17 MeV, and the atomic binding energy is only 0.69 MeV, i.e. 0.038% of the nuclear binding energy).

Expressing $\Delta M$ in a.m.u., the binding energy in MeV is calculated as:

$$W = 931.48 \cdot \Delta M. \tag{1.2'}$$

The *specific binding energy* is defined as the average binding energy per nucleon

$$\varepsilon(A,Z) = \frac{W(A,Z)}{A} \tag{1.3}$$

In Figure 1.1, the variation of $\varepsilon$ is plotted versus the mass number $A$. A maximum for the nuclides of medium mass number is apparent [1]. It follows that the nuclei of the isotopes in the mid-range of the periodic table, at about $20 < A < 180$ are the most stable. Light nuclei ($A < 20$), as well as the heavy ones ($A > 180$) are less stable, and there is a possibility that, through their transmutation into more stable nuclei, the difference in binding energy should become available as nuclear energy. For instance, if an uranium nucleus ($A$ = 238, $Z$ = 92, and $\varepsilon$ = +7.6 MeV) breaks up into two nearly equal fragments (with $\varepsilon$ = +8.5 MeV), the energy released is, accordingly, 238(8.5-7.6) MeV, or about 200 MeV.

*Nuclear fission* is the process of nuclear transformation through which a nucleus breaks into two (rarely more) nuclei of comparable mass. For heavy nuclei, fission is exoenergetic. The reverse process, of synthesis of a nucleus from two nuclei of smaller mass, is the *nuclear fusion*. For light nuclei, fusion is also exoenergetic [1-6].

The total binding energy for a nucleus with a mass number $A$ has been computed by Weizsäcker [7], who gave a half-empirical equation, that has, since then,

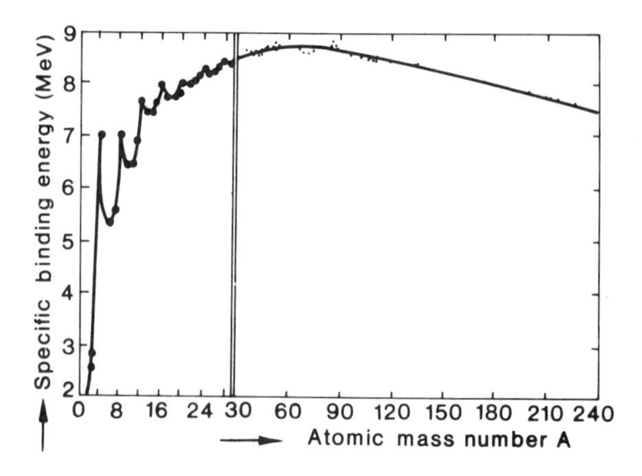


Fig. 1.1. Specific binding energy vs. the atomic mass number.

been amended and made more exact. This equation reads:

$$W(Z,A) = c_v A - c_c \frac{Z(Z-1)}{A^{1/3}} - c_s A^{2/3} - c_a \frac{(A-2Z)^2}{A} \pm c_p A^{-3/4} + S(Z,N). \quad (1.4)$$

The first term in this formula represents the binding energy of the $A$ nucleons, considered as embedded in a limitless nuclear matter. Its substantiation is based on the saturation of nuclear forces, here $c_v$ being the binding energy of a nucleon interacting with only a finite number of the other nucleons. The second term stands for the potential energy corresponding to electrostatic repulsive forces among protons. For a spherical nucleus having an uniformly distributed protonic charge $c_c = (3/5)\cdot e^2/(4\pi\varepsilon_0 R_0)$, where $\varepsilon_0$ is the vacuum dielectric constant, and $R = R_0 A^{1/3}$ is the nuclear radius, with constant $R_0 = 1.4 \times 10^{-15}$ m.

The following term takes into account the excess energy of the nucleons from the surface layer. Considering, by analogy with the liquid state, a surface tension $\sigma$, one obtains $c_s = 4\pi R_0^2 \sigma$.

The fourth term is due to the difference between the number of protons and that of neutrons. The lowest energy state, as given by the Pauli principle, corresponds to $Z = N$, but the compensation of the Coulomb repulsion, increasing with $Z^2$, calls for a faster increase of the neutron number $N$, compared to $Z$.

The term $\pm c_p A^{-3/4}$ stands only with the nuclei of even $A$, the + sign being associated to even — even nuclides, and the - sign with the odd — odd ones. It is due to the pairing energy. For odd-$A$ nuclei this term vanishes.

The last term in (1.4) is a correction introduced by the so-called "shell model" of the nucleus, being significant for nuclei having magic numbers of protons and neutrons ($Z$; $N$ = 2, 8, 20, 28, 50, 82, 126).

The notion of binding energy allows the understanding of nuclear stability. An assessment of the energy released in nuclear transformations can also be easily made from binding energy data.

For a family of isobars (different nuclei having the same mass number $A$), a certain value of $Z/A$ exists which maximizes the binding energy $W(Z,A)$. Stable nuclei are situated around this maximum (one stable isobar for $A$-odd and two stable isobars for $A$-even).

Plotting nuclear masses versus $Z$ and $N$ into a three-dimensional diagram, one obtains a Segrè "map" which shows a minimum for the stable nuclei domain (Fig. 1.2).

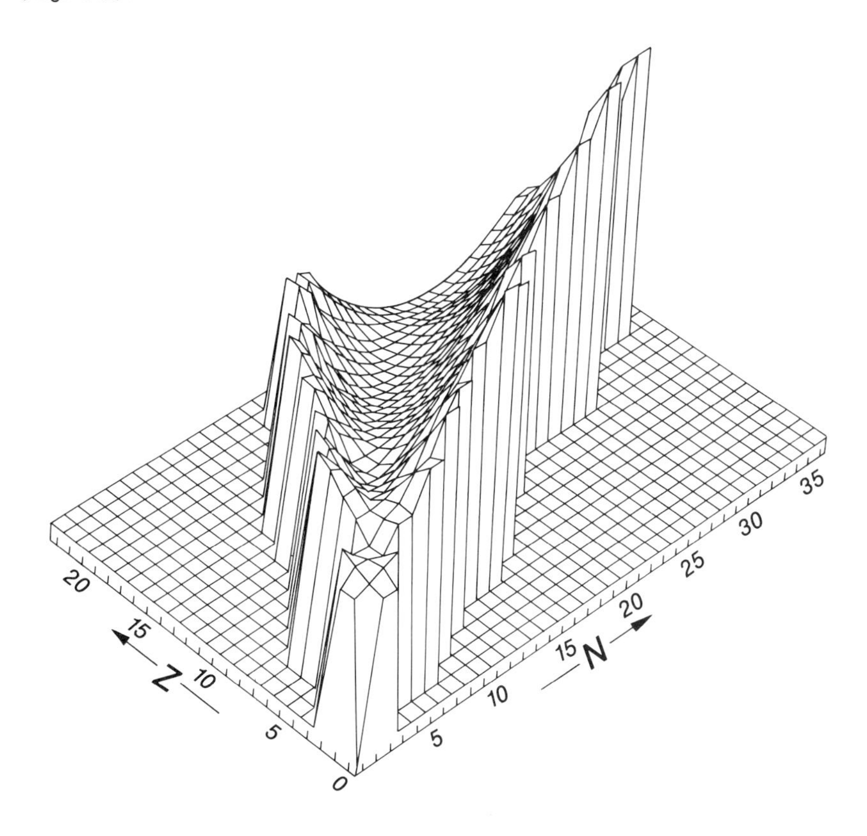


Fig. 1.2. The Segrè's chart, in the region of the light elements. The stablest nuclides are on the bottom of the valley (computer-generated picture) [8].

## 1.2. NUCLEAR TRANSFORMATIONS

Nuclear transformations are either spontaneous (radioactive decays) or due to the nuclei interacting with incident particles (nuclear reactions).

A *radioactive decay* of the nuclide ${}^{A}_{Z}X$ into the resulting nuclide ${}^{A_1}_{Z_1}Y$ can be written as

$$ {}^{A}_{Z}X \rightarrow {}^{A_1}_{Z_1}Y + {}^{A_2}_{Z_2}a. $$

The symbol "$a$" usually stands for an α-particle (α-decay), a photon or an internal conversion electron (nuclear de-excitations), or a pair β-particle + neutrino (β-decay).

In order for the transformation to be spontaneous, the decay process must release energy, i.e.

$$W_d = [M_X - M_Y - M_a]c^2 > 0. \tag{1.5}$$

This quantity is called *decay energy* and equals the share of rest energy transformed into kinetic energy of the decay products. In the laboratory frame of reference, it is written

$$W_d = T_Y + T_a - T_X . \tag{1.6}$$

Using experimental values for masses, the limit of stability for every transformation of a nucleus may thus be determined. The equations (1.5) and (1.6) are obvious consequences of the law of energy conservation. Other conservation laws are also verified in radioactive transformations, namely those for momentum, angular momentum, and, with limited validity — for parity, isospin, etc.

From a quantum mechanical standpoint, a decay can be viewed as a transition from an initial to a final state due to a perturbation. The quantity describing the transformation is the *transition probability* per unit time, $\lambda$, directly related to the *average life-time* in the unstable state, $\tau$, the *half-life*, $T_{1/2}$, and the *natural energy width* of the unstable state $\Gamma$:

$$\lambda = \frac{1}{\tau} = \frac{\ln 2}{T_{1/2}} = \hbar^{-1}\,\Gamma, \tag{1.7}$$

where $\hbar = h/2\pi$ is Planck's constant.

If there are several independent decay modes, the total probability is the sum of probabilities over all alternative processes

$$\lambda = \sum_k \lambda_k = \hbar^{-1} \sum_k \Gamma_k \tag{1.8}$$

A *nuclear reaction* is a transformation induced by the impact of colliding particles on a target nucleus

$$X + a \rightarrow Y + b.$$

The above mentioned conservation laws are also valid here, leading to very important consequences — reaction energies, selection rules, etc.

Considering each reaction process as the result of an interaction between an incident particle and a target nucleus, one may express the total number $N$ of reactions by

$$N = \sigma\Phi nVt, \tag{1.9}$$

where $\Phi$ is the flux of incident particles, $n$ — the density of nuclei in the target (i.e. the number of nuclei in the unit-volume), $V$ — the volume of the target, and $t$ — the time interval during which the reactions take place. The quantity $\sigma$ is called the reaction *cross-section*, expressed in $m^2$, or, as the nuclear physicists sometimes prefer, in barns ($1\ b = 10^{-28}\ m^2$). It depends on the type and energy of the incident particle, and of the type of the target nucleus.

The quantity

$$\Sigma = \sigma n = \sigma \rho N_0 / A \tag{1.10}$$

is called the *macroscopic cross-section* (in $m^{-1}$). Here $\rho$ is the density of the target material, $N_0$ — the Avogadro number, and $A$ — the atomic mass.

If the interactions of the incident particles with the target material take place through individual particle — target interactions, in the outgoing beam an attenuation is manifest, whose magnitude is directly connected with the macroscopic cross-section. In the occurrence of only one reaction type, the attenuation is exponential

$$\Phi = \Phi_0 \exp(-\mu x), \tag{1.11}$$

$x$ being the target thickness and

$$\mu = \Sigma = \sigma n. \tag{1.12}$$

If the target consists of several nuclear species, the total macroscopic cross-section is the sum of contributions from all the constituents; if several different processes are possible, the total cross-section, $\sigma_t$, is the sum of all contributions from the alternative processes

$$\sigma_t = \sum_k \sigma_k \tag{1.13}$$

A cross-section is labelled after the reaction it refers to, viz.: absorption cross-section ($\sigma_a$); scattering cross-section ($\sigma_s$); fission cross-section ($\sigma_f$); capture cross-section ($\sigma_c$) etc.

## 1.3. NUCLEAR FISSION

Nuclear fission was discovered in 1939 by Hahn and Strassmann. Following irradiation of a uranium sample with neutrons, they found in the material medium-weight atoms such as Ba, La, etc., which could only originate in the fragmentation of the U nucleus into two other nuclei of comparable mass — in other words, in a neutron-induced fission of the uranium nuclei. Subsequently, Flerov and Petrjak brought evidence also on a spontaneous fission of heavy nuclei. As it was learned later, the fission may be induced by other particles as well: protons, deuterons, $\alpha$-particles, $\gamma$-photons (photofission) [21].

The first model of particle-induced fission has been developed by Bohr and Wheeler [9], starting from the so-called liquid drop model of the nucleus. Consistent with this model would be that the incident particle and the target nucleus merge, after collision, into a compound nucleus in an excited state, the excess energy being derived on the account of the kinetic energy of the incident particle and of the alteration of the binding energy related to the formation of the compound nucleus itself. The excitation energy induces a deformation of the nucleus, from a spherical to an ellipsoidal shape.

A de-excitation by ejection of a gamma-photon may now be in order, in which case the nuclear reaction is only a *radiative capture*. However, if certain conditions — to be discussed later — are fulfilled, the reaction evolves, with a significant probability, towards a *fission*. In such a case the ellipsoidal compound nucleus shrinks across its central part, so that two centres of mass are created. If the resulting deformation exceeds a certain critical threshold, the Coulomb forces become prevalent and the overall process evolves towards a separation of the nucleus into distinct fragments (Fig. 1.3).

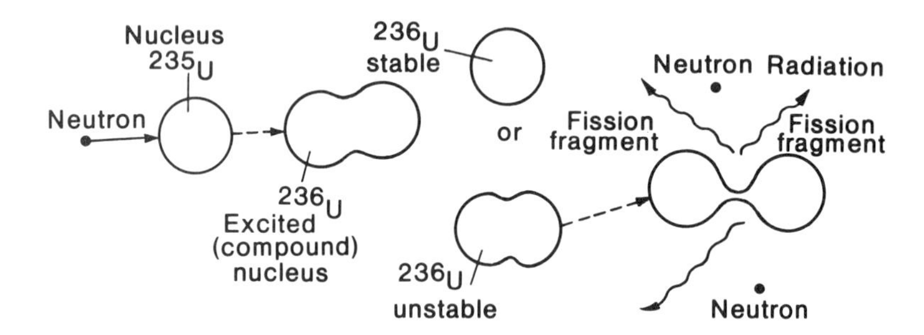


Fig. 1.3. Schematic representation of the fission of a $^{235}$U nucleus

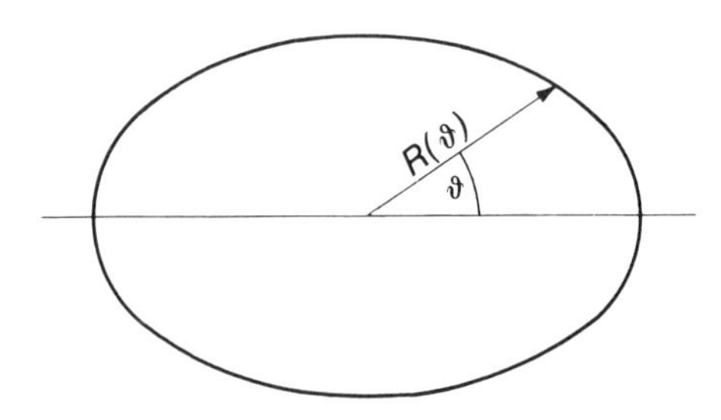


Fig. 1.4. Coordinates to describe the surface of a nucleus.

The evolution of a deformed compound nucleus is determined, in a first approximation, by the Coulomb and superficial energies described by the corresponding terms in eq. (1.4). When the nucleus becomes ellipsoidal, the Coulomb energy decreases while the superficial one increases. The evaluation of the internal energy of the compound nucleus is of utmost importance. Currently it is done on the assumption that the cylindrical symmetry is preserved during deformation. The parametric equation, $R = R(\vartheta, t)$, that describes, at a time $t$, the contour of a section through the nucleus that contains the rotation axis (Fig. 1.4) is

$$R(\vartheta,t) = R_o\lambda_o^{-1}\,[1 + \alpha_2(t)P_2(\cos\vartheta) + \sum_{n=3}^{\infty}\alpha_n(t)P_n(\cos\vartheta)], \tag{1.14}$$

where $R_0$ is the radius of the spherical nucleus, $P_n(\cos\vartheta)$ - the $n$th-order Legendre polynomial, and $\alpha_n(t)$ are time-dependent coefficients describing the dynamics of the deformation (in the sequel we shall ignore this particular dependence).

$$\lambda_o = 1 + \frac{1}{5}\alpha_2^2 + \dots ,$$

is a normalization factor, so that the volume of the incompressible nuclear fluid be kept constant.

With these accessories, one may compute both the Coulomb and the superficial energy of the deformed nucleus:

$$E_c = E_c^o(1 - \frac{1}{5}\alpha_2^2 + \dots), \qquad (1.15)$$
$$E_s = E_s^o(1 + \frac{2}{5}\alpha_2^2 + \dots),$$

where $E_c^o$, $E_s^o$ are the values for the spherical nucleus, before the beginning of the deformation.

The change in the internal energy during deformation is

$$\Delta W = V = E_c + E_s - (E_c^o + E_s^o) \qquad (1.16)$$

or, using (1.15),

$$V = \frac{2}{5} E_s^o \, [\alpha_2^2 \, (1 - x) + x \, f(\alpha_3,\alpha_4,\dots)], \qquad (1.17)$$

where

$$x = E_c^o/(2E_s^o) \qquad (1.18)$$

is known as the *fission parameter* ; $f(\alpha_3, \alpha_4,\dots)$ is a complex function of $\alpha_3$, $\alpha_4$,... In a first approximation, one may assess that the leading term in (1.17) is the first, being determined by the coefficient $\alpha_2$.

Bringing in the values known from eq. (1.4), one obtains

$$x = \frac{Z^2/A}{(Z^2/A)_{crit}} = \frac{1}{47.8} \cdot \frac{Z^2}{A}. \qquad (1.19)$$

The change in the system's energy depends on the ratio $Z^2/A$. If $x < 1$, i.e. ($Z^2/A < 47.8$), the nuclear "droplet" is stable against deformation. If $x > 1$, then $V < 0$ and the nucleus is unstable against deformation that progresses until the complete splitting of the nucleus. However, the condition $Z^2/A > 47.8$ is never fulfilled by natural nuclei; it is fulfilled only by those with $Z > 114$.

So, generally, if the induced deformation is inferior to a certain threshold, nuclei do not split. Fission takes place only when the deformation grows larger than a critical level. The sequence of nucleus deformations from the origin to the fission zone goes through those states which bring about the smallest increment in the system's energy. Figure 1.5 is a diagram in $\alpha_2 - \alpha_4$ co-ordinates, where the values of the deformation energies are plotted; the map is computed for $x = 0.76$ — the specific value for uranium. The point with the lowest energy between the origin and the fission zone is the *saddle-point*. The path that runs through this point is the easiest way to fission.

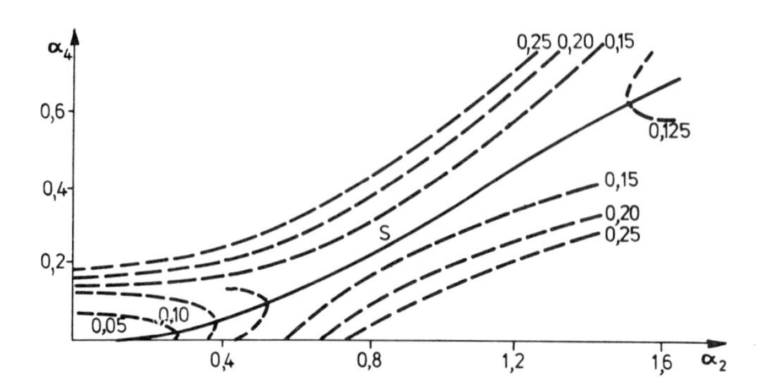


Fig. 1.5. Chart of the deformation energy in the plane $\alpha_2 - \alpha_4$, for a nucleus with $x = 0.76$.

The excitation energy of the nuclei with $x < 1$ must exceed a critical value, called *activation energy*, in order to drive them to fission. The excitation energy of a compound nucleus following a neutron capture is

$$E_{\text{excit}} = T'_{\text{n}} + S_{\text{n}} , \tag{1.20}$$

where $T'_{\text{n}}$ is the kinetic energy of the neutron in the center-of-mass system, and $S_{\text{n}}$ is the energy of separation of the neutron from the compound nucleus.

A simplified picture of the variation of the system's energy versus the distance between fragments is given in Fig. 1.6, where the activation energy $E_A$ is the height of the potential barrier.

Frenkel [10] has determined the dependence on $A$ of the activation energy, at different values of $Z$ (Fig. 1.7).

The height of the barrier was computed as a function of the fission parameter, by means of the equations [11]:

$$\begin{aligned} E_A &= 0.38\ E_s^{\text{o}}(0.75 - x), \text{ for } 1/3 < x < 2/3; \\ E_A &= 0.83\ E_s^{\text{o}}(1 - x), \text{ for } 2/3 < x < 1; \end{aligned} \tag{1.21}$$

where $E_s^{\text{o}} = c_s(1-K_sI^2)A^{2/3}$, with $I = \frac{N - Z}{A}$, $c_s$ and $K_s$ being the characteristic constants of the liquid drop model. In the case of uranium, $c_s = 17.9439$ MeV and $K_s = 1.7826$.

It is important to compare the energy of separation of the neutron from the compound nucleus with the activation energy. For the even — even nuclei, the

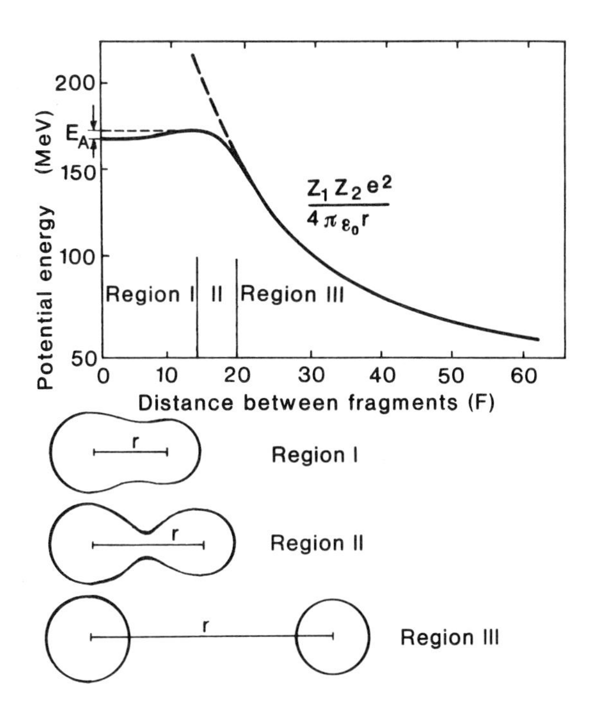


Fig. 1.6. Total potential energy (in Fermi units) vs. the distance between the centres of the fission fragments ( $Z_1$ = 38; $Z_2$ = 54 ) [2].

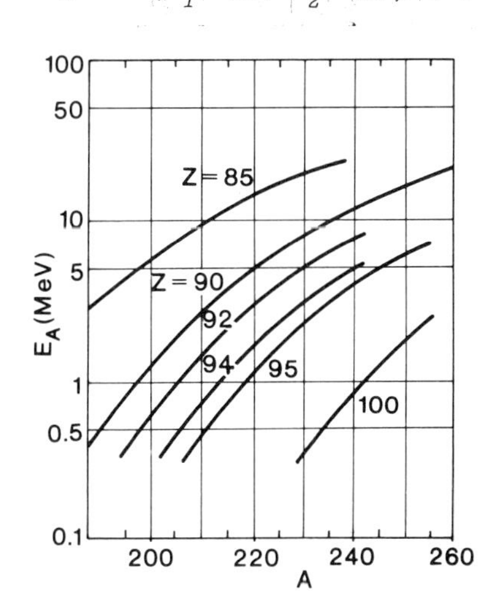


Fig. 1.7. Fission activation energy vs. atomic mass (190 ≤ $A$ ≤ 260).

extra-neutron leads to an even-odd configuration, for which $S_n$ decreases. In this case the last but one term of eq. (1.4) is missing; a high kinetic energy is then required for the incident neutron, in order to initiate fission. Such a situation occurs for the $^{238}U$ nucleus, where fission takes place with fast neutrons ($T'_n > 1.2$ MeV).

With an even — odd nucleus such as $^{235}U$, an incoming extra-neutron gives an even — even compound nucleus $^{236}U$. For the latter, $S_n$ is large (over 6.4 MeV), somewhat larger than $E_A$, so that the energy of the incident neutron can be very small. Accordingly, $^{235}U$ undergoes fission with slow neutrons. When the neutrons are in thermodynamic equilibrium with the surrounding medium, they are called *thermal neutrons*. At 293 K, a thermal neutron has a mean kinetic energy of 0.0253 eV and a most probable speed of 2200 m/s.

Uranium 235 is the only natural isotope that is fissionable with thermal neutrons; its abundance in natural uranium is 0.7113% in mass, which corresponds to an atomic abundance of 0.7204%. Other isotopes that are fissionable with thermal neutrons, and are used in nuclear power generation, are products of nuclear reactions; these are $^{239}Pu$ and $^{233}U$ — also even — odd nuclei. They are generated by neutron irradiation of natural isotopes. For instance, $^{239}Pu$ is obtained through the reaction:

$$^{238}_{92}U + ^{1}_{0}n \rightarrow ^{239}_{92}U + \gamma \xrightarrow{23.5\ \text{min}} ^{239}_{93}Np + \beta^- \xrightarrow{2.3\ \text{days}} ^{239}_{94}Pu + \beta^-$$

which implies, as seen above, two successive decays with small half-life, eventually leading to $^{239}Pu$. Plutonium 239 is long lived, having a half-life of 2.4 x $10^4$ years. The other isotope of interest, $^{233}_{92}U$, is obtained from thorium through the reaction:

$$^{232}_{90}Th + ^{1}_{0}n \rightarrow ^{233}_{90}Th + \gamma \xrightarrow{22.2\ \text{min}} ^{233}_{91}Pa + \beta^- \xrightarrow{27.4\ \text{days}} ^{233}_{92}U + \beta^-$$

Uranium 233 has a half-life of 1.6 x $10^5$ years. The three isotopes that can undergo fission when exposed to thermal neutrons, $^{235}U$, $^{239}Pu$, and $^{233}U$ are called *fissionable* or *fissile materials*, whereas the others, $^{238}U$ and $^{232}Th$ — the "raw material" for obtaining the last two fissionable isotopes — are known as *fertile materials*. Together, fissile and fertile materials make up the *nuclear fuel*.

Among fissile materials, the isotope $^{241}Pu$ is also important; it is obtained through successive capture of two neutrons by $^{239}Pu$, and is generated in nuclear reactors using plutonium fuel.

Nuclei having an even mass number, like $^{238}U$, $^{240}Pu$, as well as $^{232}Th$ and $^{234}U$ undergo fission when exposed to neutrons with energies over 1 MeV.

A significant step forward in the description and understanding of the fission process has been made with the development of a more adequate nuclear model, the *shell model*, which takes into account the distribution of nucleons into configurations called nuclear shells [12]. A thorough discussion of theories concerning nuclear fission can be found in Ref. [13], the matter falling beyond the scope of the present work.

Examining the dependence of the fission cross-section $\sigma_f$ on the energy of the incident neutron makes obvious the importance of neutron slow-down. For $^{235}U$, the dependence is plotted in Fig. 1.8 [13]. One can notice the fast increase of $\sigma_f$ at low neutron energies. In that range, the cross-section is inversely proportional to the neutron speed (the "$1/v$ law").

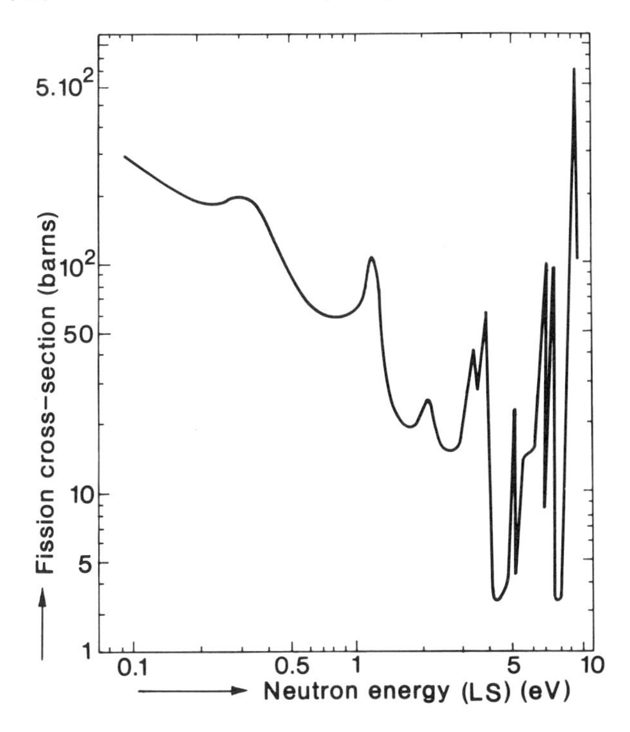


Fig. 1.8. $^{235}U$ fission cross-section vs. the energy of the incident neutron, in the laboratory system (L.S.).

Results of a fission reaction induced by either thermal or fast neutrons are two medium-mass nuclei — the *fission fragments* — as well as 2 or 3 neutrons. The splitting of a nucleus into two fission fragments can take place in a large variety of modes, all subject to the condition that the sum of the mass numbers of the two fragments comply with the nucleon number conservation law. The experimental curves, plotted in Fig. 1.9, show the percentage yield of given nuclei produced as fission fragments, versus their mass number, at the fission of $^{235}U$ and $^{239}Pu$ with thermal neutrons. Obviously, the probability of a symmetric split, i.e. into two equal fragments, is quite low; on the contrary, the emergence of fragments having a mass ratio of about 2/3, has a maximum probability. Significant yields are recorded for the fragments with $72 < A < 162$.

When $^{235}U$ splits up, the most probable fission fragments have the atomic numbers $Z = 38$ (strontium) and $Z = 54$ (xenon). One of the fission reactions of a relatively high probability goes as follows:

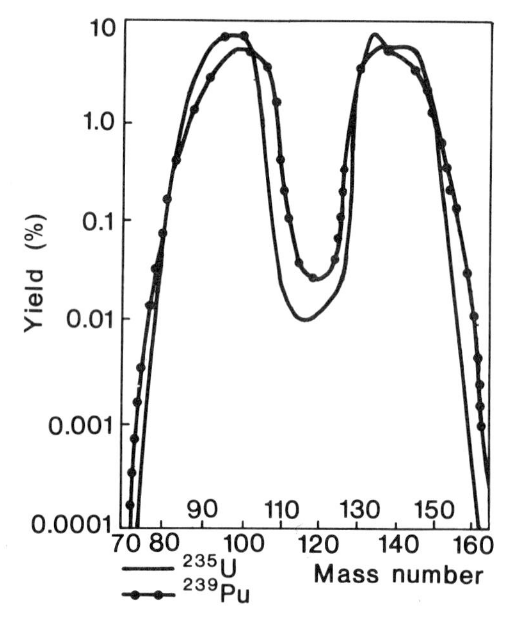


Fig. 1.9. Fission products yield.

$$^{235}_{92}U + ^{1}_{0}n \rightarrow ^{94}_{38}Sr + ^{140}_{54}Xe + 2^{1}_{0}n.$$

Possessing excess neutrons, the fission fragments are not stable. They undergo β-decays and γ-decays, turning themselves into stable nuclei only after 3 to 5 decays, on the average. For example, both fragments yielded by the reaction displayed above are radioactive (the stable isotopes of the corresponding elements being strontium 88 and xenon 136). One can see, then, that the neutrons in excess over the stable configuration is leading to a chain of β-decays that are genetically related:

$$^{94}_{38}Sr \xrightarrow{\beta^-} ^{94}_{39}Y \xrightarrow{\beta^-} ^{94}_{40}Zr \text{ (stable)},$$

or

$$^{140}_{54}Xe \xrightarrow{\beta^-} ^{140}_{55}Cs \xrightarrow{\beta^-} ^{140}_{56}Ba \xrightarrow{\beta^-} ^{140}_{57}La \xrightarrow{\beta^-} ^{140}_{58}Ce \text{ (stable)}.$$

All these processes contribute their share in the balance of energy released through fission. For the fission of $^{235}U$, this balance looks as follows:

| | |
|---|---|
| (a) kinetic energy of fission fragments | 165 MeV |
| (b) kinetic energy of neutrons | 5 MeV |
| (c) energy of prompt γ radiation | 7 MeV |
| (d) energy of β and γ radiations generated by the decay of fission products | 13 MeV |
| (e) energy of anti-neutrinos generated by the β-decay of fission products | 10 MeV |
| Sum total: | 200 MeV |

The largest share of the energy released through fission is transformed into internal energy (in the thermodynamic sense) of the surrounding medium, wherefrom it can be extracted as heat.

The neutrons are ejected either simultaneously with the fission process (*prompt* neutrons), or with a certain retardation (*delayed* neutrons). Delayed neutrons represent a small fraction of all the neutrons generated by fission, i.e. about seven such neutrons for every thousand fission events, but the part they play in the control of the fission reaction in nuclear reactors is of considerable consequence (see ch. 6). They are emitted by certain highly excited nuclei in the genetically connected decay chains of some fission fragments. The energy of the fission-generated neutrons is distributed over a continuous spectrum, similar to that of the kinetic energy of molecules in a gas. This analogy allows the definition of a nuclear "evaporation" temperature, by the equation [14]:

$$k_B T_N = 0.5 + 0.43(1 + \nu)^{1/2}, \tag{1.22}$$

where $k_B$ is the Boltzmann constant, $\nu$ is the average number of neutrons released per fission event — that is close to 2.5 (for specific examples, see Table 3.3), and the product $k_B T_N$ is expressed in MeV. In the framework of the same analogy, the energy spectrum of the neutrons is described by a Maxwellian distribution (Fig. 1.10):

$$S(E) = \frac{2}{\sqrt{\pi}} \frac{\sqrt{E}}{(k_B T_N)^{3/2}} \cdot \exp\left(-\frac{E}{k_B T_N}\right). \tag{1.23}$$

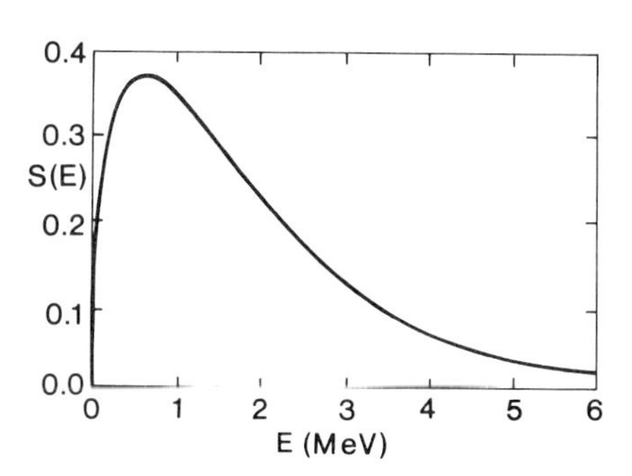


Fig. 1.10. The energy spectrum of the fission neutrons.

The average energy of fission neutrons can then be computed as:

$$\overline{E} = \int_0^\infty E\, S(E)\, dE = \frac{3}{2} k_B T_N , \tag{1.24}$$

resulting close to 2 MeV [14]. Taking into account the average number of neutrons per fission event, $\nu \simeq 2.5$, the kinetic energy transported by the neutrons amounts to about 5 MeV per fission event.

## 1.4. THE NUCLEAR REACTOR

At present, in nuclear reactors the release of nuclear energy is based on the *nuclear fission* process. The achievement and control of the self-sustained nuclear fission reaction — a notable success of nuclear physics — became possible only through an extraordinary endeavour of the science and technology of materials.

### 1.4.1. Chain Fission Reaction

A peculiar feature of the neutron-induced fission — beyond the release of significant quantities of energy — is that more neutrons are emitted than absorbed. A value higher than 1 for $\nu$ provides the requisites for the initiation of an ever higher number of fission reactions, so that, once the process has started, it spreads over the entire fissile material, and the number of neutrons and fission events grows fast. Such a process — of each fission act entailing others, while the number of reactions soars rapidly — is a *chain reaction*. Under appropriate conditions the fission chain reaction can be controlled, the number of neutrons being limited to a predetermined level. Such a steady mode of operation is achieved in the nuclear reactor.

In a chain fission reaction one must take into account the entire series of processes that a generation of neutrons undergoes, starting with their production through fission at a certain moment, and ending with their triggering new fissions, which provide further for the neutrons of the next generation. For thermal neutron-induced fission, a significant process in this series consists in diminishing the energy of the fast neutrons, which enables them to trigger new fissions; this process is called *neutron moderation*, and is achieved via collisions of the fast neutrons with nuclei of certain *moderator materials* (see ch. 5), which has a slowing-down effect. Prior, during, or after the slow-down, some neutrons are lost, either through absorption or through escaping off the medium that contains the fissile material (the multiplicative medium). The neutron balance from one generation to the next must therefore be carried over a whole cycle, from the generation to the regeneration of fast neutrons. Such a balance is outlined in Fig. 1.11, the cycle involving the following stages:

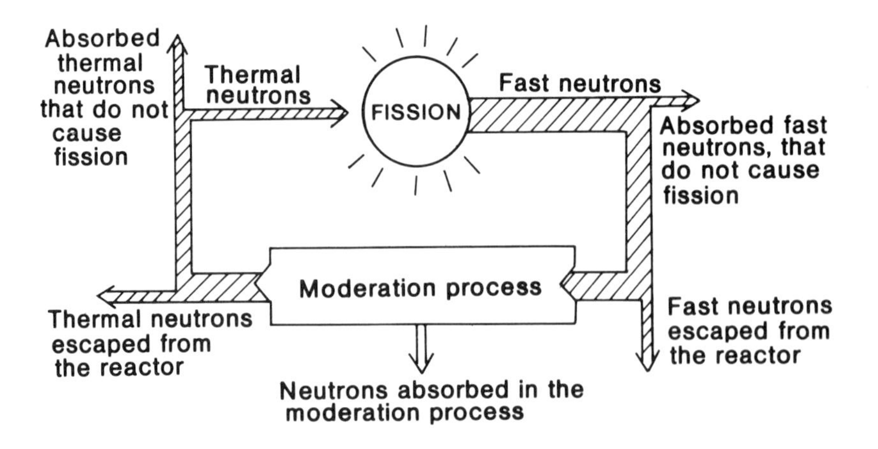


Fig. 1.11. The neutron balance, in thermal neutron reactors.

1. When interacting with nuclei of the fissile material, the fast neutrons emitted in a fission event can directly induce other fissions, thus increasing $\varepsilon$ times the number of fast neutrons; $\varepsilon$ is then the *fast neutron multiplication factor*.

2. Neutrons are slowed down till their energy reaches the region where the cross-section of radiative capture exhibits pronounced resonances (for $^{238}U$ it ranges between 9 and 41 eV). The fraction of neutrons that are thus captured is $(1-p)$, $p$ being the *probability of avoiding a resonance absorption*.

3. The number of thermal neutrons that resulted from the moderating process, and also survived the hazard of radiative capture, is further reduced through capture in moderator materials, in structural materials or in fuel diluents — by a fraction $(1-f)$, $f$ being the *thermal utilization factor*.

4. The thermal neutrons left can interact with the fissile nuclei in the fuel, which leads to captures and to new fissions, thus generating $\eta$ *fast neutrons for each thermal neutron absorbed*. This factor is the product of the ratio of the fission cross-section $\sigma_f$ to the absorption cross-section $\sigma_a = \sigma_f + \sigma_c$, by the number $\nu$ of fast neutrons released per fission event.

Summing up, one eventually obtains the ratio of the neutron numbers in two successive generations, called the *multiplication coefficient*

$$k_\infty = \varepsilon p f \eta \qquad (1.25)$$

This expression is known as the *four factor formula*.

The above equation does not take into account neutron losses (prior and after slowing down) that are due to their leaking off the multiplicative medium, thereby being suited only to an infinitely extended active medium.

In the actual case, an *effective multiplication coefficient* is defined as:

$$k_{\text{eff}} = k_\infty P_r P_t \,, \qquad (1.26)$$

where $P_r$ and $P_t$ are the probabilities for the fast and thermal neutrons respectively to remain inside the core. The evaluation of these probabilities requires the introduction of some quantities featuring the geometry of the multiplicative medium and the processes of neutron slow-down and diffusion: the geometry factor $B$, the diffusion length $L$ and the slow-down length $L_i$. The square of the slow-down length is also called the *Fermi age* $\tau = L_i^2$. With the above notations one has [14]:

$$P_r = (1 + B^2\tau)^{-1}, \; P_t = (1 + B^2L^2)^{-1}, \qquad (1.27)$$

and

$$k_{\text{eff}} = k_\infty (1 + B^2\tau)^{-1}(1 + B^2L^2)^{-1} \simeq k_\infty (1 + B^2M^2)^{-1}, \qquad (1.28)$$

where $M^2 = L^2 + \tau$ is the *migration area*.

The value of the effective multiplication coefficient determines the evolution of the chain reaction. When $k_{\text{eff}} > 1$, the number of neutrons and of fission events grows exponentially, and the chain reaction is self-amplifying; the operating conditions are said *supercritical*. When $k_{eff} < 1$, the number of neutrons and of fission events decrease, and the reaction stops; the conditions are then *subcritical*. When $k_{eff} = 1$, the chain reaction is in a steady regime — the operating conditions are *critical*. In order to achieve and maintain *criticality*, certain conditions are required, among which are a sufficient amount of fissionable material, together with an adequate moderator; their arrangement in a suitable geometry; the limitation of neutron losses, etc.; all these will be discussed further on.

Given a set of circumstances, a certain mass of fissionable material, known as *critical mass*, is required to achieve criticality. Although its magnitude is manifestly sensitive to any change in the environment of the fissionable material, the concept of critical mass has nevertheless a meaning intrinsically connected to and characterizing the fissionable material. Thus, it is often heard that plutonium's critical mass is inferior to that of uranium's, meaning that, under identical conditions, a critical chain fission reaction may be obtained using less plutonium than uranium. Data as listed in Table 1.2 [15] actually imply a warning to those manipulating fissionable materials, highlighting that *in no circumstance* quantities defined as "critical masses" should be allowed to pile up — otherwise the consequence is a possible triggering and an uncontrolled development of a chain fission reaction.

TABLE 1.2 $^{235}U$ and $^{239}Pu$ Critical Mass

| | *Metallic* (kg) | *Aqueous solution* (kg) |
|---|---|---|
| $^{235}U$ | 22.8 | 0.82 |
| $^{239}Pu$ | 5.6 | 0.51 |

Another significant point is the dependence on temperature of the effective multiplication coefficient $k_{eff}$. This is determined by two material-sensitive factors. The first is the variation with the temperature of the absorption cross-section:

$$\sigma_a(T) = \sigma_a(T_o)\cdot(T_o/T)^{1/2}, \qquad (1.29)$$

where $\sigma_a(T)$ and $\sigma_a(T_o)$ are the absorption cross-sections at the temperatures $T$ and $T_o$, respectively. The second pertains to neutron diffusion. The thermal motion of nuclei has the effect of broadening the absorption maxima in the resonance region, and, therefore of increasing the probability that the non-thermalized neutrons are captured. This alters the diffusion length $L$, the Fermi age $\tau$ and, accordingly, the migration area. A change in the geometry factor $B$ follows, which, in turn, changes the terms regarding the dimensions of the reactor core. The above requirements determine to a great extent the choice of materials that can be used in a reactor.

### 1.4.2. Structure and Characteristics of a Nuclear Reactor

The essential part of a reactor is the reaction (active) core, also known as "*the core*", containing the fuel and the moderator in either a homogeneous mixture or a heterogeneous configuration (Fig. 1.12). When the core is heterogeneous, the fuel is composed of *fuel elements* — a solution that provides for a higher overall mechanical strength and also allows the containment of the radioactive fission products. The moderator is there to diminish the energy of the fast neutrons that are generated through fission, rendering them thermal.

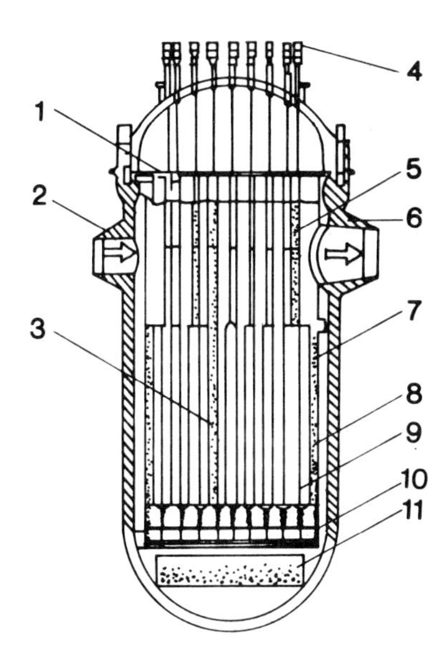


Fig. 1.12. Vertical cut through the pressure vessel of a water-cooled-and-moderated nuclear reactor: 1 - control rods, extracted from the core; 2 - cooling water inlet; 3 - control rods in the core; 4 - control rods driving mechanism; 5 - supports of upper plate; 6 - water outlet; 7 - core; 8 - core vessel; 9 - fuel assembly; 10 - core support; 11 - flow distributor.

The fuel/moderator assembly is surrounded by a *reflector*, meant to diminish the neutron leaks from the core, thus setting a lower value for the required critical mass — and for the cost of a fuel load. Removal of heat from the reactor core is achieved by means of *cooling agents*. In most of the cases a closed circuit is used, where a primary heat-transfer agent removes the heat from the reactor core, transfers it in a heat exchanger to a secondary agent, then returns to the core.

The control of the chain reaction in a nuclear reactor is performed by means of *control rods*. These steer the kinetics of the chain reaction in accordance with the fissile material consumption and the gradual accumulation of fission products, securing the critical conditions required for a normal operation of the reactor and, at the same time, ensuring its safety. The *reactor shielding* separates the intense radiation sources inside the reactor from its surroundings, diminishing radiation doses down to levels that are considered permissible, from a biological standpoint.

Both fission and fusion reactors require a *thermal protection*, which will be discussed in chapters 8 and 12.

The reactor is brought to criticality through altering the neutron flux by means of the control rods. Normally, all designs provide for supercriticality, and it is up to the control rods to either stop the reactor, or let it operate at a pre-set power level. Since during operation fission reactions generate nuclei with large absorption cross-sections for neutrons, the multiplication factor $k_{eff}$ decreases in time, and the reactor gradually loses its provisions of criticality. Eventually, the nominal power level can no longer be reached, and the fuel must be replaced. The underlying process is called *reactor poisoning*. The fission products that contribute significantly to poisoning are listed in Table 1.3, each with its main nuclear characteristics, such as the absorption cross-section, half-life, and fission yields [16].

TABLE 1.3 Fission Products with a High Neutron Absorption Cross-Section

| | *Absorption cross-section* (b) | $T_{1/2}$ | *Fission yield* (%) | | |
|---|---|---|---|---|---|
| | | | $^{233}U$ | $^{235}U$ | $^{239}Pu$ |
| Group I | | | | | |
| $^{135}Xe$ | $3.4 \times 10^6$ | 9.2 hours | 6.6 | 6.4 | 5.3 |
| $^{149}Sm$ | $6.6 \times 10^4$ | stable | 0.7 | 1.15 | 1.8 |
| $^{151}Sm$ | $1.2 \times 10^4$ | 80 years | 0.5 | 0.73 | 1.98 |
| $^{113}Cd$ | $2.5 \times 10^4$ | stable | 0.02 | 0.011 | 0.075 |
| $^{155}Eu$ | $1.4 \times 10^4$ | stable | 0.015 | 0.031 | 0.22 |
| Group II | | | | | |
| $^{153}Eu$ | 420 | stable | 0.1 | 0.15 | 0.8 |
| $^{83}Kr$ | 205 | stable | 1.2 | 0.54 | 0.05 |
| $^{115}In$ | 197 | stable | 0.02 | 0.01 | 0.05 |
| $^{143}Nd$ | 300 | stable | 5.1 | 5.9 | 6.0 |

As indicated, $^{135}Xe$ has the largest absorption cross-section. The fission products with the mass number 135 belong to the decay chain:

$$^{135}Te \xrightarrow{2\ min} {}^{135}I \xrightarrow{6.7\ h} {}^{135}Xe \xrightarrow{9.2\ h} {}^{135}Cs \xrightarrow{2\times10^4\ yr} {}^{135}Ba$$

the segment I→Xe→Cs bearing, as said above, the chief responsibility for reactor poisoning.

Let $x_1$ be the iodine concentration, $\sigma_1$ its absorption cross-section, $\lambda_1$ its decay constant, and $\gamma_1$ its fission yield. Then the rate of change of the iodine concentration is:

$$\frac{dx_1}{dt} = \gamma_1 \Sigma_f \Phi - \sigma_1 \Phi x_1 - \lambda_1 x_1 , \tag{1.30}$$

where $\Sigma_f$ is the macroscopic fission cross-section, and $\Phi$ the neutron flux.

The solution of eq. (1.30) reads:

$$x_1 = \gamma_1 \Sigma_f \Phi(\lambda_1 + \sigma_1 \Phi)^{-1} \cdot [1-\exp(-(\lambda_1 + \sigma_1 \Phi)t)]. \tag{1.31}$$

Iodine reaches its saturation concentration ($x_{1\infty}$) after roughly 4...5 half-lives ($\sigma_1 \Phi$ can be neglected as compared with $\lambda_1$):

$$x_{1\infty} = \gamma_1 \Sigma_f \Phi / \lambda_1 \tag{1.32}$$

The $^{135}Xe$ concentration rate is therefore:

$$\frac{dx_2}{dt} = \gamma_2 \Sigma_f \Phi + \lambda_1 x_1 - \sigma_2 x_2 \Phi - \lambda_2 x_2 . \tag{1.33}$$

In a similar manner, the saturation concentration of $^{135}Xe$ is found to read:

$$x_{2\infty} = (\gamma_1 + \gamma_2) \Sigma_f \Phi / (\sigma_2 \Phi + \lambda_2) \tag{1.34}$$

Reactor poisoning $Q$ is defined by:

$$Q = \sigma_2 x_2 / \Sigma_f , \tag{1.35}$$

where $\sigma_2$ is the absorption cross-section of the product with concentration $x_2$.

Merging eq. (1.34) into eq. (1.35), xenon poisoning in steady-state conditions can be expressed as:

$$Q = (\gamma_1 + \gamma_2) \sigma_2 \Phi / (\sigma_2 \Phi + \lambda_2). \tag{1.36}$$

For low, as well as for high fluxes, eq. (1.36) assumes simplified forms:

$$\begin{aligned} Q &= (\gamma_1 + \gamma_2) \sigma_2 \Phi / \lambda_2 , && \text{at low fluxes;} \\ Q &= (\gamma_1 + \gamma_2) , && \text{at high fluxes} \end{aligned} \tag{1.37}$$

Table 1.4 presents the values of the poisoning with $^{135}Xe$, corresponding to several neutron fluxes [16].

Besides the loss of neutrons through absorption in $^{135}Xe$ nuclei, xenon generates a second undesirable phenomenon. Following a reactor shut-down, the beta-decay of $^{135}I$ keeps on producing $^{135}Xe$ nuclei that are no longer destroyed through neutron capture. The amount of $^{135}Xe$ increases considerably, so that within a few hours poisoning reaches such a level as to make impossible re-starting the reactor. Solving eqs. (1.30) and (1.33) under assumptions of a shut-down reactor ($\Phi = 0$), one finds that the maximum xenon

TABLE 1.4 Xenon Poisoning of Reactors

| $\Phi$ (n/cm$^2$. s) | $10^{11}$ | $10^{12}$ | $10^{13}$ | $10^{14}$ | $10^{15}$ |
|---|---|---|---|---|---|
| $Q$ | $0.9\times10^{-3}$ | $8.3\times10^{-3}$ | $3.65\times10^{-2}$ | $5.6\times10^{-2}$ | $5.9\times10^{-2}$ |

concentration is reached by about 11 hours after shut-down (Fig.1.13). A new start-up of the reactor in that condition is impossible. Taking into account the 9.2 hour half-life of $^{135}Xe$, one has to wait 20 — 30 hours for the decay of xenon nuclei, before re-starting the reactor.

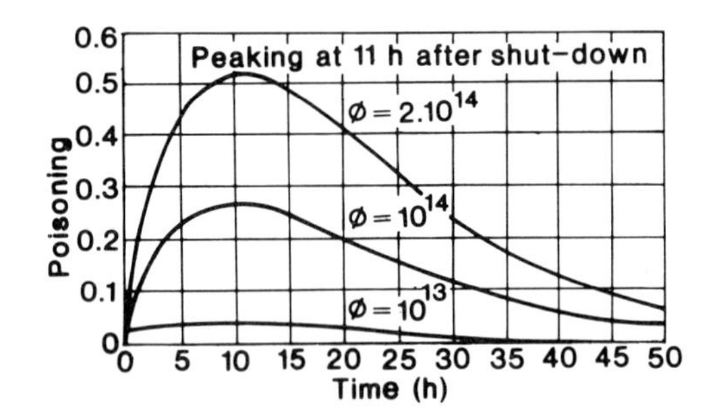


Fig. 1.13. Time variation of reactor poisoning following a shut-down, vs. the neutrons flux prior to shut-down, in n/cm$^2$ . s.

The duration of a fuel "burning" in a reactor is limited. There are two ways of expressing the degree of utilization of a nuclear fuel, namely either by means of the *integrated flux* of neutrons having irradiated the fuel element (*fluency*), or by the *burn-up factor*.

The integrated flux, or fluency is defined as:

$$\tilde{\Phi} = \int_0^t \Phi(t)dt \tag{1.38}$$

and is measured in neutrons/m$^2$.

The burn-up factor can be expressed either by the percentage of fissioned nuclei, or by the number of fissions per unit-volume of fissile material, or as specific energy generated by fission (MW day/tU). The relation between the units expressing the burn-up are presented in Table 1.5 [17].

The power of a reactor, $P$, depends on the type considered, on the neutron flux and on the mass, or volume of fuel in the core. Taking into account that about 95% of the energy released by a fission event ($E_f \simeq 200$ MeV) is transformed into thermal energy, the reactor power can be written as:

$$P = 0.95\, E_f \Sigma_f \tilde{\Phi} V_R\ , \tag{1.39}$$

TABLE 1.5 Burn-up (units)

| | | | Uranium | $UO_2$* | UC |
|---|---|---|---|---|---|
| % | $=C_1\cdot$(fluency) | $C_1=100\,\frac{N_f}{N_U}\,\sigma_f$ | $5.5\times10^{-20}\,\frac{N_f}{N_U}$ | $5.5\times10^{-20}\,\frac{N_f}{N_U}$ | $5.5\times10^{-20}\,\frac{N_f}{N_U}$ |
| Fissions.$\cdot cm^{-3}$ | $=C_2\cdot$(%) | $C_2=6.02\times10^{21}\,\frac{\rho}{M'}$ | $4.8\times10^{20}$ | $2.3\times10^{20}$ | $3.3\times10^{20}$ |
| | $=C_3\cdot$(fluency) | $C_3=6.02\times10^{23}\,\frac{\rho}{M'}\cdot\frac{N_f}{N_U}\,\sigma_f$ | $26.4\,\frac{N_f}{N_U}$ | $12.6\,\frac{N_f}{N_U}$ | $18.2\,\frac{N_f}{N_U}$ |
| MWd/tU | $=C_4\cdot$(fis/cm$^3$) | $C_4=1.85\times10^{-18}\,\frac{E_f}{A}\cdot\frac{M'}{\rho}$ | $10^{-19}\,E_f$ | $2\times10^{-19}\,E_f$ | $1.4\times10^{-19}\,E_f$ |
| | $=C_5\cdot$(%) | $C_5=1.1\times10^{4}\,\frac{E_f}{A}$ | $46.5\,E_f$ | $46.5\,E_f$ | $46.5\,E_f$ |
| | $=C_6\cdot$(fluency) | $C_6=1.1\times10^{6}\,\frac{E_f}{A}\cdot\frac{N_f}{N_U}\,\sigma_f$ | $2.5\times10^{-18}\cdot E_f\,\frac{N_f}{N_U}$ | $2.5\times10^{-18}\cdot E_f\,\frac{N_f}{N_U}$ | $2.5\times10^{-18}\cdot E_f\,\frac{N_f}{N_U}$ |

* 95% from theoretical density (T.D.)
$N_f$ = number of fissile atoms per unit-volume;
$N_U$ = number of uranium atoms per unit-volume;
$\rho$ = fuel density;
$\sigma_f$ = fission cross-section;
$E_f$ = energy per fission (MeV);
$A$ = atomic mass of uranium;
$M'$ = fuel molecular mass divided by the number of atoms per molecule

where $E_f$ is the energy released by a fission event (neutrinos' energy neglected), and $\bar{\Phi}$ is the flux averaged over the reactor volume $V_R$. For a uranium-235 reactor, a particular form of eq. (1.39) is

$$P = 3.82 \cdot 10^{-8}\,\bar{\Phi}\,m, \tag{1.40}$$

where $\bar{\Phi}$ is expressed in neutrons/cm$^2$.s, the fuel mass $m$ in kg, and the power $P$ in watts.

A quantity often dealt with in the reactor theory, also of great importance in designing its components, particularly the fuel elements,is the *specific power*. It is the thermal power released per unit-volume of the core or, alternatively, per unit-mass of fuel or per unit-length of fuel element (*linear specific power*). The latter definition is often the preferred one in reactor design. The levels of linear specific power — the higher the better — are, however, limited by the thermal and mechanical strength of materials to, typically at present, 15,000 up to 60,000 W/m [15].

The higher its rated power, the lower the cost at which a nuclear power plant generates energy. Theoretically, the power at which a reactor is rated can soar as high as one may desire; in practice, it is limited by the rate of heat removal from the reactor, so that the designed temperature be kept constant.

That argues for the necessity of materials with superior properties at high temperatures. On the other side, familiar thermodynamical considerations show that the conversion efficiency of thermal into electric energy increases with the working temperature.

It follows that lowering the generating costs of the electricity of nuclear origin is directly related to a constant raising of the maximal working temperature in the reactor. The efficiency of the thermodynamic cycle is limited by the maximum permissible temperature of the cooling fluid at the surface of the fuel element cladding. At present, it reads typically between 250 and 600 $^{0}$C, thus placing the efficiency anywhere between 25 and 42%. For natural and heavy water-cooled reactors, maintaining the water in liquid state requires a pressure up to 15 MPa. Therefore, a nuclear reactor environment is featured by relatively high temperatures and pressures, that add to the presence of an intense radiation flux.

Nuclear energy, as transformed into thermal energy in the nuclear reactor, is ultimately converted into electric energy by means of a water-vapour cycle performed on a turbo-generator set, as outlined in Fig. 1.14. In Table 1.6, several reactor types are presented, together with their main features and materials.

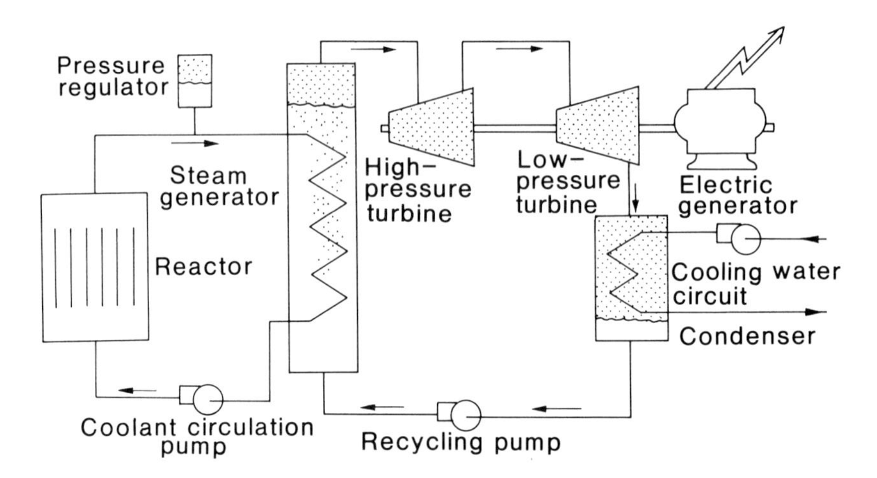


Fig. 1.14. Schematic flow-chart of a nuclear power station.

### 1.4.3. Classification of Reactors. Types of Reactors

After the advent of the nuclear reactor — the first one being conceived, built and operated in 1942 under the guidance of Enrico Fermi — many other reactors were built, of different types and for various purposes. At present, nuclear reactors are classified as follows:

(a) by the energy of the neutrons that induce fission reactions:
- thermal neutron reactors;
- fast neutron reactors;

(b) by the arrangement of the components in the core:
- homogeneous reactors;
- heterogeneous reactors;

(c) by purpose:
- research reactors;
- material testing reactors;
- power reactors;
- propulsion reactors.

TABLE 1.6 Several Reactors and their Characteristics

| *Characteristics* | *Reactor* | | | | | | |
|---|---|---|---|---|---|---|---|
| | 1 | 2 | 3 | 4 | 5 | 6 | 7 |
| Neutrons | thermal | thermal | thermal | thermal | thermal | fast | thermal |
| Power rating | - | 6 MW | 50 MW | 750 MW | - | 1200 MW | - |
| Material: | | | | | | | |
| - fissile | $UO_2$ | 5% enriched $^{235}U$ | natural U | natural $UO_2$ | $UO_2$ | 17% $PuO_2$ 83% $UO_2$ | mixture of U and Zr hydride |
| - fertile | - | - | - | - | $ThO_2$ | $UO_2$ | - |
| - moderator | graphite | graphite | graphite | heavy water | - | - | graphite |
| - reflector | - | graphite | graphite | light water | graphite | $UO_2$ | Be |
| - cooling | - | water | $CO_2$ | heavy water | He | liquid Na | liquid NaK |
| - control | cadmium | $B_4C$ | boron-steel | Cd, $D_2O$ | $B_4C$ | - | B |
| - structural | - | steel | steel | steel; zircaloy | steel; graphite | steel | steel, Nb carbide |
| - shielding | - | concrete | concrete | concrete | concrete | concrete | - |
| Temperature of the moderator | 20 °C | 600 °C | 250 °C | 55-71 °C | - | - | 530 °C |
| Neutron flux ($n/cm^2 \times s$) | | $5 \times 10^{13}$ | $1.3 \times 10^{13}$ | $1.2 \times 10^{14}$ | $1.6 \times 10^{14}$ | $6 \times 10^{15}$ | |
| Type | experi-mental | power | power | power | experi-mental | power | propulsion |

1. CP-1 (the first nuclear reactor, built by E.Fermi),Chicago,USA,1942
2. Obninsk nuclear power station, USSR, 1954
3. Calder Hall, UK, 1956
4. CANDU, Bruce, Canada, 1976
5. Dragon, UK, 1964
6. Super-Phenix, France; in project
7. SNAP-50; auxiliary power source in space applications, USA

Reactor taxonomy can also consider the kind of moderator, cooling agent, rated power, etc. More often than not, for the identification of a reactor one specifies, in initials, the cooling agent, moderator and some other feature —

all defining a type of reactor. In Table 1.7 the main reactor types are listed and grouped into classes (see also Appendix 2).

TABLE 1.7 Reactor Types

| *Type* | *Versions* | |
|---|---|---|
| LWR (Light Water Reactor) | PWR | (Pressurized Water Reactor) |
| | BWR | (Boiling Water Reactor) |
| HWR (Heavy Water Reactor) | PHWR | (Pressurized Heavy Water Reactor) the most prominent, at present, is CANDU (CANadian Deuterium Uranium) |
| | BHWR | (Boiling Heavy Water Reactor) |
| Gas-cooled, graphite moderated reactor | GCR | (Gas-Cooled Reactor) |
| | AGR | (Advanced Gas-cooled Reactor) |
| | HTGR | (High Temperature Gas-cooled Reactor) |
| | UHTGR | (Ultra High Temperature Gas-cooled Reactor) |
| TBR (Thermal Breeder Reactor) | LWBR | (Light Water Breeder Reactor) |
| | CANDU-Th | (CANDU-type reactor using Thorium as fertile material) |
| FBR (Fast Breeder Reactor) | LMFBR | (Liquid Metal Fast Breeder Reactor) |
| | GCFBR | (Gas-Cooled Fast Breeder Reactor) |
| MSR (Molten Salt Reactor) | MSBR | (Molten Salt Breeder Reactor) |
| Others | OMR | (Organic-Moderated Reactor) |
| | LMFR | (Liquid Metal Fuel Reactor) |
| | SNAP | (Systems for Nuclear Auxiliary Power) |
| | NERVA | (Nuclear Engine for Rocket Vehicle Application) |

Today, owing to the experience gained and to technical developments and improvements, the various types of reactors in nuclear power plants are able to generate electric power with increased efficiency and at high rated power. Figure 1.15, where only the most powerful plants built up by the time as specified were taken into account, is indicative for this development.

The utilization of the natural resources of nuclear fuel and their long-term planning are heavily sensitive to the types of reactors considered. The efficiency of fuel utilization can be much improved in *breeder reactors* (Fig. 1.16). That is why in the mid-future these are supposed to assume an impressive development. In a breeder reactor the fertile material is transformed, through nuclear reactions initiated with neutrons, into a fuel that is fissionable with *thermal* neutrons. That effective fuel as obtained can be either similar to the material burned in the reactor, when one has a fuel *reproduction*, or a

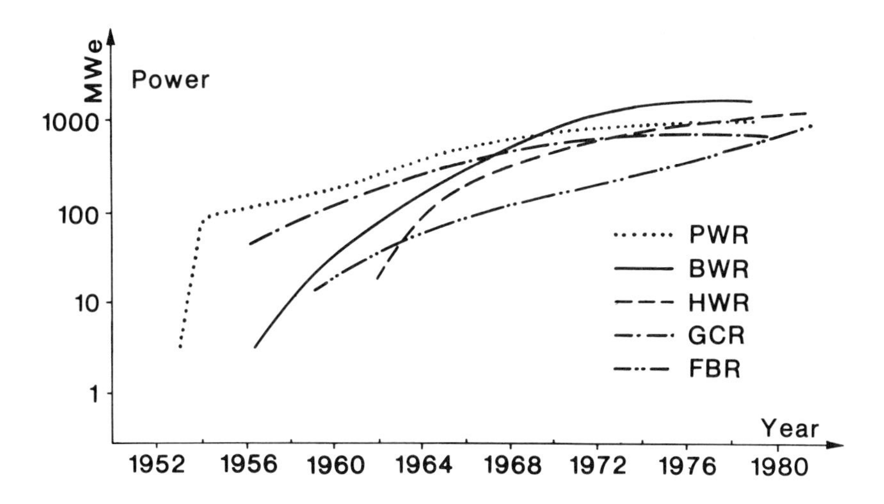


Fig. 1.15. Dynamics of unit power installed in nuclear power stations, for different types of reactors

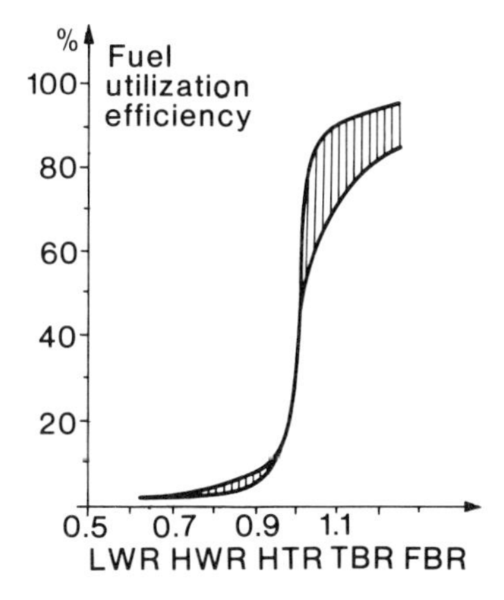


Fig. 1.16. Nuclear fuel utilization efficiency, for different types of reactors.

different fissile material, in which case one has a *conversion*. The *breeding ratio* (of either reproduction or conversion) is defined as the ratio of the amount of fuel produced to the amount consumed.

Fast neutron reactors are especially well-suited to operate in breeding conditions, using $^{239}Pu$ or $^{233}U$ as fissile materials, and $^{238}U$ or $^{232}Th$ as the fertile ones [13]. The first breeding cycle, belonging to the FBR line, is based on the following reactions:

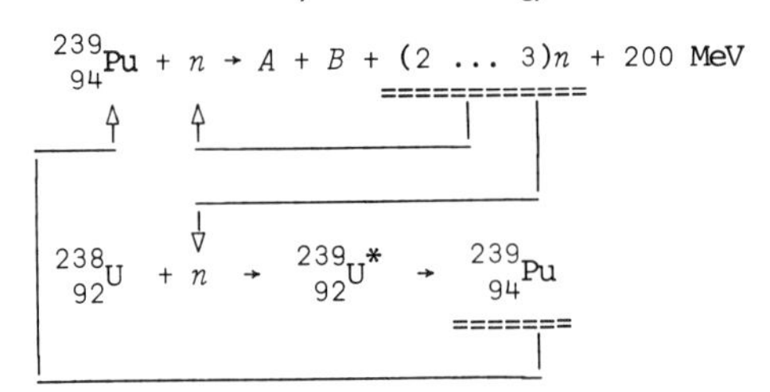


One out of every 2 — 3 neutrons emitted in a fission event contributes to the chain fission reaction of plutonium, whereas the others contribute to the transformation of uranium 238 into plutonium.

In a similar manner, the TBR line uses the cycle:

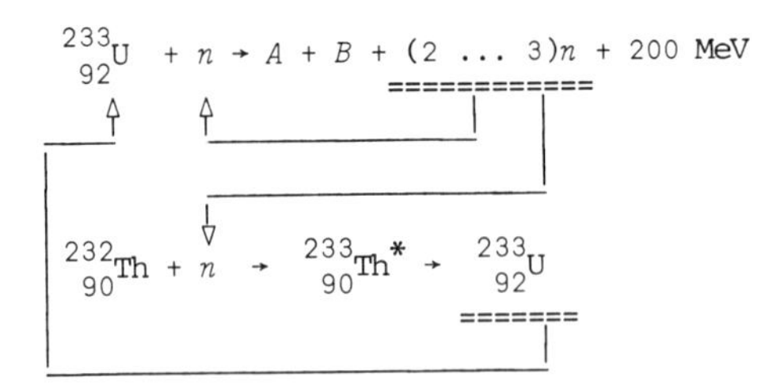


In this cycle thorium is transformed in uranium 233.

Owing to their high efficiency in utilizing natural resources of nuclear fuel, the breeder reactors are mentioned most frequently in all long-term projections on the development of fission-based nuclear power worldwide.

Overall, the selection of a reactor type to sustain the implementation of nuclear power into a national power system is by no means a simple matter. Such a selection requires a multi-dimensional analysis, carried out thoroughly and with proper responsibility, an assessment of the relative influence of many factors, among which outstanding are:

1. the goals and strategies of economic and social development;

2. the nature and magnitude of the demand for electric and thermal energy, their projected dynamics, and the possible alternatives of meeting them using conventional and/or other resources;

3. the structure and outlook of the already established power system: total installed power, rated powers, mix of energy vectors, degree of interconnection, etc.;

4. the potential of infrastructures, particularly the machine building industry; the extent and pace of possible indigenous production of nuclear power materials and equipment; resources of skilled manpower and training capability;

5. the availability of capital investments, in relation with the other priorities in that field;

6. the availability of indigenous natural uranium resources;

7. the uranium world market and its evolution: resources and deposits, price dynamics, diversification of suppliers and commercial products;

8. world production of enriched uranium and evolution of its consumption, diversification of supply sources, improvement and diversification of manufacture processes, etc.;

9. the degree of development and potential for improvements, of different thermal reactor types;

10. the outlook on the potential of the fast breeder reactors to play a part in the energy system, including the time horizon of their possible implementation;

11. the development and improvement of separation and enrichment processes, both for uranium and heavy water;

12. the possible contribution of international scientific, technological and economic co-operation to the solution of technology and supply problems.

## 1.5. ISOTOPIC ENRICHMENT OF NUCLEAR MATERIALS

Most of the reactor types involve nuclear materials having an isotopic composition that differs from that of the respective natural elements, that is — enriched in one of the constituent isotopes. For instance, the LWR type uses as fuel a variety of uranium that is enriched in $^{235}U$, whereas the HWR type uses as moderator, and (almost always) as cooling agent the heavy water (see also Appendices 2 and 3). As has already been said, $^{235}U$ represents only approximately 0.72% of the natural uranium, and deuterium is found in natural water in very small concentrations, around $2.10^{-4}$% (atomic).

Enrichment is, therefore, vital to a sound, long-term nuclear power policy. The unquestionable success of the more than 200 light water reactors (LWRs) now in operation owes much not only to the relatively low construction costs and high reliability, but also to the steady trend to cost reduction in uranium enrichment.

On the other hand, if — in order to circumvent the constraints of the uranium enrichment process — the use of natural uranium (in the form of $UO_2$) is commended, then the use of a heavy water moderator is in order, the production of which implies again isotopic enrichment techniques.

This explains why, at present, uranium ore deposits associated with ground or surface water do not turn into natural nuclear reactors. A fossil natural reactor was nevertheless discovered at Oklo (Gabon). There, in the distant past, a part of the deposit now mined has worked its way as a natural reactor. In that forgotten age, the isotopic abundance of $^{235}U$ had been significantly higher (about 3%), thus allowing the initiation of a chain reaction moderated by natural water [18,19].

A first generation of power reactors, developed in France and the U.K., were the natural uranium fueled/graphite moderated reactors. This line features a moderate burn-up factor, leading to a relatively high uranium consumption. At present, one reactor type that was commercially proved as viable and efficient, also ensuring a good utilization of the fuel, is the natural uranium/heavy water-moderated reactor. The pressure-tube reactors of this type developed in Canada are known as CANDU reactors.

It follows that for every type of reactors isotopic enrichment processes play an important part.

Both enriched uranium and heavy water are obtained through *isotope separation* processes. Every process is characterized by an *elementary separation factor* $\alpha$, defined as:

$$\alpha = \frac{c'(1 - c'')}{c''(1 - c')}\ , \qquad (1.41)$$

where $c'$ and $c''$ are the atomic fractions of the isotope of interest contained in the output and waste, respectively.

A high separation factor indicates an economically sound process, as far as plant size, and consumption of raw materials and energy. Other features of the process, such as inconvenient production rates, operating parameters, aggressiveness of the agents at work, may possibly overrun the advantage of a high separation factor.

The unit performing the elementary separation process is called a *separating element*. Several such elements, connected in parallel, constitute a *separating stage*, and several stages connected in series make a *cascade*. The way of combining several elements into stages and cascades settles the total productivity of the facility, and the overall enrichment factor of the final product.

The elementary separation factors are, as a rule, only slightly larger than unity, due to the isotopic effects being themselves small. As a consequence, separation plants often consist of a large number of elements, stages and cascades, processing vast quantities of raw materials to obtain a final product of required concentration [3,20].

Present separating technologies involve many recent advances in physics, physical chemistry, modern engineering and high technology. Potential separation processes are numerous and diverse (see chapters 3 and 5), but among them only a few were developed to a commercial, truly industrial scale. The selection of an enrichment method under some given terms implies a preliminary analysis of the comparative weight of three main factors, namely: the target output and isotopic concentration; the physical and chemical properties of the raw material (which dictates the separation factor); and, not the least, the profitability, that depends chiefly on the energy-intensiveness of the process, the capital investment per unit-output, and the operating costs. Isotope separation plants, particularly those enriching uranium or producing heavy water, are important units of today's nuclear industry.

Specific considerations on isotope separation of $^{235}U$ and of heavy water can be found in Chapters 3 and 5, respectively.

REFERENCES

1. Malearov, V. V. (1961). *Bazele teoriei nucleului atomic* (translation from Russian). Edit. tehnica, Bucharest; Evans, R. D. (1955). *The Atomic Nucleus*. McGraw Hill Book Comp., New York)
2. Enge, H. A. (1970). *Introduction to Nuclear Physics*. Addison Wesley Publ. Co., Reading (Mass.).
3. Ursu, I. (1973). *Energia atomica*. Edit. Stiintifica, Bucharest.
4. Sirokov, I. M., and Iudin, N. P. (1972). *Yadernaya Fizika*. Izd.Nauka, Moscow.
5. Spolski, E. (1974). *Atomnaya Fizika*. Izd.Nauka, Moscow.
6. Muhin, K. N. (1974). *Experimentalnaya Yadernaya Fizika*. Atomizdat, Moscow.
7. von Weizsäcker, C. F. (1935). *Z.Phys.*, 96, 431.

8. National Science Foundation (1980). *The Five-Year Outlook*, Vol. 2. NSF, Washington, D.C., USA, p. 102.
9. Bohr, N., and Wheeler, H. A. (1939). *Phys.Rev.*, 56, 426.
10. Frenkel, S. (1974). *Phys.Rev.*, 72, 914.
11. Vandenbosch, R., and Huizenga, J.R. (1973). *Nuclear Fission*. Academic Press, New York.
12. Strutinski, V.M. (1967). *Nucl.Phys.*, A95, 420.
13. Wilets, L. (1964). *Theories of Nuclear Fission*. Clarendon Press, Oxford; Meyerhof, W. E. (1967). *Elements of Nuclear Physics*. McGraw Hill Book Comp., New York)
14. Berinde, A. (1977). *Elemente de fizica si calculul reactorilor nucleari*. Edit. Tehnica, Bucharest.
15. Sauteron, J. (1965). *Les combustibles nucléaires*. Ed. Hermann, Paris.
16. Schulten, R., and Guth, W. (1976). *Fizica reactorilor* (translation from German). Edit. Tehnica, Bucharest.
17. Robertson, J. A. L. (1969). *Irradiation Effects in Nuclear Fuels*. Gordon and Breach Science Publ., New York.
18. Weber, F. and Bonhomme, M. (1975). In *The Oklo Phenomenon, Proceedings of a Symposium, Libreville, 23-27 June 1975, organized by the International Atomic Energy Agency in co-operation with the French Atomic Energy Commission and the Government of the Republic of Gabon.* International Atomic Energy Agency, Vienna. IAEA-SM-204/16, pp. 17-33.
19. Cowan, G. A. (1976). *Scientific American*, 235, 1, 36.
20. Vasaru, G. (1968). *Izotopi stabili*. Edit. Tehnica, Bucharest.
21. Ivascu, M., and Poenaru, D. N. (1981). *Energia de deformare si izomeria formei nucleelor*. Edit. Academiei, Bucharest.

# CHAPTER 2

# Structure and Properties of Materials

## 2.1. MICROSCOPIC STRUCTURE OF MATERIALS

Substances can be found in solid, liquid and gaseous states, depending on the temperature and pressure conditions they are subject to. These conditions restrain to various degrees the possible modes of motion of the constituent particles, the intensity of interaction forces between them. At a sufficiently low temperature all substances, except helium, are in solid state, featuring the fact that the microscopic distance between particles is of the same order of magnitude with their dimensions. The motion of the particles is restricted to vibrations around certain equilibrium positions. Owing to this, the solids preserve their volume and shape, and resist deformation.

In solids,the equilibrium positions of particles can assume an orderly pattern, evolving a crystalline lattice (crystalline solids); alternatively, they affect disorder (amorphous solids). The perfect crystal is an abstraction — in fact a very useful one for solid state physics. On the other hand, in practice one never finds perfect crystals. Crystals have defects, whether local or extended, due to the presence of impurities, to the preparation procedure, and also induced by thermal and mechanical processing, irradiation, erosion, corrosion, etc. As the temperature rises, the number of defects increases and, thus, the integrity of the crystalline structure can be lost. Above a certain temperature the crystalline solid melts.

In liquid state the particles or groups of particles can move relatively free, being able to depart from each other; thereby a liquid does not preserve its shape.

The liquid and solid states are known, generically, as the condensed states of the matter, in order to distinguish them from the gaseous state, in which the constituent particles are relatively independent, and from the plasma state, in which the atoms themselves lose integrity.

### 2.1.1. Structure of Crystalline Solids

The arrangement of atoms in a crystalline solid in orderly structures is a consequence of the fact that, when bringing two atoms close to one another, the rearrangement of the atomic orbitals leads, in most cases, to bound systems which, for a certain distance between atomic nuclei, achieve a stable equilibrium state [1-3]. There are four types of bonds (forces) between atoms, that give microscopically stable structures. These are:

(a) the ionic bond,
(b) the covalent bond,
(c) the metallic bond,
(d) the Van der Waals bonds.

In the *ionic bond*, the constituents of the crystalline lattice are ions of opposite charges, with fully occupied electronic shells (see Appendix 4), and interacting mostly through electrostatic forces of attraction. Electrostatic forces of repulsion, as well as quantum forces acting between electrons and nuclei, prevent an indefinite approach of the ions of opposite charge, a fact that determines the existence of an equilibrium distance. Due to the electrostatic nature of the bond, the number of neighbouring ions around a given ion depends mainly on the partners' ionic radii. This type of bond can be modelled by considering spheres of different diameters (corresponding to the constituent ions) [4].

Some simple ionic structures are represented in Fig. 2.1.

Uranium dioxide $UO_2$, for example, has a fluorite-type structure ($CaF_2$ — Fig. 2.1c) [5]. The uranium ions are positioned in the corners and the centres of faces of the cubic unit-cell, while the oxygen ions make up a cube inside this system. In this way, each uranium ion is surrounded by eight oxygen ions (co-ordination 8), while the oxygen ion has in its first co-ordination sphere 4 uranium ions. Due to this peculiar arrangement of the ions in the fluorite-type lattice, in the preparation of $UO_2$ excess anions can fill interstitial positions (in the centre of the unit-cell), which may evolve a non-stoichiometric structure:

$$UO_2 + (x/2)O_2 \rightarrow UO_{2+x}$$

Experimentally evidenced, such a situation corresponds to enlarged distances between the oxygen ions, thereby the crystalline lattice being distorted; an X-ray diffraction investigation allows a precise evaluation of such distortions.

As a rule, ionic compounds have high melting points. In solvents of high dielectric constant they dissociate. Generally, in molten state they present an electrical conductivity of the ionic type, while in solid state they are poor conductors of both electricity and heat. The ionic bond is at the origin of several specific mechanical properties: cleavage, lack of malleability (that makes frail solids) etc.

The *covalent bond* links neutral atoms that share pairs of electrons of their atomic structure. Such bonds, via pairs of electrons, are found more often in organic compounds.

In inorganic compounds covalent bonds are observed in molecules that are made of identical atoms, or in compounds made of elements that belong to the fourth column of the periodical table. Some characteristic stuctures based on covalent bonding are shown in Fig. 2.2. Featuring the covalent crystals are the directed bonds, that point to well-determined directions in space with respect to each other, and whose number corresponds to the valency of the constituent atoms. As a consequence, in compounds based on this type of bonding the covalent forces act within molecules, while the molecules themselves are bound to one another by forces of a different kind. Such is the

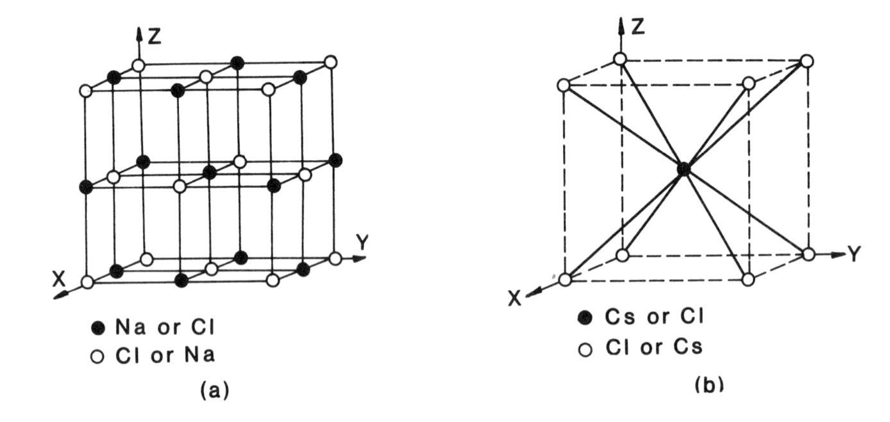


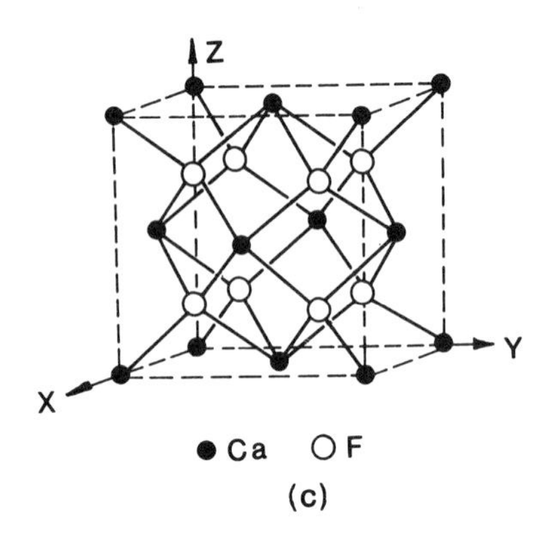


Fig. 2.1. Some typical structures for ionic crystals: (a) NaCl, (b) CsCl, (c) $CaF_2$

case of naphthalene and anthracene crystals. In graphite, through covalent bonding a stratified structure of carbon atoms is built up (planar hexagonal sublattices); forces between such sublattices are of the Van der Waals type (second-order electrostatic polarization forces). Pure covalent bonds can be found only in a limited number of compounds; among them — the diamond, whose structure is depicted in Fig. 2.2b, and AlN and AlP, having a wurtzite structure (Fig. 2.2c).

The *metals* distinguish themselves from other substances by a number of specific physical properties, like mechanical resistance, malleability, high electric and thermal conductivity, opacity for the visible spectrum, etc. In a simplified picture, the metallic structure can be regarded as an ionic lattice immersed into a quasi-free electron gas. The cohesion that develops between ions and free electrons induces the orderly arrangement of ions in a crystalline lattice, while the high mobility of electrons gives the metal the high electric and thermal conductivity. This type of structure is characterized

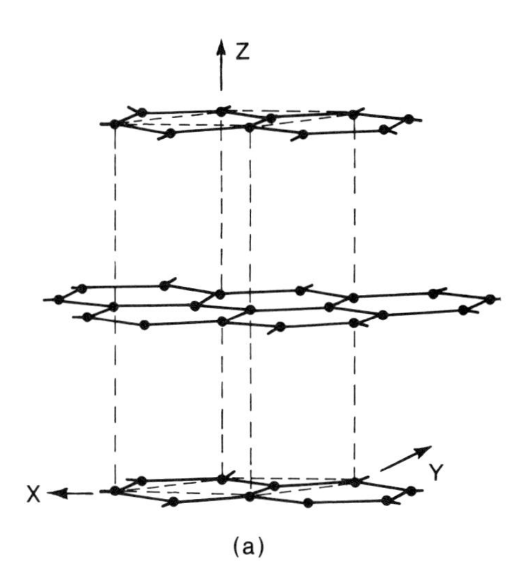


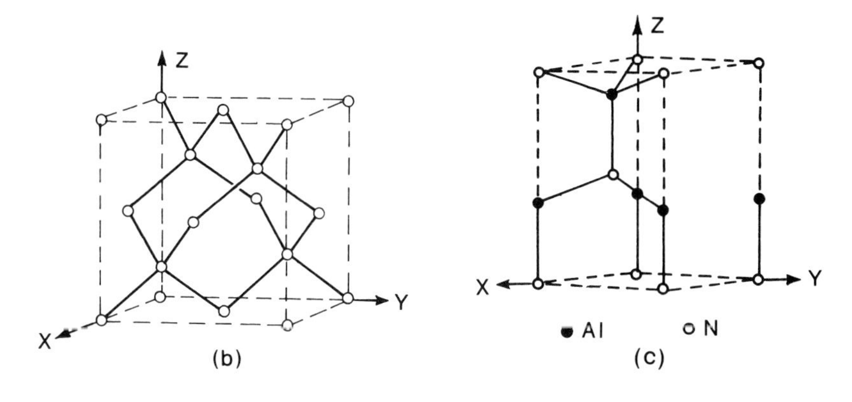


Fig. 2.2. Some typical structures for covalent crystal: (a) graphite, (b) diamond, (c) ZnS (wurtzite).

by close-packed structures of large densities. This gives the metals their remarkable properties. The different manners in which a set of rigid spheres, symbolizing the ions of the metal lattice, can be arranged in a close-packed structure are schematically depicted in Fig. 2.3 [5]. The most common crystalline structures of metals are (Table 2.1 [5]): face-centred cubic ($A_1$), body-centred cubic ($A_2$), hexagonal close-packed ($A_3$). Many of the metals exhibit polymorphism, and their stable structure at room temperature is not necessarily the simplest one. For example, uranium in normal conditions has a complex structure (X), and changes into the simple structure $A_2$ only above 1033 K.

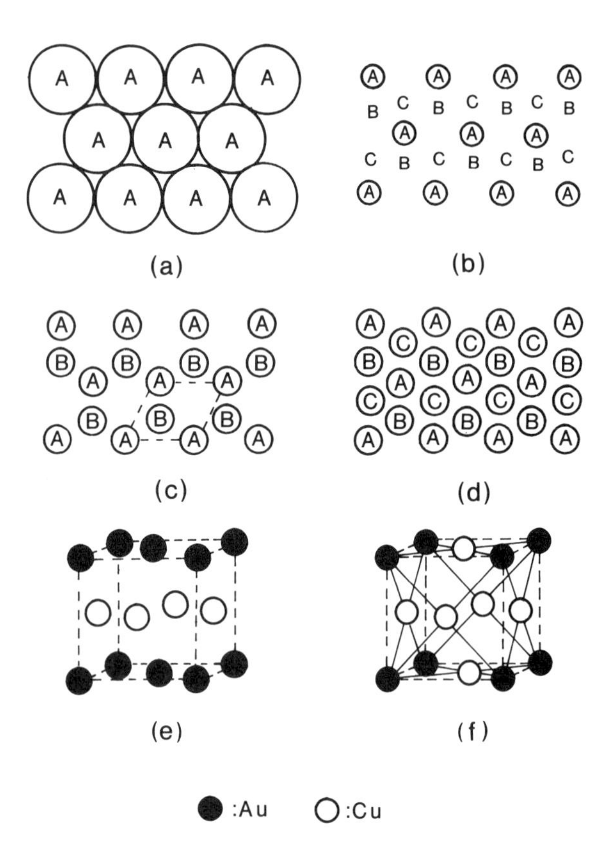


Fig. 2.3. Some typical structures of crystalline metals and alloys: (a) layer of spheres arranged in a close-packed structure; (b) a close-packed layer than can accommodate another, overlapping close-packed layer; (c) hexagonal close-packed structure; (d) cubic close-packed structure: (e) copper-gold alloy structure, with a regular phase CuAu; (f) the same alloy, with a regular phase $Cu_3Au$.

One of the particular characteristics of the metallic structures is their alloying ability. In contrast with the chemical compounds the alloys preserve, at least qualitatively, the general characteristics of the constituent elements; the usual concepts of classic chemistry concerning the valency apply with difficulty to the alloys. The alloys can be obtained in a wide range of compositions and, therefore, of properties. From this point of view a classification can be achieved by dividing the metallic elements into four classes: $T_1$, $T_2$, $B_1$ and $B_2$ (Table 2.2 [5]). The elements of the $T_1$ class are strongly electropositive. The $T_2$ class contains the transition elements (3d, 4d, 5d, 4f and 5f). Within the class B the segregation of the two categories is less clear; $B_1$ groups together elements with a more marked metallic character, while $B_2$ contains elements with a lesser metallic character. The binary alloys can be classified into four categories:

TABLE 2.1 Structures of Metal Elements

| | 1A | 2A | 3A | 4A | 5A | 6A | 7A | 8 | | | 1B | 2B | 3B | 4B | 5B | 6B |
|---|---|---|---|---|---|---|---|---|---|---|---|---|---|---|---|---|
| 2 | Li | Be | | | | | | | | | | | | | | |
| | $A_2$ | $A_2$ | | | | | | | | | | | | | | |
| | $A_1$ | $A_3$ | | | | | | | | | | | | | | |
| | $A_3$ | X | | | | | | | | | | | | | | |
| 3 | Na | Mg | | | | | | | | | | | Al | | | |
| | $A_2$ | $A_3$ | | | | | | | | | | | $A_1$ | | | |
| | $A_3$ | | | | | | | | | | | | | | | |
| 4 | K | Ca | Sc | Ti | V | Cr | Mn | Fe | Co | Ni | Cu | Zn | Ga | Ge | As | Se |
| | $A_2$ | $A_1$ | $A_1$ | $A_3$ | $A_2$ | $A_2$ | X | $A_2$ | $A_3$ | $A_1$ | $A_1$ | X | X | X | X | X |
| | | $A_3$ | $A_3$ | $A_2$ | | $A_3$ | $A_1$ | $A_1$ | $A_1$ | $A_3$ | | | | | | |
| | | X | | | | X | $A_2$ | | | | | | | | | |
| 5 | Rb | Sr | Y | Zr | Nb | Mo | Tc | Ru | Rh | Pd | Ag | Cd | In | Sn | Sb | Te |
| | $A_2$ | $A_1$ | $A_2$ | $A_3$ | $A_2$ | $A_2$ | $A_3$ | $A_3$ | $A_1$ | $A_1$ | $A_1$ | X | X | X | X | X |
| | | $A_3$ | | $A_2$ | | $A_3$ | | | | | | | | | | |
| | | $A_2$ | | | | | | | | | | | | | | |
| 6 | Cs | Ba | La | Hf | Ta | W | Re | Os | Ir | Pt | Au | Hg | Tl | Pb | Bi | Po |
| | $A_2$ | $A_2$ | $A_1$ | $A_3$ | $A_2$ | $A_2$ | $A_3$ | $A_3$ | $A_1$ | $A_1$ | $A_1$ | X | $A_3$ | $A_1$ | X | X |
| | | | $A_3$ | $A_2$ | | | | | | | | | $A_2$ | | | |
| 7 | Fr | Ra | Ac | | | | | | | | | | | | | |
| | - | - | $A_1$ | | | | | | | | | | | | | |
| Lanthanides | | | | | | | | | | | | | | | | |
| | Ce | Pr | Nd | Pm | Sm | Eu | Gd | Tb | Dy | Ho | | Er | Tm | Yb | Lu | |
| | $A_1$ | $A_1$ | X | - | X | $A_2$ | $A_3$ | $A_3$ | $A_3$ | $A_3$ | | $A_3$ | $A_3$ | $A_1$ | $A_3$ | |
| | $A_3$ | X | | | | | | | | | | | | | | |
| Actinides | | | | | | | | | | | | | | | | |
| | Th | Pa | U | Np | Pu | Am | Cm | Bk | | Cf | E | Fm | Mv | No | | |
| | $A_1$ | X | X | X | X | - | - | - | | - | - | - | - | - | | |
| | $A_2$ | | $A_2$ | | $A_1$ | | | | | | | | | | | |
| | | | | | $A_2$ | | | | | | | | | | | |

a) alloys between true metals of class T;
b) alloys between a true metal (class T) and one of the class B;
c) alloys between two metals of class B;
d) interstitial alloys, in which one component is a non-metallic element.

TABLE 2.2 Classification of Metal Elements, from the Angle of their Alloying Properties

| *Metals* | | | | | | | | | | | *B-group elements* | | | | |
|---|---|---|---|---|---|---|---|---|---|---|---|---|---|---|---|
| $T_1$ | | $T_2$ | | | | | | | | | $B_1$ | | $B_2$ | | |
| Li | Be | | | | | | | | | | | | | | |
| Na | Mg | | | | | | | | | | | Al | Si | | S |
| K | Ca | Sc | Ti | V | Cr | Mn | Fe | Co | Ni | Cu | Zn | Ga | Ge | As | Se |
| Rb | Sr | Y | Zr | Nb | Mo | Tc | Ru | Rh | Pd | Ag | Cd | In | Sn | Sb | Te |
| Cs | Ba | La | Hf | Ta | W | Re | Os | Ir | Pt | Au | Hg | Tl | Pb | Bi | Po |
| | | Lanthanides and Actinides | | | | | | | | | | | | | |

Generally one can say that, when passing over these categories from (a) to (d) the properties become progressively less metallic, and the system acquires an increasingly marked character of a chemical compound.

In Fig. 2.3 a few possible ordering structures in alloys are given. Fig. 2.3e and f show the structures of CuAu and $Cu_3Au$ compounds.

The alloys can be divided into three categories: (i) mechanical mixtures, in which the components are not mutually soluble; (ii) solid solutions, in which the components dissolve totally or partially, and (iii) intermetallic compounds. These kinds of alloys are frequently met among the materials used in nuclear reactors.

In the solid solutions the gradual replacement of one type of atom by the other is carried out statistically, accompanied by small changes in size of the elementary unit-cell, but with no noticeable modification of the structure. Thus, admitting a certain degree of disorder, the alloys can be obtained in a broad range of compositions. The variety of compositions is further widened by the possibility of including into the system some defects, like vacancies; these play a very important role in determining the alloy's properties.

In alloys the perfect order as well as a complete crystallographic disorder are two extreme cases; the shift from one extreme to the other can be achieved gradually, as the temperature changes. Nevertheless, there are some content ratios of the alloy components, and temperature ranges, for which an ordered structure can be achieved over the entire volume of the alloy. In such conditions, the analysis of the physical quantities that characterize the alloy indicates a phase transition of the order — disorder type.

An accurate quantum-mechanical treatment shows that the crystalline structure and the corresponding phase of the alloy depend ultimately on the ratio of the number of electrons to the number of atoms; more exactly, what matters is *the number of free electrons per atom.* In phases with a defectless crystalline structure, there is a direct relationship between the two ratios; in phases with structural defects the number of electrons per unit cell is of essence.

A rather different situation is encountered with the *interstitial alloys*, where the atoms of the alloying element are distributed in the interstices of the base-metal array (host lattice). Commonly, this one is a transitional metal, and the alloying elements are hydrogen, carbon, boron and nitrogen. Chief features of such alloys are: a high melting temperature; metallic shine and optical opacity; a good electric conductivity; and a wide-ranging composition.

The metallic hydrides — alloys of metals with hydrogen — exhibit properties somewhat different from those of other interstitial alloys.

Examples of interstitial alloys are given in Table 2.3.

TABLE 2.3 Elements that Match in Interstitial Alloys

| | | |
|---|---|---|
| Zr — H | Th — N | Th — C |
| Ta — H | Zr — N | Zr — C |
| Ti — H | U — N | Hf — C |
| | Nb — N | U — C |
| | Ti — N | Nb — C |
| | W — N | Ti — C |
| | Mo — N | W — C |
| | Mn — N | Mo — C |
| | Fe — N | V — C |
| | | Fe — C |

Taking into account the above-mentioned properties, these alloys cannot be regarded as simple interstitial solutions. In these situations it is necessary to consider the nature of the bonding forces between components, that brings these substances closer to the plain chemical compounds. In general, such bonds involve *sp* hybridized orbitals. A notable stability is obtained for some cubic structures, because these obey simultaneously the requirements of metal and non-metal bonding configurations.

The steel structure is an example of such an alloy, and one of the greatest technical importance. The iron — carbon system is of a great complexity, on the account of the small size of the iron atoms; the ratio of iron to carbon ionic radii is ~1.6, which allows the formation of either interstitial-type alloys, or of even more complex structures. On the other hand, complexity results also from the iron's polymorphism. Thus, up to 910 °C iron has the body-centred cubic structure called $\alpha$Fe, between 910 and 1400 °C it displays a face-centred cubic one called $\gamma$Fe, and up to 1539 °C (the melting point) it has again a body-centred cubic arrangement called $\delta$Fe. Moreover, the physical properties of steel depend, besides the number and nature of co-existing phases, on the existence and mutual arrangement of the microcrystallities (alloy microstructure). Therefore, different varieties of steel can be prepared, on a wide range of properties.

## 2.1.2. Phase Transitions

The aggregation phases of the substance depend primarily on temperature and pressure, but for more complex substances such as the alloys, other parameters, like composition, also play a decisive role.

An overview on this complex relationship can be obtained from the *phase diagrams* , that express the relation between temperature, pressure and the system's physical state. A phase diagram for a single-component system is given in Fig. 2.4. The curves (a), (b), (c) mark the limits of three zones, corresponding to the three physical phases: solid (S), liquid (L) and gaseous (G). Curve (b), separating the solid and liquid phases, expresses the pressure dependence of the melting (crystallization) temperature. The curve (c), which separates the liquid and gaseous phases, corresponds to the vapour equilibrium pressure at evaporation or condensation. The curve (a) which separates the solid and gaseous states corresponds to the equilibrium pressure at sublimation. The three curves meet in a single point, called the *triple point* (A). For $P = P_0$ and $T = T_0$ the three phases coexist. Point B at the upper end of the curve (c), is the *critical point*; for $P = P_c$ and $T = T_c$ one can not distinguish the boiling liquid and the saturated vapours as separate phases.

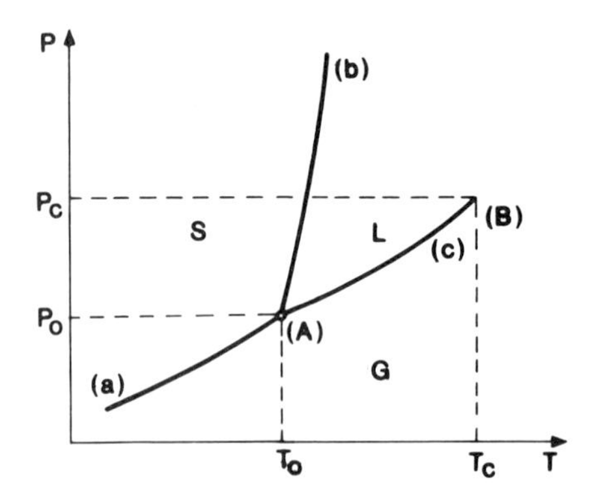


Fig. 2.4. Equilibrium (phase) diagram for a single-component substance.

For multicomponent systems the most common representations are the *equilibrium diagrams* (analogous to the phase, or state, diagrams) in which the relation between temperature, composition and physical state of the system in equilibrium conditions is graphically expressed. In such diagrams the pressure is supposed to be constant. Equilibrium diagrams highlight the changes in the state of aggregation, or the phase transformations of the system. These diagrams can be derived from the *cooling curves*. An example is the well-known iron — carbon diagram.

The term "phase transformation" (transition) is more general than "state transformation", because it includes all transformations from one phase to another that are characterized by a sudden change in a physical property. If the phase transformation is accompanied by a latent heat, it is said to be of the first kind. Examples are transformations of the aggregation state. If the transformation does not involve latent heat it is said of the second kind. First-kind phase transitions feature discontinuities in some thermodynamic quantities (internal energy, enthalpy, etc.), whereas second-kind transitions entail discontinuities in the first-order derivatives of thermodynamic potentials (jumps in the specific heat, etc.). Examples of second-kind phase transitions are the transformations ferromagnetic — paramagnetic; conducting — superconducting; and order — disorder.

In Fig. 2.5 the equilibrium diagram of the bismuth — cadmium system is shown. It is characteristic for the case (i) of paragraph 2.1.1. By increasing the cadmium concentration, the solidification temperature of the mixture decreases gradually along the curve AB. A "*eutectic point*" at 144 °C and a concentration of bismuth of ~ 45 at % can be singled out in the diagram. It indicates the separation, from the liquid phase, of several phases at the lowest solidification temperature of the system. The metallographic constituent resulting from the solidification is called, in this particular instance, a *eutectic* ;if, on similar terms, it results through transformations in solid state, it is called a *eutectoid*. As the Cd concentration increases, the solidification temperature slides along the curve BC. The AB and BC curves represent the beginning of solidification; above them, the system is in liquid state. The horizontal line passing through B represents the end of the solidification process; at temperatures below this line the system is solid.

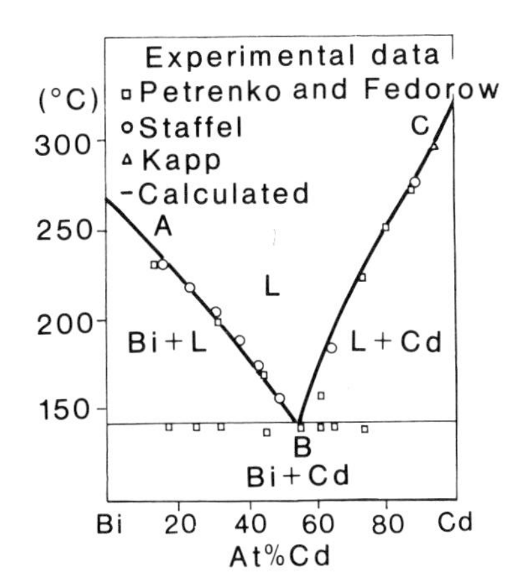


Fig. 2.5. Equilibrium diagram of the Bi — Cd system.

A representative example for the case (ii) in paragraph 2.1.1 is the U — Zr system (Fig. 2.6) [7]. Above 900 °C $Zr_{\beta}$ and $U_{\gamma}$ are completely soluble, in any proportions, evolving a solid solution or, above the melting temperature, a liquid that is atomically homogeneous. Below 900 °C, due to Zr to U phase transformations (Zr passes into its α-phase and U into its α- or β-phase) there is only a partial solubility in solid state and the system becomes more complex. In this category enters also the Zr — Ti alloy, that has a large mechanical strength at high temperature (Fig. 2.7 [3]). Below 545 °C an solid solution is formed, regardless of the concentration of the components, whereas in the 890 — 1580 °C temperature range a β solid solution is obtained. A different situation is encountered with the U — Th alloys (Fig. 2.8 [3]). Here, a partial solubility in liquid phase is evidenced.

Finally, the case (iii) in 2.1.1 is illustrated by the U — C system [7]; one region of its phase diagram is given in Fig. 2.9 [3]. The presence of the UC, $UC_2$ or $U_2C_3$ is noted. A similar behaviour is exhibited by the $UCl_4$ — KCl system (Fig. 2.10 [3]). This diagram can be decomposed into two diagrams of type (i) (see for example Fig. 2.6) — the first for the system KCl — $K_2UCl_6$ (the part ABC),and the second for $K_2UCl_6$ — $UCl_4$ (the portion CDE). For a composition of 33 mol % $UCl_4$, the compound $K_2UCl_6$ emerges. It does not decompose before melting.

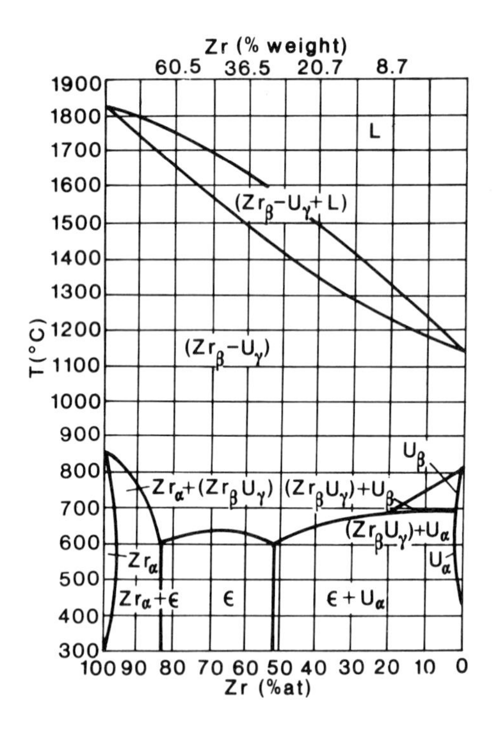


Fig. 2.6. Equilibrium diagram of the U − Zr system.

All these diagrams, as determined experimentally from the cooling curves, indicate, at the same time, the phase transformations that the system can undergo.

## 2.2. LATTICE DEFECTS

### 2.2.1. Types of Defects

In given conditions of temperature and pressure one may consider as crystalline solids those structures which exhibit an order in the positionning of the constituent atoms, that expresses in the very existence of the crystalline lattice, and preserve their shape and properties over a long time.

A series of studies have pointed out that the properties of materials depend to a considerable extent on the defects that exist in the crystalline lattice. This remark says not only that, on occasions, defects would worsen the physical properties of materials but, on the other hand, that some properties of genuine interest are largely determined by the very defects, their type, movement over the lattice and the way they have been generated.

One may say that, in actuality, no material is ideal. For example, a metal is a conglomerate of microcrystals (a mosaic), where each microcrystal is, in turn, a set of monocrystals of nearly the same orientation. Not even these monocrystals are perfect (ideal), since they also have certain inner defects.

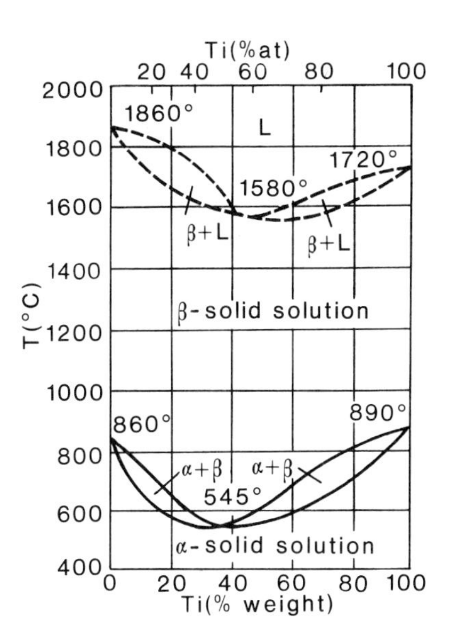


Fig. 2.7. Equilibrium diagram of the Zr — Ti alloy.

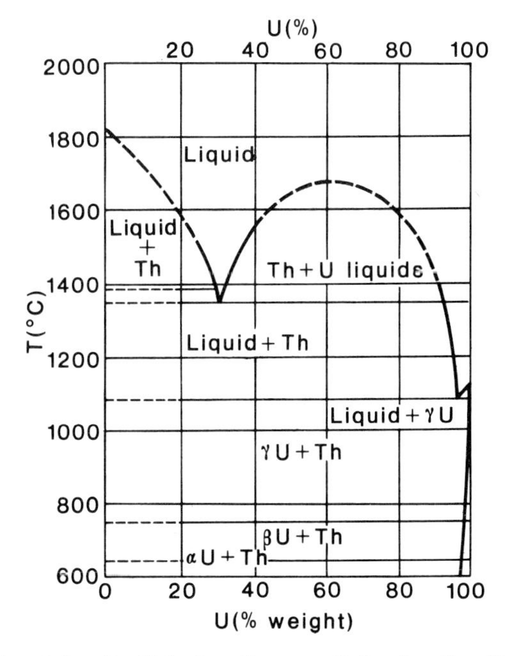


Fig. 2.8. Equilibrium diagram of the U — Th alloy.

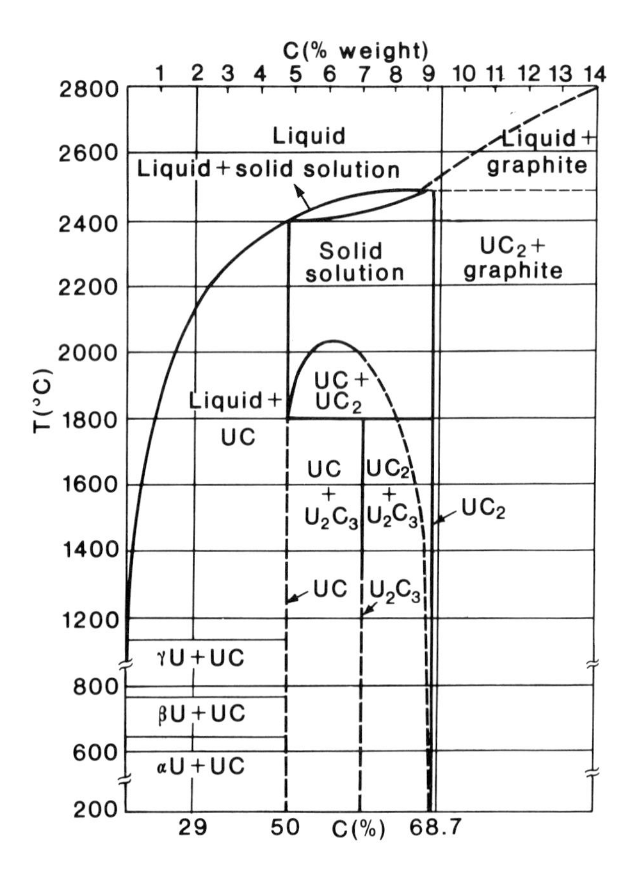


Fig. 2.9. Equilibrium diagram of the U — C system.

Defects may be classified in several ways: they may be *point-*, *linear-*, *planar-* and *volume*-defects. Defects can also be connected to *lattice imperfections*, *impurities* , *disorder* of the constituents over equivalent lattice points, as well as to *valency-disordered structures* (different valency states of similar ions), to global or local *non-stoichiometry*, etc.

As to the mechanical properties of solids, the most important defects are the *vacancies, interstitial atoms, substitutional impurities* (point defects); *dislocations*(linear defects); *grain boundaries* (planar defects); *microfissures, pores, voids, texture, inhomogeneities* (volume defects). As a result of mechanical, thermal and irradiation stresses, defects can be generated, displaced, clustered or eliminated. At present the dynamics of these processes is, to a considerable extent, elucidated, although one cannot say that all problems are completely solved.

Let us pause first upon the way in which the mechanical properties of the materials can be expressed,and then analyse the effect of radiations upon them.

The elastic — plastic properties of materials are mainly determined by the following mechanism: external mechanical stresses, as well as stresses induced

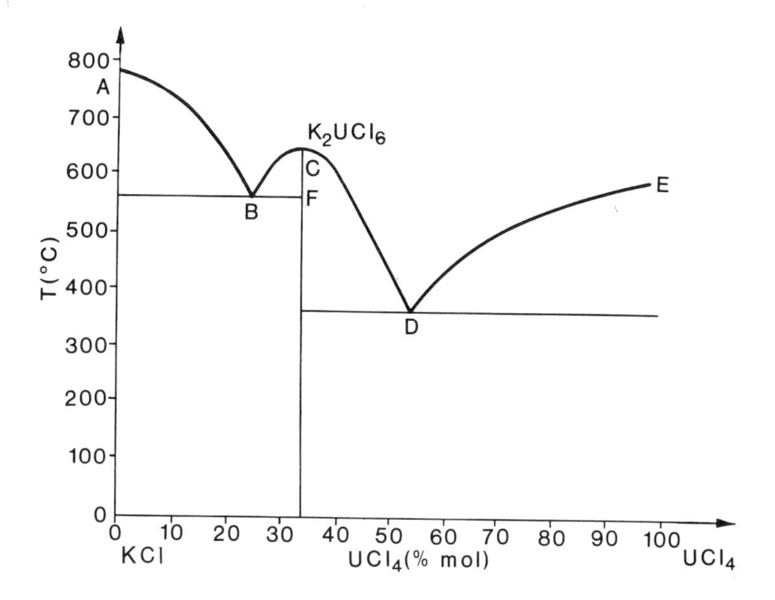


Fig. 2.10. Equilibrium diagram of the $UCl_4$ — KCl system.

by thermal and radiation fields cause the displacement of point defects and dislocations. Under the effect of the same factors these defects can multiply. By a specific process of interaction defects can cluster, migrate and increase in dimensions, leading gradually to a inhomogeneity, on a microscopic scale, of the chemical bonds between atoms. Then the stresses will concentrate gradually upon some zones of the material, determining, at the beginning, the activation of some atomic slip-planes, continuing with generation of micro-fissures and ending up with internal fractues and their gradual expansion. The stage of inelastic, permanent, strain may become manifest in the phenomenon of creep (viscous flow) at small stress, or in plastic strains at stresses that go beyond the limit of elastic strain.

The generation, development, and clustering of defects is a process specific to any given material and the understanding of these mechanisms allows finding of appropriate solutions to improve the qualities of materials.

### 2.2.2. Point Defects

The atoms that form the crystalline lattice are in equilibrium in certain sites. Leaving these sites requires jumping over a barrier of potential and, therefore, a certain energy (called *activation* energy). At thermal equilibrium the most frequent point defect in metals is the vacancy, also known as a "Schottky defect" (the absence of an atom from a site of the crystalline lattice), because its generation requires the least increase in the free energy as compared to other types of defects. When the atom is in an interstitial site, the pair vacancy — interstitial atom is a "Frenkel defect". Because vacancies and interstitials are defects in thermal equilibrium, their number increases with temperature. Point defects may also be introduced by plastic deformation, lack of stoichiometry, quenching from high-temperature, particle-beam irradiation, etc. These control the mobility of the atoms in solids. Point defects can be thermally annealed, either by vacancy — interstitial annihilation or by migration of defects up to the outer surface of the crystal and to the grain boundaries. Defects can cluster, forming voids, or cavities.

A real crystal, having $N$ ordered atoms in a perfect lattice and $N_v$ vacancies has an energy given by:

$$E_o = N\varepsilon_o + N_v\varepsilon_v , \tag{2.1}$$

where $\varepsilon_o$ stands for the average energy per atom in an ideal lattice, and $\varepsilon_v$ is the average energy for the generation of a vacancy. At thermal equilibrium their number will be obtained by minimizing the Gibbs free energy. The fraction of unoccupied sites in the lattice is given by:

$$\frac{N_v}{N_v + N} = \left(\exp\left(\frac{S_v}{k_B}\right)\right) \cdot \exp\left(-\frac{\varepsilon_v}{k_B T}\right) , \tag{2.2}$$

where $k_B$ is Boltzmann's constant,and $S_v$ is the excess-entropy at the generation of a vacancy. Typically, $\exp(S_v/k_B)$ has a value somewhere between 5 and 50. The energy to generate a vacancy is of the order of (1...2)eV(~2 . $10^{-19}$ J). Illustratively, it turns out that for $\varepsilon_v = 1$ eV, $T = 1270$ K and $S_v \simeq 0$ the vacancy fraction is ~$10^{-4}$.

The same type of evaluation may be applied to the formation of interstitials out of the same type of atoms which constitute the lattice. The energy of formation of an interstitial is somewhat greater than that for the formation of a vacancy, and the equilibrium concentration is very small. The concentration can increase due to the generation of *Frenkel pairs* as, for example, is the case with the irradiation of graphite, the strtucture of which provides a fairly good stability at ambient temperature. Removing these defects can be achieved by annealing above 200 $^{o}$C (see section 2.6.4). Another type of interstitials are the foreign atoms, for example, xenon atoms generated at the fission of uranium.

If, originally, the volume concentration of point defects is not the same over the entire bulk of the sample, then their migration will ultimately lead to an equalization of the concentration. This process of diffusion (migration) can be described by *Fick's* second law:

$$\frac{\partial c}{\partial t} = D\nabla^2 c + Q , \tag{2.3}$$

where $D$ is the diffusion coefficient of the given species, $c$ the volume concentration, $Q$ the rate of formation of defects in the unit-volume.

Such a process is witnessed when the fission products formed within the fuel migrate. The diffusion coefficient depends on temperature according to the law of thermally activated processes:

$$D = D_o \exp(-E/(k_B T)) , \tag{2.4}$$

where $E$ is the activation energy for a jump of the defect from one site of the lattice to another. Depending on the diffusion mechanism associated to a given defect, the diffusion activation energy may be chosen either as the migration energy of the defect, as, for example, is the case with foreign atoms and defects induced by irradiation, or as the sum of migration and formation energies, as in the case of vacancies in equilibrium with the lattice or in the case of self-diffusion. Experimental measurements have revealed that the activation energy for the diffusion of hydrogen in metals is of ~10 kJ/mol, and for uranium diffusion in $UO_2$ is ~500 kJ/mol. In comparison to the data in Table 2.4, these values represent extreme cases [8], corresponding to the two examples above.

TABLE 2.4 Diffusion Parameters, in Several Cases of Interest

| *Material* | | *Diffusion coefficient* | *Activation energy* | |
|---|---|---|---|---|
| | | $D_0$ ($cm^2$/s) | $E$(eV) | $E$(kJ/mol) |
| Fe (bcc) | self-diffusion | $1.0 \cdot 10^{-2}$ | 3.0 | 288.8 |
| Cu | self-diffusion | $2.0 \cdot 10^{-5}$ | 2.05 | 197.2 |
| Ta | self-diffusion | $2.0 \cdot 10^{-4}$ | 4.78 | 460.5 |
| Nb | self-diffusion | $1.3 \cdot 10^{-4}$ | 4.13 | 397.7 |
| Ge | self-diffusion | $7.8 \cdot 10^{-4}$ | 3.0 | 288.8 |
| αU | self-diffusion | $4.5 \cdot 10^{-4}$ | 1.83 | 176.6 |
| βU | self-diffusion | $2.8 \cdot 10^{-3}$ | 1.92 | 185.2 |
| γU | self-diffusion | $1.19 \cdot 10^{-3}$ | 1.16 | 111.9 |
| γPu | self-diffusion | $2.1 \cdot 10^{-5}$ | 0.72 | 69.5 |
| δPu | self-diffusion | $4.5 \cdot 10^{-3}$ | 1.03 | 99.4 |
| $\delta_2$Pu | self-diffusion | $1.6 \cdot 10^{-2}$ | 1.14 | 110.0 |
| εPu | self-diffusion | $2.0 \cdot 10^{-2}$ | 0.80 | 77.2 |
| αZr | self-diffusion | $2.1 \cdot 10^{-7}$ | 1.17 | 112.9 |
| βZr | self-diffusion | $4.2 \cdot 10^{-5}$ | 1.04 | 100.3 |
| Cu in Al | | $2.3 \cdot 10^{-4}$ | 1.45 | 140.2 |
| Th in W | | $1.0 \cdot 10^{-4}$ | 5.4 | 519.1 |
| C in Fe (bcc) | | $2.0 \cdot 10^{-5}$ | 0.9 | 85.8 |
| Zn in Cu | | $3.4 \cdot 10^{-5}$ | 1.98 | 190.1 |
| Fe in Nb | | $1.5 \cdot 10^{-4}$ | 3.38 | 325.2 |
| Co in Nb | | $0.7 \cdot 10^{-4}$ | 3.06 | 295.1 |
| P in Ge | | $2.0 \cdot 10^{-4}$ | 2.5 | 240.7 |
| B in Ge | | $4.0 \cdot 10^{-4}$ | 4.5 | 439.5 |
| U in V (1373-1773 K) | | $1.0 \cdot 10^{-7}$ | 2.66 | 256.6 |
| U in Nb (1993-2373 K) | | $5.0 \cdot 10^{-6}$ | 3.33 | 321.3 |
| U in Ta (1873-2423 K) | | $7.6 \cdot 10^{-5}$ | 3.66 | 353.1 |
| U in Mo (2073-2373 K) | | $1.3 \cdot 10^{-6}$ | 3.28 | 316.5 |
| U in W (1973-2473 K) | | $2.0 \cdot 10^{-3}$ | 4.49 | 433.2 |

The diffusion process, which is random and isotropic, can be altered if a gradient of potential is manifest in the material. This gives the defects a directional movement which overlaps with the isotropic diffusion. Such situations may be encountered when there is a temperature gradient in the sample, or under a non-uniform stress of the sample, etc. Due to a thermal gradient, a diffusion process may appear, preventing, for example, the concentration of a two-component system becoming uniform [9].

As an example of diffusion, one may mention the migration of hydrogen toward the cooler regions in the Zircaloy cladding of a fuel rod, in light water reactors. Thus,zones of the cladding at lower temperatures become more brittle. Another example is the migration of plutonium toward the warmer central regions of the fuel rod, in fuel elements made of a mixture of $PuO_2$ — $UO_2$. This migration affects in an undesirable manner both the neutron flux distribution and the thermal characteristics of the fuel element [10].

The movement of defects under the influence of a field of internal stresses may be illustrated by the process of trapping vacancies or interstitials on impurities, dislocations, internal surfaces or volume defects. The types of lattice defects create around them a region of high stress which acts as a trapping centre. As soon as a vacancy or an interstitial comes across such a region, it moves right down to the centre of the defect. As a consequence there comes a gradual accretion of imperfections which can either grow, or move. On this property, one of the methods of identifying the dislocations has been founded: the *method of decoration*. To this purpose impurity atoms are dissolved into the material; later they would precipitate preferentially on lattice defects. An examination in visible light or infrared allows the direct observation of the precipitation result — thus, of the regions with defects.

Trapping of gas bubbles (xenon) on the dislocation lines represents another example. At temperatures around 1900 K the bubbles migrate easily over distances of microcrystallite dimensions, and over intervals of days or months. If on their way they encounter microfissures, they may escape from the material.

### 2.2.3. Dislocations and Creep

Among the other types of defects, the most important from the point of view of mechanical properties of materials are dislocations and grain boundaries [11,12]. Unlike the point defects, they are not in a state of thermodynamic equilibrium, because their formation energy is much too large to be overtaken by the bound energy that implies the configurational entropy with which they contribute to the free energy. These appear in the process of crystallization (as the solid evolves from the liquid phase), or as a consequence of stress, or work done upon the material. Such defects cannot be completely annealed.

Plastic properties of materials can be largely explained on the basis of these two types of defects.

The dislocation, as a structural defect in crystals, has been introduced by the necessity to explain the great difference between the actual magnitude of the fracture strength and the one calculated with a perfect crystal model. Thus, the theoretical fracture strength of a crystal could reach magnitudes up to $10^4...10^5$ MPa, whereas actually measured values are approximatively three orders of magnitude lower.

Because the fracture generally appears in a crystalline plane (as it happens with the near-perfect crystals, for instance, the cleavage of NaCl crystals, etc.) the lowering of the fracture limit can only be caused by the imperfections existing at microscopic level. Such impurities are, for example, the dislocation lines. The most important types of dislocations are the *edge dislocation* and the *screw dislocation*. Groups of these can evolve as, for example, *loop dislocations*.

The edge dislocation may be seen as a crystalline half-plane inserted between the crystalline planes of the perfect crystal (Fig. 2.11). The edge of the half-plane is called the dislocation line. The order of atoms in the lattice, some lattice constants away from the dislocation line, does not reveal any imperfection. Yet, if one surrounds a dislocation line with a geometric circuit, in a plane perpendicular to it, in such a way as to pass through an equal number of lattice sites along the opposite edges of the loop, unlike in the case of a perfect lattice one will not reach the same site one has started with. The difference has been called the *Bürgers vector* (Fig. 2.11). The symbol for the dislocation line in this case is ⊥. The edge dislocation is characterized by a dislocation line that is perpendicular to the Burgers vector. A shear stress (denoted by $\sigma$ in Fig. 2.11) will determine a slip of atomic planes along a slip plane. Because such a slip requires an amount of energy much smaller than the displacement of a whole atomic plane, the presence of the dislocation reduces by several orders of magnitude the maximum shear stress, equal to 0.1 $G$ for perfect crystals ($G$ = the shear modulus).

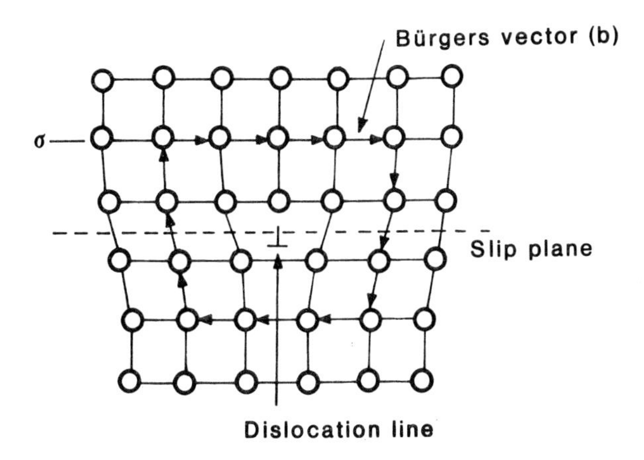


Fig. 2.11. Edge dislocation.

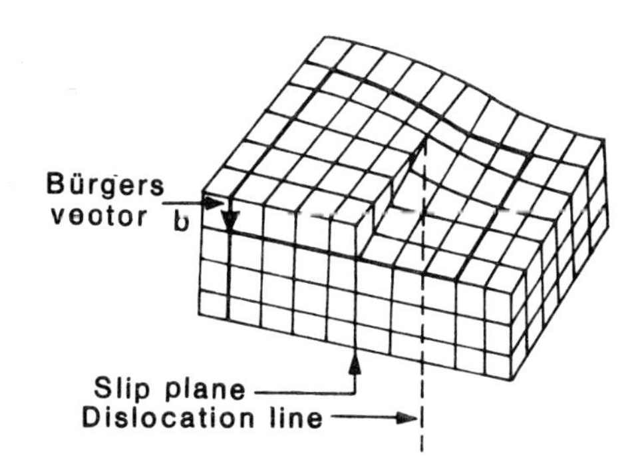


Fig. 2.12. Screw dislocation.

In the case of screw dislocations the Bürgers vector is parallel to the dislocation line (Fig. 2.12).

For both kinds of dislocations, the dislocation line can move along a perpendicular direction, the line lying in the slip plane. The shear stress in the case of this displacement is much lower than the yield point for perfect crystals. Apart from the movement of the dislocation caused by shear stresses in the slip plane, a dislocation can also move by "climbing". This mechanism consists either in a diffusion of atoms next to the dislocation line towards the vacancies, which causes the supplementary half-plane of the edge dislocation to shrink, or in the diffusion of atoms toward the dislocation line, in which case the half-plane expands. This process of dislocation movement is important in order to understand the phenomenon of stationary creep in metals, in which case the activation energy is just the energy for atomic self-diffusion in pure metals.

In crystals there are dislocations which do not exhibit either a pure screw or a pure edge character. This is the case of dislocation loops, where the two kinds of dislocation coexist and pass smoothly from one another, as long as one follows the dislocation line (Fig. 2.13). Unlike the simpler kinds, where the dislocation lines end up on the crystal surface, dislocation loops are confined entirely inside the crystal.

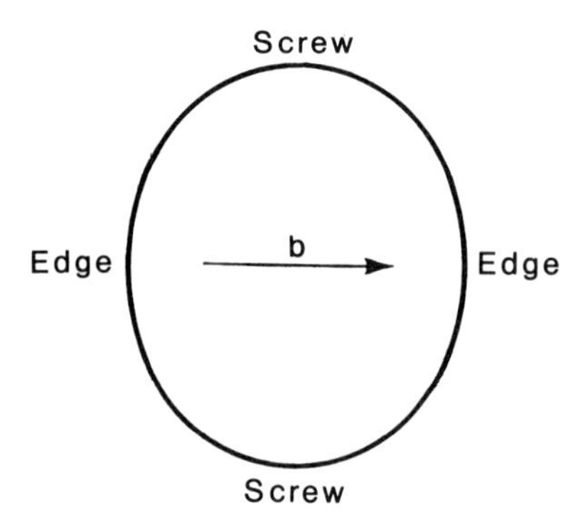


Fig. 2.13. Dislocation loop in a slip plane.

The strain field that a dislocation creates is uniformly distributed along the dislocation line. This makes the energy density associated with the strain of the medium along the dislocation line constant. It may be shown that, irrespective of the dislocation type, the total strain energy per unit-length of dislocation line is [11]

$$W_{\tau} \sim Gb^2 , \qquad (2.5)$$

where $G$ is the shear modulus and $b$ is the Bürgers vector.

Generating, stretching, deforming or displacing a dislocation requires appropriate work. Yet the movement of dislocations and, therefore, the ability of a material to deform plastically is limited by a series of phenomena. For example, in a material severely strained the moving dislocations intersect each other. Therefore, dislocation steps may appear, which diminishes their mobility. Likewise, the steps lengthen the dislocations and thus such crossing needs supplementary energy. A stongly cold-worked metal, containing a large density of dislocations, becomes less plastic — it hardens up. This phenomenon is known as work-hardening. For the material to become plastic again it should be recrystallized. Recrystallization needs a temperature of ~ 0.4 $T_f$, where $T_f$ is the metal's absolute melting temperature.

In the presence of impurities the movement of dislocations is dramatically slowed down. An impurity which yields a strain field in its neighbourhood may block an edge dislocation. The energy of the impurity — edge dislocation system decreases as long as the two approach each other, their ensemble thus forming a stable system.

Compact crystallographic structures (fcc, hcp) may be described as being formed out of superposed atomic planes; each plane may be obtained from the neighbouring one by a translation that is either perpendicular, or parallel to the plane (Fig. 2.14a and 2.14b).

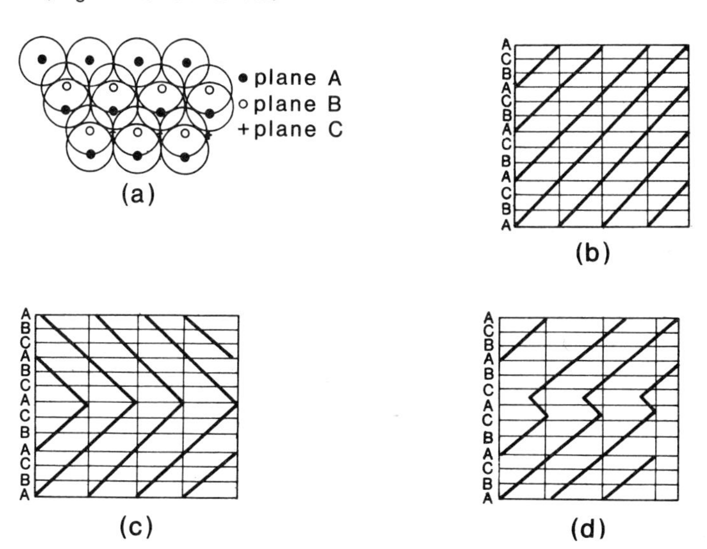


Fig. 2.14. Stacking faults: (a) and (b) the method of generating atomic planes in a crystal; (c) and (d) stacking faults.

For example, a pattern like ABCABC... is characteristic of the fcc structure, whereas a pattern that reads ABAB... is characteristic of the hcp structure.

The sequence of atomic planes may be modified, for example to read ABCABCACBACB... This is illustrated in Fig. 2.14c for the fcc structure. The alteration of the ABCABC... pattern of the perfect crystal is called a *stacking fault*. An example of such a defect is given in Fig. 2.14d.

If the stacking fault is confined only to the inside of the crystal, its edge may be considered a dislocation with a Bürgers vector shorter than the lattice translation vector, which explains why this type of dislocation is sometimes called *partial* or *imperfect* dislocation.

A stacking fault would thus be an extensive region limited by partial dislocations, that presents a packing order different from the regular one.

Normally, a metal consists in a compact collection of microcrystals (crystallites, grains), each microcrystal being a formation which contains point defects and dislocations. Microcrystals are separated by zones where the crystalline structure is strongly distorted. Such a region has a thickness of about 1 nm (Fig. 2.15a). In between microcrystals, there could be boundaries which are much less distorted (Fig. 2.15b), made of dislocations which form a separation surface between two regions of an ordered structure. The orientation of crystallites in a solid is never perfectly aligned to a given direction. If the departure from the chief orientation is small ($\vartheta < 10^{0}$) then the defect which appears is called *small-angle*, or *tilt, boundary*. At sufficiently high temperatures the dislocations may arrange themselves in regular arrays, in patterns like those presented in Fig. 2.15b. Prolonged annealing will modify this mosaic structure, because larger microcrystals will grow in dimensions.

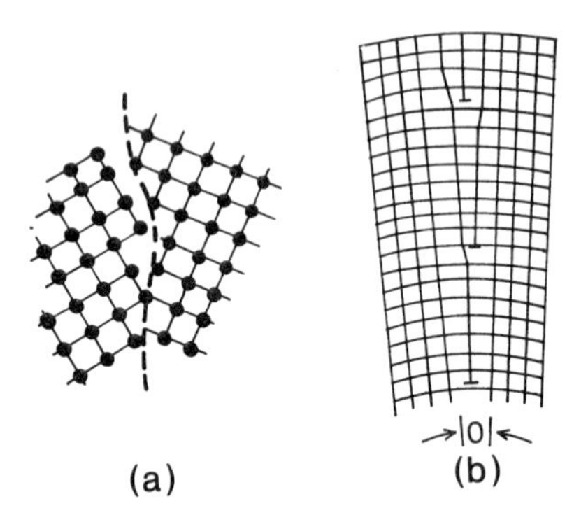


Fig. 2.15. Grain boundaries: (a) large-angle grain boundary; (b) small-angle grain boundary.

Mechanical properties which come out from these facts are important. Thus, because of the different orientations of microcrystals, to induce a plastic deformation much more effective slip planes are required by comparison to the case of monocrystals. If only a small number of slip planes are active (Fig. 2.16a), then the random orientation will in some microcrystals make the shear stress exceed the critical values, whereas in some others this would not happen. As a consequence, the plastic strain will only be produced in the moment when the component of the shear stress along an active slip plane exceeds the critical value for all the orientations relative to the applied stress (Fig. 2.16b).

When subjecting a body for a long time at constant stress another phenomenon occurs, which is referred to as *creep*. Creep is a continuous, irreversible deformation that occurs in samples under stress over a longer time. A typical creep curve is given in Fig. 2.17.

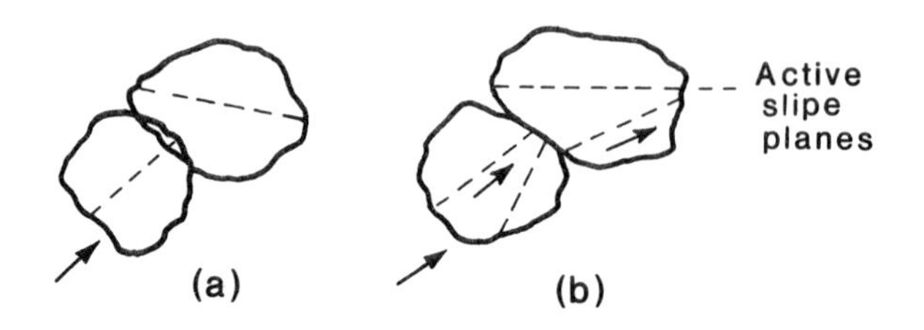


Fig. 2.16. Influence on the plastic strain of the lack of orientation of microcrystals: (a) small number of active planes: (b) a larger number of active planes.

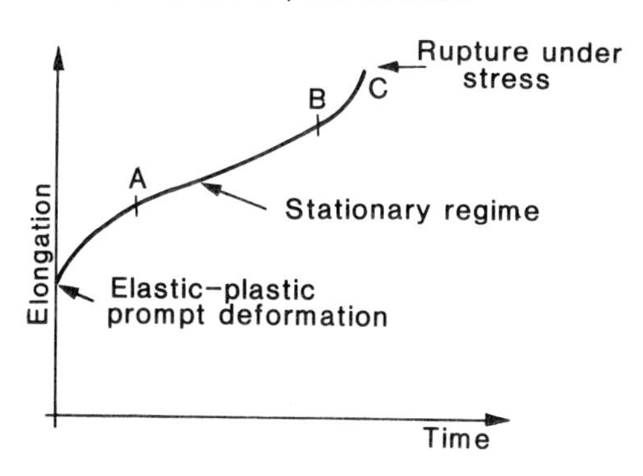


Fig. 2.17. Typical creep curve.

Experimentally, creep presents three stages: primary creep, during which the creep rate slows down (up to A); secondary, or steady-state, creep (from A to B) where the creep rate stays essentially constant; and tertiary creep (from B to C) on which the creep rate soars shortly before failure occurs, which is called stress-rupture.

As a phenomenon of slow, irreversible plastic deformation, creep is a result of thermal fluctuations which make those dislocations and slip planes not normally active at a particular time to become gradually free, by thermal activation. The probability per unit time of supplying the energy ($E$) needed to get a dislocation moving is proportional to the Boltzmann factor $\exp(-E/(k_B T))$. Hence the creep rate exhibits a very pronounced temperature dependence.

A quantitative analysis gives the creep rate ($d\varepsilon/dt$) during the regime of secondary creep:

$$\frac{d\varepsilon}{dt} = \text{const}.\sigma^n \exp(-E/(k_B T)). \tag{2.6}$$

where $\varepsilon = \Delta l/l$ is the strain, $\sigma$ is the applied shear stress, $n$ is a constant equal to 4 for creep rates governed by dislocation displacement, $T$ is the absolute temperature, and $E$ is the creep activation energy.

Creep may also occur by sliding of adjacent crystallites along grain boundaries, or by diffusion of vacancies from one side of a grain to the other. This last case is called the *Nabarro – Herring creep mechanism*.

## 2.3. INFLUENCE OF PHYSICAL AND CHEMICAL AGGRESSIVENESS ON MATERIAL PROPERTIES

In reactors, materials are permanently subject to two types of stresses, apart from the mechanical and thermal ones. The first stems from their environment, and is typical for all equipment operated in chemically and physically aggressive media. The other is specific, and results from irradiation; it is related to the generation and diffusion of defects and impurities in the material. In the first category are, for example, *physical erosion* appearing in the pipes under forced fluid circulation; *chemical and electrochemical corrosion* under the effect of the atmospheric agents, resulting from a direct contact of chemically reactive materials. In the second category fall the same kinds of aggressive effects, but enhanced due to the presence of intense radiation fields that feature the environment of a nuclear reactor, which may induce peculiar processes such as surface *blistering*, *creep*, modification of lattice dimensions (change of shape), formation of *bubbles* and *voids*. In this paragraph the first set of topics will be discussed; in other sections the effects specific to the reactor environment will be tackled, at an appropriate level of details (viz. section 2.6).

The normal physical and chemical action on materials is either due to physical processes of atoms entering their structure, or is induced by chemical reactions. Penetration effects are generally slow and, therefore, do not play a significant role unless some physical conditions favour them. Such conditions materialize when one has high temperature gradients (which is usually the case inside fuel rods), a strong irradiation regime, etc.

Apart from these conditions, the chemical reactions are in control of the changes, in time, of the properties of substances.

The main sources of chemical corrosion — as the chemical effect of the aggressive agent is known — are the air and industrial gases that contain oxygen, hydrogen, nitrogen, ammonia, water vapours, carbon oxides, etc. [13-15]. This gaseous environment induces such phenomena as *oxidation*, *decarburation*, *embrittlement*, *nitruration*, etc. Effects are worse when the medium is damp, and the temperature high. The effect of other impurities, contributing to the speeding up of the corrosion process, adds to the compound. Examples are the impurities that are chemically inactive (e.g. carbon), but have instead a great capacity of absorbing gases from the atmosphere, as well as the inactive, nonabsorbent particles which favour, instead, the phenomenon of capillary condensation (e.g. the sand).

Several cases may provide a better appraisal on the corrosive effects of fluid media. The oxidation of metals at high temperature goes, in general, by the reaction:

$$nM + (m/2)O_2 \rightleftharpoons M_nO_m.$$

As the metal covers with an oxide film, this may inhibit further corrosion, as is the case with aluminium. In the case of iron, at high humidity (> 60%) a spongy film of oxides ($xFeO \cdot yFe_2O_3 \cdot 2H_2O$) forms on the surface of the metal; yet it does not protect the metal from further oxidizing by the environment.

Oxidation leads, generally, to a worsening of the mechanical properties of materials, both by the fact that it diminishes in time the geometrical sizes of the components, and by the process known as *cracking corrosion* under stress. Under this type of corrosion the metal, or the alloy, is destroyed by formation of cracks, that result from the combined effects of the corrosive medium and of the residual, or external, static stresses applied upon the material. This kind of corrosion affects particularly the austenitic, ferritic, martensitic stainless steels, the lightly alloyed steels, as well as the aluminium alloys, nickel, copper, titanium alloys, etc.

*Alkaline (caustic) embrittlement* is another variety of corrosion under stress that appears in steel in the presence of alkaline media at high temperatures and pressures. This is extremely dangerous in the case of steam boilers, where there is a large concentration of salts. The salt crust, lining the water pipes of the boilers, consitutes a region of high alkalinity that attacks the metal:

$$Fe + 2NaOH + 2H_2O \longrightarrow Na_2Fe(OH)_4 + H_2.$$

Hydrogen, at high pressure and temperature, corrodes the steel causing embrittlement. Hydrogen diffuses in the bulk of the metal and reduces the iron carbide at the level of crystallite boundaries according to the reaction:

$$Fe_3C + 2H_2 \longrightarrow CH_4 + 3Fe.$$

The resulting methane accumulates in the regions rich in defects, creating local internal stresses that facilitate cracking. This kind of corrosion may be diminished by alloying with titanium, chromium, zirconium, tungsten and vanadium – all elements that form stable carbides.

Figure 2.18 indicates the temperatures at which the embrittlement of steel alloyed with stabilizing elements appears. The corrosion behaviour in different gases of several metals is given in Table 2.5 [15].

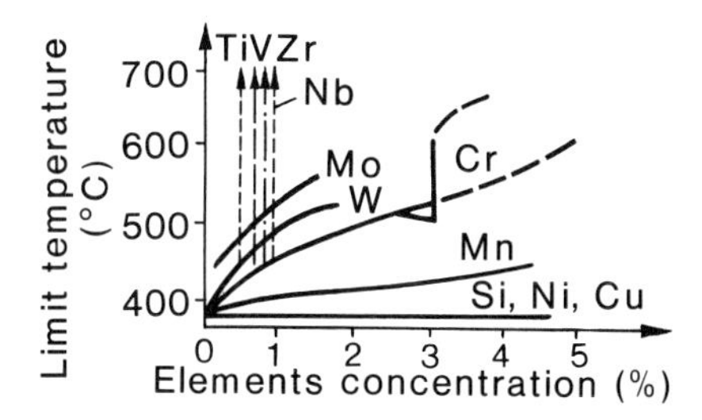


Fig. 2.18. Influence on stainless-steel embrittlement, of alloying elements.

TABLE 2.5 Corrosion Behaviour of Several Metals, in Gaseous Media

| *Metal* | *Temperature* °C | *Weight increment* [ . $10^{-2}$ $kg/m^2$] in 24 hours | | | |
|---|---|---|---|---|---|
| | | $O_2$ | $H_2O$ | $CO_2$ | $SO_2$ |
| Fe | 700 | 51.10 | 62.20 | 50.70 | 35.40 |
| Cr | | 0.47 | 0.05 | 0.27 | 0.16 |
| Ni | | 0.96 | 0.03 | 0.39 | 92.30 |
| Co | | 3.70 | 0.80 | 2.70 | 59.80 |
| Cu | | 12.00 | 3.30 | 6.50 | 0.13 |
| W | | 41.20 | 2.10 | 13.90 | 47.00 |
| Fe | 900 | 124.30 | 57.50 | 113.30 | 500.00 |
| Cr | | 2.20 | 1.20 | 1.30 | 3.20 |
| Ni | | 2.80 | 1.40 | 3.60 | 83.70 |
| Co | | 93.10 | 25.90 | 44.30 | 163.30 |
| Cu | | 43.90 | 15.30 | 12.30 | 0.20 |
| W | | 376.20 | 179.20 | 13.90 | 29.00 |

*Electrochemical corrosion* [15,16] is characterized by the fact that it is accompanied (and influenced) by the flow of an electric current through the corroded metal. The metal is in contact with an electrolyte in which solvated ions or complex combinations are formed. As a consequence, at the surface of the metal two types of reactions take place, that develop in parallel and conserve the electric charge:

1. the anode reaction of metal oxidation,generating positive ions and electrons

   $M \rightarrow M^{Z+} + Ze^-$;

2. the cathode reaction of reduction, by which the electrons left in the metal following the metal solvation are taken over by an acceptor in the solution. The cathode process may develop as follows:

(a) in an acid environment: $2H^+ + 2e^- \rightarrow H_2$;

(b) in an alkaline environment: $O_2 + 2H_2O + 4e^- \rightarrow 4(OH)^-$;

(c) in a neutral environment: $Fe^{3+} + e^- \rightarrow Fe^{2+}$ (in oxygenated solutions).

The effect of these processes upon metals depends on the behaviour of the outcome of the metal ionization. A soluble corrosion product would not protect the metal, while a less-soluble product will or will not protect the metal, depending on whether the film deposit is compact or spongy. In all these cases the pH of the solution is of particular consequence.

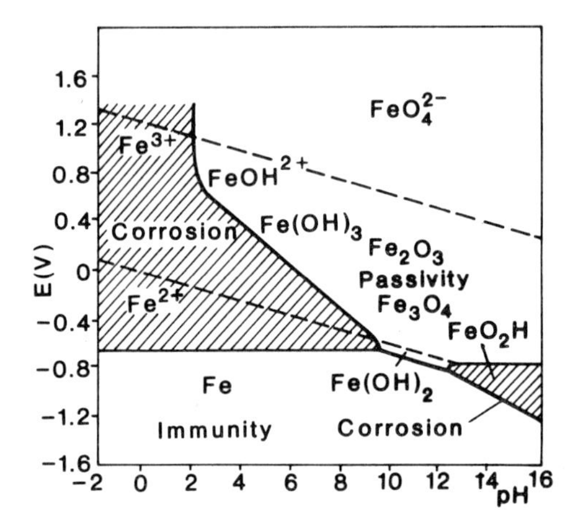


Fig. 2.19. Voltage — pH diagram (simplified) for the iron — water system, at 298 K.

A convincing example is drawn from the (simplified) diagram of thermodynamic stability under iron — water corrosion at 298 K (Fig. 2.19 [15]). On the abscissa is the pH, and in the ordinate is the potential. Areas separated by full lines are the regions of stability for the chemical species produced. In the immunity area the metal ionization reaction is not thermodynamically possible. In the passivity area the metal corrosion is thermodynamically possible, but it is prevented by the formation of a protecting film on the metal surface. In the corrosion area the solution of metal is accompanied by the following reactions:

$Fe \rightarrow Fe^{2+} + 2e^-$ (anode process);

$O_2 + 2H_2O + 4e^- \rightarrow 4(OH)^-$ (cathode process).

Corrosion products are $FeO$ and $Fe(OH)_2$.

The process of chemical corrosion is influenced by the purity and homogeneity of the metal, the mechanical quality of the surface (rugosity), and the temperature.

Thus, corrosion is a very complex process, hard to prevent, and sensitive to the actual local conditions.

## 2.4. MATERIAL CONSTANTS

The engineering and utilization of materials require a good knowledge of their physical properties. These are expressed as *material constants* — a term saying that, for pure substances, in given conditions, they have well-determined magnitudes. As a rule, when external conditions change these values vary only a little. The magnitude of the constants depends on the nature of the material and may, sometimes, be calculated on the basis of a microscopic model. Yet, in general, the material constants are determined from measurements. Because pure substances have constants that differ from those of mixtures, measuring the material constants also allows the characterization of the purity of the material under investigation.

Important constants that qualify the physical properties of materials are the *mechanical* constants (compressibility, elastic moduli, etc.); those featuring the *crystalline structure* (density, lattice constants, etc.), the *phase transformations* (melting temperature, boiling temperature, temperatures of polymorphic and magnetic transitions, etc.); the *thermodynamic, electric, optical* constants, etc. (see also Appendix 5).

Let us follow briefly the microscopic bases of the main properties of materials, expressed through material constants. One of the simplest microscopic models of a solid is the *two-atom model*, in which, under the action of binding forces, two different atoms form a molecule. The binding force depends on the manner in which the potential energy of interaction between atoms varies as a function of their mutual distance (Fig. 2.20). Two features are outstanding; namely the presence of a minimum in the potential energy at some distance $R_o$, and the asymmetry of the curve around this particular point. Knowing the function $E_p = f(r)$ one can infer the force required to either compress or stretch out this "ensemble":

$$F = -\frac{dE_p}{dr}. \tag{2.7}$$

In Figure 2.20 the dependence on distance of the force $F$ is indicated by a broken line. The minimum of $F(r)$, at $r = R_1$, corresponds to the inflexion point of the curve $E_p(r)$. The absolute magnitude of this minimum represents the maximal effort that the model-"solid" can withstand at stretching.

The aspect of the curve $E(r)$ indicates that, for $r > R_o$ mainly attraction forces are in the act, while for $r < R_o$ repulsion forces prevail. Such a behaviour can be modelled by choosing $E_p(r)$ of the form:

$$E_p(r) = -a/r^m + b/r^n, \tag{2.8}$$

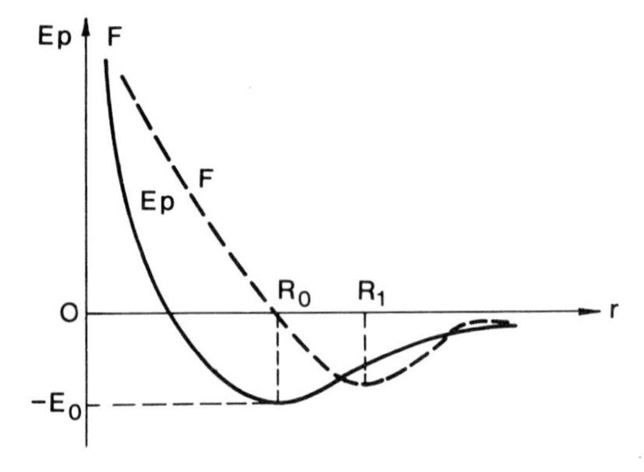


Fig. 2.20. Potential energy ($E_p$) and interaction force ($F$) versus the distance between two atoms.

where $m < n$, and $a$ and $b$ are constants. Using the condition $dE_p/dr = 0$ to identify the minimum, one gets the distance $r = R_0$ that corresponds to the equilibrium position of the two atoms:

$$R_0 = (nb/ma)^{1/(n-m)}. \qquad (2.9)$$

From knowing the constants $m$, $n$, $a$ and $b$, $R_0$ follows immediately; reciprocally, knowing $R_0$ and three other constants one can determine the fourth [12].

Now, assuming that this concept of mutual positioning of two different atoms stands valid to be generalized in the three-dimensional space, one gets a simple cubic crystalline structure, in which one takes that each atom is elastically bound to all its neighbours (Fig. 2.21).

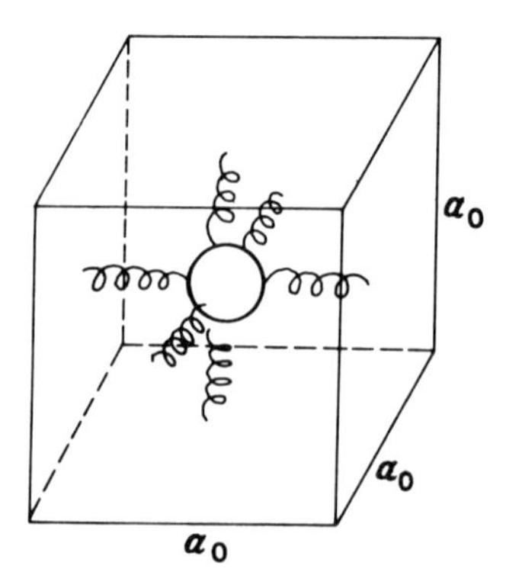


Fig. 2.21. One-atom elementary cell model [8].

On the assumption that the extrapolation to three dimensions does not modify in any essential way the equilibrium distance $R_0$ between atoms, one has

$R_0 = a_0$, $a_0$ being the cubic lattice constant. So, one comes at expressing the lattice constant as a function of the constants implied in the expression of the potential energy which, in turn, reflects the electronic structure of the atoms. If $a_0$ is the length of the edge of the (simple) cube that contains one atom of the lattice, the *density* of the solid is:

$$\rho = m_A / a_0^3 \tag{2.10}$$

where $m_A$ is the mass of the constituent atom.

In this way one gets the first material constant as a function of the parameters which characterize the interaction between the constituent atoms of the solid. Evidently this model is oversimplified; yet it is indicative of the manner in which a crystalline structure can be analysed.

Let us assume now that in a monoatomic "solid" a deviation $\Delta R$ from the equilibrium position $R_0$ appears, changing it into $R_0 + \Delta R$. In this situation,

$$F(r) = \frac{nb}{(R_0 + \Delta R)^{n+1}} - \frac{ma}{(R_0 + \Delta R)^{m+1}} . \tag{2.11}$$

Because, for $R = R_0$, the force $F$ vanishes, and taking into account that $\Delta R \ll R_0$, by expanding $F(r)$ in series of powers of $\Delta R / R_0$ one gets the simple expression:

$$F(\Delta R) = (m-n) \frac{nb}{R_0^{n+1}} \cdot \frac{\Delta R}{R_0} \tag{2.12}$$

Considering $N$ atoms, these will form a chain of a length $NR_0 = L_0$. The total deformation will be $N\Delta R = \Delta L$, and the force implied:

$$F(\Delta R) = \frac{(m-n)nb}{R_0^{n+1}} \cdot \frac{\Delta L}{L_0} = A \frac{\Delta L}{L_0} . \tag{2.13}$$

This is, obviously, *Hooke's law* for the elastic deformation of a homogeneous and isotropic body. The constant $A$ is proportional to the *Young modulus*, which can thus be determined from an analysis of the microscopic structure of the material. Actually, the values obtained for the Young modulus on the basis of such a model are far away from the experimental values, especially for metals.

In a similar manner, the *bulk modulus*, defined as:

$$K_b = -V \left( \frac{\mathrm{d}p}{\mathrm{d}V} \right)_T \tag{2.14}$$

can be determined, where $V$ is the volume of the solid and $p$ is the pressure exerted upon it. At the temperature of absolute zero, the entropy variation vanishes and one can write $\mathrm{d}U = -p\mathrm{d}V$, so that

$$K_b = -V \frac{\mathrm{d}^2 U}{\mathrm{d}V^2} \tag{2.15}$$

As in these conditions the kinetic energy may be neglected, the energy $U$ of the crystal is given by the potential energy of all interacting constituent atoms $E_{tot}$, represented in Fig. 2.20.

For the various types of crystals, the interaction energy may be expressed as follows:

(a) for *inert gas crystals* (Van der Waals forces):

$$U(r) = -a/r^6 + b/r^{12} \quad \text{(Lennard – Jones potential);} \tag{2.16}$$

(b) for *ionic crystals* (electrostatic forces):

$$U(r) = \lambda \exp(-r/p) \pm q^2/(4\pi\varepsilon_0 r), \tag{2.17}$$

where the signs ± consider the ions of the same and opposite charges, respectively;

(c) for *metals*:

$$U(r) = -A/r + B/r^2. \tag{2.18}$$

The attraction forces are, in this case also, of electrostatic type. The repulsive term is determined by the kinetic energy of the free electrons.

Several constants that characterize the mechanical properties of some materials are given in Table 2.6.

In the case of isotropic and homogeneous bodies the elastic material constants are interdependent. Using the notations in Table 2.6 one gets:

(a) the *shear modulus*:

$$G = \frac{E}{2(1 + \mu)}\,; \tag{2.19}$$

(b) the *longitudinal velocity of sound*:

$$v_l = (E/\rho)^{1/2} \tag{2.20}$$

(for cylindrical rods of a diameter much inferior to the sound wavelength);

(c) *coefficient of compressibility* $\quad \varkappa = -\frac{1}{V}\left(\frac{\partial V}{\partial p}\right)_T = \frac{1}{K_b}$

$$\varkappa = 3(1-2\mu)/E. \tag{2.21}$$

The above considerations are valid for strains not too large and at temperatures sufficiently low. At larger deformations the *asymmetry* of the potential curve with respect to the equilibrium position $R_0$ becomes relevant (Fig. 2.20). In general it is much easier to stretch a body than to compress it. Therefore, at large stresses the stretching modulus differs from the

TABLE 2.6 Some Mechanical Constants of Materials

| *Material* | *Density* (ρ) ($10^3$ kg/m$^3$) | *Elasticity modulus (E)* (Pa) | *Shear modulus (G)* (Pa) | *Poisson coefficient*, μ | *Longitudinal sound velocity*, $v_l$ (m/s) |
|---|---|---|---|---|---|
| Aluminium | 2.69 | $0.73 \cdot 10^{11}$ | $0.26 \cdot 10^{11}$ | 0.34 | $5.1 \cdot 10^3$ |
| Carbon steel | 7.7 | $(2-2.1) \cdot 10^{11}$ | $0.81 \cdot 10^{11}$ | 0.24-0.28 | $5.1 \cdot 10^3$ |
| Copper | 8.96 | $1.2 \cdot 10^{11}$ | $0.45 \cdot 10^{11}$ | 0.35 | $3.9 \cdot 10^3$ |
| Cadmium | 8.85 | $0.51 \cdot 10^{11}$ | $0.2 \cdot 10^{11}$ | 0.30 | $2.3 \cdot 10^3$ |
| Molibdenum | 10.2 | | $1.45 \cdot 10^{11}$ | | |
| Nickel | 8.9 | $1.99 \cdot 10^{11}$ | $0.78 \cdot 10^{11}$ | 0.3 | $5.0 \cdot 10^3$ |
| Tantalum | 16.6 | $1.89 \cdot 10^{11}$ | | | $3.4 \cdot 10^3$ |
| Wolfram | 19.2 | $3.55 \cdot 10^{11}$ | $1.33 \cdot 10^{11}$ | 0.17 | |
| Zinc | 7.3 | $(0.4-1.3) . 10^{11}$ | $(0.3)-0.5) . 10^{11}$ | 0.2-0.3 | $3.7 \cdot 10^3$ |
| Lead | 11.34 | $0.17 \cdot 10^{11}$ | $0.08 \cdot 10^{11}$ | 0.45 | $1.3 \cdot 10^3$ |
| Tin | 7.29 | $0.54 \cdot 10^{11}$ | $0.18 \cdot 10^{11}$ | 0.33 | $2.6 \cdot 10^3$ |
| Glass | 2.6 | $0.7 \cdot 10^{11}$ | $0.2 \cdot 10^{11}$ | 0.2 | $5.0 \cdot 10^3$ |
| Wood | 0.3-1.1 | $0.1 \cdot 10^{11}$ | $0.01 \cdot 10^{11}$ | | $(3-4) . 10^3$ |

compression modulus. In Table 2.6 the value of $\rho$, $E$, $G$, $\mu$, and $v_l$ are given, for some representative materials.

In turn, the fracture strength is linked to the maximum value of the tensile stress (opposing the pressure) which would cause the rupture of the material. This is given by:

$$-p_{max} = \frac{dU_{tot}}{dV} \tag{2.22}$$

on the condition that $d^2U_{tot} / dV^2 = 0$. As a result of stretching, the distance between two atoms increases from $R_0$ to $R_1$ (Fig. 2.20). Calculations based on this model lead to *fracture strengths* which, in the case of metals, exceed many times those measured experimentally. As was pointed out earlier, this discrepancy is especially due to imperfections in the lattice. Several values for the strength of steel and pig iron are given in Table 2.7.

High pressures compress the materials with the effect of diminishing the interatomic distances, so that $R < R_0$ (Fig. 2.20). In Fig. 2.22 isothermal compressibility curves for NaCl, aluminium and antimony are displayed [17]. If

TABLE 2.7 Steel and Pig Iron Tensile Strength

| *Material* | *Tensile strength* (MPa) | *Yield* (%) |
|---|---|---|
| Steel (OL 60) | 500 — 700 | 20 — 10 |
| Cast steel | 360 — 600 | 20 — 10 |
| Pig iron | 280 — 360 | |

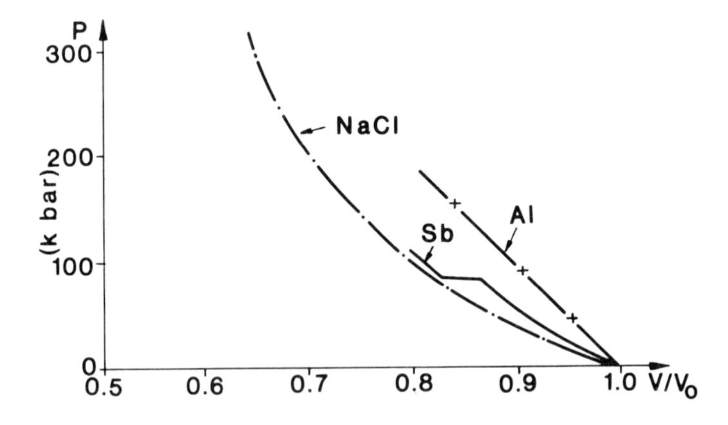


Fig. 2.22. P — V curves for some solids.

changes in the crystalline structure occur (phase transitions), they manifest themselves by flat lines on the P — V curve (as is the case, for example, with antimony for $p$ = 83 kbar). Phase changes which appear at high pressures make some of the crystalline structures stable only at such pressures and, sometimes, at high temperatures. This aspect can be illustrated on the carbon phase diagram, the knowledge of which has been instrumental in the obtaining of artificial diamonds (Fig. 2.23) [17].

Since, obviously, solids are studied at temperatures above absolute zero, the *thermal movement* of the lattice atoms has to be taken into account. As long as temperature increases, the atoms vibrate with a larger and larger amplitude around the position $R_0$ (Fig. 2.20). Because of the asymmetry of the potential curve near $R_0$, larger deviations from $R_0$ must be described by anharmonic terms. A series expansion around $R_0$ of the potential energy of interaction gives:

$$E_p = A(r-R_0)^2 + B(r-R_0)^3 + \ldots \tag{2.23}$$

The term in $(r-R_0)$ does not appear because it has as a coefficient the first derivative of the potential energy with respect to $r$, and, as we are around a minimum (equilibrium) point, this derivative vanishes. The first anharmonic term is $B(r-R_0)^3$ and it shows that, at higher temperatures, the vibrations will no longer be symmetric with respect to the initial position of equilibrium

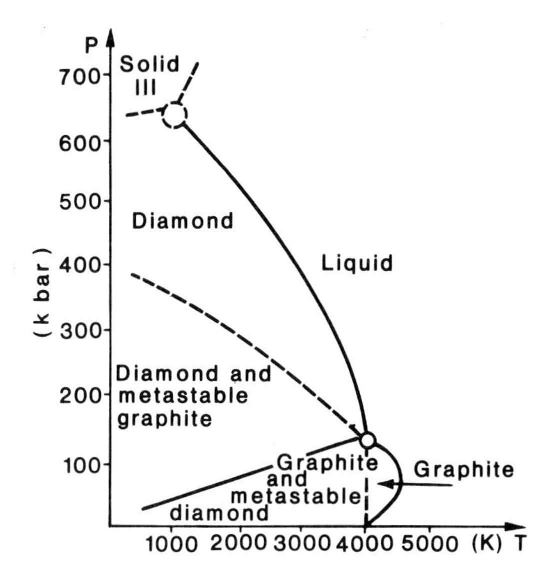


Fig. 2.23. Carbon's phase diagram [17].

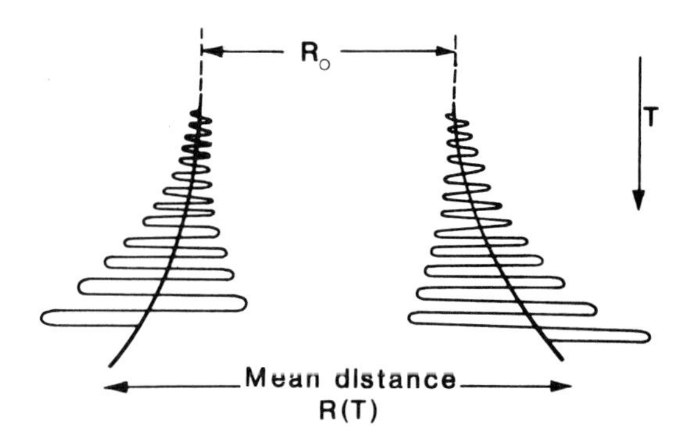


Fig. 2.24. A model for thermal expansion.

$R_o$, because $E_p$ itself is no longer symmetric. Consequently, the average distance between the atoms (the lattice constant, $a$) increases with temperature, as Fig. 2.24 suggests. Thus, as the temperature increases a change in the dimensions of the body will appear — an expansion. To describe this effect, a material constant called *coefficient of thermal expansion*, ($\alpha$), is introduced; it is defined in thermodynamics as:

$$\alpha = \frac{1}{V} \cdot \left( \frac{\partial V}{\partial T} \right)_p \tag{2.24}$$

The *thermal expansion coefficient* expresses the variation with temperature of one of the body's dimensions. In the case of homogeneous and isotropic bodies it is $\alpha_l = \alpha/3$. Usually, $\alpha$ depends on temperature. In many cases the linear thermal expansion coefficient is given as an average value over the interval between two temperatures $T_1$ and $T_2$:

$$\overline{\alpha}_l = \frac{1}{T_2 - T_1} \cdot \int_{T_1}^{T_2} \alpha_l \mathrm{d}T. \tag{2.25}$$

In general, thermal properties of solids depend on the manner the atoms vibrate around the equilibrium position. In the approximation of small (harmonic) oscillations, featured by:

$$E_\mathrm{p} = A(r-R_\mathrm{o})^2 = Ay^2, \tag{2.26}$$

the movement of each atom is approximately described by the equation:

$$m \frac{\mathrm{d}^2 y}{\mathrm{d}t^2} = F = -\frac{\mathrm{d}E_\mathrm{p}}{\mathrm{d}y} = -2Ay, \tag{2.27}$$

where $m$ is the mass of the atom.

The solution of this equation is of the form:

$$y = y_\mathrm{o} \sin\omega t, \tag{2.28}$$

where $y_\mathrm{o}$ is the amplitude of the oscillations, and

$$\omega = 2\pi\nu = (2A/m)^{1/2}. \tag{2.29}$$

The frequency $\nu$ is known as *Einstein's frequency* and characterizes the lattice dynamics.

This vibration frequency can be evaluated by estimating the constant $2A$. The latter has a value of roughly 100 N/m. For medium mass atoms (Fe, Cu, etc.) one gets $\nu \simeq 10^{13}$ s$^{-1}$.

However valuable for a straightforward understanding, this model of lattice vibrations — known as the *Einstein model* — is still oversimplified. For a more realistic model of a crystal, that would take into account the interactions between atoms, a better approximation assumes that the vibration frequencies are spread over a range between zero and a maximal value, $\nu_{max}$. This model is known as the *Debye model*. Calculations show that, in the Debye model, the temperature-dependent energy of the crystal is [12]

$$U = \frac{9Nk_\mathrm{B}T^4}{\theta^3} \cdot \int_0^{\theta/T} \frac{x^3}{\exp x - 1} \mathrm{d}x. \tag{2.30}$$

where $N$ is Avogadro's number, $k_B$ is Boltzmann's constant, $x = h\nu/k_BT$, $T$ is the absolute temperature, and $\theta$ is the *Debye temperature* defined as

$$\theta = h\nu_{max}/k_B. \tag{2.31}$$

Here $h$ is the Planck constant.

Knowing the temperature-dependence of the internal energy $U$ of the lattice (the caloric equation) one can calculate another characteristic of the material — the *heat capacity* defined as:

$$c_\alpha = \left(\frac{\partial U}{\partial T}\right)_\alpha \tag{2.32}$$

where $\alpha$ is a state parameter, that is kept constant in the process. Applying eq. (2.32) to a mole of substance one can define the *molar heat at constant pressure*, $C_p$, and, the *molar heat at constant volume*, $C_v$, respectively.

In the low-temperature approximation, the molar heat at constant volume is

$$C_v = 234\ Nk_B(T/\theta)^3 \quad [\text{J/mol.K}]. \tag{2.33}$$

The dependence on temperature of the (molar) heat capacity in the Debye model for $T/\theta < 1.4$ is plotted in Fig. 2.25.

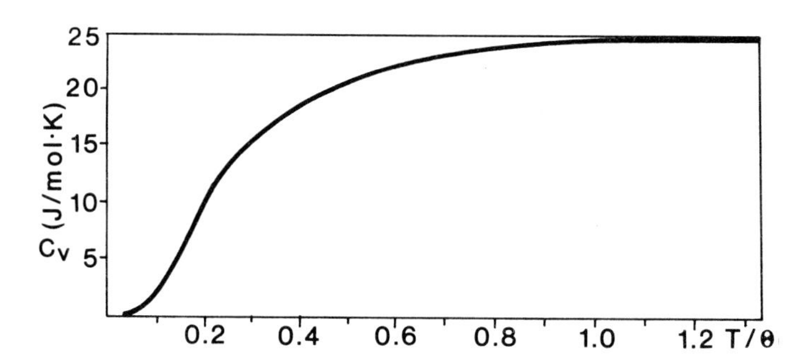


Fig. 2.25. Molar heat variation vs. temperature, in the Debye model.

The various material constants that characterize substances from the point of view of their thermal properties are interdependent. Thus, the expansion coefficient $\alpha$ may be expressed as:

$$\alpha = (\gamma C_v/3V)\varkappa, \tag{2.34}$$

where $\gamma$ is the *Grüneisen constant*, that characterizes anharmonicity of the atomic vibrations; in theory it is equal to $(m+n+3)/6$, where $m$ and $n$ are the powers which appear in the expression (2.8) of the potential energy, $\varkappa$ is the compressibility, and $V$ is the atomic volume.

Several material constants that thermally characterize some substances are given in Table 2.8.

TABLE 2.8 Thermal Constants of Materials

| *Material* | *Debye temperature* ($\theta$) (K) | *Grüneisen constant* ($\gamma$) | *Molar heat at 293 K* ($C_p$) (J/mol . K) | *Linear expansion coefficient* $\alpha_l \cdot 10^6$ ($K^{-1}$) | *In the temperature range* (K) |
|---|---|---|---|---|---|
| Al | 426 | 13.6 | 24.02 | 23.8 | 273 – 373 |
| Be | 1160 | 2.22 | 17.91 | 13.3 | 293 – 473 |
| C (diamond) | 2240 | | 5.65 | 0.91 | 273 – 373 |
| C (graphite) | 2065 | | 8.04 | 1.5-1.7 | 293 – 373 |
| Cd | 188 | 6.27 | 26.16 | 33 | 293 – 373 |
| Cu | 348 | 7.44 | 24.27 | 16.5 | 273 – 373 |
| Fe | 464 | 50.2 | 25.24 | 12.3 | 273 – 373 |
| Mo | 470 | 21.1 | | 5.2 | 273 – 373 |
| Pb | 108 | 33.6 | 27.28 | 28.8 | 273 – 373 |
| Ti | 430 | 35.5 | | 8.5 | 273 – 373 |
| U | 200 | 109 | | | |
| W | 405 | 12.1 | 24.74 | 4.5 | 273 – 373 |
| Zr | 310 | 30.3 | | | |

The *specific (molar) heat* of a mixture can be determined from the sum of the specific (molar) heats of the components (the Kopp and Neumann law). For example, the specific heat of glass may be calculated starting with the specific heats of the constituent oxides. Yet, if the material undergoes a phase transition the specific heat changes; this behaviour is rendered evident in Fig. 2.26. The specific heat of alloys may also be obtained approximately by using the Kopp and Neumann law (for mixtures).

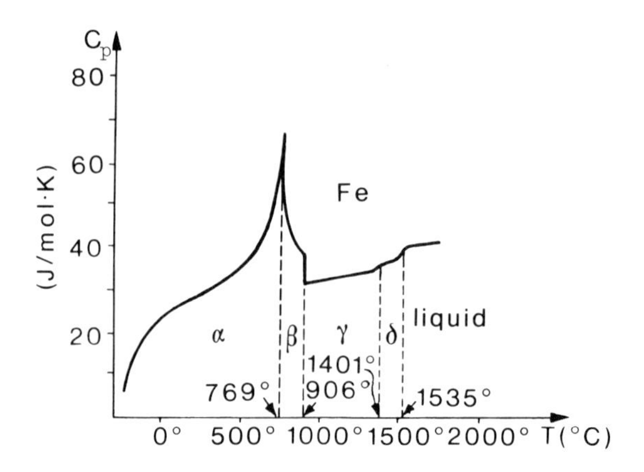


Fig. 2.26. Iron's specific heat, in the temperature range -50 – 1700 °C.

For metals, apart from the heat capacity of the lattice one must also take into account the heat capacity of the electrons. The free electrons of the metal have an energy distribution that depends on temperature. This one contributes a supplementary term to the total heat capacity, but, at ambient temperature, this contribution is sufficiently low to be neglected.

Another material constant is connected to the *thermal conductivity*. The heat flux that establishes in a body between two regions of different temperature also depends on the nature of the material. The constant which characterizes this property has been called the *coefficient of thermal conductivity* or *thermal conductivity* ($K$), and appears in the transport equation:

$$\frac{dW}{dS} = -K \cdot \frac{\partial T}{\partial n} \quad , \tag{2.35}$$

where $dW/dS$ is the elementary heat flux that flows through the unit-area due to the temperature gradient $\partial T/\partial n$ — normal to the surface.

For a stationary regime, the heat flux ($W$) across a plane wall of thickness $d$ and area $S$ is:

$$W = KS(T_2 - T_1)/d, \tag{2.36}$$

$T_1$ and $T_2$ standing for the temperatures on each side of the wall, respectively.

The heat transport may be performed (i) by displacements or vibrations of atoms (molecules) of the fluid, or solid material, respectively, and (ii) by means of electrons — in metals.

For the thermal conductivity of gases the kinetic theory gives the expression:

$$K = \frac{1}{3}\,\rho\bar{v}\bar{\lambda}C_v; \tag{2.37}$$

where $\rho$ is the density, $\bar{v}$ and $\bar{\lambda}$ are the average velocity and the mean free path of the molecules, respectively.

Gases and vapours are the least thermally conductive materials. Their thermal conductivity increases markedly with temperature.

As a rule, non-metallic liquids have a low thermal conductivity. Among them, however, water shows a slightly better performance in conducting heat.

Among solid materials, the best thermal conductors are the metals. The chief mechanism of thermal conductivity of metals is the conduction through quasifree electrons. This is why thermal conductivity of metals may be correlated to their electric conductivity, which is determined by quasifree electrons as well. Indeed, at a given temperature, the ratio of the *thermal conductivity* ($K$) to the *electric conductivity* ($\sigma$) is a constant. This ratio, expressed in ($V^2/K$), increases linearly with temperature (the Wiedemann — Franz law):

$$K/\sigma = (\pi^2/3) \, . \, (k_B/e)^2 \, . \, T = 2.45 \, . \, 10^{-8} \, T. \tag{2.38}$$

Here e stands for the electron charge.

Another mechanism of thermal conduction in solids accounts for the vibrations of the crystalline lattice. Normally these vibrations, being essentially elastic, propagate through the crystal at sound speed. Yet, it is a fact that heat propagates much slower; this is due to both the anharmonicity of atomic vibrations around the equilibrium sites, and to the influence of the impurities.

Overall, the total conductivity will be the sum of lattice conductivity ($K_1$) and electronic conductivity ($K_e$). Because $K_e >> K_1$, the conductivity of non-metallic materials is much inferior to that of metals.

The thermal conductivity of several materials is given in Table 2.9, and in Fig. 2.27 the variation with temperature of the thermal conductivity of water and water vapours is plotted.

Among non-metallic materials, a large conductivity is exhibited by the saphire crystal. The influence of defects on thermal conductivity can be easily followed in Fig. 2.28, where the dependence on temperature of the thermal conductivity of the synthetic saphire, as a function of the defects induced by the fast neutron irradiation, has been displayed.

For non-homogeneous materials — the spongy ones, for example — the conductivity can be calculated assuming that the pores do not perturb the structure of the temperature field, and also that the pore itself has a conductivity $K_p$ that differs from that of the compact solid $K_s$. In this case, defining the *volume porosity* $P$ as the ratio of the pore volume to the total volume, and assuming that the pores have similar shapes and sizes, the conductivity in a given direction may be calculated by considering a model in which the pores are connected "in series" to form "tubes", whereas the tubes are connected in parallel (*Loeb's equation*):

$$K = K_s(1-P). \qquad (2.39)$$

This expression has been obtained assuming $K_p << K_s$ [18]. The porosity $P$ can be estimated by knowing the theoretical density of the compact body ($\rho_s$ = T.D.) and the density of the spongy body ($\rho$):

$$P = 1 - \rho/\rho_s. \qquad (2.40)$$

Such measurements render evident small deviations from the theoretical expression of $K$, a fact that calls for introducing an adjusting parameter, $\alpha$; thus, the conductivity equation reads:

$$K = K_s(1-P)/(1+(\alpha-1)P), \qquad (2.41)$$

where $\alpha$ has values between 2 and 3.

For uranium dioxide, a semiempirical equation, valid between 150 $^oC$ and 1000 $^oC$ is [18]:

$$K/K_s = 1-(2.58 - 0.58 \cdot 10^{-3}\, T)P, \qquad (2.42)$$

(see eq. (9.2), and also Appendix 11).

TABLE 2.9 Thermal Conductivity of Several Materials

| *Material* | *Temperature* (*K*) | *K* (W/m . K) |
|---|---|---|
| Helium | 273 | 0.1436 |
| Neon | 273 | 0.0464 |
| Argon | 273 | 0.01616 |
| Krypton | 273 | 0.00872 |
| Xenon | 273 | 0.00511 |
| $H_2$ | 273 | 0.1733 |
| $O_2$ | 273 | 0.0245 |
| $N_2$ | 273 | 0.0243 |
| NO | 273 | 0.02372 |
| CO | 273 | 0.02302 |
| $CO_2$ | 273 | 0.0143 |
| Dry air | 273 | 0.0243 |
| Methane | 273 | 0.0302 |
| Glycerin | 293 | 0.2849 |
| Engine Oil | 293 | 0.1396 |
| Transformer Oil | 293 | 0.1314 |
| Water | 293 | 0.5815 |
| Aluminium (liquid) | 973 | 30.714 |
| Tin (liquid) | 573 | 32.564 |
| Lead (liquid) | 673 | 16.282 |
| Mercury (liquid) | 293 | 8.722 |
| | 373 | 10.647 |
| Aluminium (commercial) | 293 | 209.34 |
| Beryllium | 293 | 167.47 |
| Silver | 293 | 420 |
| Copper | 293 | 390 |
| Iron (Armco) | 293 | 63 |
| Nickel (98% pure) | 293 | 58 |
| Cadmium | 293 | 93.04 |
| Molibdenum | 293 | 383.79 |
| Lead | 293 | 34.77 |
| Tantalum | 293 | 54.66 |
| Wolfram | 293 | 168.63 |
| Duralumin | 293 | 165.15 |
| Constantan | 293 | 22.60 |
| Monel | 293 | 22.1 |
| Electron | 293 | 116.3 |
| Nichrome | 293 | 14 |
| Granite | 293 | 2.9 |
| Graphite | 273 | 168.6 |
| Glasswadding | 293 | 0.042 |
| Quartz glass | 273 | 1.35 |
| Crown glass | 273 | 1.16 |
| Paraffin | 293 | 0.25 |
| Plexiglass | 273 | 0.19 |
| KF | 293 | 7.1 |
| NaCl | 293 | 7.1 |
| KCl | 293 | 7.1 |
| $CaF_2$ | 293 | 12 |
| AgCl | 293 | 1.1 |

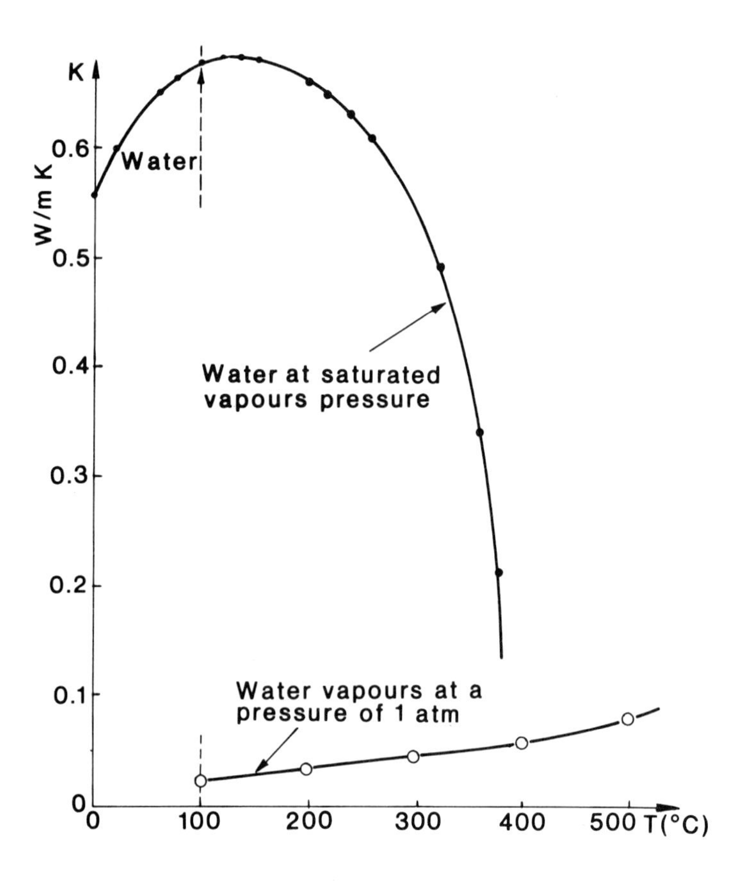


Fig. 2.27. Thermal conductivity of water and water vapours vs. temperature.

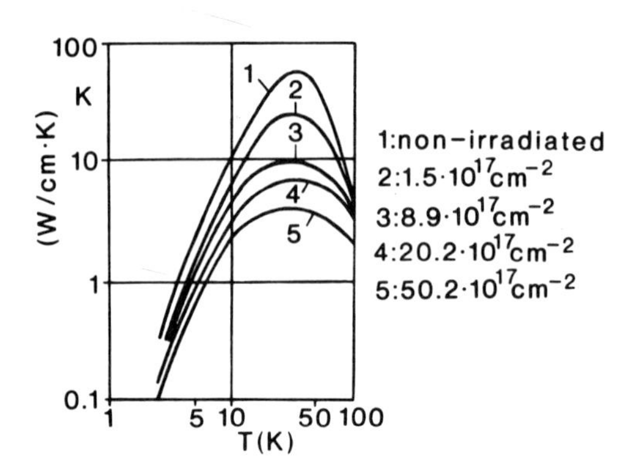


Fig. 2.28. Thermal conductivity of non-irradiated, and fast neutron-irradiated sapphire vs. temperature.

## 2.5. INFLUENCE OF PROCESSING ON MATERIAL PROPERTIES

Materials used in nuclear engineering are obtained through various chemical and metallurgical procedures, being then mechanically and thermally processed to have them brought to the desired shape and characteristics. All these operations may, more or less, alter the properties of materials. As a consequence, a good knowledge of the influence of processing and working on the finite material is of interest, either in order to conduct the operations in such a way as to obtain superior qualities, or to impart new properties. Such processes can be: *crystallization* by cooling a liquid phase, *annealing*, *pressing* at high temperature, *briquetting*, *pelleting*, *sintering*, *mechanical alloying*, etc.

Later in this book, some of these processes, of interest in nuclear materials' technology, will be discussed.

Solidification appears at a characteristic temperature, when the free energy ($f$) of the system corresponding to the solid state is lower than that corresponding to the liquid state [19]. From a microscopic point of view this process starts with the spontaneous formation, as the liquid cools down, of atomic clusters called *crystallization germs*. Such germs set in by gradual extension of the short-range order pre-existent in the liquid, the long-range order having not yet been dynamically stable. The dynamic stability of the clustering atoms which make up the germs as well as their dimensions increase as temperature decreases. Initially, the germs have variable sizes due to statistical fluctuations. Growth of those larger among them will prevail. The subsequent growth of a crystallization germ depends on reaching a critical size, which brings about a decrease of the free energy.

The transition of a given volume $V$ into solid state (the germ volume) is accompanied by a decrease $\Delta f$ of the free energy per unit-volume. Meanwhile, the formation of internal boundaries raises the superficial energy. Overall, the free energy will vary with $-V\Delta f + S\sigma$, where $S$ is the germ's area and $\sigma$ is the superficial tension at the liquid — solid interface. Thus, only by reaching a critical dimension the germ may become a dynamically stable phase and may expand further. The energy necessary for the formation of a germ of critical (or larger than critical) size is provided by thermal fluctuations that exist near the solidification temperature. This is why any solid impurity in the liquid phase will be a preferred site for solidification, corresponding to an energy less than that required in the formation of the germ out of the liquid phase. Once a germ which exceeds the critical dimension has been formed, it grows ever further. The lattice defects of such a crystallite represent points of preferential growth. Then the crystallization goes on by multiplication of germs and growth in their dimensions. As long as they grow up free from one another, the germs exhibit regular geometric shapes; but as soon as the distinct crystallites come into close contact with each other they undergo mutual perturbations; the faces which have come into contact will no longer change, the growth continuing on the directions left free. So, the shapes and volumes of the crystallites in the bulk of the solid become irregular (Fig. 2.29). Summing this up, crystallization is determined by two factors:

(a) the rate of germ formation ($n'$);

(b) the rate of germ growth ($v_g$).

Both $n'$ and $v_g$ depend on temperature and nature of material. Since microcrystallite sizes are decisive in determining the mechanical properties of materials, the control of their dimensions is extremely important. For example, the dependence on microcrystallite dimensions of the yield point is given by the *Hall — Pach equation* :

Fig. 2.29. Crystallization stages, for a metal; as the process progresses microcrystallites are formed.

$$\sigma_c = \sigma_o + K/d^{1/2}, \tag{2.43}$$

where $d$ is the microcrystallite dimension, and $\sigma_o$ and $K$ are constants for the considered metal.

The crystallization starts usually at the walls of the vessel containing the melt (ingot mould, crystallizer) because, on one hand, this is cooler and, on the other hand, the wall is an "impurity" which induces the formation of the first crystallites. The crystallites grow from the wall to the centre (in the direction of the thermal gradient) and branch out, this process being called *dendritic growth*. The arms of the dendrites are isolated from one another by extremely thin layers of insoluble impurities and very fine pores that appear at solidification as a result of volume contraction.

In general, a cross-section through a metallic ingot presents three different structural zones, the sizes of which depend on the way the cool-down took place. The wall region which cools down in the first place determines the formation of relatively fine, *equiaxial* crystallites. The next region, which is much broader, has a structure of oblong dendrites and is called the region of *columnar* crystals. Fast cooling makes this region extend down to the centre. Yet, in general, in the centre of the ingot a new zone of equiaxial crystals is formed.

If the solidification of an alloy develops nonhomogeneously, then in various zones of the ingot different compositions are obtained — a phenomenon which is called *segregation*. The same effect corresponds to the separation from the molten material of some impurities, during slow cooling. This process can be used to eliminate impurities from a material (zone melting), or to refine certain metals in the process of extraction. The non-uniform and rough structure which appears at casting can be altered by warm and cool work or by annealing. As a result one gets a fine, uniform structure which allows one to obtain the prescribed properties.

There is a series of cases in which primary materials are such that their alloying is impossible. Such cases are illustrated, for example, by *metalloceramic* materials, that are increasingly employed in nuclear engineering. A similar problem occurs when one tries to alloy metals with extremely different melting temperatures. In such cases, even if the very high temperatures necessary for melting the component of the higest melting point are reached, the process of segregation at cool-down will prevent the alloy being formed. Modern techniques of *powder metallurgy* [20], would then provide for the obtaining of the desired system. This process requires transformation of the two components into powder, mixing the powders mechanically, then pressing and *sintering*. The last two stages can be effected separately or simultaneously [21-23].

Another case in which powder metallurgy is extremely efficient is the production of nuclear oxide fuels. By chemical procedures a fine powder results which, having a low density as compared to the theoretical one (< 30% T.D.), does not comply with the thermal conductivity requirements. In these cases processes of densifying and sintering are in order.

To get mechanical rigidity and a required density the powder has to be subject to such a treatment as to achieve a very large contact surface between particles. By straightforward pressing, an increase of the contact interface due to the coming closer, clustering and permanent deformation of powder's particles is accomplished [20]. Of particular importance is the granulometric distribution which, given the best conditions, allows a close contact and a complete filling of the voids (maximum density). A variant of this process is *mechanical alloying*, by which the powders one wants to alloy, or from which formation of a composite is sought, are mixed together in a ball-mill where they are worked upon for a sufficiently long time. With the process of milling, a mutual compression of the particles is achieved, so that, from a certain moment on, the components can no longer be distinguished from one another [23].

*Sintering* is a metallurgical procedure in which thermally activated densification and recrystallization of some powder conglomerates, with or without small quantities of smelt added, are achieved [20]. Sintering is a complex process in which a transport of material which accomplishes the "welding" of particles at the level of their contact points and surfaces takes place. One or more mechanisms can contribute to the realization of this transport as, for example, evaporation and condensation, surface and volume diffusion, viscous (plastic) flow of material. During sintering, processes of recrystallization and grain growth, of reduction of oxide films on the surface of the particles, and of elimination of absorbed gases may also take place. Because the materials obtained by sintering, unlike those obtained by casting, exhibit porosity, as a main characteristic, the characterization of sintered materials is usually done by measuring the density and, sometimes, thermal and electric conductivities – both sensitive to this parameter.

The chief advantage of the sintering method is the possibility of producing metallic materials with very special properties.

## 2.6. IRRADIATION EFFECTS ON MATERIALS

A nuclear reactor is an intense source of high energy radiations that can alter the physical, chemical and mechanical properties of materials entering its structure. The radiations in question have a very broad energy spectrum. Fission neutrons have energies up to 10 MeV, most of them in the range of 0.5 — 2 MeV. Thermal neutrons have the most probable energy at about 0.03 MeV. The energy spectrum of the gamma radiations is also very broad, the average energy lying around 1 MeV. The intensity of these radiations may reach up to $10^{15}$ n/cm$^2$ . s, and $10^{13}$ photons/cm$^2$ . s, respectively.

The interaction of radiation with matter is a complex phenomenon. Each type of radiation existent in the nuclear reactor (fast and thermal neutrons, $\alpha$, $\beta$, $\gamma$ radiations, fission fragments, etc.) has several peculiarities, and each type of material will respond in a specific manner to the interaction with this radiation.

The mechanisms of technological relevance which lead to an alteration of the physical properties, as a consequence of the interaction of materials with radiations, are:

1. *The interaction of fission fragments with lattice atoms of the fuel material*; this reigns mainly at the level of the fuel, owing to the very short path (approximately 10 μm in $UO_2$) of the fragments in the bulk of the fuel [24,25].

2. *Elastic collisions of neutrons with lattice atoms*; this may come up at greater distances from the places where neutrons are generated; so this effect may touch upon the majority of materials which make up the reactor.

3. *Ionizations caused by fission fragments, and $\beta$ and $\gamma$ radiations*; the ionization has less important effects on the solids that are employed in the construction of nuclear reactors [26].

### 2.6.1. Neutron Effect

As electrically neutral particles, neutrons are not affected in their movement across the matter by the electric fields of the nuclei and electrons, their interaction with matter consisting in direct collisions with the nuclei. A fission-born fast neutron has an average kinetic energy of approximately 2 MeV, which it dissipates by elastic and inelastic collisions. In a process of *elastic collision*, the energy transferred by a neutron to an atomic nucleus can be calculated classically, on grounds of energy and momentum conservation. For a neutron of mass $M_0$ and kinetic energy $E_0$ the maximum energy transferred to a nucleus of mass $M_1$ (in the case of a head-on collision) is:

$$E_{max} = 4M_oM_1E_o/(M_o + M_1)^2. \tag{2.44}$$

In Table 2.10 some typical values for the energy transferred in such a process are given.

If the energy transferred to an atom is greater than a certain energy $E_d$ (the *displacement energy*) then the atom may be knocked out from its position in the lattice. The value of $E_d$ is about four times greater than the binding energy of the atom, and may be anywhere between 25 and 50 eV, for most materials.

The atom that is knocked on by a fast neutron receives an energy much larger than that required to extract it from its site in the lattice, and behaves

TABLE 2.10 Maximum Transferred Energy, at Collision with a Neutron

| *Target nucleus* | $E_{max}$ (MeV) |
|---|---|
| H | 2.0 |
| Be | 0.72 |
| C | 0.56 |
| Al | 0.28 |
| Cu | 0.12 |
| U | 0.033 |

subsequently as a fast ion (*primary knock-on atom*), moving at velocities greater than the velocity of sound in the respective material. Experiment has shown that these primary knock-on atoms collide elastically as well as inelastically with other atoms of the lattice, generating *secondary*, *tertiary*, etc. *knock-on atoms* (Fig. 2.30).

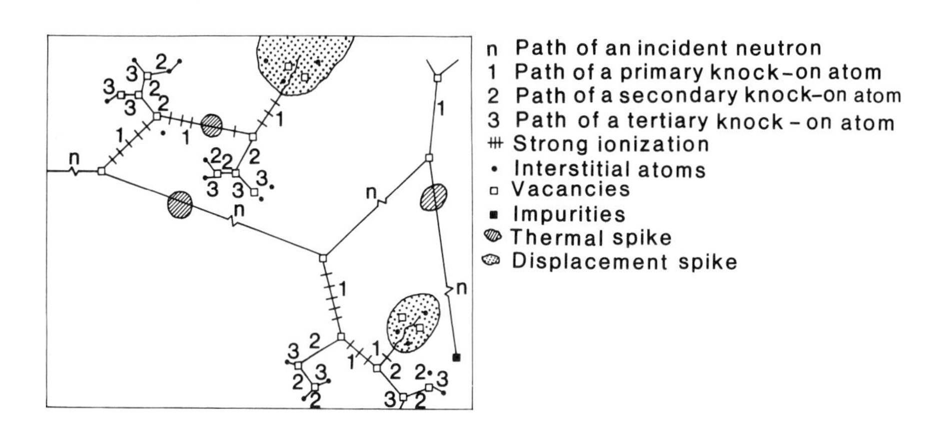


Fig. 2.30. Effect of neutron irradiation on a solid.

The knock-on atoms abandon their sites in the lattice leaving behind empty places (vacancies — Schottky defects) and finally occupying *interstitial* positions. Vacancy — interstitial pairs thus introduced in the lattice affect the properties of the material (its density, electric resistivity, etc.) as well as the rates of the processes which imply mass transport (diffusion, nucleation, etc.).

In the process of *inelastic collisions* the primary knock-on atoms dissipate their energy over the matter by interactions with electrons. The electronic excitation is more probable when the velocity of the incident ion exceeds the velocity of the orbital movement of the electrons in the knock-on atom. Therefore, inelastic collisions will prevail for the primary knock-on atoms (of high energy), while elastic interaction will be more frequent for the secondary, tertiary, etc., knock-on atoms (of lower energy).

The threshold energy of ionization in insulators, $E_i$ (eV) is given by:

$$E_i = \frac{1}{8} \cdot \frac{M_o}{m} I, \tag{2.45}$$

where $M_o$ is the mass of the primary knock-on atom, $m$ is the electron mass and $I$ is the ionization energy for the knock-on atom.

In the laboratory system, when a particle of mass $M_o$, velocity $v$ and electric charge $Z_oe$ passes at a distance $b$ by a particle at rest, of mass $M_1$ and charge $Z_1e$ , the energy $\Delta E$, transferred at small deflection angles to the particle at rest, is

$$\Delta E = M_o Z_o^2 Z_1^2 e^4 / ((4\pi\varepsilon_o)^2 M_1 b^2 E), \tag{2.46}$$

where $E = (1/2) M_1 v^2$ . If the particle at rest is an electron, it has a much lighter mass than that of the incoming nucleus, so that the absorption of energy from the primary atom will be much larger ( $\sim 10^3$ times) than in the case of an interaction with a nucleus as a target.

If the primary ion is of high energy it might penetrate the electronic shells of the knock-on atom (a Rutherford-type collision). If the primary ion has not enough energy its electronic cloud will not penetrate profoundly that of the atom at rest, and the interaction process will be similar, in many ways, to the elastic spheres collision.

The energy transferred to an atom in a collision of the Rutherford type is at most equal to $E^*$, where:

$$E^* = 4E_R^2 Z_o^2 Z_1^2 (Z_o^{2/3} + Z_1^{2/3}) M_o / (M_1 E_o), \tag{2.47}$$

where $E_R = 2.17 \times 10^{-18}$ J (13.6 eV — the ground state energy of the hydrogen atom) and $E_o$ is the kinetic energy of the incident ion. For primary knock-on atoms, $E^*$ is much higher than the displacement energy $E_d$, and, therefore, Rutherford collisions will cause the displacement of $\sim 100$ atoms that happen to come across the trajectory of the incident ion.

### 2.6.2. Thermal and Displacement Spikes

Seitz and Koehler [27] have figured out a mechanism of "disordering" a crystalline lattice through the effect of radiations, called "thermal spike". Such a phenomenon appears when an incident particle knocks on an atom of the lattice, transferring to it an amount of energy smaller than that necessary to displace it permanently from its equilibrium position. This energy is dissipated by vibrations, the knock-on atom transferring the received energy to the neighbouring atoms which, in turn, may be set in a state of intense vibration. In this way, in a small region around the atom a temperature burst above 1000 K, for a very short duration ($10^{-13}$ — $10^{-11}$ s),takes place. After the heating up and, possibly, the sudden melting of the material entailing a local redistribution of the atoms, a fast cool-down and, thus, the solidification of the substance follows.

At a point situated at a distance $r$ away from the thermal spike, after a lapse of time $t$ from the initiation of the spike, the temperature is given by the equation:

$$T(r,t) = T_0 + \frac{Q}{(4\pi)^{3/2} cd(Dt)^{3/2}} \cdot \exp\left(-\frac{r^2}{4Dt}\right), \qquad (2.48)$$

where $Q$ is the energy conveyed to the knock-on atom; $c$ is the heat capacity of the material; $d$ is its density; and $D$ is its thermal diffusion coefficient.

Under the action of internal and external perturbations, featuring an activation energy $E'$ the atoms in the crystal lattice sites may also perform — apart from the vibrational movement — jumps off their normal sites. Let us suppose that, under the action of thermal spikes, the atoms in a material perform such "jumps" (hops) in interstitial positions, or replace themselves in the lattice, at an average rate of the process given by:

$$\nu = \nu_0 \exp(-E'/k_B T), \qquad (2.49)$$

where $T$ is the local temperature, $E'$ is the activation energy of the process, and $\nu_0$ is the effective frequency which includes the entropic factors.

Suppose the sample has $n_0$ atoms per unit volume. The occurrence of the thermal spike in the lattice will cause a number $n$ of hops that may be calculated by integrating the expression (2.49) over the spatial and temporal variables implicit in the temperature $T$, while requiring that temperature depends upon these variables according to eq. (2.48). The additional number of hops $n$ may be inferred by subtracting from the number of hops which occur in the substance, at temperature $T_0$, in the absence of the thermal spike:

$$\Delta n = n_0 \nu_0 \int_0^\infty 4\pi r^2 dr \int_0^\infty dt \,\{\exp[-E'/k_B T(r,t)] - \exp[-E'/k_B T_0]\}. \qquad (2.50)$$

If there are $N_0$ thermal spikes/cm$^3$ . s in the material then the increment in the hopping rate will be $N_0 \Delta n$ (hop/cm$^3$ . s).

The integral in the expression (2.50) is difficult to evaluate. Seitz and Koehler [27] have studied a simpler case, at $T_0 = 0$. In the end, the following expression for the number of extra-hops was obtained:

$$\Delta n = 0.016(\nu_0 r_s^2/D)(Q/E)^{5/3}, \qquad (2.51)$$

where $r_s = (3/4\ \pi n_0)^{1/3}$.

Among the effects of the thermal spikes are *disordering* processes in originally ordered alloys; the stimulation of *local diffusion*; *phase transitions*; and other processes activated by a local and short burst of heating.

Brinkman [27] came up with the conjecture that, for lighter elements, the mechanism of defect generation by irradiation consists in the creation of vacancy — interstitial pairs, whereas for heavier elements disordering is due to *displacement spikes* (Fig. 2.31). To explain displacement spikes one admits that along the trajectory of ionizing particle a local heating of the substance is produced. As a result of this heating, part of the atoms in the substance evaporate, to condense in other regions, thus occupying interstitial sites or replacing atoms pre-existing in the lattice (Fig. 2.31). Displacement spikes cause the displacement of ~ $10^4$ atoms and a local heating up to ~$10^4$ K.

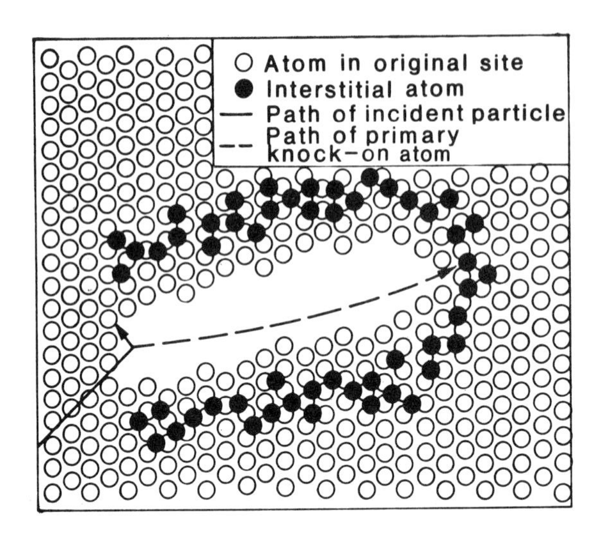


Fig. 2.31. Generation of a displacement spike.

After the heating and melting, solidification follows; as a result, in the lattice a disordering appears, and along the displacement spikes dislocation loops are introduced.

Brinkman's model suggests, for heavy metals, an acceleration of diffusion at the atomic level. Let $c(x)$ be the concentration of an atomic species in a one-dimensional crystal [26]. If the spike occurs in the position $x'$, the composition is altered immediately over a distance $l$ (the length of the spike). The resulting concentration, $c'(x)$, is (Fig. 2.32):

$$c'(x) = \begin{cases} \bar{c}(x'), & \text{for } x' - l/2 < x < x' + l/2, \\ c(x), & \text{otherwise,} \end{cases} \tag{2.52}$$

where $\bar{c}(x')$ is the average of $c(x)$ over the interval centred on $x'$ (Fig. 2.32):

$$\bar{c}(x') = \frac{1}{l} \int_{x'-l/2}^{x'+l/2} c(x)\,dx. \tag{2.53}$$

Suppose the spikes occur at random, in different places and moments, yet with an average rate $N_0$ (spikes/cm . s). The probability that a spike centred between $x'$ and $x' + dx'$ be generated during the time interval from $t$ to $t + dt$ is $N_0 dx' dt$. When such a spike appears, the concentration around the point $x$ changes by $\Delta c(x,t) = \bar{c}(x',t) - c(x,t)$, when $|x-x'| < l/2$, where $\bar{c}(x',t)$ is given by eq. (2.53). Evidently, $\Delta c(x,t) = 0$ for $|x-x'| > l/2$. In these conditions, the average variation in time of the concentration is:

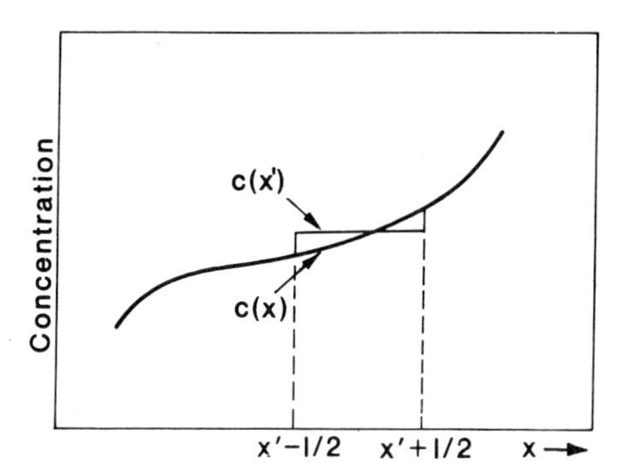


Fig. 2.32. Concentration of an atomic species in the region of a displacement spike.

$$\left\langle \frac{\partial c(x,t)}{\partial t} \right\rangle = N_0 \int_{x-l/2}^{x+l/2} dx' \; [\bar{c}(x',t) - c(x,t)]. \tag{2.54}$$

One may evaluate the average concentration $\bar{c}(x',t)$ by expanding in series $c(x)$ of eq. (2.53) around a point $x$ within the integration interval:

$$\bar{c}(x',t) = \frac{1}{l} \int_{x'-l/2}^{x'+l/2} \sum_{n=0}^{\infty} \frac{1}{n!} c_n(x,t)(\xi - x)^n d\xi, \tag{2.55}$$

where
$$c_n(x,t) = \frac{\partial^n c(x,t)}{\partial x^n} .$$

Solving the integral in (2.55) gives:

$$\bar{c}(x',t) = \sum_{n=0}^{\infty} \frac{c_n(x,t)}{l(n+1)!} \left[ (x'-x+\frac{l}{2})^{n+1} - (x'-x-\frac{l}{2})^{n+1} \right]. \tag{2.56}$$

If the result (2.56) is used in eq. (2.54), by integration one gets:

$$\left\langle \frac{\partial c(x,t)}{\partial t} \right\rangle = \sum_{n=1}^{\infty} \frac{2N_0 l^{2m+1}}{(2m+2)!} c_{2m}(x,t), \tag{2.57}$$

where one can admit $c_{2m}$ to be the derivative of the order $2m$ of $\langle c(x,t) \rangle$.

Because $l$ is very small and the concentration depends but weakly on $x$, from the sum in the right-hand side of (2.57) one may retain only the first term. Therefore,

$$\frac{\partial}{\partial t} \langle c(x,t)\rangle \simeq \left(\frac{N_0 l^3}{12}\right) \frac{\partial^2}{\partial x^2} \langle c(x,t)\rangle . \tag{2.58}$$

The expression (2.58) is a *diffusion equation* that describes the time-evolution of the concentration in agreement with Fick's law, with a (one-dimensional) diffusion coefficient $N_0 l^3/12$. The diffusion coefficient is strongly dependent both on the linear dimension of the spike and on the number of displacement spikes per unit-length of the path.

### 2.6.3. Other Types of Interaction

A peculiar case of irradiation is the *self-irradiation* of radioactive materials. The most interesting and best-studied case is the self-irradiation of plutonium with its own $\alpha$ radiations of 5 MeV [25]. The recoil nucleus, $^{235}U$ (which turns out from the $\alpha$ decay of $^{239}Pu$) with an energy of 87 keV will cause a displacement spike similar to those produced by the fission fragments.

The *neutron absorption* by the nuclei of materials (the main process being the K-capture) introduces impurities (nuclei formed on the spot) in the basic material, which can alter its properties.

The *electromagnetic radiation*, by its interactions with matter (photoelectric effect, Compton effect, pair generation) may induce supplementary ionizations and displacements of the atoms from the normally occupied sites of the lattice.

The irradiation has an important effect on the corrosion of materials in aqueous media, via the *radiolythic decomposition of water*, when ions and free radicals of large chemical activity are generated. Likewise, the irradiation acts upon the transport phenomena in solids, the phase transitions, the reaction rates, etc. For example, diffusion in solids is accomplished by migration of vacancies and interstitials; as we have previously seen, the irradiation increases the natural concentration of these types of defects, which leads to an increase in the rate of diffusion.

### 2.6.4. Mobility and Annihilation of Defects

The defects introduced into materials by irradiation have a considerable mobility and can interact with one another; they may even annihilate each other. During irradiation a dynamic equilibrium sets in, between generation and annihilation of defects. In this process temperature plays a very important role. To quench the induced defects, the irradiation must be done at very low temperature.

The process of defect annihilation is referred to as *annealing*, and has two components: *thermally-activated* annealing and the *irradiation-activated* annealing [28,29]. When some properties of non-irradiated and irradiated materials are measured and compared, one notes differencies in the characteristic values. The irradiated material, heated up at a certain temperature, then tested at ambient temperature, will present a change in properties, as compared to the sample not recovered in this way, due to the thermally-activated annealing. In Fig. 2.33 the results of differential thermal analysis of an irradiated sample are shown. The energy given out by

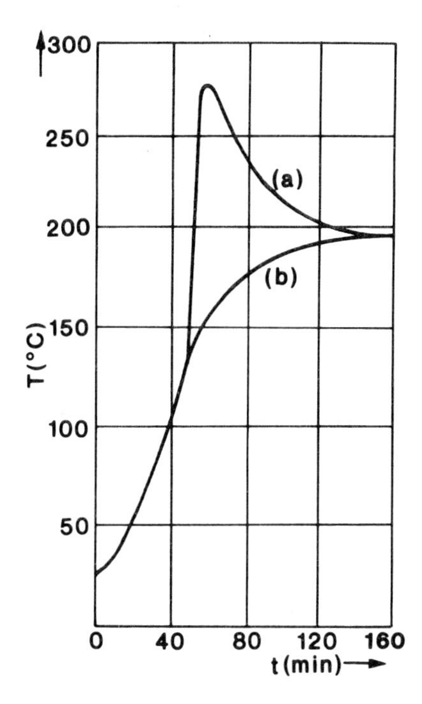


Fig. 2.33. Temperature changes of a graphite sample due to annihilation of defects generated by irradiation at normal temperature [26]: (a) irradiated sample, in an oven at 200 °C, (b) the same sample again in the oven, after its first cool-down.

the annihilation of the internal defects which causes an increase in the sample temperature is rendered evident. This effect does not appear at subsequent recovery.

The defects induced by irradiation can be annihilated due to a continuous bombardment with various types of particles during the irradiation (the active component of the irradiation).

Extinction of the irradiation-induced defects in a material may follow from their *diffusion* through the material till they get stuck in traps; from *annihilation* of vacancies and interstitial atoms, either by recombination, or by absorption on surfaces, grain boundaries or dislocations; from *pinning* of vacancies and interstitials on substitutional impurities as a result of the local strain of the lattice etc. [27,28].

As an example, we shall discuss briefly the recovery of a material by the diffusion of defects in the lattice. The distribution in a crystal of a particular type of defect may be characterized by its concentration $c(\vec{r},t)$. The diffusion of defects in a homogeneous environment, as an effect of a concentration gradient, is described by the diffusion equation:

$$\frac{\partial c}{\partial t} = D\nabla^2 c. \tag{2.59}$$

The coefficient of diffusion $D$ is proportional to the frequency with which a defect hops to an adjacent site:

$$D = \gamma a^2 \nu_j, \tag{2.60}$$

where $a$ stands for the lattice constant and $\gamma$ is a factor depending on the geometry of the lattice in the neighbourhood of the defect. The hopping frequency $\nu_j$ is given by an Arrhenius-type equation:

$$\nu_j = \nu_o \exp(-E_m/(k_B T)), \tag{2.61}$$

where $E_m$ is the activation energy for the displacement of the defect and $\nu_o$ is the effective frequency ($\sim 10^{13}$ s$^{-1}$).

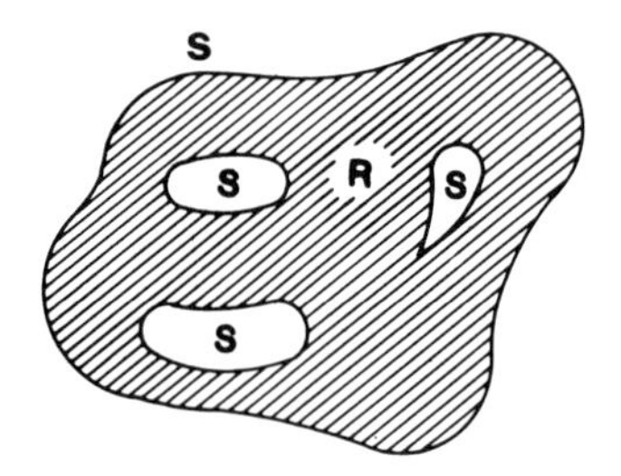


Fig. 2.34. Model of internal trap region, where diffusion of defects takes place.

The problem of material recovery is thus translated into the diffusion of defects of various types on internal traps. Internal traps consist of discontinuities, or imperfections, in the crystalline lattice — dislocations, grain boundaries, clusters of point defects, voids, etc. Such traps are conceived as being contained into imaginary, closed surfaces (Fig. 2.34) on which the concentration of defects is always nil. This is a classical problem in the theory of diffusion; solutions are quite well known [30].

Let us consider a region $R$ (Fig. 2.34) bounded by various absorbent surfaces $S$ on which $c = 0$. For this region of arbitrary shape one may always define a set of functions $\varphi_i(\vec{r})$ such that:

$$\nabla^2 \varphi_i(\vec{r}) + \lambda_i \varphi_i(\vec{r}) = 0, \tag{2.62}$$

and $\varphi_i = 0$ on the surfaces $S$. The eigenvalues $\lambda_i$ of the equation (2.62) depend on the size and shape of the considered region. A complete set of orthonormal eigenfunctions $\varphi_i(\vec{r})$ may be found, so that:

$$\int_R \varphi_i(\vec{r}) \varphi_j(\vec{r}) d\vec{r} = \delta_{ij}, \tag{2.62a}$$

where $\delta_{ij} = 1$ if $i = j$, and $\delta_{ij} = 0$ if $i \neq j$.

The general solution of the equation (2.59) may be written as a linear combination of functions in this set, that reads:

$$c(\vec{r},t) = \sum_{i=0}^{\infty} a_i \varphi_i(\vec{r})\exp(-\lambda_i Dt), \tag{2.63}$$

where

$$a_i = \int_R \varphi_i(\vec{r})c(\vec{r},0)d\vec{r}.$$

The concentration at any point $\vec{r}$, and, therefore, the total amount of substance that leaves the region $R$ decreases exponentially in time, $D$ being the factor which characterizes the process. After a sufficiently long time the evolution of the process is determined by the lowest eigenvalue $\lambda_0$, the terms corresponding to the other eigenvalues becoming negligible. Thus the expansion (2.63) may be written as:

$$c(\vec{r},t) \simeq a_0\varphi_0(\vec{r})\exp(-\lambda_0 Dt). \tag{2.64}$$

In many practical situations the expression (2.64) is valid for any time interval. This time dependence resembles the one which describes the kinetics of chemical reactions, where the number of molecules (here defects) $n$ that have not entered chemical reactions changes with time according to the law:

$$\frac{dn}{dt} = -Kn. \tag{2.65}$$

Here $K$ stands for a rate constant, independent of $n$. The equation (2.65) indicates the fact that $n \sim \exp(-Kt)$. Comparing eqs. (2.65) and (2.64) one obtains:

$$K = \lambda_0 D. \tag{2.66}$$

Considering a crystal of linear dimension $L$, which does not contain internal, but only external, traps (its very surface), $\lambda_0$ will be of the order of $1/L^2$; in this case the asymptotic "decay" constant will be:

$$K = \pi^2 D/L^2. \tag{2.67}$$

In the same conditions, for a sphere of radius $R$ one gets:

$$K \simeq \pi D/R^2, \tag{2.68}$$

and for a parallelepiped of sizes $A$x$B$x$C$:

$$K = \pi^2(1/A^2 + 1/B^2 + 1/C^2)D. \tag{2.69}$$

Thus, if one of the sizes is small, it will have a dominant effect on $K$.

For a crystal that does contain internal traps, for example spherical traps (of radius $r_1$), made out of clusters of vacancies of radius $r_o(r_1 >> r_o)$ one gets:

$$K \simeq (3r_o/r_1^3)D = 4\pi r_o N_o D, \quad (2.70)$$

where $N_o$ is the number of traps per unit volume ($N_o = (4\pi r_1^3/3)^{-1}$).

Another kind of sink for the accumulation of point defects by diffusion is represented by dislocations. Assuming the dislocations to have a cylindrical form of radius $r_o$ and infinite length, contained in reflecting cylinders of radius $r_1$ and having a density $N_o$ of dislocation lines per unit-surface of the crystal, then:

$$K = 2\pi N_o D/\ln(r_1/r_o) \quad (2.71)$$

The diffusion of defects toward dislocations at intermediate temperatures is greatly influenced by the field of mechanical stresses induced by the dislocation, so that the equation (2.71) turns inadequate. In such circumstances, one may use instead a model suggested by Cottrell and Bilby [31].

### 2.6.5. Methods to Limit the Irradiation Effects

There are several means of general nature to limit the irradiation effects on materials:

1. *Changing the irradiation temperature* : if one finds, for example, that a strong hardening of the material appears as a result of an irradiation at a certain temperature, then an increase in temperature during irradiation will substantially alleviate this occurrence.

2. *Using isotropic materials with simpler bonds* : materials used in various applications have to have, as far as possible, simple structures, with a high degree of symmetry and to be relatively isotropic; materials with metallic bonds are the most resistant to the effects of irradiation, followed by those with ionic and covalent bonds.

3. *Admixture of some alloying elements* : this considerably reduces the instability of phases and the effects of anisotropy.

A more complete and detailed review on these methods may be found in Bullough and others [32].

### 2.6.6. Alterations in Material Properties

Changes in the crystalline lattice structure of materials and the generation of defects at irradiation lead to alterations of the mechanical and thermal properties of materials. Sometimes the electric, magnetic and optical properties are strongly distorted too.

The formation of defects and the generation of new elements by irradiation leads to a gradual modification of the crystalline lattice both as a result of crystallographic disordering, and by the structural transformations involved.

The alteration of the crystalline lattice parameters of quartz when irradiated with neutrons is an example. From Fig. 2.35 one can see that, as the neutron fluency increases, the ratio $c/a$ decreases. If the irradiation fluency exceeds

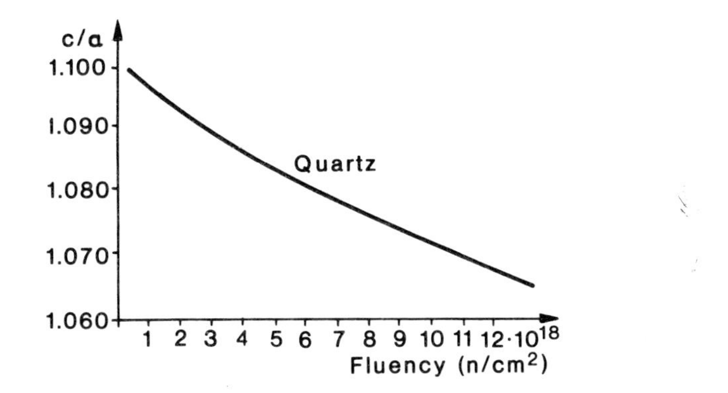


Fig. 2.35. Variation of the lattice constant ratio ($c/a$) for the crystalline quartz, versus neutron fluency.

1.2 x $10^{20}$ n/cm$^2$, the long-range order in the monocrystal is destroyed and α-quartz turns into a vitreous state.

The density of the crystalline quartz irradiated, in the reactor, with neutrons decreases. For a fluency of 1.5 x $10^{20}$ n/cm$^2$, the alteration in density reaches 14.5%. The quartz glass presents a density increase of up to 3% of the initial value, at a fluency of 5 x $10^{19}$ n/cm$^2$ (Fig. 2.36) [33].

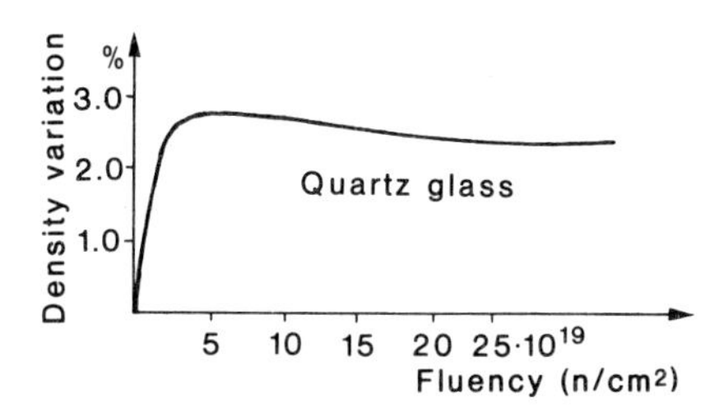


Fig. 2.36. Quartz glass density variation versus neutron fluency.

A similar example is provided by the irradiation with neutrons of graphite. The $c$ constant of the elementary cell increases with the neutron fluency (Fig. 2.37a) and the volume expansion varies correspondingly (Fig. 2.37b) [34]. Likewise, the mechanical resistance of graphite changes (Fig. 2.38) [35]. By irradiation, the thermal properties of graphite change.Thus neutron irradiation causes the thermal conductivity to decrease (Fig. 2.39). Electric properties are also affected by irradiation. So, the disorder induced in the lattice, the generation of defects, and the creation of traps at irradiation with neutrons lead to a decrease in both carriers' mobility and concentration. Consequently, the resistivity of metals increases. This effect is illustrated in Fig. 2.40, on zinc,nickel,and copper, subject to a flux of 6 x $10^{11}$ n/cm$^2$ . s at 20 K [40]. The fact that the increment in resistivity is proportional to the duration of irradiation is connected to the fact that, on the average, the rate of generation of irradiation defects is constant. The situation is different with the irradiation of semiconductors, where the presence of more than one

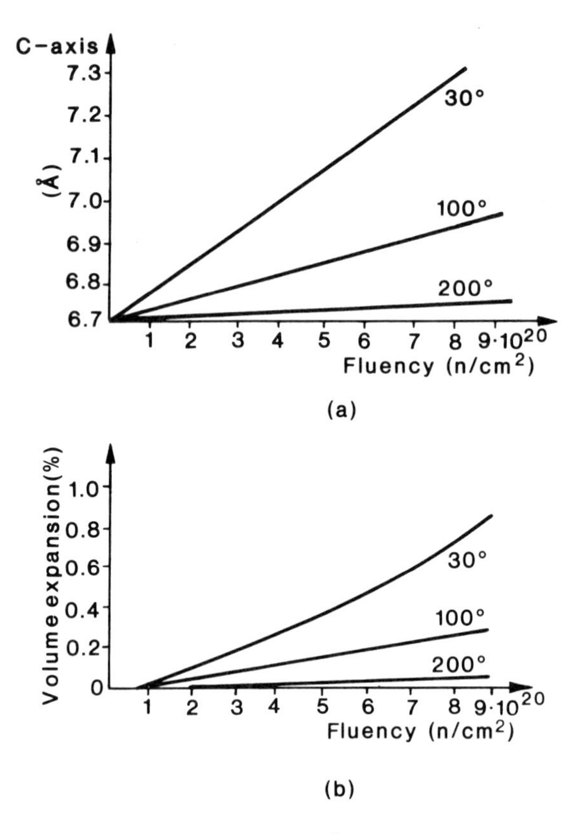


Fig. 2.37. Lattice constant $c$ (Å) variation, (a), and graphite expansion, (b), versus neutron fluency.

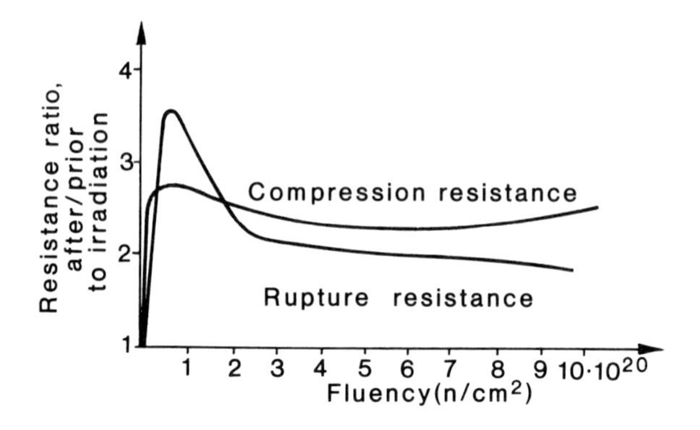


Fig. 2.38. Variation of mechanical properties of graphite with the neutron fluency.

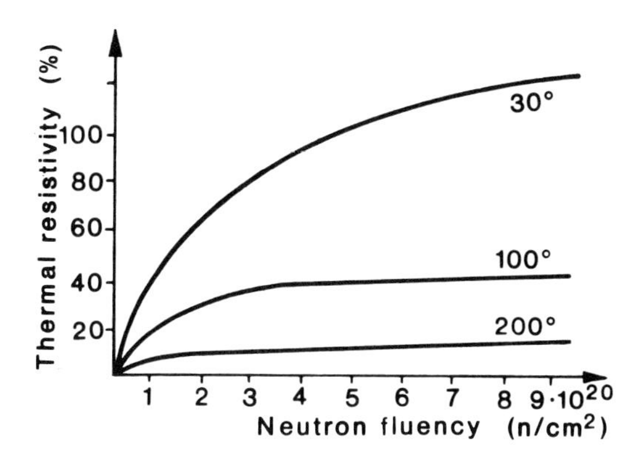


Fig. 2.39. Dependence on neutron fluency of the thermal resistivity change of graphite.

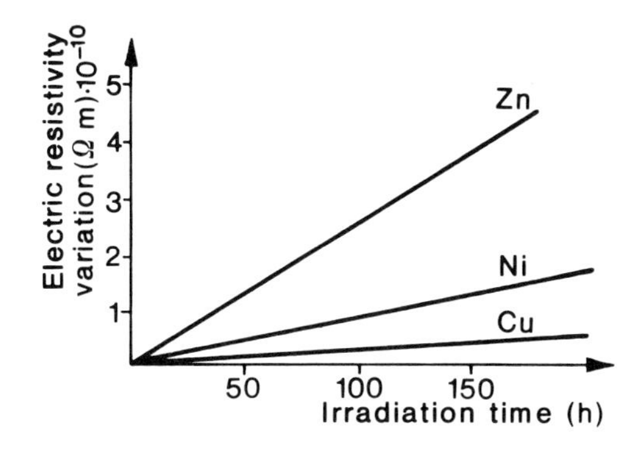


Fig. 2.40. Variation of the electric resistivity of Cu, Ni and Zn, at neutron irradiation.

type of carrier breaks the linear connection of the resistivity with the neutron fluency. This behaviour is evidenced in Fig. 2.41, on the case of a germanium sample irradiated with fast neutrons [36]. The experimental data may be analysed on the basis of a model which assumes three types of charge carriers, as it results from Hall effect measurements.

An interesting case is the variation of the resistivity of metallic uranium under neutron irradiation [37]. In Fig. 2.42, $\tau$ stands for the ratio of the number of atoms which have already fissioned to the total number of atoms (the burn-up factor). One may see that, after causing a fast variation of the resistivity, the irradiation leads to a saturation zone. The simplest model which can explain this saturation effect is based on the assumption that the irradiation generates certain defects while annihilating others. Thus, at each moment, the number of defects created is equal to the number of defects annihilated. The model takes into account the formation of the displacement spikes due to irradiation and, therefore, allows inferring of some of the specific parameters of the process. For example, the calculations give a value of $2.1 \times 10^{-22}$ $m^3$ for the volume of a displacement spike in uranium.

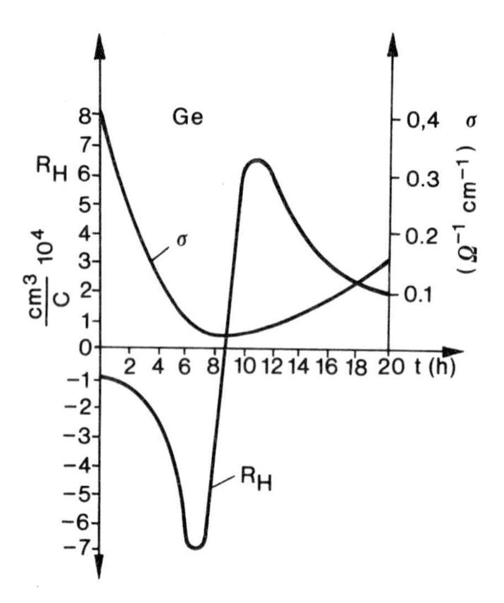


Fig. 2.41. Change with time of electric conductivity (σ) and Hall constant ($R_H$), for fast neutron-irradiated germanium.

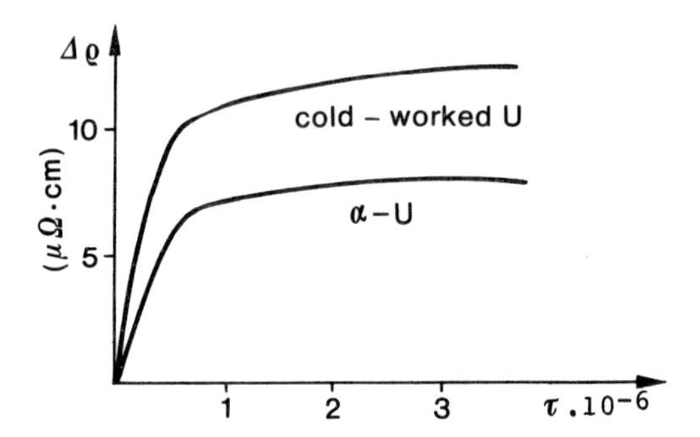


Fig. 2.42. Variation of electric resistivity of metallic uranium (in α-phase and cold worked), at fast neutron irradiation.

A similar case is encountered with the self-irradiation of plutonium [38]. Since plutonium is α-active, emitting α particles of 5.1 MeV, the metal in time changes its electric resistivity, the number of defects per unit-time created by self-irradiation being constant. Thus the tipical variation per unit-time of the resistivity for metallic plutonium in α-phase at 5 K, for low fluencies is $6.0 \times 10^{-2}$ Ω . cm/h [38]. In this case, also, the variation of the resistivity has a saturation tendency, described by a classical law of the form:

$$\frac{d\rho}{dt} \sim (\rho_{sat} - \rho). \tag{2.72}$$

The irradiation of insulator materials leads invariably to an increase in the electric conductivity. This is due to the fact that, at not too high temperatures, the conductivity is of ionic type and, generally, is favoured by the existence of lattice defects. Such a behaviour is illustrated in Fig. 2.43, for some ionic crystals [39].

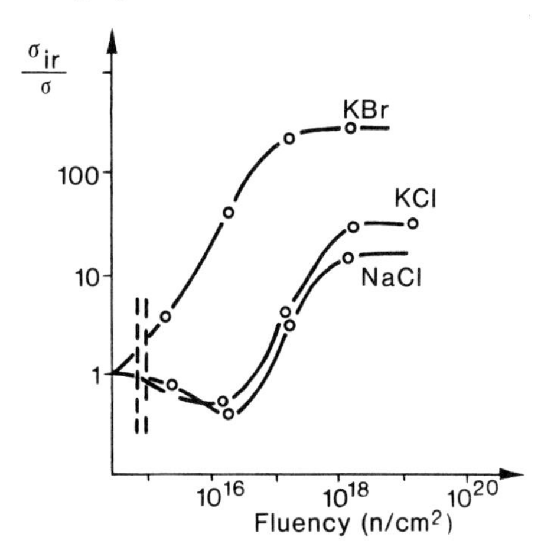


Fig. 2.43. Variation of electric conductivity of some neutron-irradiated alkali halides.

Also, the optical properties are affected by irradiation. Thus, after neutron irradiation the reflectivity of a ceramic plate shows a strong dependence on the wavelength in the region 500 — 650 nm (Fig. 2.44) [40]. Another example is the increment, in a certain wavelength domain, of the light absorption coefficient, for irradiated alkali halides. In LiF crystals the effect is proportional to the irradiation dose (Fig. 2.45) [41].

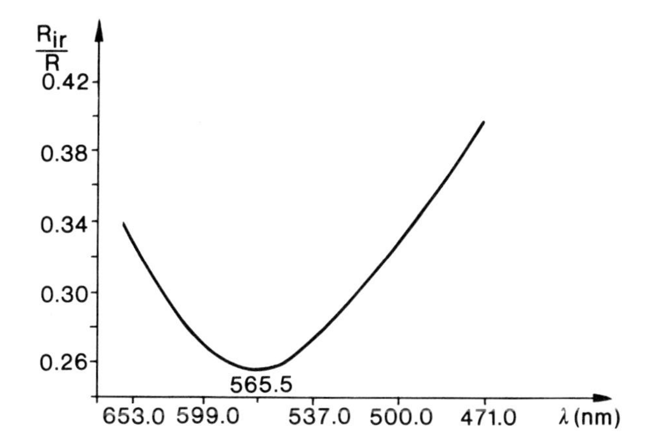


Fig. 2.44. Variation of reflectivity of a neutron-irradiated ceramic plate (fluency: $10^{18}$ n/cm$^2$).

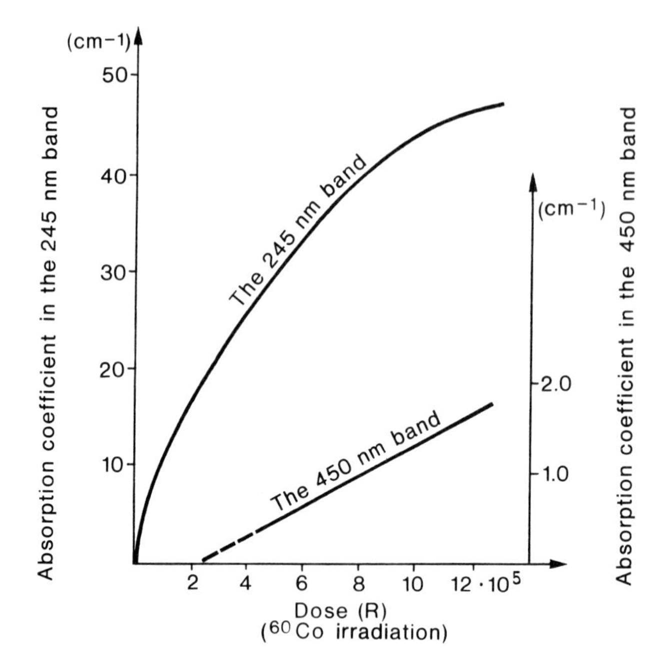


Fig. 2.45. Changes in the optical absorption by LiF in the 245 nm and 450 nm bands, at γ-irradiation.

Materials with special magnetic properties are extremely sensitive to irradiation. These may, at irradiation, present both an increase and a decrease of the magnetic permeability. In Fig. 2.46 the relative variation at irradiation of the permeability for two types of ferrites is displayed [42].

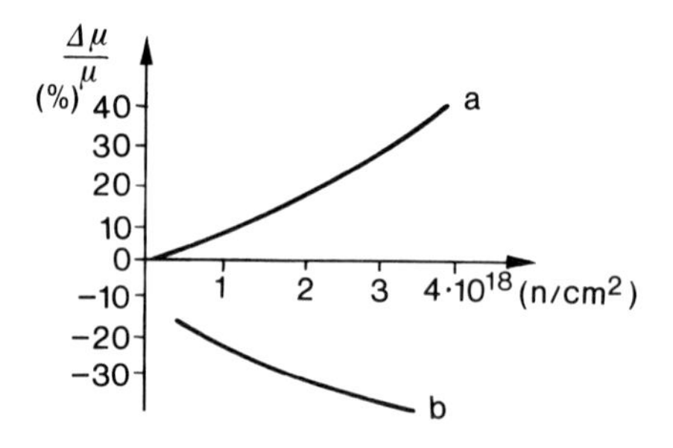


Fig. 2.46. Relative variation with neutron fluency of the magnetic permeability of two types of ferrites.

The effects of irradiation on materials are extremely complex and, generally, have considerable practical consequences. It is for this reason that the study of irradiation effects on substances of technological interest is a topic of great importance to nuclear materials technology.

As far as the operation of a nuclear reactor is concerned, the electric, optical, and magnetic properties are relatively less important. Yet in several cases these become essential. Thus, some insulating materials may turn into conductors, and so may alter the accuracy of whatever electrical measurements are carried out in the reactor core. Some transparent materials may gradually become opaque, and may get tinted under irradiation. On irradiation, magnetic materials with special qualities may lose these to a considerable degree. Likewise, the semiconductor devices may be strongly affected by high-dose irradiations, which may lead to defective functioning of control instruments.

The influence of irradiation on the mechanical properties of materials of nuclear interest will also be discussed in the subsequent sections (Figs. 4.7, 4.11, 4.12, and 5.2).

To conclude this section let us summarize the principal effects of the irradiation on those materials that play an important part in the operation of the reactors.

(a) Thermal neutron reactors. The most important effects of irradiation on stainless steel and zircaloy are embrittlement, creep, increase of hydrogen absorption, and corrosion due to water radiolysis. Their potential/actual impact on the pressure vessel of the reactor is to be carefully assessed/ monitored, because of the special role of the vessel in the overall protection system of the nuclear unit (its insulation from the environment). In the particular case of HTGR there seem to be no special problems with the irradiation of the metallic parts; the effects are felt mainly by graphite.

(b) Fast neutron reactors. The conditions in this case (high temperature, intense neutron flux and, sometimes, a corrosive environment) make problems much more numerous and difficult. Most affected are the fuel elements casings. Here one may encounter swelling and peeling, creep, embrittlement and irradiation-activated diffusion. So far, there is a conspicuous lack of good theoretical models to help solve these problems.

(c) Fusion reactors. All differences considered — as regards physical properties in fusion reactors as compared to those in the fission — the problems with the irradiation of materials are rather similar. Taking into account that, in this case, the energy of the neutrons is much higher (~ 14 MeV), it is believed that the most difficult problem will be the wall of the burning chamber which will be strongly embrittled due to helium generation in very large quantities (see also Chapter 12). A pulsed operation of the reactor would probably be the cause of a greater "fatigue".

At present, in this domain there still is a wide fan of problems that have not yet been satisfactorily attended. Some of them are discussed by McHarque and Scott [43].

### 2.6.7. Testing Materials at Irradiation

The best method of testing nuclear materials at irradiation is to place them inside the nuclear reactor, in conditions as close as possible to those corresponding to their future uses, and to analyse the changes that happen with them. To that purpose, special testing reactors are being built. These assure high irradiation doses, difficult to obtain over convenient lapses of time in normal reactors. Also, they have adequate auxiliary equipment to handle the samples, both inside and outside the core, and are provided with "hot" laboratories that enable handling of strongly irradiated materials.

Among the testing reactors are: (a) on thermal neutrons: MIR in the USSR (100 MW, $\Phi_{th} = 5 \times 10^{14}$ n/cm$^2$ . s) and ATR in the USA (250 MW, $\Phi_{th} = 10^{15}$ n/cm$^2$ . s); (b) on fast neutrons: EBR II and FFTF in the USA, BOR-60 in the USSR, DFR in the UK, Rapsodie in France, etc. [44].

These reactors allow irradiation with neutrons the energy of which does not exceed 2 MeV. For the higher energies that are required for testing materials to be used in fusion reactors ( ~ 14 MeV), different neutron sources have been suggested, based on stripping reactions[1] (d,n) the deuterons being obtained from particle accelerators.

Samples irradiated in testing reactors, in conditions that replicate, as faithfully as possible, those in the power reactors are then subject to various tests; first and foremost of mechanical resistance. Testing devices are specially designed to enable working with activated samples (see also Appendix 6).

To study the irradiation defects one may use irradiation with charged particles (heavy and light ions, electrons). In this case the main technique of observing the defects is transmission electron microscopy. One may study and simulate the processes of creating voids in materials (with heavy ions); changes in the mechanical properties, embrittlement, creep under irradiation (with light ions: p, d, $\alpha$); the dynamics of producing the chief varieties of defects — voids, dislocations, complex clustering, etc.

In parallel, the influence of irradiation is studied from the point of view of the local structure of the crystalline lattice (point defects, electronic structure, changes in the chemical binding, etc.). To this effect, methods now considered as classic are employed,such as the *electron paramagnetic resonance*, *Mössbauer effect* [45], or channelling techniques [46]. Methods are also used by which one studies the variation of a physical parameter in proportion to the defect concentration, such as measurements of electrical and thermal conductivity; lattice constants; stored energy; elasticity moduli, etc., or of a physical parameter that is sensitive to the type of defects, such as magnetoresistance, Hall coefficient, thermoelectric effect, X ray and neutron scattering amplitude (see also Chapter 11).

For specific problems concerning the irradiation effects upon materials and methods of testing their properties the reader is referred to Refs. 47 and 48.

## 2.7. SELECTING MATERIALS OF NUCLEAR INTEREST

When selecting materials for nuclear applications one must take into account those general and special merits enabling superior performances in the severe operational conditions of a nuclear reactor. Among the desirable general features are: mechanical resistance, ductility, stability, corrosion resistance and adequate heat transfer properties, to which one has to add accessible manufacturing technologies and bearable costs. Moreover, observance of special requirements concerning the characteristics related to neutron absorption, sensitivity to activation, resistance to irradiation effects, etc. is in order.

The first step in selecting materials for a certain application consists in a realistic definition of mechanical, thermal and irradiation conditions to which the respective material will be exposed in the reactor. The desired *mechanical resistance* of the material must be inferred from the stresses and strains it will be subject to during the operation of the reactor. *Ductility* is connected to resistance, as it is known that, as a rule, a superior ductility may be achieved only to the detriment of resistance.

---

[1]The stripping reaction (d,n) consists in the detachment of the proton in the incident deuteron and its capture by the target nucleus. The neutron left out continues on its way.

The *stability* of materials is closely linked to temperature and to the duration of exposure to a given temperature. Temperature affects mechanical strength and creep, and is also determinant in the generation of microstructural alterations in the material, with direct effects on its properties.

*Assurance of an adequate thermal transfer* is required, among other things, by the necessity of limiting the temperature that certain parts may reach.

The *neutron absorption* in materials which make the core of the reactor has to be as low as possible. A loose observance of a proper economy of neutrons forces the design into a superfluous oversizing of the core in order to reach criticality. On the other hand, reactor control and protection requires the choice of strongly neutron-absorbent materials.

In Table 2.11 the absorption cross-sections for thermal neutrons are given, for several materials [49]. Materials in group I, having a small cross-section, are adequate for use in the core of natural uranium reactors. In the second group are materials that can be used with enriched uranium which, on the account of a greater excess of reactivity, allows the use of materials with a somewhat larger cross-section. To the third group belong materials with a large cross-section, which are avoided in reactor construction, apart from the case when they are sought for the adjusting and shut-down of the reactor (see also Appendix 7).

TABLE 2.11 Absorption Cross-Sections for Thermal Neutrons

| *Group I* | | *Group II* | | *Group III* | |
|---|---|---|---|---|---|
| *Material* | *Cross-section* *b* | *Material* | *Cross-section* *b* | *Material* | *Cross-section* *b* |
| $CO_2$(1 at) | 0.004 | Zn | 1.06 | Mn | 12.6 |
| O | < 0.0002 | Nb | 1.1 | W | 19.2 |
| $D_2O$ | 0.0019 | Sr | 1.16 | Ta | 21.3 |
| C | 0.003 | Ba | 1.17 | Cl | 31.6 |
| He | 0.0068 | N | 1.78 | Co | 34.8 |
| Be | 0.009 | Mo | 2.5 | Ag | 62 |
| F | < 0.010 | Stainless steel | ~ 2 ... 3 | Li | 67 |
| Bi | 0.032 | Fe | 2.53 | Re | 84 |
| Mg | 0.063 | Cr | 2.9 | Au | 98 |
| Si | 0.13 | Cu | 3.59 | Hf | 105 |
| Pb | 0.17 | Inconel | ~ 4.1 | In | 190 |
| Zr | 0.18 | Monel | ~ 4.2 | Hg | 380 |
| P | 0.19 | Ni | 4.6 | Ir | 430 |
| Al | 0.215 | V | 5.1 | B | 750 |
| Al (2S) | 0.26 | Ti | 5.6 | Dy | 1100 |
| H | 0.33 | Sb | 6.4 | Cd | 2550 |
| Al (3S) | 0.36 | | | Eu | 4600 |
| Ca | 0.43 | | | Gd | 46,000 |
| Na | 0.49 | | | | |
| S | 0.49 | | | | |
| $H_2O$ | 0.66 | | | | |

It is desirable that the neutron radiative capture reactions in materials lead to elements with short lifetimes, and which emit low energy radiations.

Choosing materials which satisfy this desideratum simplifies the maintenance operations and eases handling and storage of radioactive wastes.

The compound of these requirements is further enlarged with other technical and economic criteria. Among them, the *manufacturing procedures* influence the choice, to the effect that such materials must be chosen which can be obtained from available raw materials at reasonable costs, and using standard manufacturing techniques.

The criterion of resistance to irradiation effects provides for a proper, comprehensive definition of the required properties and manufacturing procedures, that would warrant confidence in the stability and maximal performance of nuclear materials in the demanding environment of a reactor. A nuclear reactor (with thermal neutrons) is a complex system articulating, in principal, the following parts: *fuel materials, moderator, reflector, reactivity control devices, cooling agents, structural and protection parts.* Each of these has special assignments in ensuring a controlled, smooth and safe operation of the reactor.

## REFERENCES


1. Vainstein, B. K., Cernov, A. A., and Suvalov, L. A. (1979, 1980). *Sovremennaya krystallografyia*, Vol. 1, 2 and 3. Izd. Nauka, Moskow; Vainstein, B. K., and Cernov, A. A. (1975). *Problemy sovremennoy krystallografyi*. Izd. Nauka, Moskow.
2. Jeludev, I. S. (1968). *Fizika krystalliceskyh dielektrikov*. Izd. Nauka, Moskow.
3. Dobrescu, L. (1960). *Tehnica reactoarelor nucleare*. Edit. Academiei, Bucharest.
4. Greenwood, N. N. (1970). *Ionic Crystals; Lattice Defects and Nonstoichiometry*. Butterworths, London.
5. Evans, R. C. (1966). *An Introduction to Crystal Chemistry*. Cambridge Univ. Press.
6. Kubaschewski, O. (1962). In *Proc.Symp.Thermodynamics of Nuclear Materials*, IAEA, Vienna. p. 219.
7. Hansen, M. (1958). *Constitution of binary alloys*. McGraw-Hill, New York; Elliott, P. R. (1965). *Constitution of binary alloys*, first supplement. McGraw-Hill, New York; Shunk, F. A. (1969). *Constitution of binary alloys*, second supplement. McGraw-Hill, New York.
8. Wert, C. A., and Thomson, R. M. (1964). *Physics of Solids*. McGraw-Hill, New York.
9. Nenitescu, C. D. (1979). *Chimie generala*. Edit. didactica si pedagogica, Bucharest.
10. Fedorov, G. B., and Smirnov, E. A. (1978). *Diffuzyia v reaktornyh materialah*. Atomizdat, Mokcow.
11. Hull, D. (1965). *Introduction to Dislocations*. Pergamon Press, Oxford.
12. Kittel, C. (1971). *Introducere in fizica corpului solid* (translation from English). Edit. Tehnica, Bucharest.
13. Shreir, L. L. (1963). *Corrosion*, vol.1, 2. Newnes, London.
14. Rosenfeld, I. L. (1970). *Korrozya; zaschita metallov*. Izd. Metallurgyia, Moskow.
15. Constantinescu, M. (1977). *Protectia anticorosiva a metalelor*. Edit. didactica si pedagogica, Bucharest.
16. Oniciu, L. (1977). *Chimie fizica; electrochimie*. Edit. didactica si pedagogica, Bucharest.
17. Sorohan, M. (1977). *Introducere in fizica si tehnica presiunilor inalte*. Edit. Tehnica, Bucharest.
18. Olander, D. R. (1976). *Fundamental Aspects of Nuclear Reactor Fuel Elements*. Nat.Techn.Inf.Service, US Dept. of Commerce, Springfield, TID-26711-P1.

19. Frenkel, I. I. (1953). *Introducere in teoria metalelor* (translation from Russian - 1950). Edit. Tehnica, Bucharest; Lakhtine, I. (1978). *Metallographie et traitements thermiques des metaux*, MIR, Moskow.
20. Labusca, E. (1957). *Introducere in metalurgia pulberilor*. Edit. Academiei, Bucharest.
21. Dixon, T. H. R., and Clayton, A. (1971). *Powder Metallurgy for Engineers*. The Machinery Publ., London.
22. Sims, T. C., and Hagel, W. C., (Eds.) (1972). *The Superalloys*. John Wiley and Sons, London.
23. Benjamin, J.S., and Volin, T.E. (1974). *Metallurgical Transactions*, 5,1929.
24. Robertson, J. A. L. (1969). *Irradiation Effects in Nuclear Fuels*. Gordon and Breach Science Publ., New York.
25. Leteurtre, J., and Quere, Y. (1972). *Irradiation Effects in Fissile Materials*. North-Holland Publ. Co., Amsterdam.
26. Dienes, G. S., and Vineyard, G. H. (1957). *Radiation Effects in Solids*. Interscience Publ., New York.
27. Seitz, F., and Koehler, J. S. (1956). Displacement of Atoms during Radiation, in *Solid State Physics*, Vol. II. Academic Press, New York, p. 307; Brinkman, J. A. (1956). *Am.J.of Physics*, 24, 246.
28. Stoneham, A. M. (1975). *Theory of Defects in Solids*. Oxford Univ.Press.
29. Tikanov, B. A., and Davydov, E. F. (1977). *Radiatsionnaya stoykosti teplovidelyaiuschyh elementov iadernyh reaktorov*. Atomizdat, Moskow.
30. Carslow, H. S., and Jaeger, J. C. (1959). *Conduction of Heat in Solids*. Clarendon Press, Oxford.
31. Cottrell, A. H., and Bilby, B. A. (1949). *Proc.Phys.Soc.(London)*, A62, 49.
32. Bullough, R., Eyre, B. L., and Kulcinsky, G. L. (1977). *J.Nuclear Materials*, 68, 168.
33. Primak, W. (1957). *Phys. Rev.*, 119, 1240.
34. Wirtz, K. (1955). Production and Neutron Absorption of Nuclear Graphite. In *Proc.Intern.Conf.Geneva*, 8(P), p. 1132.
35. Woods, W. K., Bupp, L. P., and Flecher, J. E. (1955). Degats infliges par l'irradiation au graphite artificiel. In *Comptes Rendus Intern.Conf. Geneva*, 8(P), 746.
36. Buras, B. and others (1963). In *Radiation Damage in Solids*, Vol. 3. IAEA, Vienna, pp. 55-64.
37. Quere, Y. (1963). *J.Nuclear Materials*, 9, 290.
38. Olsen, C. E., Elliot, R. O. (1962). *J.Phys.Chem.Solids*, 23, 1225; Griffin, C. S., Mendelssohn, K., and Mortimer, M. J. (1968). *Cryogenics*, 8, 110.
39. Budilin, B. V., and Vorobiov, A. A. (1962). *Deistvie izlucenyia na ionye struktury*, Gosatomizdat, Moskow.
40. Motoc, C. (1964). *Actiunea radiatiilor nucleare asupra corpului solid*. Edit. Academiei, Bucharest.
41. Kubo, K., and Ozawa, K. (1963). In *Radiation Damage in Solids*, Vol. 3, IAEA, Vienna, p. 121.
42. Labusca, E., Mirion, I., Andreescu, N., and Biscoveanu, I. (1965). In *Non-Destructive Testing,Nuclear Technology*, Vol.2, IAEA,Vienna, pp.75-106.
43. McHarque, C. J., and Scott, J. L. (1978). *Metall.Trans.*, 9A, 151.
44. Ullmaier, J. H., and Schilling, W. (1980). Radiation Damage in Metallic Reactor Materials, In *Physics of Modern Materials*, Vol. 1, IAEA, Vienna.
45. Wertheim, G. K., Haussmann, A., and Sander, W. (1971). *The Electronic Structure of Point Defects*. North-Holland Publ.Co., Amsterdam.
46. Swanson, M. L., Howe, L. M., and Quenneville, A. F. (1975). In *Proc.Int. Conf.on Fundamental Aspects of Radiation Damage in Metals*. Gatlinburg, US Dept.of Commerce, Springfield.
47. Ibragimov, Sh. Sh. (Ed.) (1978). *Radiatsionnyie defecty v metallitcheskyh kristallah*. Nauka, Alma-Ata.
48. Kostiukov, N. S., Antonova, H. P., Zilberman, M. I., and Aseev, N. A. (1979). *Radiatsionnoe electromaterialovedenye*. Atomizdat, Moskow.
49. Stephenson, R. (1954). *Introduction to Nuclear Engineering*. McGraw-Hill Book Co., Inc., New York.

## CHAPTER 3

# Fuel Materials

The "heart" of a nuclear reactor is its core, which contains the *fuel* that cradles the fission reactions. A nuclear fuel must contain fissionable nuclei. There are only four fissile isotopes that are of practical interest: $^{235}U$, $^{233}U$, $^{239}Pu$ and $^{241}Pu$. Isotope $^{235}U$ is the only natural nuclide that is fissionable by thermal neutrons; the others are offsprings of neutron-induced transmutations.

Nuclear fuel materials undergo a long series of processing steps, both prior to their loading into the reactor and after being used. To make fuels, a fabrication process must be designed and implemented, beginning with uranium ore mining, continuing with its conversion into a suitable chemical form, followed by recurrent chemical purifications and, when appropriate, isotopic enrichment; the process ends by bringing the fuel material to the required shape and size. Following the utilization in the reactor, spent fuel processing, extraction of recyclable components, waste storage and reutilization of recovered materials are in order. Overall, such a sequence constitutes the *nuclear fuel cycle* [1]. One variety of this cycle, that provides for plutonium recovery, is outlined in Fig. 3.1.

The fuel that nuclear reactors use may be delivered alternatively as a metal; a ceramic compound (oxides, carbides, nitrides, etc.), a solid dispersion (metallic, metallo-ceramic, carbo-ceramic, ceramic, vitro-ceramic) or a fluid (gases, liquids). The *metallic fuels* present certain merits, as a result of some suitable thermal and nuclear properties; however, they do not behave satisfactorily under long-term irradiation conditions. In comparison, the *ceramic fuels* (oxides, carbides, U-, Pu-, Th-nitrides) rate lower as far as nuclear and thermal properties, offering instead the advantage of a greater stability under irradiation and at high temperatures. The *dispersed fuels* have been created for the purpose of improving reactor performance, especially as far as in-core energy density and operating temperature of cooling agents. *Fluid fuels* offer, in principle, advantages as to the fabrication costs and operation performance, though problems with corrosion and handling tend to downgrade their overall merits.

### 3.1. URANIUM

Uranium is found in the earth's crust down to a depth of 16 km, at an average abundance of $2 \times 10^{-5}\%$, which is more than the abundance of such metals as

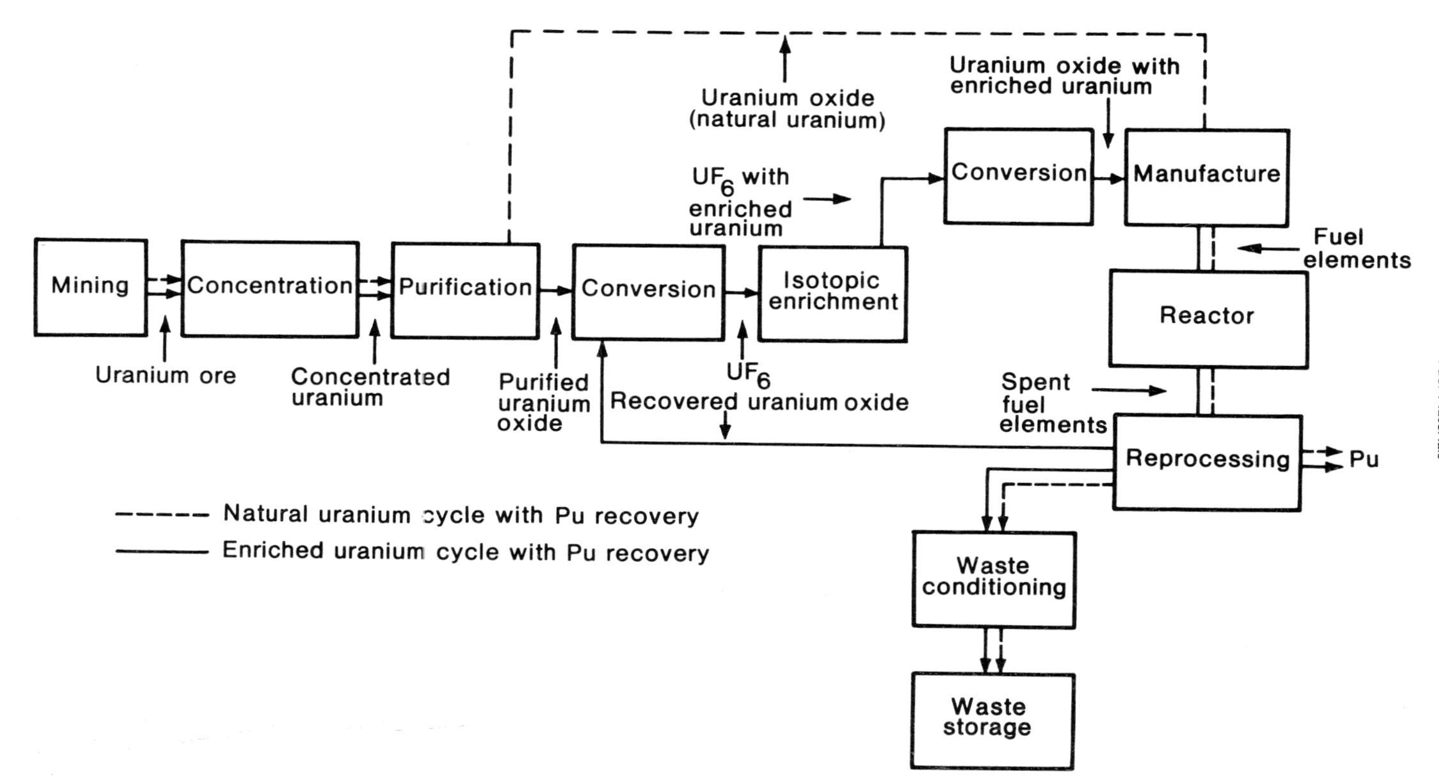


Fig. 3.1. Flow-sheet of nuclear fuel cycles with plutonium recovery.

mercury, silver, bismuth or cadmium. In sea and ocean waters uranium is to be found as soluble salts, at concentrations between $0.4 \times 10^{-7}$ and $23 \times 10^{-7}$ g/l.

There are three types of rocks that may contain uranium. Two of them contain primary and, respectively, secondary uranium minerals, while the third one contains uranium as an impurity embedded in host crystal lattices; as yet, the last type is of no economic relevance. Of great interest are the primary minerals, generated in the depths of the earth when the crust was formed. Secondary minerals evolved at the surface as a result of chemical transformations (most commonly oxidation) of primary minerals. In the primary minerals uranium is captured in oxides, either as such or in association with other metals. In secondary minerals uranium is to be found in a hexavalent state. Here are some examples:

1. primary minerals: uraninite, $UO_2$; pitchblende, $U_3O_8$; torianyle $(Th,U)O_2$;

2. secondary minerals: carnotite, $K[UO_2/VO_4]\cdot 1\frac{1}{2}H_2O$; calcolite, $Cu[UO_2/PO_4]\cdot 8H_2O$; antumite, $Ca[UO_2/PO_4]\cdot 8H_2O$; becquerelite, $2UO_3\cdot 3H_2O$.

As a rule, all these minerals are mixed with other oxides ($PbO$, $ThO_2$, $ZrO_2$, $CaO$, $MgO$, $Al_2O_3$, ... ) so that the chemical composition may vary from one sample to another [2].

The content in disperse uranium of some rocks is given in Table 3.1. [3].

TABLE 3.1 Content in Dispersed Uranium of Some Rocks

| *Rock* | *Content* | |
|---|---|---|
| | *$SiO_2$* (% weight) | *U* (g/t) |
| Granite | 70 | 9.0 |
| Granodiorite | 66 | 7.7 |
| Diorite | 60 | 4.0 |
| Basalt | 50 | 3.0 |
| Gabbro | 50 | 2.4 |
| Peridotite | 43 | 1.5 |
| Dunite | 40 | 1.4 |

The use of uranium in nuclear power generation is by far the chief utilization of this element. Figure 3.2 gives a projection of the estimated uranium consumption for the manufacturing of fuel meant to feed various reactor types, over the years 1975 — 2025 [4]. Resource savings are notable in the case of fast breeder reactors (FBR) as compared to light water reactors (LWR). The trends also indicate limits to proven reserves, as well as to estimated and potential resources.

Uranium may also be utilized in the preparation of solid laser-active crystals (calcium fluoride-trivalent uranium — $CaF_2:U^{3+}$); luminiphore materials (systems containing $U^{6+}$); as an active ingredient in glass colouring, etc.

Obtaining uranium from raw minerals requires concentration, refining and reduction. Through *concentration* one separates the uranium from rocks. The rock is ground, then treated with uranium-dissolving reactives (leaching). The next step is uranium recovery from solution, leading to the product known as

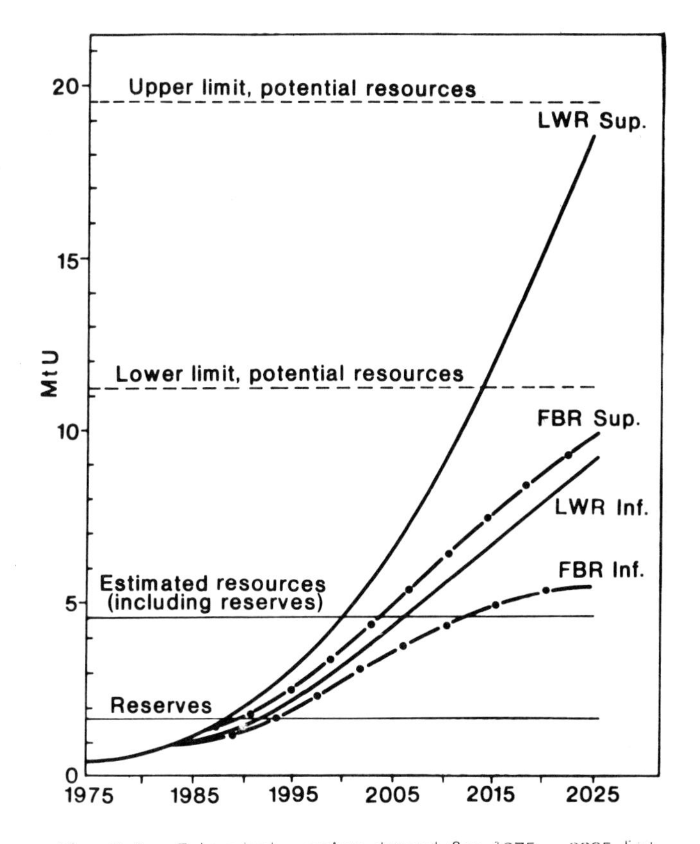


Fig. 3.2. Estimated uranium demand for 1975 — 2025 [4].

"yellow cake", which contains 60 — 85% $U_3O_8$. *Refining* is the process through which uranium-carrying compounds are upgraded to attain a high purity, in compliance with nuclear requirements. During this phase one aims at eliminating particularly those impurities that have a large cross-section for neutron absorption. In the process, $UO_3$ is converted to $UO_2$ and then to $UF_4$ — if one aims at obtaining metallic uranium, or to $UF_4$ and then to $UF_6$ — if isotopic enrichment is next envisaged. $UF_6$ is a volatile compound that sublimes at 56 $^{\circ}C$; it is almost exclusively used for isotopic enrichment.

Metallic uranium can be obtained either by *chemical reduction* of tetrafluoride or uranium dioxide, or by *electrolytic reduction* of uranium halogenides or of uranium dioxide. More details regarding concentration, refining and reduction, and the technologies for obtaining nuclear purity uranium from ore, are available from Appendix 13.

### 3.1.1. Properties

The ground state of the uranium atom is $^5L_6$ and corresponds to the electronic configuration (Rn) $5f^36s^26p^66d^17s^2$. The first excited states (of successively greater energy) of the atom are:

$$f^3d^2s,\ f^4s^2,\ f^2d^2s^2,$$
$$f^3s^2p,\ f^3dsp,\ f^4ds,\ f^4sp,\ \text{etc.}$$

Metallic uranium comes in three allotropic states (α, β and γ). The α-U phase crystalizes in the orthorombic system (Fig. 3.3) and is stable up to 663 °C. The phase β-U, of a tetragonal structure, subsists in the temperature range 663 — 764 °C, while the phase γ-U, presenting a body-centred cubic structure, remains stable from 764 °C up to the melting point. Table 3.2 displays some of the physical properties of uranium [7,8,11].

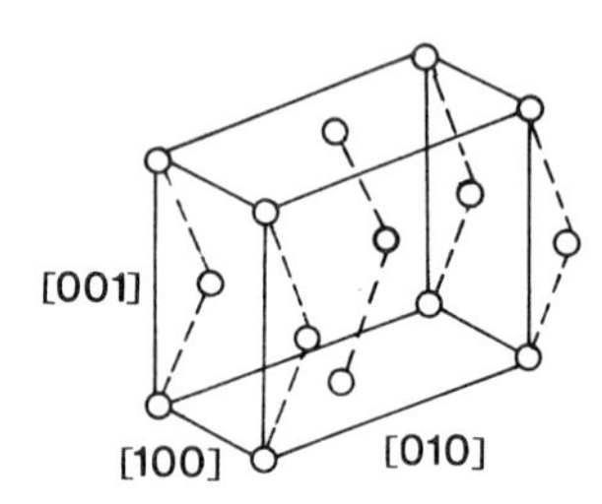


Fig. 3.3. The orthorhombic unit-cell of α-uranium.

TABLE 3.2 Physical Properties of Uranium

| | |
|---|---|
| Density | 19,130 kg/m$^3$ |
| Allotropic transformations: | |
| α → β | 663 °C |
| β → γ | 764 °C |
| Crystal structure: | |
| α-U – orthorombic | $a$ = 2.852 Å ; $b$ = 5.865 Å; $c$ = 4.955 Å |
| β-U – tetragonal | $a$ = 10.759 Å ; $c$ = 5.656 Å |
| γ-U – b.c.c. | $a$ = 3.524 Å |
| Melting point | 1133 ± 1 °C |
| Boiling point | 3813 °C |
| Thermal expansion coefficient (between 25 and 125 °C): (°C)$^{-1}$ | |
| in direction [100] | 21.17 . 10$^{-6}$ |
| in direction [010] | -1.5 . 10$^{-6}$ |
| in direction [001] | 23.2 . 10$^{-6}$ |
| Bulk expansion coefficient | 45.8 . 10$^{-6}$ |
| Thermal conductivity: (W/m . K) | |
| at 350 K | 25.95 |
| at 670 K | 32.65 |
| Electric resistivity (25 °C) | (2-4) . 10$^{-7}$ Ω m |

The analysis of the mechanical and physical properties of α-U reveals its pronounced anisotropy. Table 3.2 gives the values of the coefficient of thermal expansion along various crystallographic directions, while Fig. 3.4. draws the temperature dependence of the lattice parameters.

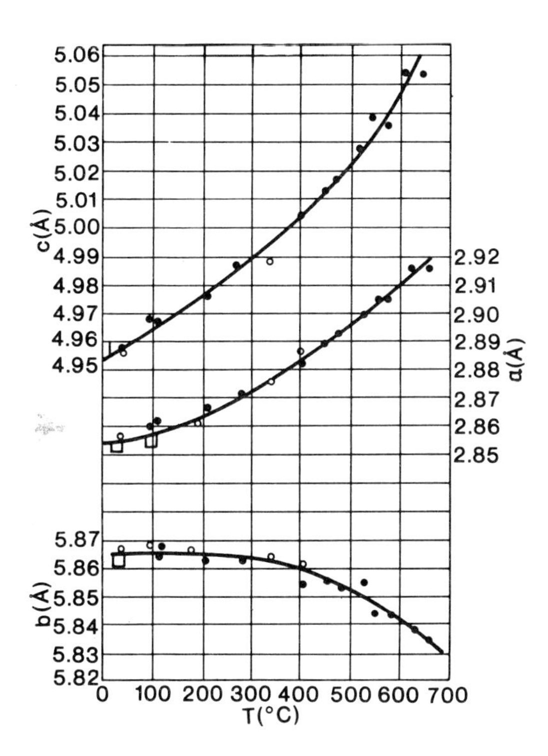


Fig. 3.4. Temperature effect on the lattice cell parameters of uranium.

The temperature dependence of the thermal conductivity of metallic uranium is illustrated in Fig. 3.5 [9]. Table 3.3 outlines certain nuclear properties of several of the uranium isotopes.

Due to anisotropy, the mechanical properties of α-U depend strongly on the directions of the applied stresses. The degree of ordering of the crystallites is determined by the fabrication procedures and by thermal treatments. As an example, in Table 3.4 some mechanical properties of uranium obtained through lamination at 300 °C and 600 °C respectively, then submitted to various types of thermal treatments are presented [10]. The data indicate a decrease in tensile strength and an increase in ductility, as the test temperature increases.

The mechanical properties are strongly altered by irradiation.

Thus, following an irradiation of a uranium sample at 120 °C (0.035% burn-up), the tensile strength diminishes by 27% as compared to the initial value, the flow resistance doubles and the sample is no longer ductile (Table 3.5) [10]. A thermal treatment further performed on irradiated samples

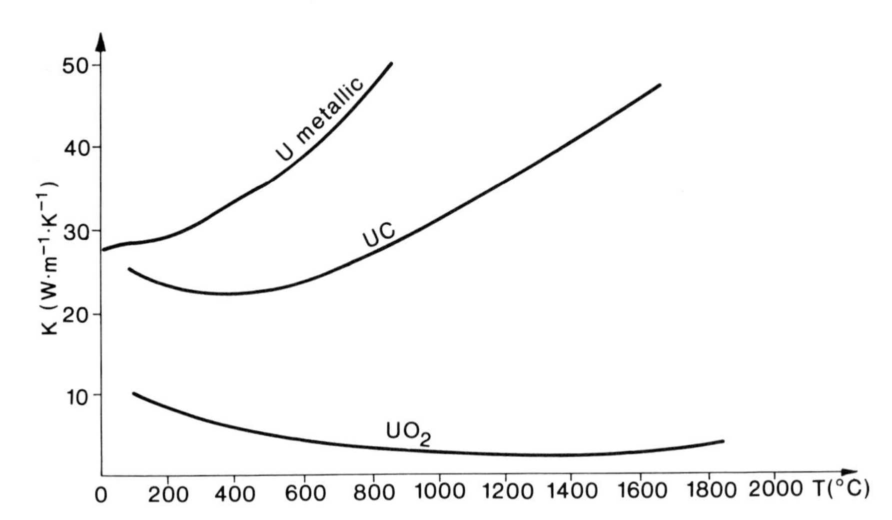


Fig. 3.5. Thermal conductivity of metallic uranium, uranium carbide and uranium dioxide.

TABLE 3.3 Nuclear Constants of Uranium (Thermal Neutrons)

| *Isotope* | $^{233}U$ | $^{235}U$ | $^{238}U$ | *natural U* |
|---|---|---|---|---|
| Microscopic fission cross-section (b) | 532 | 582 | - | 4.18 |
| Absorption cross-section (b) | 588 | 694 | 2.73 | 7.68 |
| Average number of fast neutrons emitted per fission act | 2.52 | 2.43 | - | 2.46 |
| Number of fast neutrons per absorbed thermal neutron | 2.28 | 2.07 | - | 1.37 |

(at 400 °C) provides for a continued reduction in tensile strength, whereas the ductility remains practically the same.

During irradiation a hardening of the material is also recorded. An α-U sample obtained through lamination has a Vickers hardness of 220 ± 10; after irradiation at a burn-up of 0.10% the hardness increases to 315 ± 20.

The main valence states of uranium are U(III), U(IV) and U(VI). In aqueous solution the most stable among them is U(VI), usually expressed in the form of the uranyl ion $(UO_2)^{2+}$. Metallic uranium is extremely reactive, comparable in many respects with magnesium. It reacts with nitrogen at high temperatures, and with hydrogen gives a hydride ($UH_3$) that decomposes at 350 °C.

The effect of temperature on the uranium corrosion rate in certain coolants is given in Table 3.6 [11,12]. Uranium does not react with liquid sodium or potassium.

TABLE 3.4 Mechanical Properties of Uranium

| *Sample* | *Testing temperature* (°C) | *Tensile strength* (MPa) | *Yield resistance* (MPa) | *Elongation* (%) |
|---|---|---|---|---|
| Rolling at 300 °C | 25 | 777 | 301 | 6.8 |
| Annealed in α-phase [1] | 300 | 245 | 123 | 49.0 |
| | 500 | 78 | 36 | 61.0 |
| Annealed in β-phase [2] | 25 | 448 | 172 | 8.5 |
| | 500 | 74 | 50 | 44.0 |
| Rolling at 600 °C | 25 | 620 | 182 | 13.5 |
| Annealed in α-phase [1] | 300 | 224 | 133 | 43.0 |
| | 500 | 78 | 36 | 61.0 |
| Annealed in β-phase [2] | 25 | 434 | 175 | 6.0 |
| | 500 | 189 | 109 | 33.0 |

[1] Annealing in α-phase, 12 h at 600 °C, slow cooling
[2] Annealing in β-phase, 12 h at 700 °C, slow cooling

TABLE 3.5 Effect of Irradiation on Tensile Properties of Uranium

| | *Tensile strength* | | *Yield strength* | | *Elongation* | |
|---|---|---|---|---|---|---|
| | daN/m$^2$ | *Change,%* | daN/m$^2$ | *Change,%* | m/100 m | *Change,%* |
| Control specimen | 7170 | 0 | 2275 | 0 | 17 | 0 |
| Irradiated at 120 °C at a burn-up of 0.035% | 5420 | -27 | 4930 | +117 | 0.36 | -97.9 |
| Irradiated, annealed for 15 h at 400 °C in vacuum | 4480 | -38 | 3585 | +58 | 0.54 | -96.8 |

The corrosion rate of irradiated uranium is much larger than that of non-irradiated uranium.

Uranium, especially as a powder, is handled with difficulty, both because of its feature of being highly pyrophoric and because it is highly toxic. Uranium strikes mainly the kidneys. The uranium concentration that is considered hazardous for the human body is 0.05 mg/m$^3$. In terms of radioactivity, through alpha emission and its fixation in bones, liver, spleen and kidneys, uranium is even more harmful. The maximum permitted amount in air corresponds to an activity of $1\times10^{-8}$ – $3\times10^{-8}$ μCi/l.

TABLE 3.6 Corrosion Rate, for Uranium

| *Temperature* (°C) | *Corrosion rate* ($10^{-2}$ kg/m$^2$ . h) | | |
|---|---|---|---|
| | *In air* | *In water* | *In liquid sodium* |
| 50 | | 0.066 | |
| 70 | | 0.45 | |
| 90 | | 1.00 | |
| 100 | 0.56 | 2.7 | |
| 138 | 1.08 | | |
| 183 | | 139 | |
| 200 | 10.9 | | 1.11 |
| 300 | 21.8 | | |
| 400 | 118 | | |
| 500 | 157 | | 1.39 |
| 600 | | | 4.3 |

### 3.1.2. Processing Methods

Processing of metallic uranium is made difficult by certain factors, among which are fast oxidation in contact with the air, the tendency to absorb hydrogen, and the anisotropy of the properties. Among the main processing methods are *lamination*, *casting*, *extrudation*, *forging* [13], *sintering* of metallic powders, etc. Casting is in order to obtain slabs (bar strips), in view of further processing; lamination, which may be applied in any of the three phases ($\alpha$, $\beta$, $\gamma$) is performed especially through hot-rolling in the $\alpha$-phase of uranium.

Since uranium powders are highly pyrophoric, the conventional methods of powder metallurgy have to be amended, to ensure adequate safety with handling and processing. Uranium parts may be obtained from powders in several ways, e.g. through cold-compacting followed by sintering; cold-compacting, sintering, re-pressing, final thermal treatment; hot-pressing.

Choosing among alternatives in uranium processing depends markedly on temperature. In the $\gamma$-phase, uranium presents a great plasticity, but processing is still difficult since it requires high temperatures. Phase $\beta$-U is generally harder and more brittle than the other two phases; however, by carefully controlling the processing temperature, lamination and extrudation become feasible. To process uranium in the $\alpha$-phase a wide range of techniques are available.

### 3.1.3. Dimensional Stability

In reactor conditions, metallic uranium undergoes significant dimensional modifications as a result of repeated thermal cycles (thermal cycling) and of the impact fission fragments have on its structure. Dimensional alterations are induced also by the action of fast neutrons, though to a much smaller extent.

As an effect of thermal cycling, the uranium dimensional stability is affected both by *anisotropic modification of sizes*, and by surface *ridging*. The amplitude of size modifications depends on the material features (texture, crystallite sizes) and on cycling conditions. The swelling rates attain maximal values for polycrystalline, textured materials, whereas the $\alpha$-U monocrystal neither gets swelled nor ridged.

The *swelling* and *ridging* phenomena which occur during thermal cycling are also recorded with the material irradiation by fission fragments. The crystallite size is not so important in the process of irradiation-induced swelling, but it has a significant bearing on surface ridging. Texture, in exchange, is a chief factor in determining the dimensional effects of the irradiation. The texture effect expresses itself in a reduction of the sample size along the direction that is parallel to the axes [100] and an increase in size along the direction of the axes [010], while in the direction corresponding to the axes [001] the size remains practically unchanged; thereby, the volume variation is negligible.

Dimensional modifications as recorded experimentally can be analysed by means of the equation [14,15]:

$$L = L_0 \exp(Gf) \tag{3.1}$$

where $L_0$ is the initial length, $L$ is the sample length after a fraction $f$ of the nuclei have been split, and $G$ is the *growth constant*, or coefficient. From eq. (3.1) one may obtain:

$$G = [\ln(L/L_0)]/f \tag{3.2}$$

or, for small distortions, $G$ may be expressed as:

$$G = \frac{\text{relative growth}}{\text{burn-up}} = \frac{\Delta L}{L_0 f} \tag{3.3}$$

Thus, growth is a macroscopic measure of the atoms transport at microscopic level as a result of fission acts. Starting from a monocrystal with $N_0$ uranium atoms, let us suppose that, during a fission act, $q$ atoms are transfered from the planes (100) into the planes (010) and, consequently, the monocrystal dimensions will increase in the direction [010] by $\Delta L / L_0 = q / N_0$. After $N_0 f$ fission acts, that correspond to a burn-up $f$, the relative growth will be $\Delta L/L_0 = qf$. The growth coefficient $G$ will be $G = \Delta L/(L_0 f) = q$. At macroscopic scale this is reflected by a roughly 100% elongation of the sample in the direction [010] at a burn-up of 0.2% (1850 MW day/ton).

Table 3.7 gives the growth coefficients of uranium irradiated at 100 °C up to a burn-up of 0.1% [10]. For comparison, the values of the thermal expansion coefficients are also presented.

TABLE 3.7 Coefficients of Growth under Irradiation and Expansion Coefficients

| *Crystallographic direction* | *Growth coefficient* | *Thermal expansion coefficient* [ . $10^{-6}/^{o}C$] | | |
|---|---|---|---|---|
| | | *25 – 125 °C* | *25 – 325 °C* | *25 – 650 °C* |
| [100] | -420 ± 20 | 21.7 | 26.5 | 36.7 |
| [010] | +420 ± 20 | -1.5 | -2.4 | -9.3 |
| [001] | 0 ± 20 | 23.2 | 23.9 | 34.2 |

Limitations in the growth effects determined by thermal cycling or by irradiation are achieved through a choice of such manufacturing procedures and thermal treatments that should avoid texturing, leading to a material with fine grain structure and random grain orientations.

Dimensional stability is also affected by uranium swelling. A volume modification, swelling is a consequence of fission products occurrence in the uranium mass. As stated above, among the fission products there are inert gases, xenon and krypton, which are insoluble in the uranium lattice. For instance, 1 ml of uranium generates 4.73 ml of inert gases, on a burn-up of 1%. The gas as generated diffuses in the fuel volume, agglomerates and collapses on various types of defects (cavities, inclusions, irradiation-induced defects) producing bubbles. In time, these expand causing fuel swelling (a decrease in density). If one admits that all the gas generated through fission is stored in the spherical bubbles occurring in the material volume, then, according to the law of ideal gases, the pressure $p$ in the pores vs. the volume relative modification is:

$$p\Delta V/V = Bk_B T/a^3 \tag{3.4}$$

where $k_B$ is Boltzmann's constant, $B$ is the average number of gas atoms in the volume $a^3$, of the elementary cell of the lattice, and $T$ is the absolute temperature.

A more accurate analysis of gas behaviour in the bubbles starts from the Van der Waals equation [16]. In the case of very small bubbles it is necessary to consider also the Laplace pressure:

$$p_{\text{Laplace}} = 2\gamma/r \tag{3.5}$$

where $\gamma$ is the surface tension and $r$ the bubble radius; this adds to the external pressure, to obtain a more realistic value of the pressure, to be used in the law describing gas behaviour.

Studies of electron microscopy on the interface of material and bubble did not reveal any significant density of dislocations — a fact that suggests that the material has not been plastically distorted as a result of the bubble inner pressure, and thus — that the bubble inner pressure is balanced by the surface pressure. This balance stands under conditions of continued fission gas accumulation in the bubble, owing to a diffusion of an appropriate number of vacancies from the neighbouring zone.

On these considerations, a relation can be established between bubble radii and the distance between their centres, $L$, or their number per unit volume, $n$ [14]. The material is seen as a group of "cells", each with a bubble in the middle. The number of gas atoms of each cell is $L^3B/a^3$. If the whole amount of gas is stored in the bubble, then pressure is given by:

$$p = (3Bk_B L^3 T)/(4\pi a^3 r^3) \tag{3.6}$$

If $p$ equals the pressure generated by the surface tension (thereby neglecting the external pressure) it results:

$$r^2 = L^3(3Bk_B T)/(8\pi\gamma a^3) = [(3Bk_B T)/(8\pi\gamma a^3)] \cdot \frac{1}{n} \tag{3.7}$$

In this way a relation is obtained between the density of bubbles and their radii, the proportionality constant depending on the irradiation conditions (temperature, burn-up) and on the physical properties of the material.

On similar assumptions, the *macroscopic swelling* can be estimated from the relation:

$$\frac{\Delta V}{V} = \frac{4\pi}{3}\left(\frac{r}{L}\right)^3 = \frac{Bk_BT}{2\gamma a^3}\cdot r = \frac{\sqrt{3}}{2\sqrt{\pi}}\left(\frac{Bk_BT}{2\gamma a^3}\right)^{3/2} \cdot \left(\frac{1}{n}\right)^{1/2} \tag{3.8}$$

This indicates that small volume modifications take place for small bubbles, relatively close to one another.

In time, the fission gas bubbles in the bulk of the material grow, either as a result of the increase in inner gas concentration, or through pores migration, followed by their mutual merging. When an atom located on the bubble surface performs — under the action of an external perturbation — a jump corresponding to a lattice constant $a$, with a frequency $\nu_D\psi\ \exp(-E_s/k_BT)$ ($\nu_D$ = the Debye frequency; $\psi$ = an entropic factor; $E_s$ = the activation energy) then the bubble as a whole moves to a distance of $3a^4/(4\pi r^3)$ [17]. Since there exist $4\pi r^2/a^2$ atoms on the bubble surface, the jump frequency is $4\pi(r/a)^2.\ \nu_D\psi\exp(-E_s/k_BT)$ and, consequently, the diffusion coefficient of the bubble $D_p$ is given by:

$$D_p \simeq \frac{3a^6}{8\pi r^4}\nu_D\psi\ \exp(-E_s/k_BT) = \frac{3}{8\pi}\left(\frac{a}{r}\right)^4 D_s \tag{3.9}$$

where $D_s$ is the surface diffusion coefficient. Eq. (3.9) illustrates the great mobility of small bubbles and predicts an exponential dependence on temperature of the mobility.

If the bubble is located in a temperature gradient or is under the action of the strain field in an area of dislocations or of grain boundaries, then a force $F$ will act upon it, casting a directional motion of a velocity $v$ over its random motion. According to Nernst — Einstein relation one has

$$v = [D_p/(k_BT)]\ .\ F \tag{3.10}$$

and thus:

$$v \simeq \frac{3}{8\pi}\frac{D_s}{k_BT}\left(\frac{a}{r}\right)^4 F \tag{3.11}$$

For the cases described above, the forces $F$ may be expressed by:

$$\begin{aligned} F &= \frac{4\pi k_B}{3a^3}\frac{dT}{dt}r^3, && \text{(temperature gradient)} \\ F &= \mu b^2 && \text{(dislocations)} \\ F &= \pi\gamma_1 r && \text{(grain boundaries)} \end{aligned} \tag{3.12}$$

Here $\mu$ denotes the material shear modulus; $b$ is the Bürgers vector of dislocation while $\gamma_1$ is the grain boundary energy.

Knowing the bubble migration rate as a result of random and directional motion enables one to calculate the rate of their dimensional growth. The complexity of this calculation considered, we shall only sketch below the model describing bubble growth as a consequence of their merging by random motion. As a first step, the average time of migration $t_m$ until the joining of two bubbles becomes probable is evaluated. The bubble diffuses over a distance equal to its diameter in a time $4r^2/(6D_p)$. The distance between bubbles being $L$, each bubble gets $(L/2r)^3$ available places where it could migrate. Thus, the probability of merging with another bubble is $Z(2r/L)^3$, where $Z = 6D_p/4r^2$ represents the number of places explored by the bubble in a unit of time. Then the migration time $t_m$ is

$$t_m = \frac{1}{Z\left(\frac{2r}{L}\right)^3} = \frac{4r^2}{6D_p}\left(\frac{L}{2r}\right)^3 = \frac{L^3}{12D_p r}. \tag{3.13}$$

By substituting eq. (3.9) for $D_p$ and eq. (3.6) for $L^3$ and using (3.5) one obtains:

$$t_m = \frac{16\pi^2}{27} \cdot \frac{\gamma}{aBk_B TD_s} \cdot r^5 \tag{3.14}$$

Suppose that initially all bubbles had a radius $r_o$ and that they join in pairs; then, after a time $t_m$, their number drops to half, and each radius becomes:

$$r_1^2 = 2r_o^2 \tag{3.15}$$

Since $t \sim r^5$, the merging time for the next set of bubbles (of $r_1$ radius) will be $2^{5/2}$ times longer; after $m$ such steps the bubble radius will grow from $r_o$ to $r_o(\sqrt{2})^m$ in a time:

$$t = t_m[1+2^{5/2}+(2^{5/2})^2+ \dots + (2^{5/2})^{m+1}] = (t_m/4.7)[(2^{5/2})^m-1] \tag{3.16}$$

Consequently,

$$t \simeq t_o(r^5-r_o^5)/(4.7\, r_o^5), \tag{3.17}$$

where $t_o = \dfrac{16\pi^2}{27}\dfrac{\gamma}{aBk_B TD_s} r_o^5.$

The average bubble radius after the lapse of time $t$ is

$$r \simeq \left[\frac{27 \cdot 4.7}{16\pi^2}\frac{aBk_B TD_s t}{\gamma} + r_o^5\right]^{1/5}. \tag{3.18}$$

### 3.1.4. Uranium Alloys and Compounds

To avoid some of the phenomena that limit the use of metallic uranium as fissionable material, uranium alloys or compounds are employed. They have a better dimensional stability, are corrosion-resistant, etc. Three types of uranium alloys that can be used as fuel materials are known [9]:

(a) alloys with a low content of alloying elements;
(b) alloys with a high content of alloying elements;
(c) metallic dispersions.

A low content of alloying elements preserves the crystalline structure of the metallic uranium, while also providing for a stabilization of its morphology and upgrading it with suitable mechanical and physical features. These alloys are used in the case of the natural uranium fuel element and, consequently, the alloying elements should have small cross-sections for thermal neutron capture; such elements may be carbon, aluminium, iron, chromium, silicon, zirconium or niobium. An example is the "Sicral" alloy (SiCrAl), with a fine crystal distribution, which does not allow the occurrence of large bubbles containing fission gas. Another example is the alloy with 0.5 – 1% weight Mo – an ingredient that raises the flow limit of the material.

To the second type belong alloys designed for operation at high temperatures. In order to maintain the crystal structure stability within a wide range of temperatures a higher concentration of alloying constituents is required, which limits the utilizations to enriched uranium only. The ingredients provide for maintaining a crystal structure similar to phase $\gamma$ within the range of temperatures 20 — 800 °C. The most employed ingredient is molybdenum. It can be used at concentrations of up to 10% in weight. In the EBR-2 (USA) fast neutron experimental reactor a uranium + "fissium" alloy has been employed (consisting of 50% Mo, 40% Ru, 5% Rh, 3% Pd, 2% Zr, 0.1% Nb — a composition that corresponds approximately to the proportions of fission products that occur in the fuel after a longer operation). As a rule, through alloying the corrosion resistance is improved — a consequence of the presence of the $\gamma$-phase.

A third type of alloys are the fine dispersions under the form of uranium grains or uranium compounds in a host metal known as the *matrix*. The matrix is crucial in determining the mechanical and thermal characteristics of the nuclear fuel; in this case the uranium that is used has to be highly enriched.

Aluminium, zirconium and stainless steel are the metals most frequently used as matrices.Some varieties of uranium metallic dispersions are given in Table 3.8. Table 3.9 gives physical data regarding dispersions more frequently employed.

The equilibrium diagram of the AlU system (Fig. 3.6) provides more insignt on the way the intermetallic compounds are derived [12]. The equilibrium diagrams of U-Zr and U-C alloys are given in Figs. 2.6 and 2.9, respectively. The effect of alloying on the thermal conductivity $K$ is shown in Fig. 3.7.

Summing up all these, the metallic fuels based on uranium and its compounds do present certain advantages, such as an adequate thermal conductivity, a good neutron economy, and good resistance to thermal shocks. Such qualities kept metallic fuels in use, especially in the field of research reactors. On the other hand, their rather weak corrosion resistance, the phase transformations and the dimensional instability at deep burn-up irradiations, have determined, in the field of nuclear power reactors, a preference for *ceramic fuels*. These have an adequate mechanical resistance at high temperatures, a high corrosion resistance, and also a good thermal stability under irradiation. The uranium ceramic compounds of prevailing use nowadays in the fabrication of fuel elements for power reactors are introduced in Table 3.10, through some of their physical and nuclear properties [11].

TABLE 3.8 Uranium in Metallic Dispersions

| *Matrix* | *Compound* | *Percentage of U* | *Mixture type* |
|---|---|---|---|
| Al | $UAl_4$ (inter-metallic compound) | 5 — 30% | Metallurgical dilution |
| | $UO_2$<br>$U_3O_8$ | | "Cermet" |
| Zr | γ-U | ≤ 2 — 3% | Metallurgical dilution |
| Stainless steel | $UO_2$ | 20 — 25% | "Cermet" |

TABLE 3.9 Characteristics of Uranium Compounds that can be Dispersed in Metallic Matrices

| *Compound* | *Density* ($10^3$ kg/m$^3$) | *Melting point* (°C) |
|---|---|---|
| $UAl_2$ | 8.10 | 1590 |
| $UAl_3$ | 6.70 | 1320 |
| $UAl_4$ | 6.00 | 730 |
| UC | 13.63 | 2370 |
| $UC_2$ | 11.68 | 2470 |
| $UO_2$ | 10.97 | 2880 |

Among the uranium ceramic compounds, *uranium dioxide* is now the most frequently used as a nuclear fuel. Uranium dioxide has a high melting point (see also Fig. 12.13). Its physical properties are described in Refs 18, 19, and 40. Table 3.10 gives some indications to this effect. The $UO_2$ thermal conductivity dependence upon temperature is shown in Fig. 3.8 [20]. The thermal expansion coefficient $\alpha$ depends on the stoichiometry and on the presence of other oxides; its variation with temperature is given in Fig. 3.9.

As noted already, the $UO_2$ crystal lattice is that of fluorite, of the f.c.c. type (Fig. 2.1c). The fact that the central positions in the lattice are not occupied by uranium ions but by the oxygen makes it possible that $UO_2$ present deviations from stoichiometry.

The main type of structural disorder that occurs in $UO_2$ is the presence of interstitials in the oxygen sub-lattice. This kind of disorder is noted both in the stoichiometric compound and in the non-stoichiometric one ($UO_{2+x}$). In order to secure electrical neutrality in the non-stoichiometric compound, part of the uranium ions have to be pentavalent if $x > 0$, or trivalent for $x < 0$. A region in the equilibrium diagram of the U — O system is depicted in Fig. 3.10 [20].

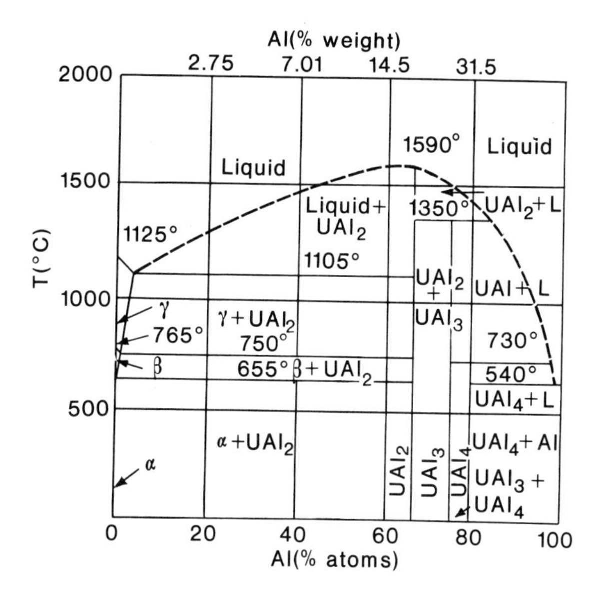


Fig. 3.6. Equilibrium diagram of U -Al alloys.

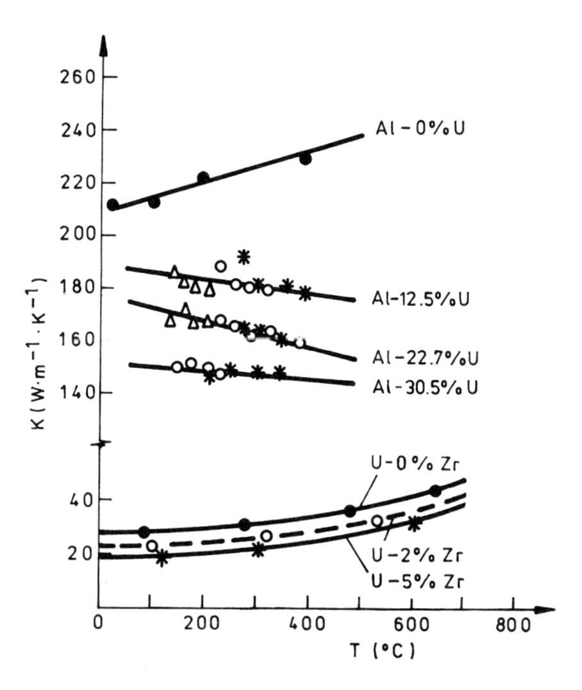


Fig. 3.7. Temperature dependence of thermal conductivity, for U, some U – Zr and U -Al alloys.

TABLE 3.10 Some Properties of the Uranium Compounds

| Compound | *Structure* Lattice type | *Structure* Cell constant ($10^{-10}$ m) | *Melting point* (°C) | *Density* ($10^3$ kg/m$^3$) | *Nuclear properties (at 100% TD)*[1] Macroscopic cross-section ($cm^{-1}$) Fission | Capture |
|---|---|---|---|---|---|---|
| $UO_2$ | cubic ($CaF_2$) | $a$ = 5.704 | 2880 | 10.97 | 0.102 | 0.187 |
| UC | cubic (NaCl) | $a$ = 4.961 | 2370 | 13.63 | 0.137 | 0.252 |
| $UC_2$ | Tetragonal | $a$ = 3.524<br>$c$ = 5.999 | 2470 | 11.68 | 0.115 | 0.211 |
| UN | cubic (NaCl) | $a$ = 4.889 | 2850 | 14.32 | 0.143 | 0.327 |
| USi | Orthorhombic | $a$ = 5.66<br>$b$ = 7.66<br>$c$ = 3.92 | 1600 | 10.40 | 0.098 | 0.184 |

[1]TD = theoretical density

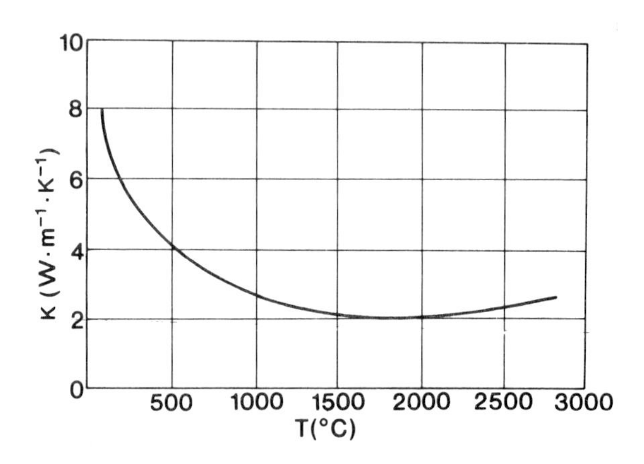


Fig. 3.8. Temperature dependence of thermal conductivity for sintered $UO_2$ (0.95 TD).

As a nuclear fuel, uranium dioxide may come in a variety of forms: pellets (bulk or hollow), "islands" dispersed in a metallic matrix; very small (800 microns) graphite-coated spherical particles, etc. To obtain uranium dioxide one starts with uranium ore concentration, followed by concentrate purification, preparation of uranium trioxide, and its conversion to $UO_2$ powder.

The powder is then subject to a densification process, i.e., pressing-sintering, melting-casting or vibration compacting. Thus, a material is obtained with a density close to the theoretical value (TD). For example, the

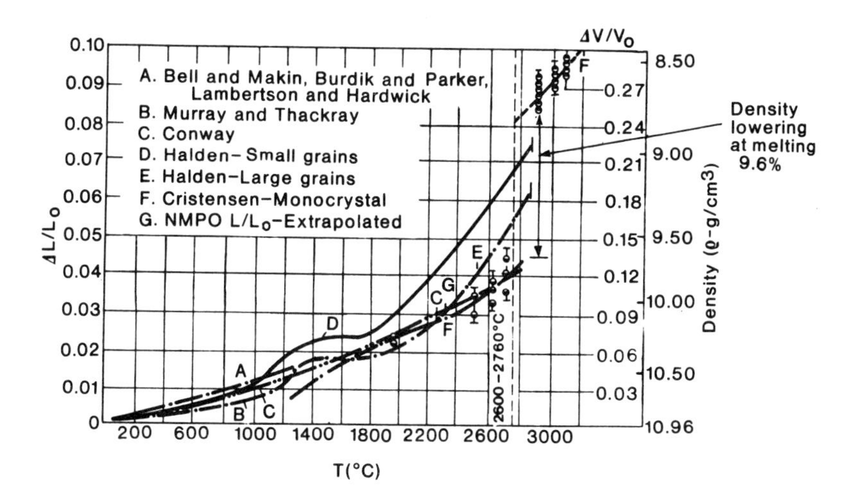


Fig. 3.9. Temperature dependence of the expansion coefficient of $UO_2$.

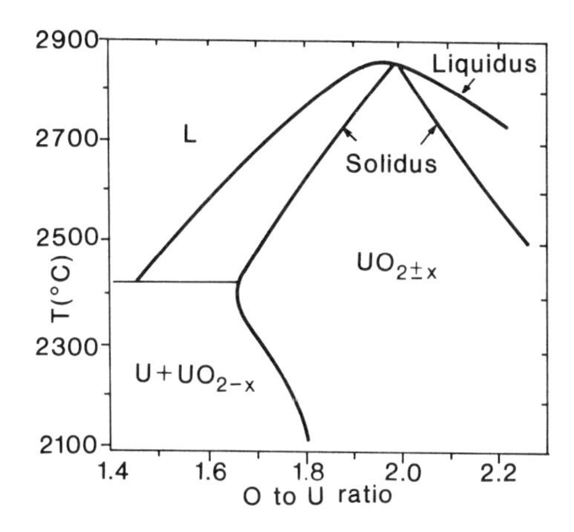


Fig. 3.10. High-temperature region of equilibrium diagram for $UO_{2+x}$.

pellets obtained from sintering in a hydrogen environment at temperatures of 1700 $^{\circ}C$ for 8 – 10 hours have densities of about 93 – 96% TD [21,22].

The $UO_2$ tensile strength is about 71 MPa and the elasticity modulus is about $2.26 \times 10^5$ MPa. Tensile strength increases with temperature. Over 1400 $^{\circ}C$ uranium dioxide turns plastic; some of its physical properties are presented in Appendix 11.

It has been found experimentally that $UO_2$ is stable in distilled water up to temperatures of about 300 °C. A good corrosion resistance in liquid metals is also observed; for example, samples of sintered $UO_2$, of a great density, corrosion-tested in a mixture of molten sodium and potassium at 600 °C have lost only 0.1 mg/cm$^2$ in weight, over 72 hours.

In reactor conditions uranium dioxide has a very good dimensional stability [23 — 25]. On the other hand, due to a low thermal conductivity (about 3 W/m x °C at 1000 °C) very high temperatures and temperature gradients settle in the fuel — a fact that may contribute to fuel cracking, densification, fission gas release, structure modifications, etc. These effects will be dealt with in more detail in Chapter 9.

### 3.1.5. Uranium Enrichment in $^{235}U$

The factors of isotopic separation of $^{235}U$ from $^{238}U$ are very small, so that the separation processes are very costly. Most of the enriched uranium produced nowadays worldwide is obtained through the *gaseous diffusion process*. Essentially, it is based on the preferential penetration of a porous membrane, by the lighter molecules of a gas under pressure in contact with the membrane. In the case of uranium enrichment, the gas is a mixture of $^{235}UF_6$ and $^{238}UF_6$. The separation factor of the process is given by

$$\alpha = (M_2/M_1)^{1/2} \tag{3.19}$$

where $M_1$ and $M_2$ are the molecular weights of $^{235}UF_6$ and $^{238}UF_6$, respectively.

The extremely small value of $\alpha$, ($\alpha = 1.0043$) indicates the need of cascading a great number of separation elements, of a rather sophisticated structure — Fig. 3.11. Thus, in order to reach a 3% concentration of $^{235}U$ starting from the natural concentration (0.72%) a cascade of 1250 steps is in order. The depleted product still has a concentration of 0.2% in $^{235}U$. An enrichment up to 90% requires multiplying the number of steps up to more than 4000 [26,27]. The diffusion chamber, the membrane (whose micropores are measured in nanometres) and the compressor raise difficult engineering problems. All components are to be made of corrosion-resistant materials, for which the action of both $UF_6$ and of fluorine hydride have to be duly attended. In addition, one has to avoid formation of solid uranium compounds that might obstruct the membrane pores. The specific electric energy consumption in the gaseous diffusion industrial plants amounts to 2400 kWh/kg — separation work. Capital investment and operating costs are considerable; as related to the unit output, they depend strongly on the plant capacity, in such a way that only facilities of very large capacity are cost-effective.

Today, the gaseous diffusion supplies 98% of the world production of enriched uranium; it is a mature technology that ensures a high reliability and relatively low costs when compared to other processes. Its merits could be improved by at least 20%, through better porous barriers and other materials, as well as through a partial recovery of the compression heat released from the diffusion stages.

One potential way to improve and optimize this process relies on automation. In this sense an important role could be ascribed to the potential shown by a continuous on-line monitoring of the uranium concentration, of its isotopic composition at different check-points in the plant, free of the need of sample extraction. It is believed that, some time in the future, the magnetic resonance will be among the methods able to meet this requirement (see Chapter 11).

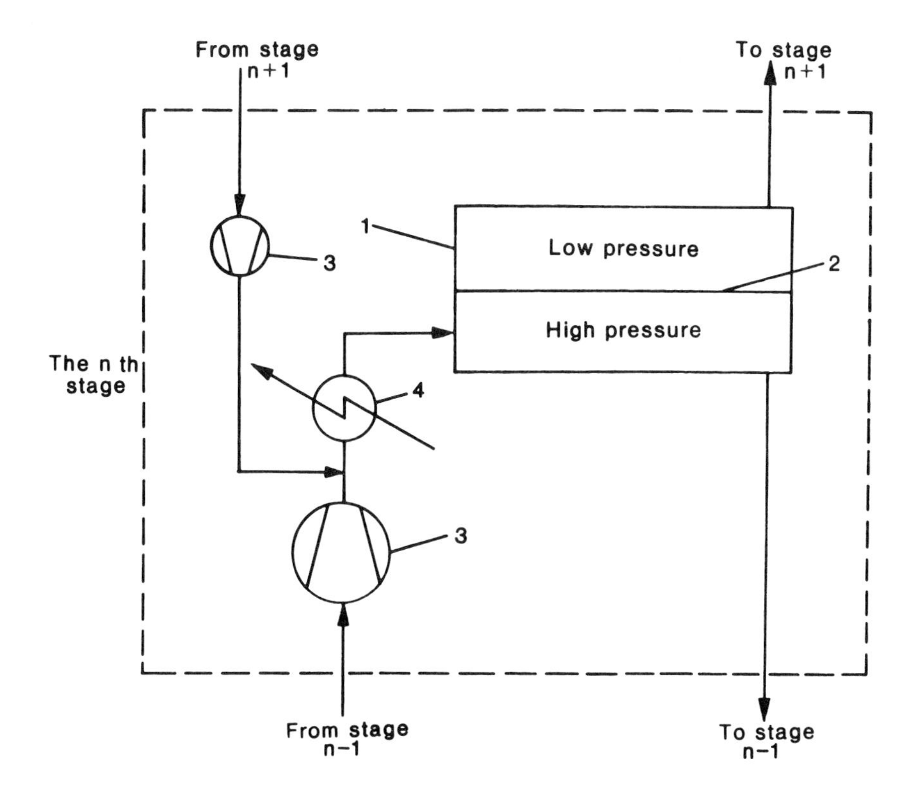


Fig. 3.11. Flow-sheet of a gaseous diffusion stage.
(1) diffusion chamber; (2) membrane;
(3) compressor; (4) heat exchanger.

$UF_6$ *centrifuging* is another process, characterized by a much larger separation factor, that enjoys today an increasing interest. The separation factor is given by the equation:

$$\alpha = \exp[(M_1-M_2)\omega^2 r^2/(2RT)], \tag{3.20}$$

where $\omega$ is the circular frequency, $r$ the centrifuge radius, $R$ the ideal gas constant, $T$ the absolute temperature, while $M_1$ and $M_2$ are the molecular weights as above.

It is interesting to note that, since the separation factor depends on the mass *difference* and not on the absolute molecular masses as such, one may choose for separation less corrosive, thus easier to handle, products.

The process in a separation element is presented in Fig. 3.12. This implies feeding close to the rotor spindle and extraction of the product and wastes at the centrifuge ends, close to the wall. For peripheral speeds in the range 300 — 350 m/s separation factors of 1.06 up to 1.16 are obtained. The operating requirements, especially the rotation speed, make binding the use of high technology and equipment. As compared with diffusion, the necessary number of separation stages is substantially smaller, and the specific energy consumption is only 300 — 350 kWh/kg separation work. Such advantages are counterbalanced by the small processing capacity of a centrifuge and by a specific capital investment larger than that required for gaseous diffusion. Overall, the process of separation through centrifuging make competitive average-size production capacities. The automatic control of the flow-chart and a permanent striving towards optimization constitute ways to increasing the efficiency of this separation process.

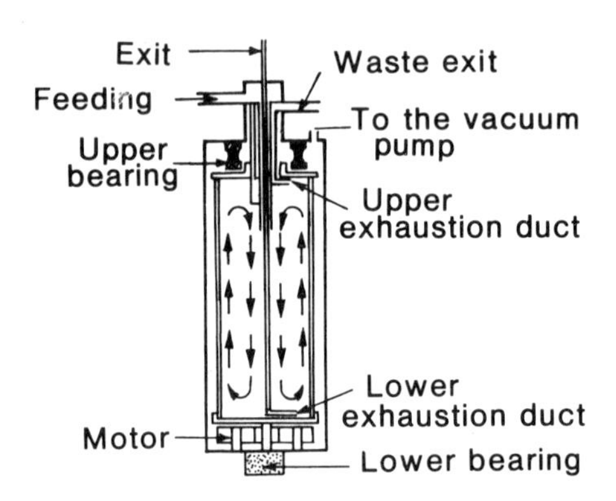


Fig. 3.12. Centrifugal separation unit.

The selection of any one of the methods of uranium enrichment to be used on an industrial scale is conditioned by technological and economic criteria. Thus,

1. gaseous diffusion enjoys an advanced and well-developed technology, while centrifuging is more of a novel process;

2. although with centrifuging the separation factor is larger, considering the fact that the flow in every separation element is very low, one ends up with the finding that, at equal outputs, the advantage of centrifuging over diffusion — of steep enrichment through a small number of stages — is counteracted by the need to increase the number of elements set in parallel;

3. centrifuging requires equipment and materials of truly fine chemical and mechanical merits, having in view the high stresses caused by the huge centrifugal force that exerts on a rotor operating in the aggressive environment of uranium hexafluoride;

4. on the other hand, gaseous diffusion, beside posing difficult problems with obtaining performant porous membranes, also presents the disadvantage that the respective enrichment plants must be inherently very large enterprises, whereas those using centrifuging can be made smaller, since these do not need the compelling utilities of the gaseous diffusion [28-30].

The *nozzle separation* is performed on a device that is schematically presented in Fig. 3.13. The gas — a mixture of 5% $UF_6$ and 95% $H_2$ or He, is blown in the nozzle, under pressure; following a curved path the jet travels at supersonic speed. In the process, a splitting of the gas flow into two fractions takes place, the inner fraction being richer in the light component. For an admixture of 5% mol $UF_6$ in 95% mol He and an expansion from 600 to

150 torr, at a jet fractioning ratio of 1/3, a separation factor of roughly 1.015 is obtained, which places the process in an intermediate position between diffusion and centrifuging as to the required number of stages. Recent estimations indicate a specific investment close to that in the diffusion plants, but an electric energy consumption considerably larger — of about 4000 kWh/kg — separation work. Still, technological improvements are expected that are believed to have the potential of increasing the competitiveness of this process.

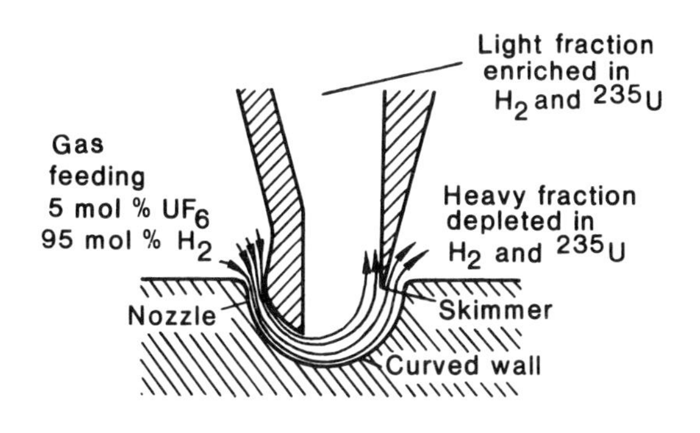


Fig. 3.13. Nozzle separation unit.

*Membrane-jet separation* tries to combine the advantages of jet (nozzle) separation with those of membrane (diffusion) separation. Yet present-day developments with this process do not go beyond the laboratory scale.

An avenue now under investigation and assessment, which presents interesting prospects for uranium isotopic separation, is *laser separation* [31-33]. The process is based on selective excitation by means of power, frequency-tunable lasers, of uranium vapours or of molecules of some uranium volatile compounds, such as $UF_6$. In principle, three methods may be used:

(a) *Deflection of excited atoms.* The atomic beam that forms the isotope mixture (generated by heating and evaporation of the material at high temperatures) passes through a laser beam capable of exciting only one of the two isotopes. If the excited atoms have electric or magnetic properties that differ from those of non-excited atoms, then they may be extracted from the beam by means of electric or magnetic methods.

(b) *Recoil of dissociating molecules.* The molecular beam containing the isotopes is, again, obtained by heating at relatively high temperatures. The molecules that contain one of the isotopes will have absorption bands different from those of the molecules containing the other isotopes. The laser beam frequency may be thus chosen so that it excites and dissociates only the molecules containing a certain isotope. As a consequence of momentum conservation, the lighter molecular fragments will be deflected more markedly than the heavy ones, thereby enabling their separate collection.

(c) *Radiation pressure.* Assume that one of the isotopes is selectively excited by laser radiation. At the de-excitation of this isotope a transversal component of the momentum occurs. By repeating this excitation — de-excitation process the ions of the isotope in question are taken off the ion beam and may be collected, thus achieving isotopic separation.

The first procedure has been experimentally tested to separate the isotopes $^{235}U$ and $^{238}U$. To this effect two lasers have been used, one on xenon ions (tuned on 378.1 nm), the other on krypton ions (tuned either on 350.7 nm or on 356.4 nm). The xenon laser excites the $^{235}U$ atoms bringing them to a metastable state. These metastable-state atoms are then ionized by the second laser beam (on krypton). The $^{235}U$ atoms are subsequently deflected by means of an electric field that does not act upon the $^{238}U$ atoms.

The laser isotopic separation offers the advantage of a large separation factor. One may obtain uranium up to about 60% rich in $^{235}U$ with the atomic method, and up to 1 — 10% with the molecular method, on using $UF_6$, at a low energy consumption and with investments substantially smaller than those required by the classical methods. The main disadvantage, however, of the atomic method, is the need to work at very high temperatures (around 2500 K), which raises difficult technical problems with the corrosive action of liquid uranium and its vapours.

Other separation processes are also being studied, but available scientific information on these is rather scarce at the moment. Here one includes *chemical processes*, based on reactions of two uranium compounds in different phases; the separation factor as achieved does not overtake the one of the diffusion process.

Outstanding among the recent research directions in uranium isotopic separation is the separation through polymer membranes, using two-membrane cells, one being permeable for the light component, the other for the heavy component. Cascading double-cell, polymer membrane separation units indicates attractive values of the separation factor. It is believed that, some time in the future, this process will become economically sound [34].

## 3.2. PLUTONIUM

Plutonium was first obtained in 1940 by the team of researchers led by G. T. Seaborg (USA), through irradiating $^{238}U$ with accelerated particles [35]. Reactions available for plutonium production are:

$$^{238}_{92}U + ^{2}_{1}d \rightarrow ^{238}_{93}Np + 2^{1}_{o}n$$
$$^{238}_{93}Np \rightarrow ^{238}_{94}Pu + \beta^- \quad (T_{1/2} = 2.1 \text{ days})$$
$$^{238}_{92}U + ^{2}_{1}d \rightarrow ^{239}_{93}Np + ^{1}_{o}n$$
$$^{239}_{93}Np \rightarrow ^{239}_{94}Pu + \beta^- \quad (T_{1/2} = 2.35 \text{ days})$$
$$^{238}_{92}U + \alpha \rightarrow ^{240}_{94}Pu + 2n$$
$$^{238}_{92}U + \alpha \rightarrow ^{241}_{94}Pu + n.$$

Since the half-life of plutonium isotopes is relatively short (see Table 3.13) it cannot be encountered as such in our natural environment. And yet, owing to the capture reactions $(n,\gamma)$ or to $(\alpha,n)$ reactions in uranium, plutonium may be found in minute amounts associated with uranium ores, such as pitchblende or carnotite. Plutonium production through the capture reaction $(n,\gamma)$ follows the chain:

$$^{238}_{92}U + ^{1}_{o}n \rightarrow ^{239}_{92}U + \gamma$$
$$^{239}_{92}U \rightarrow ^{239}_{93}Np + \beta^- \quad (T_{1/2} = 23.5 \text{ min.})$$
$$^{239}_{93}Np \rightarrow ^{239}_{94}Pu + \beta^- \quad (T_{1/2} = 2.35 \text{ days})$$

This is the very process that provides for plutonium production in nuclear reactors.

The $^{239}Pu$ isotope that is obtained is fissionable by thermal neutrons. On its fission, about three neutrons are generated, which makes this process efficient for fuel conversion (fertile isotope conversion into fissionable isotopes).

### 3.2.1. Properties

Plutonium is a white-silvery metal that rapidly darkens in contact with the air, because of oxidation. Metallic plutonium presents six allotropic states. The transitions between them, and the associated volume modifications, are displayed in Table 3.11 [7,10,11]. The first transition occurs at 122 °C; up to the melting temperature other five allotropic transformations take place. Phase $\delta$ exists only within a temperature span of about 30 °C. Two of the transformations show negative volume variations. Table 3.11 gives some data regarding structural features and a few physical properties for each of the six phases. Table 3.12 gathers certain nuclear characteristics of plutonium isotopes of practical interest [10].

The mechanical properties of plutonium are strongly affected by the concentration of impurities, intrinsic defects, as well as by residual stresses, the anisotropy of the expansion coefficient, etc. Here are some characteristic values at room temperature [36,37]:

| | |
|---|---|
| elastic modulus | $8.9 \times 10^4$ MPa |
| shear modulus | $3.7 \times 10^4$ MPa |
| Poisson ratio | 0.21 |
| tensile strength | 350 MPa |
| compression strength | 220 MPa |
| elongation (at 25 °C) | 0.068% |
| melting temperature | 640 °C |

The thermal and electrical properties of metallic plutonium are most unusual for a metal. The expansion coefficient of phases $\delta$ and $\delta'$ is negative, i.e. by heating plutonium shrinks.

Like uranium, plutonium powder is strongly reactive and pyrophoric. At room temperature it reacts easily with oxygen and hydrogen, but slower with water. The main valence states of plutonium are: (II), (III), (IV), (V), (VI) and (VII), the state Pu(IV) being very stable [38]. Plutonium is extremely toxic, both chemically and due to its strong radioactivity. It deposits itself in bones, thus being highly dangerous. The biological half-life of inhaled radioactive material is about 200 years.

Radioactive properties of plutonium isotopes are given in Table 3.13.

### 3.2.2. Processing Methods

Plutonium processing applies to procedures currently used for other metals, i.e., melting and casting, rolling, extruding [38,39]. Yet the high toxicity of this element compounds considerably the performance with these conventional procedures. In terms of biological protection assurance, plutonium melting and casting seem to be the most advantageous, due to plutonium's low melting point, high fluidity, high density, small volume modification during solidification, etc. Since specific properties vary from one phase to another, distinct processing methods are used. $\alpha$-plutonium is brittle and difficult to process other than by forging — by means of pressure. Phase $\delta$ is ductile and may be processed by any of the usual techniques. In case of processing at

TABLE 3.11 Crystalline Structure, Physical Properties and Phase Transformations of Metallic Plutonium

| Phase | Crystalline structure | Cell parameters ($10^{-10}$ m) | Density ($10^{-3}$ kg/m$^3$) | Average thermal expansion coefficient ($^{\circ}$C)$^{-1}$ | Electric resistivity ($10^{-8}$ ($\Omega$m)) | Temperature of transition to the next phase ($^{\circ}$C) | Volume change at transition (%) |
|---|---|---|---|---|---|---|---|
| α | Monoclinic | $a$ = 6.1835<br>$b$ = 4.8244<br>$c$ = 10.973<br>β = 101.81°<br>(at T=294 K) | 19.816<br>(298 K) | $55 \cdot 10^{-6}$ | 150 at 25 °C | 122 | 8.9 |
| β | Body-centred monoclinic | $a$ = 9.284<br>$b$ = 10.463<br>$c$ = 7.859<br>β = 92.13°<br>(at T=463 K) | 17.82<br>(406 K) | $35 \cdot 10^{-6}$ | 110 at 200 °C | 205 | 2.4 |
| γ | Face-centred ortho-rhombic | $a$ = 3.1587<br>$b$ = 5.7682<br>$c$ = 10.162<br>(at T=506 K) | 17.14<br>(508 K) | $36 \cdot 10^{-6}$ | 110 at 300 °C | 318 | 6.7 |
| δ | Face-centred cubic | $a$ = 4.6371<br>(at T = 593 K) | 15.92 | $-21 \cdot 10^{-6}$ | 102 at 400 °C | 452 | -0.4 |
| δ | Face-centred tetragonal | $a$ = 4.701<br>$c$ = 4.489<br>(at T=738 K) | 16.01<br>(720 K) | $-65.6 \cdot 10^{-6}$ | | 479 | -3.0 |
| ε | Body-centred cubic | $a$ = 3.631<br>(at T=763 K) | 16.48<br>(783 K) | $4 \cdot 10^{-6}$ | 120 at 500 °C | 640 | ~ 0.1 |

TABLE 3.12 Plutonium Nuclear Constants (Thermal Neutrons)

| *Constants* | $^{238}Pu$ | $^{239}Pu$ | $^{240}Pu$ | $^{241}Pu$ |
|---|---|---|---|---|
| Fission cross-section (b) | 18 | 738 | 4 | 971 |
| Absorption cross-section (b) | 468 | 1025 | 350 | 1336 |
| Average number of neutrons emitted per fission act | - | 2.91 | 3 | 3 |

TABLE 3.13 Radioactivity of Plutonium Isotopes

| *Isotope* | $T_{1/2}$ (years) | *Type of decay* |
|---|---|---|
| $^{238}Pu$ | 86.4 | α |
| $^{239}Pu$ | $2.4 \cdot 10^4$ | α |
| $^{240}Pu$ | $6.6 \cdot 10^3$ | α |
| $^{241}Pu$ | 13.2 | $\beta^-$ |
| $^{242}Pu$ | $3.8 \cdot 10^5$ | α |

high temperatures, subsequent cooling entails allotropic transitions that modify both the shape and the properties of the original part.

The ceramic powders based on plutonium, as for instance the mixtures $(U+Pu)O_2$, can be compacted in order to obtain pellets. Also it is considered that the compacting technologies on vibrations and die forging provide for better qualities of the final product.

### 3.2.3. Dimensional Stability

Thermal cycling from normal temperatures up to the temperatures corresponding to phases δ or ε, leads to progressive decrease of the material density. During the $\alpha \rightleftarrows \beta$ cyclings, voids are induced in the material, at microscopic scale, that expand in the case of $\alpha \rightleftarrows \beta \rightleftarrows \gamma$ cyclings. During $\gamma \rightleftarrows \varepsilon$ cyclings, the structural modifications are less important and voids are not induced. The $\alpha \rightleftarrows \varepsilon$ cycling of high purity plutonium induces a growth in sizes (length and volume); the sample diameter increases over the whole length, reaching maximal values at the ends and at the centre.

### 3.2.4. Corrosion Behaviour

From a chemical standpoint, plutonium is more reactive than uranium; it oxidates in air at room temperature, generating a layer of $PuO_2$ powder. The phenomenon is precipitated by environmental moisture. Thus, a plutonium sample exposed in dry air at 65 °C for 200 hours presents a loss in weight of about 0.015 mg/cm². When relative humidity increases to 5% the weight loss attains 1 mg/cm² [39]. When the material is heated to a temperature corresponding to the transition $\beta \rightarrow \gamma$, the oxidation rate decreases.

In aqueous environments the corrosion implies diffusion of oxygen and $OH^-$ ions through the pre-existent oxide layer, which leads to the formation of oxides and hydrates. This kind of corrosion constitutes a major problem in cladded fuel elements on metalic plutonium, where clad failure may entail formation of corrosion products (mixtures of plutonium oxides and hydrates) that ask for a much bigger volume than the original material.

### 3.2.5. Plutonium Compounds

Metallic plutonium is a fissionable material too "concentrated", thereby requiring "dilution" in order to be used as a fuel. Moreover, its chemical reactivity, structural complexity and peculiar physical and mechanical properties make unattractive its utilization in a pure state.

Experience shows that plutonium has an alloying behaviour similar to that of uranium. Plutonium has an even stronger tendency to form intermetallic compounds. Alloying this metal facilitates fuel fabrication, improves dimensional stability (by eliminating phase transformations) and increases corrosion resistance. As alloying elements, Al, Ga, Mo, Th, and Zr are frequently used. A fuel containing Pu in phase δ, stabilized by 3.59% Ga, has a corrosion resistance in moist air much superior to that of the pure metal: the weight loss is diminished to only 0.1 $mg/cm^2$ for an exposure of 27,000 hours.

Among the plutonium *ceramic compounds*, the *oxide* and the *carbide* are of primary interest. These are very similar to the respective uranium compounds, with which they may form solid solutions in the whole range of concentrations.

Plutonium gives three oxides, but only the dioxide ($PuO_2$) is employed in the manufacturing of uranium — plutonium mixed fuels. The plutonium dioxide has a density of 11,460 $kg/m^3$ and the melting point at 2280 °C. It can be obtained either by thermal decomposition of plutonium nitrate, or by precipitation and calcination of plutonium hydroxide or of plutonium oxalate. The resulting $PuO_2$ powder is mechanically mixed with $UO_2$ powder; the mixture is used to obtain compact pellets, through customary procedures in powder metallurgy. The mixtures of uranium and plutonium oxides with a content of up to 20% Pu are used as fuels in the fast breeder reactors.

Several physical properties of nuclear materials that are of interest in conducting analyses of the behaviour under irradiation of LWR fuel rods made of mixed $(UPu)O_2$ fuels are gathered in Appendix 11 (see also Refs 40 and 41).

## 3.3. THORIUM

Thorium was first discovered and extracted as a metal by Berzelius (1830). Its radioactive properties were recognized by Marie Curie. The thorium isotopes and some of their characteristics are presented in Table 3.14 [36]. Natural thorium is monoisotopic, consisting entirely in $^{232}Th$.

The chief source of thorium is *monazite*, a mineral containing phosphates of rare earths and — in variable proportions — thorium. It may also be found in *thorite* $ThSiO_4$ and *thorianite* $ThO_2$ [42,43].

Thorium is a ductile, white metal, similar to platinum [39]. For a long time it has been used as an alloying element: small quantities of thorium in magnesium, for instance, considerably improve stretch and creep properties at high temperatures.

TABLE 3.14 Nuclear Properties of Thorium Isotopes

| *Nuclide* | $T_{1/2}$ | *Cross-sections* | | *Resonance integral* | |
|---|---|---|---|---|---|
| | | *Capture* $\sigma_\gamma$ (b) | *Fission* $\sigma_f$ (b) | *Capture* $I_\gamma$ (b) | *Fission* $I_f$ (b) |
| 227 | 18.72 days | | 200 | | |
| 228 | 1.913 years | 123 | 0.3 | 1013 | |
| 229 | 7340 years | 54 | 30.5 | 1000 | 464 |
| 230 | 77,000 years | 23.2 | 0.0012 | 1010 | |
| 232 | $1.4 \cdot 10^{10}$ years | 7.40 | 0.000039 | 85 | |
| 233 | 22.2 min | 1500 | 15 | 400 | |
| 234 | 24.1 days | 1.8 | 0.01 | | |

Natural thorium is not fissionable by thermal neutrons, but its isotope $^{232}Th$ generates — through neutrons radiative capture and subsequent beta desintegration — the isotope $^{233}U$, that, in turn, is fissionable. The interest raised by thorium in the nuclear power realm, comes precisely from this possible conversion, thorium utilization in reactors allowing savings of available uranium resources.

### 3.3.1. Properties

As compared with uranium, thorium has an isotropic structure, higher mechanical strength, a melting point ~ 600 $^{o}C$ higher. Table 3.15 presents several physical properties of thorium [7]. Like uranium dioxide, thorium dioxide has a face-centered cubic structure; its melting point is 500 $^{o}C$ higher than that of $UO_2$; together they form solid solutions.

TABLE 3.15 Physical Properties of Thorium

|  |  |
|---|---|
| Density | 11,710 $kg/m^3$ |
| Allotropic transformation | |
| $\alpha \rightarrow \beta$ | 1400 ± 25 $^{o}C$ |
| Crystaline structure: | |
| $\alpha$-Th — fcc | $a_o$ = 5.087 Å |
| $\beta$-Th — bcc | $a_o$ = 4.11 Å |
| Melting point | 1690 ± 10 $^{o}C$ |
| Boiling point | > 3000 $^{o}C$ |
| Thermal expansion coefficient ($^{o}C^{-1}$) between: | |
| 30 — 100 $^{o}C$ | $11.5 \cdot 10^{-6}$ |
| 30 — 1000 $^{o}C$ | $12.5 \cdot 10^{-6}$ |
| Thermal conductivity (W/m . K): | |
| at 100 $^{o}C$ | 38 |
| at 650 $^{o}C$ | 45 |
| Electric resistivity (20 $^{o}C$) | $1.8 \cdot 10^{-7}$ Ωm |
| Cross-sections for thermal neutrons (b): | |
| scattering | 12.6 |
| absorption | 7.4 |

The mechanical properties of thorium are strongly affected by its content in impurities. Figure 3.14 outlines the effect of some impurities such as carbon, oxygen and nitrogen on tensile and ductile strength of the material obtained by arc melting. While oxygen and nitrogen do not significantly affect the tensile strength, a carbon content of 0.13% enhances the resistance from 135 MPa to 312 MPa.

Thorium has a single state of oxidation in aqueous solution: +4, of chemical properties similar to those of zirconium and cerium. Thorium is an extremely electropositive element. In powder form it is pyrophoric; in bulk it is stable in air. Because of its long half-life ($1.45 \times 10^{10}$ years) its handling is considerably less hazardous than that of uranium.

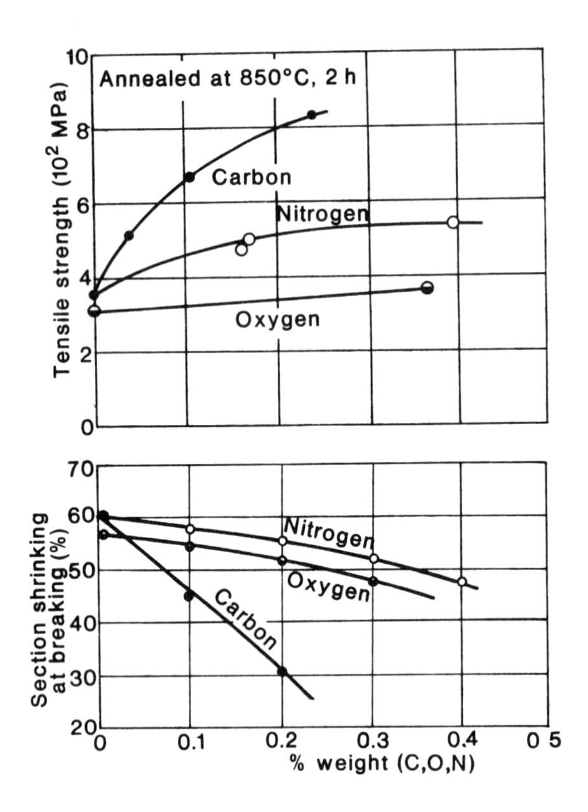


Fig. 3.14. Effects of impurities on some mechanical properties of thorium [7].

### 3.3.2. Processing Methods

Due to its high melting point and strong chemical reactivity,thorium processing by melting and casting is difficult: the operation should be performed in vacuum or inert atmosphere to prevent contamination with oxygen or nitrogen.

Parts made of thorium with a low content of O, N, Si, and Al can be obtained by extrudation (480 — 1000 °C), hot rolling, cold rolling, etc. Current procedures in powder metallurgy may be successfully applied also on thorium powders. Thorium powder may conveniently be cold-compacted, up to densities of about 95% of the theoretical density. Hot pressing (650 °C) in vacuum, at 1.4 – 2.1 kPa, allows one to obtain metallic thorium of high density.

### 3.3.3. Dimensional Stability under Irradiation

Its cubic structure gives thorium a greater dimensional stability under irradiation, as compared with uranium. Thorium is free of any anisotropic growth effects and of the associated intercrystalline tensions. A growth in hardness and a doubling of flow resistance has been noted when the material has been exposed to a neutron fluency of about $10^{19}$ n/cm$^2$. The thorium-based fuels do not present excessive swelling on internal voids formation, as with uranium, in the 400 – 550 °C temperature range.

### 3.3.4. Corrosion Behaviour

Fresh-cut thorium is shiny, silvery, but it darkens when exposed to air, due to corrosion. Up to about 350 °C the oxide layer has a protective role, preventing further, in-depth oxidation of the material. At higher temperatures the oxide layer cracks and oxidation progresses linearly with temperature, up to about 450 °C. From roughly 1100 °C the oxidation rate assumes a parabolic dependence on temperature, due to the sintering that occurs in the protective oxide layer.

Thorium has a relatively low resistance at aqueous corrosion, irrespective of temperature. At 100 °C the corrosion rate is lower, with the formation of an additional oxide layer. In the range 178 — 200 °C the oxide layer rapidly grows, cracks and falls off the material; at 315 °C the reaction is particularly rapid.

Thorium shows a satisfactory behaviour in corrosive environments consisting of liquid metals, with the exception of aluminium, at temperatures below 900 °C.

### 3.3.5. Thorium Alloys and Compounds

To increase mechanical and corrosion resistance, several alloying elements (e.g. Zr) have been associated with the pure metallic thorium. Thorium — uranium alloys have the potential of bringing together the chief property of uranium of being fissionable and the capital property of thorium of being fertile. As seen from Table 3.16, mechanical properties of thorium are also improved by alloying with uranium [44].

TABLE 3.16 Mechanical Properties of Th-U Alloys

| *Uranium content* (% in weight) | *Yield strength* (MPa) | *Tensile strength* (MPa) | *Elongation* (%) |
|---|---|---|---|
| Pure Th | 280 | 452 | 46 |
| 1.0 | 365 | 548 | 37 |
| 5.0 | 391 | 604 | 38 |
| 10.2 | 428 | 642 | 35 |
| 20.6 | 440 | 680 | 38 |
| 30.9 | 515 | 797 | 24 |
| 40.6 | 550 | 891 | 28 |
| 49.1 | 621 | 950 | 10 |

Thorium dioxide is one of the most stable oxides. Its melting temperature is 3300 °C, and the density is 10,000 kg/m$^3$.

Thorium dioxide may be either crystalline or amorphous. The amorphous variety is derived from thorium hydroxide, by calcination in vacuum at temperatures of about 340 °C. At calcination temperatures in excess of 1000 °C, crystalline thorium dioxide is obtained, that may be used in nuclear fuels manufacturing.

The composition of interest in PHWR low-burnup thorium fuel cycles is 97.5% $ThO_2$ vs. 2.5% $UO_2$. Pellets are pressed at 70 MPa and sintered at 1800°C; a density of 96% of the theoretical value is thereby obtained [45].

## REFERENCES


1. International Atomic Energy Agency (1979). *International Nuclear Fuel Cycle Evaluation, Final Report*, Vol. 1-2, IAEA, Vienna.
2. Ambartsumian, T. L., and others (1961).*Termicheskye isledovanyia uranovyh i uransoderjaschih mineralov*. Gosatomizdat, Moskow.
3. Tataru, S. (1968). *Uraniul*. Edit. Stiintifica, Bucharest.
4. Taylor, D. M., and Cameron, J. (1980). Uranium Deposits of the Future. In *Proc.Int.Symp.on the Pine Creek Geosyncline*. IAEA, Vienna, p.743.
5. Maxim, I. V. (1969). *Materiale nucleare*. Edit. Academiei, Bucharest.
6. Galkin, N. P., and Tihomirov, V. B. (1961). *Osnovnye protsessy i apparaty tehnologhyi urana*.Gosatomizdat, Moskow.
7. Weinstein, E., Boltax, A., and Lanza, G. (1964). In *Nuclear Engineering Fundamentals*, Vol. 4, *Nuclear Materials*. McGraw-Hill Book Co., New York.
8. Nakamura, J., and Takahashi, Y. (1980). *J.Nuclear Materials*, 88, 64.
9. Sauteron, J. (1965). *Les combustibles nucleaires*. Hermann, Paris.
10. Smith, C. O. (1967). *Nuclear Reactor Materials*. Addison, Wesley Publ. Co., Reading (Mass.).
11. Gerasimov, V. V. (1965). *Korrozya urana i ego splavov*, Atomizdat, Moskow; Gerasimov, V. V., and Monahov, A. S. (1982). *Materialy iadernoi tehniki*. Energoizdat, Moskow.
12. Tipton, C. R. (1960). *Reactor Handbook*, Vol. 1, New York.
13. Linard M. and others (1977). *J.Nuclear Materials*, 71, 44.
14. Robertson, J. A. L. (1969). *Irradiation Effects in Nuclear Fuels*. Gordon and Breach Science Publ., New York.
15. Leteurtre, J., and Quere, Y. (1972). *Irradiation Effects in Fissile Materials*. North-Holland Publ. Co., Amsterdam.
16. Barnes, R. S. (1964). *J.Nuclear Materials*, 11, 135.
17. Greenwood, G. W., and Speight, M. V. (1963). *J.Nuclear Materials*, 10, 140.
18. Ohse, R. K. and others (1979). *J.Nuclear Materials*, 80, 232.
19. Koike, T., and Kirihara, T. (1978). *J.Nuclear Materials*, 73, 217.
20. Asamoto, R. R., Anselin, F. L., and Conti, A. E. (1969). *J.Nuclear Materials*, 29, 67; Olander, D. R. (1976). *Fundamental Aspects of Nuclear Fuel Elements*, Nat. Techn. Inf. Service, US Dept. of Commerce, Springfield, TID-26711-PI.
21. Timmermans, W. and others (1978). *J.Nuclear Materials*, 71, 256.
22. Soni, N. C., and Moorthy, V. K. (1979). *J.Nuclear Materials*, 79, 312.
23. Lastman, B. (1964). *Radiatsionnye iavlenyia v dvuokysi urana*. Atomizdat, Moskow.
24. Nakae, N. and others (1978). *J.Nuclear Materials*, 74, 1.
25. Lindman, N. (1977). *J.Nuclear Materials*, 71, 73.
26. Vasaru, G. (1968). *Izotopi stabili*. Edit. Tehnica, Bucharest.
27. Chemla, M., and Perie, J. (1974). *La séparation des isotopes*, Presses Universitaires de France, Vendome.
28. Vasaru, Gh. (1971). *Separarea izotopilor uraniului*, Report IIS-U-1.
29. British Nuclear Energy Soc. (1976). *Proc.Intern.Conf.on Uranium Isotope Separation*. London, 1975
30. Villani, S. (Ed.) (1979). *Topics in Applied Physics*, Vol. 35, *Uranium Enrichment*. Springer Verlag, Berlin, Heidelberg, New York.
31. Ursu, I. and others (1977). *Imbogatirea izotopica cu laseri* invited lecture at the UNESCO International School "Scientific Research and Society's Needs", Bucharest.
32. Popescu, I. M. and others (1979). *Aplicatii ale laserilor*, Edit. Tehnica, Bucharest.
33. Yare, R. N. (1977). *Scientific American*, 236, 2, 86.

34. Constantinescu, D. M., Dimulescu, A., Isbasescu, G., Peculea, M. and Ursu, I. (1979). *Studiu comparativ asupra cascadelor de separare cu celule din membrane din polimeri*, The 5th Scientific Session of the Rm. Vilcea Research Centre, Rm. Vilcea.
35. Glasstone, S. (1967). *Sourcebook on Atomic Energy*, D.van Nostrand Co., Princeton (New Jersey).
36. Konobeevsky, S. T. (1959). *Doklad sovetskih uchenyh TZM.* Atomizdat, Moskow, 396
37. Konobeevsky, S. T. (1962). *Atomnaya Energhyia*, 12, 550.
38. Milyukova, M. S. and others (1965). *Analiticheskaya himya plutonia.* Izd. Nauka, Moskow,
39. Coffinberry, A. S., Schonfeld, F. W., Waber, J. T., Kelman, L. R., and Tipton, C. R. (1960). Plutonium and its alloys. In *Reactor Handbook*, 2nd edition, Vol. I, Materials Interscience Publ.Inc., N.Y.
40. Mc.Donald, P. (1976). *MATPRO — Version 9 - A Handbook of Nuclear Material Properties for Use in Analysis of Light Water Reactors Fuel Rods Behaviour*, Report — TREE — NUREG — 1005.
41. USAEC (1973). *Neutron Cross Sections*, 3rd edition, Vol. I, Brookhaven National Laboratory BNL 325.
42. Runion, T. C., Rogers, B. A., and Paine, S. H. (1960). Thorium. In *Reactor Handbook*, 2nd edition, Vol. I, Materials Interscience Publ.Inc., N.Y.
43. Kaplan, G. E. and others (1960). *Toryi, ego sirievye resursy, himyia i tehnologhyia*.Atomizdat, Moskow.
44. Adams, R. E., Berry, W. E., Milko, J. A., Paine, S. H., Peoples, R. S., and Pay, H. A. (1960). Thorium alloys. In *Reactor Handbook*, 2nd edition, Vol. I, Materials Interscience Publ. Inc., N.Y.
45. Loch, L. D., and Quirk, J. F. (1960). Ceramics. In *Reactor Handbook*, 2nd edition, Vol. I, Materials Interscience Publ. Inc., N.Y.

## ADDENDUM

The assessment of the technical and economic efforts required by the uranium enrichment enterprise is largely based on the evaluation of two parameters: the specific consumption of natural uranium (NU); and the separation (enrichment) work (SW).

The *specific consumption of natural uranium*, expressed in kg of natural uranium required in order to obtain 1 kg of enriched uranium, is given by the equation:

$$NU = \frac{x_p - x_w}{x_f - x_w} .$$

where $x_p$ is the concentration of the isotope $^{235}U$ in the enriched uranium; $x_f$ is the concentration of the said isotope in the uranium that feeds the enrichment facility (as a rule, natural uranium); and $x_w$ is its concentration in the depleted uranium at the finish of the cascade.

Normally, one takes $x_f = 0.00711$ and $x_w = 0.0020$ (though many plants operate nowadays at $x_w = 0.0030$). A drop in $x_w$ below 0.0020 is predicted for the end of this century, when it is believed that the ratio of the price of natural uranium vs. the price of the unit-separation work will sore.

The *separation* work, also expressed in kg per 1 kg of enriched uranium, is given by the equation:

$$SW = V_p + \frac{x_p - x_f}{x_f - x_w} V_w - \frac{x_p - x_w}{x_f - x_w} V_f ,$$

where $V_p$, $V_w$, and $V_f$ are the *separation potentials* associated to the enriched, depleted and feedstock-uranium, respectively.

The separation potentials are evaluated via the general equation

$$V_i = (2x_i - 1) \ln \frac{x_i}{1 - x_i}$$

where $i$ stands for $p$, $w$ and $f$ successively.

As an example, on the assumption that $x_p = 0.03$ and $x_w = 0.002$ one gets NU = 5.479 kg/kg and SW = 4.306 kg/kg.

CHAPTER 4

# Structural Materials

## 4.1. GENERAL

A nuclear reactor is a complex mechanical structure articulating a variety of subassamblies that concur in making possible a self-sustained chain fission reaction, as well as the exchange of power between the core and the outer systems of conversion and utilization of the heat of nuclear origin. There is no question that each and every material that is part of such a contraption has a "structural" function as well. However, a certain class of materials stands out in the sense that their part in, and bearing upon the development and control of the nuclear reaction is of little consequence; these are conventionally described as "structural materials".

Structural materials are, for instance, those that make the pressure vessel; pressure tubes through which the cooling agent is circulated; the clads hosting the nuclear fuel; the parts assembling, driving and steering fuel elements and control rods; the fittings and supports, the reactor resistance structure itself, etc. As one can easily notice, their main functions are of sealed enclosure; support; handling and transport; enabling effective and safe operation of various subassemblies; preventing loss of control over radioactive materials.

Depending on their location in the reactor, structural materials may be classified as *in-core* and *off-core* materials, respectively. Structural materials in the reactor's core must be able to withstand considerable mechanical stress, high temperatures, irradiation and the corrosive action of the cooling agent; also they must absorb as few neutrons as possible, and be relatively immune to activation. Likewise, off-core materials must meet special quality and reliability requirements, generically known as "nuclear quality". Considering the part they have to play as well as the demanding conditions created by the operation of the reactor, structural materials should have the following essential characteristics:

(a) a small neutron absorption cross-section (for thermal reactor in-core materials);

(b) mechanical properties favouring manufacture of reliable parts of high strength in spite of small sizes and a conservative use of materials;

(c) stability under irradiation and at high temperatures, with special emphasis on shape and ductility preservation;

(d) good endurance of the corrosive action of the cooling agents;

(e) chemical compatibility with the other materials;

(f) high thermal conductivity.

Among structural materials, those used for the nuclear fuel cladding hold a place of special importance. Their functions are:

(a) to prevent passage of the fission products from the fuel into the cooling agent;

(b) to avoid fuel corrosion by the cooling agent, which might lead to fuel fragmentation and its dispersion into the cooling circuit;

(c) to ensure the mechanical integrity of the fuel element during handling and operation, and also to allow heat transfer from fuel to cooling agent.

Table 4.1 presents several structural materials and some of their characteristics [1].

## 4.2. ALUMINIUM

Aluminium and its alloys have been used for nuclear fuel cladding and in the manufacture of passages travelled by the cooling agent and the control rods. Aluminium and its alloys are mostly used in the reactors on metallic natural or lightly enriched uranium operating at relatively low temperatures, in certain reactors designed for plutonium production, in experimental and research reactors.

### 4.2.1. Processing Methods

One of the advantages of using aluminium as a structural material is that it can be easily processed. Pure or alloyed aluminium parts may be obtained in a large variety of shapes. When casting aluminium, certain precautions are in order, in view of avoiding porosity; it results from the fact that hydrogen, although insoluble in solid aluminium, is soluble in the liquid metal, the quantity of absorbed hydrogen increasing with the temperature. The porosity, as an effect of hydrogen precipitation during solidification, can be prevented by blowing nitrogen through the melt before casting.

### 4.2.2. Properties

Table 4.2 presents certain physical properties of aluminium. Note the very good thermal conductivity as well as the small neutron absorption cross-section qualities that make it recommendable for applications in thermal reactors.

The 99.1%-purity aluminium has a fracture strength of 185 MPa and an elongation of 30 — 35%. Cold processing of the material can lead to an increase in tensile strength up to 340 MPa, accompanied by a diminishing in elongation by approximately 5%. The mechanical strength of aluminium can be substantially improved by alloying. The composition of certain typical aluminium alloys is given in Table 4.3 [1].

Figure 4.1 presents the tensile strength of certain aluminium alloys at room temperature and at temperatures of 200 °C and 310 °C [2]. The alloys 2S,

TABLE 4.1 Several Structural Materials. General Properties

| *Material* | *Absorption cross-section for thermal neutrons* | *Density at 20 °C* | *Melting point* | *Specific heat at 20 °C* | *Linear expansion coefficient at 20 °C* | *Thermal conductivity at 20 °C* | *Electric resistivity* | |
|---|---|---|---|---|---|---|---|---|
| | (b) | (kg/m$^3$) | (°C) | (kJ/kg . K) | ($10^{-6}$/°C) | (W/m . K) | (nΩ . m) | (°C) |
| Elements | | | | | | | | |
| Aluminium | 0.215 | 2699 | 600.2 | 0.9 | 23.8 | 209 | 26.55 | 20 |
| Beryllium | 0.009 | 1850 | 1315 | 1.8 | 11.6 | 159 | 59 | 0 |
| Carbon (graphite) | 0.0045 | 2220 | 3700 | 0.69 | 0.6; 4.3 [3] | 23.8 | 13750 | 0 |
| Chromium | 2.9 | 7190 | 1890 | 0.46 | 6.2 | 67 | 130 | 28 |
| Iron | 2.43 | 7870 | 1539 | 0.46 | 11.7 | 75.3 | 97.1 | 20 |
| Magnesium | 0.059 | 1740 | 650 | 1.0 | 26.1 | 155 | 44.6 | 20 |
| Nickel | 4.5 | 8900 | 1455 | 0.44 | 13.3 | 92.1 | 68.4 | 20 |
| Vanadium | 4.7 | 6100 | 1710 | 0.53 | 8.3 | 29.3 | 250 | 20 |
| Zirconium | 0.18 | 6500 | 1845 | 0.29 | 5.5 | 23.9 | 420 | 20 |
| Stainless steels | | | | | | | | |
| Austenitic | Depending on composition | | | 0.5 | 16.5-18 (at 500 °C) | 14.6 [1] | 720-900 | 20 |
| Martensitic | Depending on composition | | | 0.46 | 11.2-11.6 (at 600 °C) | 25.1 [1] | 570-670 | 20 |
| Ferritic | 2.6-3.2 | Depending on composition | | 0.46 | 10-10.5 (at 600 °C) | 21 [1] | 670 | 20 |
| Alloys | | | | | | | | |
| Al – U (12.5 w% U) | | | | | 23.5 | 183 [2] | | |
| 2S | | 2710 | 640-660 | | 25.6 | | | |
| Inconel | | 8510 | 1395-1425 | 0.45 | 11.5 (25-100 °C) | 15.0 [1] | 981 | 20 |

[1] At 100 °C; [2] At 200 °C; [3] On the *a* and *c* directions, respectively.

3S and 52S have better mechanical characteristics, as a result of their hardening in solid solution, but they cannot be hot-treated; their strength can be increased only by cold processing.

An aluminium alloy with a high mechanical strength at temperatures over 400°C can be obtained using powder metallurgy techniques.

### 4.2.3. Irradiation effects

The vacancies and the interstitial atoms generated in the metal lattices by neutron irradiation have a minor effect within the temperature range in which these defects may be annihilated. The defects induced by irradiation in the pure aluminium lattice are annealled at room temperature, and consequently the irradiation does not affect to any serious extent the physical characteristics of this metal. Still, it should be emphasized that impurities, even at very low concentrations, may induce the material into a higher annealing temperature

TABLE 4.2 Physical Properties of Aluminium

| | |
|---|---|
| Density ($kg/m^3$) | 2700 |
| Crystalline structure | fcc |
| Lattice constant (Å) | 4.0489 |
| Melting point (°C) | 660.2 |
| Boiling point (°C) | 2327.0 |
| Thermal expansion coefficient ($^oC^{-1}$) | |
| 20 — 100 °C | $23.8 \times 10^{-6}$ |
| 20 — 400 °C | $26.7 \times 10^{-6}$ |
| Thermal conductivity (W/m . K) | |
| 20 °C | 210.6 |
| 100 °C | 210.6 |
| 400 °C | 228.6 |
| Electric resistivity (nΩ . m) | |
| 20 °C | 26.6 |
| 100 °C | 38.6 |
| 400 °C | 80 |
| Absorption cross-section for thermal neutrons at $v$ = 2200 m/s (b) | 0.215 |

TABLE 4.3 Chemical Composition of Certain Aluminium Alloys

(% in weight)

| *Alloy* | *Si* | *Cu* | *Mn* | *Mg* | *Cr* | *Fe* | *Zn* | *Ti* |
|---|---|---|---|---|---|---|---|---|
| 2 S | * | 0.20 | 0.05 | - | - | * | 0.10 | |
| 3 S | 0.60 | 0.20 | 1.5 | - | - | 0.70 | 0.10 | |
| 17 S | 0.80 | 4.50 | 1.0 | 0.80 | 0.10 | 1.0 | 0.10 | |
| 24 S | 0.50 | 4.90 | 0.90 | 1.8 | 0.10 | 0.50 | 0.10 | |
| 52 S | ** | 0.10 | 0.10 | 2.8 | 0.35 | ** | 0.10 | |
| 61 S | 0.80 | 0.40 | 0.15 | 1.2 | 0.35 | 0.70 | 0.20 | 0.15 |
| 63 S | 0.60 | 10.10 | 0.10 | 0.85 | 0.10 | 0.35 | 0.10 | 0.10 |

* 1% Si + Fe; ** 0.45% Si + Fe

and, therefore, they intensify the irradiation effects [3]. These effects become apparent as a hardening of the material — a phenomenon similar to that induced by cold processing. During an irradiation at a fluency of $1.26 \times 10^{21}$ $n/cm^2$, an increase in the mechanical strength (Table 4.4 [2]) accompanied by a decrease of the material ductility has been noticed in aluminium samples at 60 °C.

As regards dimensional stability, irradiation does not bring about important changes; also, it has a minor effect on density and expansion coefficient.

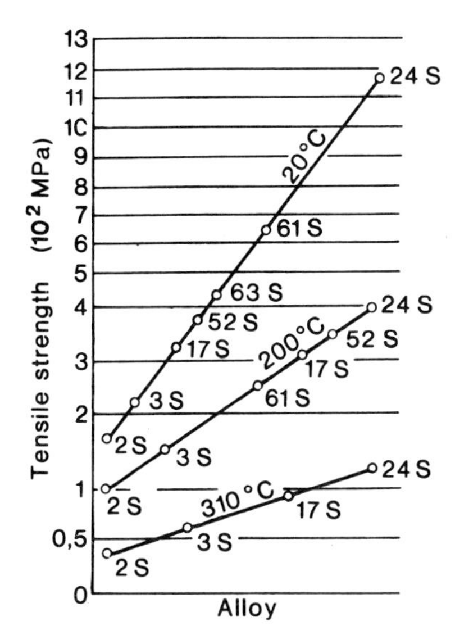


Fig. 4.1. Tensile strength of certain alluminium alloys at different temperatures (arranged in the order of increasing tensile strength).

TABLE 4.4 Irradiation Effects on Mechanical Properties of Aluminium Alloys

| *Alloy* | *Tensile strength* (MPa) | | *Elongation* (%) | |
|---|---|---|---|---|
| | *Before irradiation* | *Post irradiation* | *Before irradiation* | *Post irradiation* |
| Annealed material | 93 | 186 | 38 | 21 |
| Cold-worked material | 373 | 392 | 8.8 | 12.5 |
| Aged material | 304 | 343 | 17 | 16 |

4.2.4. Corrosion Resistance

Pure aluminium has a very good resistance to the corrosive action of the natural environment and of aqueous media. The corrosion rate for the aluminium 2S alloy increases if the solution becomes either acid (pH < 5) or basic (pH > 7) and as the temperature increases. Aluminium is resistant to corrosion in liquid metals too (Na or NaK) up to 200 °C. In water, the corrosion is accelerated by neutron irradiation.

Alloying and precipitation hot treatments reduce aluminium's resistance to corrosion; this is why, when selecting the material, a compromise should be struck between the good corrosion resistance of the pure metal and the superior mechanical characteristics of its allyos.

## 4.3. ZIRCONIUM

Zirconium is a metal with a shiny-grey aspect. In chemical compounds it is to be found in the bi-, tri- or tetravalent state, the most frequent valence being +4. Metallic zirconium has a great capacity of retaining gases ($H_2$, $N_2$, $O_2$).

Zirconium has a very small neutron absorption cross-section, a fracture resistance in alloyed state close to that of steels (see Tables 4.5 and 4.6) and a very good corrosion resistance in many environments,water included [1,2]. Due to these qualities, zirconium alloys are among the most used structural materials in nuclear reactors with thermal neutrons, moderated and cooled with water.

TABLE 4.5 Physical Properties of Zirconium

| | |
|---|---|
| Density ($kg/m^3$) | 6500 |
| Crystalline structure, at ambient temperature | Hexagonal close-packed |
| Lattice constant, at ambient temperature (Å) | $a_o$ = 3.2321<br>$c_o$ = 5.1474 |
| Melting point (°C) | 1845 ± 25 |
| Thermal expansion coefficient, at 250 °C (°C$^{-1}$) | |
| $c$ -axis | 6.15 x $10^{-6}$ |
| $a$ -axis | 5.69 x $10^{-6}$ |
| Thermal conductivity (W/m . K) at ambient temperature | 20.9 |
| Electric resistivity (nΩ . m) | |
| at 20 °C | 420 |
| at 100 °C | 590 |
| at 400 °C | 1000 |
| Absorption cross-section for thermal neutrons, at 2200 m/s (b) | 0.18 |

TABLE 4.6 Mechanical Properties of Zirconium

| *Properties* | *Annealed* | *Cold-worked* |
|---|---|---|
| Tensile strength ($10^2$ MPa) | 4.28 – 5 | 11.42 – 12.85 |
| Yield strength ($10^2$ MPa) | 1.42 – 2.14 | 9.28 – 10.71 |
| Elongation (%) | 30 – 40 | 15 |
| Rockwell hardness | 35 – 45 | 87 – 98 |

Metallic zirconium undergoes an allotropic transition at 865 °C. Below this temperature it crystallizes in the hexagonal close-packed system (zirconium α); above this temperature, it changes to a body-centred cubic lattice (zirconium β). The alloying elements can expand or restrict the domains of existence of these phases.

### 4.3.1. Manufacturing Techniques

Zirconium manufacturing implies a chemical stage and a metallurgical stage [4-6].

The *chemical stage* covers operations that start from the mined ore (the most widespread being a zirconium silicate called zircon), to end with the refined zirconium sponge. Here is, in brief, one alternative used at industrial scale:

1. reduction of zirconium ore with carbon in an electric arc furnace, leading to zirconium carbo-nitride;

2. carbo-nitride chlorination;

3. chloride solution in water and hafnium separation by extraction with an organic solvent;

4. precipitation and thermal decomposition to $ZrO_2$;

5. zirconium oxide chlorination, to zirconium tetrachloride, in the presence of high purity graphite at 800 °C;

6. reduction of zirconium chloride, in vapour state, with liquid magnesium at 800 °C, to zirconium sponge;

7. purification of zirconium sponge, by removing magnesium and magnesium chloride inclusions through vacuum distillation.

Zirconium sponge thus obtained must meet the exceptionally severe requirements of nuclear applications. The content in iron, carbon and oxygen is limited to a few parts per million (ppm), and other elements such as B, Cd, and Li, must fall under 1 ppm. A variety of analysis methods are employed, including X-ray fluorescence, emission optic spectroscopy, etc.

The *metallurgical stage* covers a series of operations starting with the obtaining of the alloyed ingot and ending with the finite product (pipes, bars, wire, stripes).

The ingot is obtained by melting in vacuum arc furnaces or vacuum electron beam furnaces of a sample obtained by pressing the zirconium sponge, with zirconium waste and alloying elements (either pure or alloyed).

The hot plastic deformation of the ingot is performed through conventional forging, rolling and extruding that aim at giving the product predetermined shapes, as well as adequate grain structures. Special precautions are in order, to avoid impurification with gases or other elements during processing.

By cold plastic deformation and hot treatment one envisages obtaining the final shape of the product, a special quality of the surface as well as a microstructure that should provide the mechanical characteristics required during operation (mechanical strength, stability during operation, corrosion resistance). Quality control of products has an extremely important part to play in manufacturing technology.

In order to illustrate the consequence of the manufacturing "history", the effects of the material cold processing on its mechanical characteristics [1,2] are given in Fig. 4.2. The effect of temperature materializes in a decrease of the material strength. The tensile strength of zirconium obtained from a zirconium sponge by electric arc melting decreases from 700 MPa (at room temperature) to 145 MPa (at 500 °C), while the elongation increases from 43% to 100%, as shown in Fig. 4.3 [2].

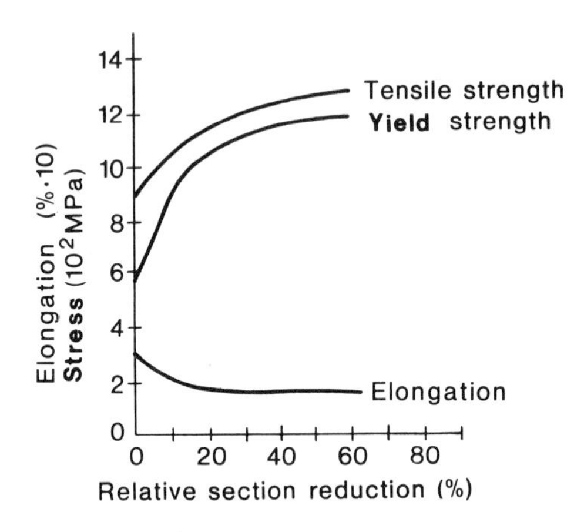


Fig.4.2. Cold-working effect on certain mechanical properties of zirconium.

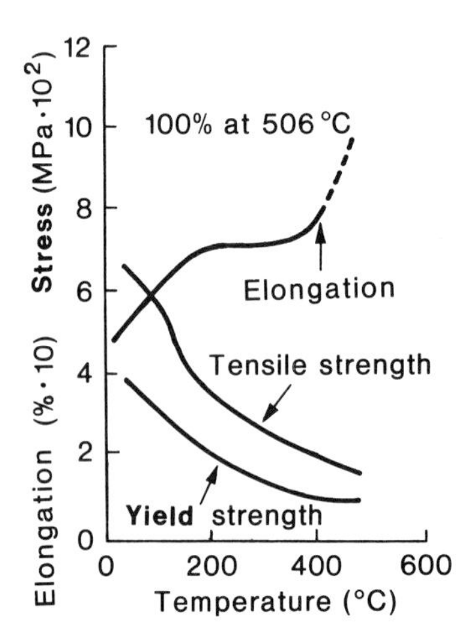


Fig. 4.3. Modification of mechanical properties of zirconium, as function of temperature.

Table 4.7 shows the influence of oxygen on the mechanical characteristics of zirconium [6 — 8]. One can see that, owing to the increase in the oxygen content, the yield and tensile strength increase simultaneously, whereas the elongation decreases.

TABLE 4.7 Oxygen Effect on Zirconium's Mechanical Properties

| *Oxigen content* (%) | *Tensile strength* (MPa) | *Yield strength* (MPa) | *Elongation* (%) | *Rockwell hardness* |
|---|---|---|---|---|
| 0.0 | 200 | 48 | 32 | 20 |
| 0.5 | 557 | 371 | 16 | 36 |
| 1.0 | 714 | 483 | 7 | 46 |
| 1.5 | 842 | 600 | 5 | 52 |
| 2.0 | 1000 | 700 | 4 | 57 |
| 2.5 | 1300 | 800 | 3 | 60 |

The hydrogen that exists in the material in small quantities has an embrittling effect stronger than oxygen or nitrogen. The zirconium — hydrogen reaction results in zirconium hydride, that precipitates as platelets in the bulk of the material (Fig.4.4 [16]). The embrittlement,as a result of the presence of 10 ppm of precipitated hydrogen, substantially reduces the material's ductility at low deformation rates.

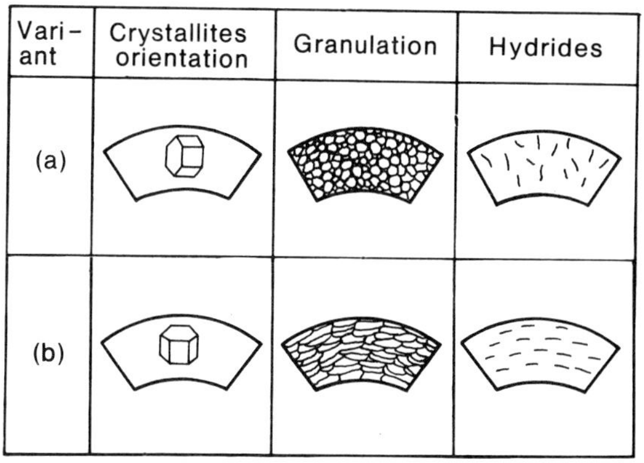

| Vari-ant | Crystallites orientation | Granulation | Hydrides |
|---|---|---|---|
| (a) | | | |
| (b) | | | |

Fig. 4.4. Crystallites and hydrides orientation in a fuel pipe-clad.

### 4.3.2. Corrosion Resistance

Zirconium has an excellent corrosion resistance in water, acid and alkaline solutions as well as in gases, at moderately high temperatures [2]. Pure zirconium is relatively resistant to corrosion in water at high pressure and temperature. Small quantities of impurities significantly reduce the corrosion resistance. The corrosion rate of zirconium varies within very large limits, from a few mg/dm$^2$ x year to the total destruction of the material in a few days (Fig.4.5) [2]. The curves of oxidation kinetics in Fig.4.5 indicate that, after a time, at temperatures over 360 $^{o}$C, there is a sudden increase in the

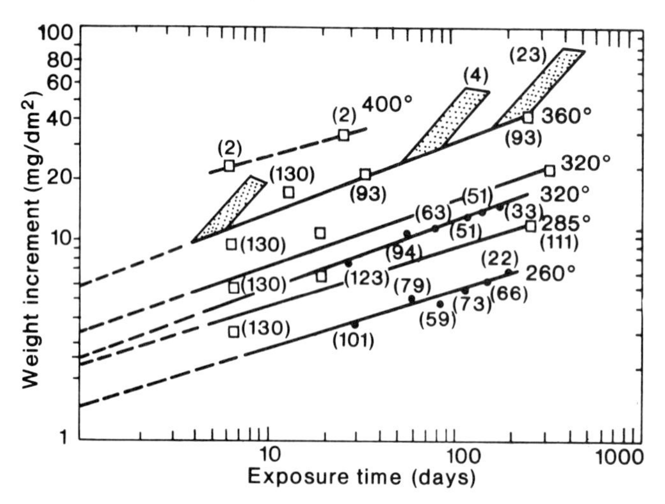


Fig. 4.5. Zirconium corrosion rate at different temperatures; the number of tested samples is specified in the brackets.

corrosion rate (represented by dark areas). The higher the temperature, the faster the recurrent change.

In an oxidizing environment the surface of zirconium gets covered with a compact film of monoclinic $ZrO_2$. The sudden jumps in the corrosion rate are explained either by the zirconium dioxide transition from the cubic or tetragonal state to the monoclinic state, or by the fracturing of the oxide film as a result of the internal stresses due to the more rapid growth of the film in the area of the grain boundaries [9,10]. In water, the corrosion of zirconium and its alloys takes place according to the reaction:

$$Zr + 2H_2O \rightarrow ZrO_2 + 2H_2 \qquad (4.1)$$

After a short initial period, the increase in weight ($\Delta W_1$) as a result of the oxide film formation observes an equation of the type:

$$\Delta W_1 = \Delta W_o + k_1 t^{1/3}, \qquad (4.2)$$

where $\Delta W_o$ is the increment in weight in the initial period, $t$ is the exposure time, and $k_1$ is the constant characterizing the process.

After reaching an extra weight of $\sim 3 \times 10^{-3}$ kg/m$^2$ the oxidation rate begins to depend linearly on time

$$\Delta W_2 = \Delta W_1 + k_2 t \qquad (4.3)$$

The *analysis of the corrosion process of the fuel elements* calls for studying the process in realistic conditions of thermal operation.

Consequently, the corrosion rate must be correlated with the heat flux at the outer surface of the fuel clad.

The corrosion rate depends on the temperature of the corrosive environment according to an Arrhenius-type equation:

$$\frac{d(\Delta W)}{dt} = k_o \exp\left(\frac{-Q}{RT}\right) \tag{4.4}$$

where $k_o$ is the rate constant, $Q$ is the activation energy per mol, $R$ is the constant of the ideal gases and $T$ is the absolute temperature.

The oxide film, acting as a thermal barrier, has a thickness [11] given by

$$S_{ox} = \gamma \Delta W, \tag{4.5}$$

where $\gamma$ is a constant.

The temperature variation $\Delta T$ — a consequence of the presence of the oxide film — is

$$\Delta T = S_{ox} q'/K = (\gamma q'/K)\Delta W, \tag{4.6}$$

where $q'$ is the thermal flux and $K$ is the thermal conductivity of the cladding material.

In the presence of the oxide film, eq. (4.4) becomes:

$$d(\Delta W)/dt \simeq k_o \exp[-Q/R(T_o + \Delta T)], \tag{4.7}$$

where $T_o$ is the temperature at the clad surface. As $\Delta T \ll T_o$, eq. (4.7) can be approximated by:

$$\frac{d(\Delta W)}{dt} \simeq k_o \exp(-Q/RT_o)\exp(Q\Delta T/RT_o^2) = k_o \exp(-Q/RT_o)\exp(\gamma Q q'\Delta W/RT_o^2 K) \tag{4.8}$$

By integrating this equation it results that:

$$\Delta W = \frac{RT_o^2 K}{\gamma Q q'} \cdot \ln \left[1 - \frac{\gamma Q q'}{RT_o^2 K} \; k_o t \, \exp(-Q/RT_o)\right]^{-1} \tag{4.9}$$

For various values of the thermal flux the function in (4.9) is deployed in Fig. 4.6. [11]. An acceleration in the corrosion process is noticed, directly related to the increase in the thermal flux. The corrosion process is not significant as long as the oxide film is thin, but it gets very fast for thick $ZrO_2$ films. The acceleration of the corrosion process is a consequence of the dependence on temperature of the corrosion rate (eq. 4.4). It is possible that the form of this dependence may change during irradiation, as an effect of the microstructure changes in the material, especially in the case of BWR reactors, where the cooling of the fuel elements is achieved in an oxidizing environment [11].

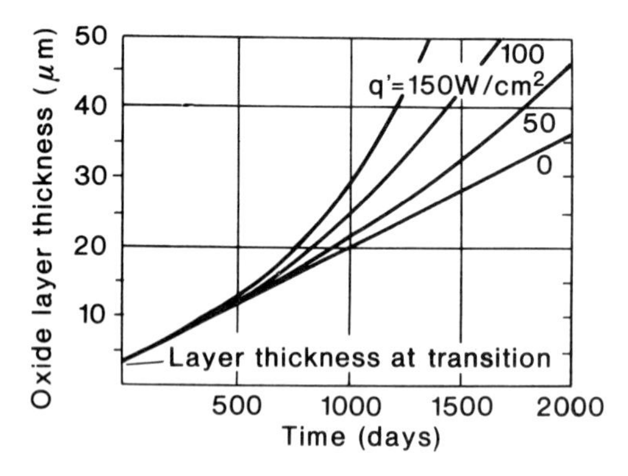


Fig. 4.6. Corrosion process acceleration, as an effect of temperature dependence of the weight increase rate.

In extreme operation conditions, for instance in case of a loss of pressure of the cooling agent, the fuel element clad made of alloyed zirconium is exposed to the corrosive action of the overheated steam ($T > 1000$ °C). The rate of corrosion of zirconium is then described by an equation of the type [12,13]:

$$\frac{dW_{Zr}}{d\sqrt{t}} = \sqrt{K_p} \tag{4.10}$$

where the constant $K_p$, characterizing the isothermal oxidation process, has the form [13]:

$$K_p = 3.10 \cdot 10^5 \exp(-33370/(RT)). \tag{4.11}$$

The values of $K_p$ are given in $(mg \times cm^{-2})^2 \cdot s^{-1}$. Integration of (4.10) leads to:

$$W_{Zr} = (K_p t)^{1/2} \tag{4.12}$$

In the equation above $W_{Zr}$ is the quantity of corroded zirconium. A similar equation may be obtained for the increment in weight:

$$W_{O_2} = (K'_p t)^{1/2} \tag{4.13}$$

The $K'_p$ constant can be obtained from the value of $K_p$, taking into account the fact that for the stoichiometric zirconium dioxide, one has:

$$W_{O_2} = 0.35\, W_{Zr}. \tag{4.14}$$

Figure 4.7 presents the temperature dependence of the reaction constant, calculated according to eq. (4.12), in comparison with the experimental results [14].

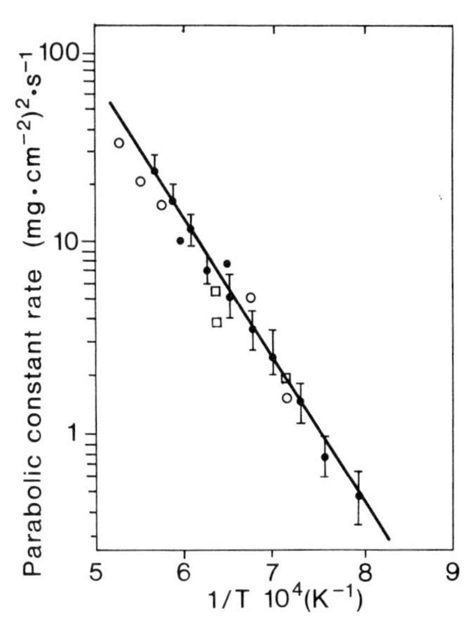


Fig. 4.7. Oxidation's parabolic rate constant in steam, of zircaloy-4, as function of temperature.

### 4.3.3. Zirconium Alloys

The ordinary industrial zirconium cannot reach the degree of purity required in order to exhibit a satisfactory corrosion resistance and appropriate mechanical characteristics. If special precautions are not observed, in the process of obtaining zirconium certain impurities may get caught into the material, the most harmful being gases (oxygen, nitrogen and hydrogen) and carbon. Nitrogen and carbon reduce corrosion resistance, hydrogen leads to embrittlement, while oxygen hardens zirconium, so that it poses problems with processing — see Table 4.7. Impurification with carbon and hydrogen may be avoided by special measures taken during the obtaining and processing of the material. To prevent all undesirable effects, zirconium must be alloyed. Zirconium alloys obtained with tin, iron, chromium and nickel are known under the generic name of zircaloy (Table 4.8). They have a much better corrosion resistance; a superior mechanical strength [2,15]; sufficient ductility; and nuclear characteristics close to those of pure zirconium, as shown in Table 4.9 [2]. The values presented in this table are strictly indicative, since the mechanical characteristics of zirconium and its alloys depend very much on the fabrication process (see Table 4.6).

TABLE 4.8 Chemical Composition of Certain Zirconium Alloys

| *Alloy* | *Content in alloying elements (%)* | | | | |
|---|---|---|---|---|---|
| | *Sn* | *Fe* | *Cr* | *Ni* | *Nb* |
| Zircaloy-2 | 1.5 | 0.14 | 0.1 | 0.06 | - |
| Zircaloy-3 | 0.25 | 0.25 | 0.05 | 0.05 | - |
| Zircaloy-4 | 1.5 | 0.17 | 0.12 | - | - |
| Ozenite | 0.25 | 0.1 | - | 0.1 | 0.1 |

TABLE 4.9 Properties of Zirconium Alloys

| *Material* | *Temperature* ($^{\circ}$C) | *Rupture strength* (MPa) | *Yield strength* (MPa) |
|---|---|---|---|
| Zr pure annealed | 25 | 354 | 110 |
| | 260 | 171 | 65 |
| Zircaloy-2 | 25 | 1180 | 945 |
| | 250 | 868 | 559 |
| Zircaloy-3 | 25 | 907 | 564 |
| | 260 | 422 | 251 |

Another class of zirconium alloys are those with Nb (1% or 2.5% in weight). They have better mechanical characteristics and corrosion resistance than zircaloy-2, especially at high temperatures.

The way in which certain physical characteristics of zirconium alloys are assessed is given in Appendix 11.

### 4.3.4. Irradiation Effects on Zirconium and its Alloys

Figure 4.8 presents the dependence on the neutron fluency of the fracture strength and elongation of a zirconium sample [16]. Relative elongation decreases substantially for fluencies up to 3 x $10^{21}$ n/cm$^2$ and then remains constant. Fracture strength increases from 70 x $10^7$ Pa to 115 x $10^7$ Pa. At fluencies higher than 3 x $10^{21}$ n/cm$^2$, the fracture strength becomes practically independent on fluency.

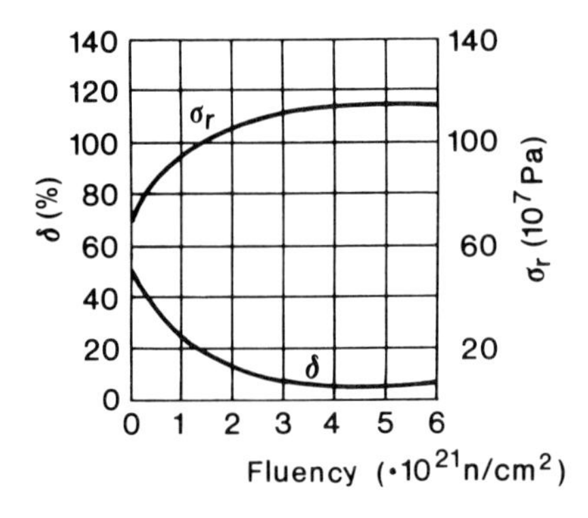


Figure 4.8. Modification of mechanical properties of zirconium, as function of fluency.

The effect of irradiation on zirconium and its alloys appears macroscopically as an altering of their mechanical characteristics under the effect of the fast neutrons and by the embrittlement of the material, owing to the absorption of hydrogen and its precipitation as zirconium hydride.

On the interaction of the fast neutrons ($E > 1$ MeV) and zirconium and its alloys, clusters of defects and dislocation loops are generated, their sizes and distribution in the bulk of the material depending upon the integrated neutron flux (fluency), the irradiation temperature and the composition of the material.

Investigation by transmission electron microscopy of zircaloy-2 irradiated samples have revealed, for instance, that the defects and dislocation loops reach a saturation density of $\sim 3 \times 10^{16}$ cm$^{-3}$ for a fluency $< 5 \times 10^{20}$ n/cm$^2$ at 573 K [18]. These microstructural changes determine an increase in tensile strength [19], a decrease in resilience [20] and an increase in irradiation-induced creep [21]. Bement [22] established a connection between the yield strength of annealed zirconium at 573 K and fluency, using the saturation equations [20] for the hardening induced by irradiation in copper and nickel:

$$\Delta\sigma_c = A[1 - \exp(-B\tilde{\Phi})]^{1/2}, \tag{4.15}$$

where $A$ is a saturation value of $\Delta\sigma_c$, $B$ is a constant and $\tilde{\Phi}$ is the fast neutrons fluency (at energies over 1 MeV).

The constant $B$ in eq. (4.15) is the number of "traps" across the propagation of the deformation due to the interaction with neutrons, multiplied by their effective volume $v$. $B$ depends on the neutron flux, varrying from $2.9 \times 10^{21}$ cm$^2$ ($\Phi = 2 \times 10^{13}$ n/cm$^2$ . s) to $0.58 \times 10^{21}$ cm$^2$ ($\Phi = 7 \times 10^{13}$ n/cm$^2$ . s) [23].

Fleischer [24] has established a relation between the hardening of the material (the increase in yield strength $\sigma_c$) and the number and distribution of the defects induced by fast neutrons:

$$\sigma_c = (\mu b/3.7)(\sum_i n_i^v d_i)^{1/2}, \tag{4.16}$$

where $\mu$ is the shear modulus, $b$ is the Bürgers vector and $n_i^v$ is the number of disc-shaped defects with a diameter $d_i$.

Another effect of the interaction of fast neutrons with zirconium and its alloys consists in changes of their initial shape and sizes [25]. Zirconium and its alloys have a crystalline structure of a hexagonal close-packed type which favours an uneven distribution on the prismatic and basal planes of the vacancies and interstitial atoms generated by the interaction with the fast neutrons.

The forming of the dislocation loops and the clustering of point defects on these lead to changes in the shape of the crystal. If all the interstitial atoms and all vacancies induced by irradiation precipitate on the dislocation loops, then the shape of the crystal changes (Fig. 4.9) while the material density remains constant.

Defects induced by irradiation that remain isolated in the lattice, i.e. do not adhere to the dislocation loops, are felt at the macroscopic level as variations in the dimensional parameters, i.e. as changes in the volume of the material.

In most anisotropic polycrystalline materials, these variations at macroscopic level are due to texture (preferential orientation of crystallites). The

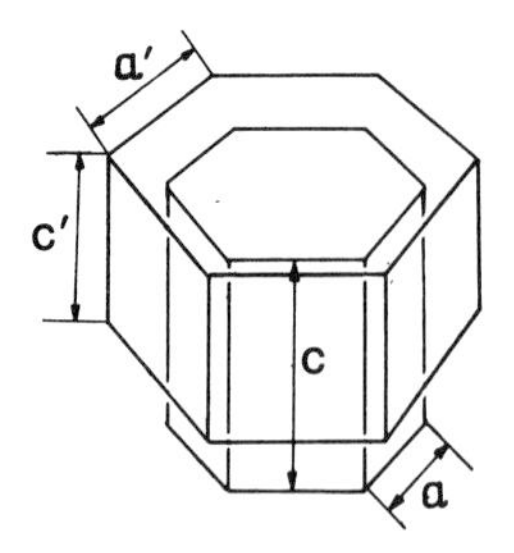


Fig. 4.9. Modification of the ideal lattice unit-cell of a zirconium alloy crystal, as a result of irradiation ($a$, $c$; $a'$, $c'$ original and distorted lattice constants, respectively).

texturing phenomenon expresses itself in an increase of the anisotropy of the physical properties of the material.

The measuring of the texture degree is usually done by means of X-ray diffraction. Thus Refs 22 and 26 report texture measurements on samples of zircaloy used in the manufacture of pressure tubes.

If, at the interaction of a fission neutron with an atom in the lattice, $G$ atoms from the basal plane are extracted and then distributed between the planes parallel to the $c$ axis, and if the vacant place in the lattice is anihilated, then an increase in length in a given direction is registered in the polycrystalline material.

Table 4.10 gives the values of the growth coefficient under irradiation [27]. Between 78 K and 553 K it is not really affected by temperature. Around the temperature of 550 K the growth coefficient depends on fluency $\tilde{\Phi}$, according to the equation:

$$G \sim (\tilde{\Phi})^{-1/3}. \tag{4.17}$$

TABLE 4.10 Growth Coefficients under Irradiation, as Function of Temperature and Fast Neutron Fluency

| *Temperature* (K) | *Fluency* | *Growth coefficient* | | *Number of samples* |
|---|---|---|---|---|
| | | *Instantaneous* | *Average* | |
| 468 | $9.7 \cdot 10^{22}$ | - | 11.6 ± 2.5 | 4 |
| 313 | $8.2 \cdot 10^{22}$ | - | 19.0 ± 1 | 7 |
| 353 | $5.2 \cdot 10^{22}$ | - | 0.77 ± 0.06 | 7 |
| 553 | $2.2 \cdot 10^{21}$ | 9 | - | 1 |
| 553 | $1 \cdot 10^{22}$ | 5 | - | 1 |
| 553 | $1 \cdot 10^{23}$ | 3 | - | 1 |
| 553 | $3 \cdot 10^{23}$ | 1.5 | - | 1 |

4.3.5. Hydrogen Absorption and Diffusion

As indicated above, under normal conditions of operation in the reactor, the diffusion of hydrogen in zirconium and its alloys in phase $\alpha$, determines the precipitation of the zirconium hydride platelets, which leads to the embrittlement of the material. In a sample of zirconium or zirconium alloy, subjected to a temperature gradient, the hydrogen dissolved in the lattice diffuses from the hot area to the cold area of the material [28]. The diffusion of hydrogen in $\alpha$-zirconium is described by the equation:

$$J_\alpha = -D_\alpha\left(\frac{dN_\alpha}{dx} + \frac{Q^*_\alpha N_\alpha}{RT^2} \cdot \frac{dT}{dx}\right), \tag{4.18}$$

where $J_\alpha$ is the hydrogen flux, $D_\alpha$ the diffusion coefficient, $N_\alpha$ the hydrogen concentration, $Q^*_\alpha$ the transport heat, $T$ the local temperature, $R$ is the constant of the ideal gases and $dT/dx$ is the temperature gradient. The diffusion coefficient, $D_\alpha$, has the following dependence on temperature:

$$D_\alpha = D^o_\alpha \exp(-Q_\alpha/(RT)), \tag{4.19}$$

where $D^o_\alpha$ is the frequency factor and $Q_\alpha$ is the activation energy of the process.

In case of a cylindrical sample, eq. (4.18) becomes:

$$J_\alpha = -D_\alpha\left(\frac{dN_\alpha}{dr} + \frac{Q^*_\alpha N_\alpha}{RT^2} \cdot \frac{dT}{dr}\right). \tag{4.20}$$

When the hydrogen concentration falls below the solubility limit, for a certain temperature it will reach a stationary level ( $J_\alpha = 0$):

$$N_\alpha = N_o \exp(Q^*_\alpha/(RT)), \tag{4.21}$$

where $N_o$ is a constant.

When the concentration of hydrogen exceeds the solubility limit, the quantity of hydrogen dissolved in zirconium in phase $\alpha$ adjusts to this solubility limit:

$$N_s = N^{(o)} \exp(-\Delta H/(RT)), \tag{4.22}$$

where $\Delta H$ is the dissolving energy and $N^{(o)}$ is a constant.

The distribution of hydrogen after a time $t$ is given by:

$$N = \frac{C^2 D^{(o)}_\alpha N^{(o)} (\Delta H + Q^*_\alpha)}{RT^4 r^2} \cdot \left[(r^2 - r_i^2)^2 \left(\frac{\Delta H + Q_\alpha}{R} - 2T\right) - \frac{2T^2 r^2}{C}\right] t \exp\left(-\frac{\Delta H + Q_\alpha}{RT}\right) + N_I \tag{4.23}$$

where

$$C = Q_\alpha/(2\pi k(r_o^2-r_i^2)), \qquad (4.24)$$

and $r$ is the current radius describing the position of the clad area under investigation, $r_i$ is the clad inner radius, $r_o$ is the tube outer radius, $R$ is the constant of the ideal gases, $N_I$ represents the initial hydrogen concentration and $k$ is a constant.

Experimental data [29] point out that the temperature dependence of the diffusion coefficient (in $cm^2/s$) is given by:

$$D_\alpha = 2.17 \times 10^{-2}\exp(-8380/(RT)). \qquad (4.25)$$

The solubility limit varies with temperature, as per the equation:

$$N_s = 9.9 \times 10^4\exp(-8250/(RT))\text{ppm} \qquad (4.26)$$

On this model one can see that hydrogen diffuses towards the outer, colder surface of the sample (Fig. 4.10) where it accumulates. If this accumulation exceeds the local solubility limit, zirconium hydride precipitates as platelets.

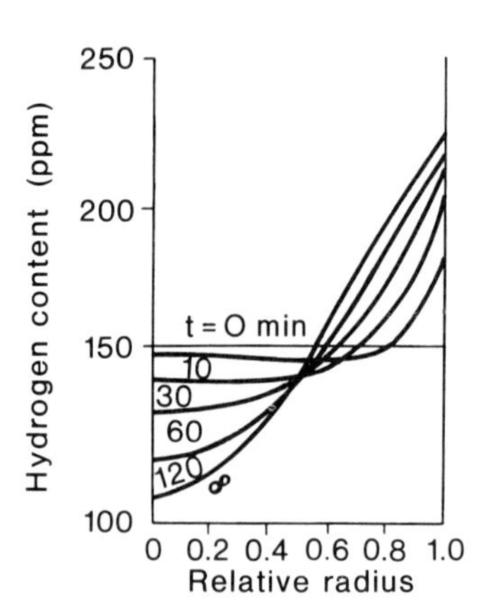


Fig. 4.10. Modification of hydrogen radial distribution in a zircaloy sample.

As soon as the hydride platelets precipitate in the material, the monophasic model presented above is no longer valid for the description of hydrogen diffusion in zirconium. A "biphasic" model is then in order, that enables also the description of hydrogen diffusion through precipitated hydrides [30].

## 4.4. STAINLESS STEELS

### 4.4.1. Properties

Stainless steels exhibit very good mechanical properties and great corrosion resistance, even at high temperatures; yet, being avid neutron absorbents, they have only limited in-core applications to thermal neutron reactors. Stainless steels are, on the other hand, chief off-core structural materials to thermal reactors, and essential materials to the entire structure of fast neutron reactors. Among stainless steels, the most used are the austenitic ones. Table 4.11 contains data indicative of the chemical composition of some varieties of steels employed in building reactors. Table 4.12 presents some physical properties of stainless steels.

TABLE 4.11 Chemical Composition of some Stainless-Steels (%)

| *Trade mark*[1] | *C* | *Si* | *Mn* | *Cr* | *Ni* | *Mo* |
|---|---|---|---|---|---|---|
| AISI 304 | 0.08 | 1.0 | 2.0 | 18-20 | 8-12 | - |
| AISI 304L | 0.03 | 1.0 | 2.0 | 18-20 | 8-12 | - |
| AISI 316 | 0.08 | 1.0 | 2.0 | 16-18 | 10-14 | 2-3 |
| AISI 316L | 0.03 | 1.0 | 2.0 | 16-18 | 10-14 | 2-3 |

[1]As per the US AISI standards.

TABLE 4.12 Physical Properties of Stainless-Steels

| | |
|---|---|
| Density (kg/m$^3$) | 7600 — 7980 |
| Melting point (K) | 1643 — 1783 |
| Thermal conductivity (W/m . K): | |
| at 373 K | 13.4 — 36.4 |
| at 773 K | 16.7 — 33.5 |
| Electric resistivity (nΩ . m): | |
| at 293 K | 450 — 790 |
| at 673 K | 880 — 1050 |
| at 1073 K | 1150 — 1210 |

The thermal conductivity of stainless steels is rather low, comparable to that of zirconium, but their higher mechanical strength allows the manufacture of fuel clads with thinner walls.

In order to illustrate the extent to which temperature influences the mechanical properties of stainless steels, Fig. 4.11 provides the values of the fracture strength $\sigma_r$, elongation $\delta$ and $\sigma_{0.2}$ respectively, in the range 0 — 800 $^{o}$C [17]. The mechanical strength of stainless steels decreases with the increase in temperature; however, these changes are not extensive, over a temperature scale up to 600 $^{o}$C. This behaviour contrasts to that of other structural materials, the use of which is limited to the range of relatively low temperatures.

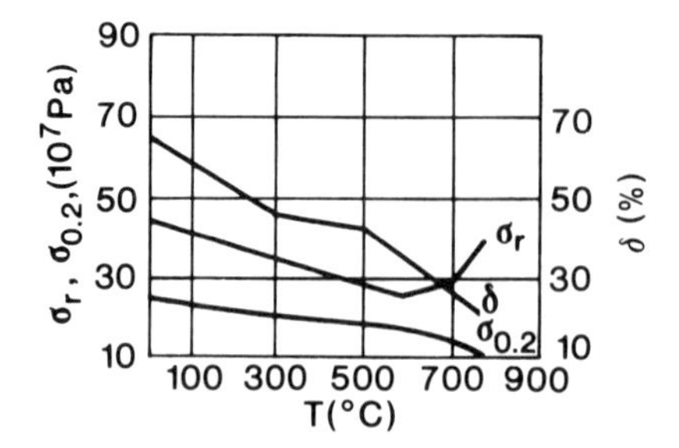


Fig. 4.11. Variation of stainless-steel strength and elongation as function of temperature.

## 4.4.2. Irradiation Effects

To illustrate the effects of irradiation on the mechanical behaviour of an alloyed (stainless) steel, Fig. 4.12 shows the change in fracture strength $\sigma_r$ and elongation $\delta$ for fluencies up to $8 \times 10^{19}$ n/cm$^2$ [16]. The behaviour is similar to that of zirconium (Fig. 4.8). Elongation decreases with the increase in fluency, while fracture strength increases. For fluencies higher than ~ $4 \times 10^{19}$ n/cm$^2$ both $\delta$ and $\sigma_r$ exhibit relatively small varriations.

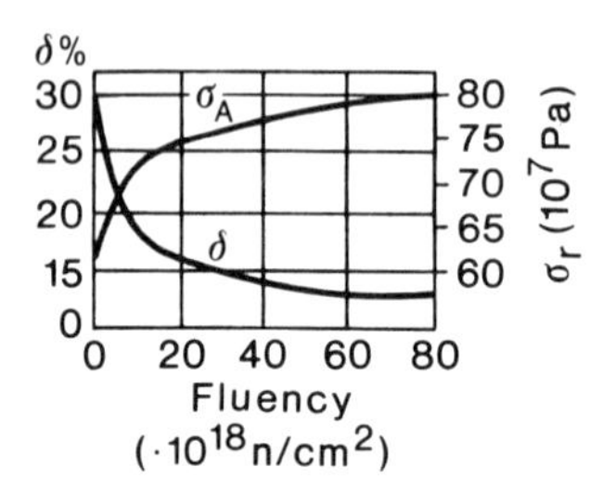


Fig. 4.12. Dependence of stainless-steel strength and elongation as function of fluency.

Besides hardening, another drawback created by irradiation is the swelling of stainless steels, following the accumulation of helium that results from nuclear reactions of neutrons with several ingredients in the material [31]. Investigations by transmission electron microscopy [32] evidenced that "swelling" is due to the occurrence of cavities inside the metal grains, of sizes up to ~ 7000 Å. The degree of swelling depends on the fast neutron fluency. After a period of "incubation" up to a fluency $\tilde{\Phi}$ of ~ $10^{22}$ n/cm$^2$, where swelling is negligible, a period in which swelling $(\Delta V/V)$ grows with $(\tilde{\Phi})^n$ follows, where $n$ is a constant greater than 1. The data accumulated so far show that the fuel element clads register, for most LMFBR reactors, relative volume growths of 5 — 10% [33].

The collisions of the fast neutrons with the atoms of the crystalline lattice generate a large number of interstitial — vacancy pairs. Part of them recombine and the rest migrate across the material before being eventually caught by "traps", the most efficient being the dislocations. The dynamic equilibrium established between created and annihilated point defects maintains the concentration of vacancies and interstitials much above the level that corresponds to the thermal equilibrium regime of the non-irradiated system.

The agglomerations of interstitial atoms introduce additional planes into the lattice, which determines a growth of the material [34]. Agglomeration of vacancies on dislocation loops induces a contraction of the lattice in the area adjacent to the loop, that compensates the swelling due to the interstitial atoms agglomeration. This contraction does not appear when the vacancies are accumulated in voids, and consequently the expansion introduced by the agglomeration of interstitial atoms is no longer compensated.

The relative stability of the voids and dislocation loops can be evaluated by comparing the energy of an agglomeration containing $m$ vacancies to that of the perfect lattice. For a void this energy is:

$$E_{void} = 4\pi R^2\gamma, \quad (4.27)$$

where $\gamma$ is the superficial tension of the solid and $R$ is the void radius. The number of vacancies $m$ can be expressed by the equation:

$$m = 4\pi R^3/(3\Omega), \quad (4.28)$$

where $\Omega$ is the atomic volume. Therefore, the energy associated to the void is:

$$E_{void} = 4\pi\gamma(3\Omega m/(4\pi))^{2/3}. \quad (4.29)$$

The energy associated to a dislocation loop made up of $m$ vacancies placed on a disc of a radius $R_1$ is:

$$E_b = (2\pi R_1)\tau_b + \pi R_1^2\gamma_{d_i}, \quad (4.30)$$

where $\tau_b$ is the energy per unit of length (linear stress) of the dislocation ( ~ $Gb^2$, where $b$ is the Bürgers vector) and $\gamma_{d_i}$ is the energy per unit of surface associated to the loop's packing defect (see Chapter 2).

In metals that have a face-centred cubic structure, with the lattice constant $a_o$, the dislocation loops are formed on the planes (111), where the area corresponding to an atom is $3^{1/2}\, a_o^2/4$. The radius of the vacancy loop that is generated when extracting $m$ atoms from (or depositing $m$ vacancies on) a plane is:

$$R_1 = (3^{1/2}\,\Omega m/(\pi a_o))^{1/2}, \quad (4.31)$$

where $\Omega = a_o^3/4$ is the atomic volume for a f.c.c. structure. Therefore the energy associated to a vacancy loop is:

$$E_b = 2\pi Gb^2(3^{1/2}\,\Omega m/(\pi a_o))^{1/2} + 3^{1/2}\,\Omega\gamma_{d_i} m/a_o \quad (4.32)$$

The evaluation of the energy associated to the voids (4.29) and dislocation loops (4.32) — with a view to assessing the stability of the two types of vacancy agglomerations — is compounded by the fact that the value of the stress in the dislocation line is not fully known. It seems that the voids are more stable for small vacancy agglomerations (small $m$) and that, when $m$ increases, the arrangement of vacancies on the dislocation loops is favoured energetically.

Given the complexity of the mathematical modelling of metal swelling, the following is just a brief account of the method to determine the increase in volume of the irradiated material, based on the information supplied by studies of transmission electron microscopy. Such studies allow the determination of the void distribution function, $N(R)\mathrm{d}R$, per unit-volume, with radii between $R$ and $R+\mathrm{d}R$. The total number of voids is:

$$N = \int_0^\infty N(R)\mathrm{d}R, \tag{4.33}$$

and their estimated average dimension is:

$$\overline{R} = N^{-1} \int_0^\infty R\, N(R)\mathrm{d}R. \tag{4.34}$$

Similarly, the swelling of the material $\Delta V/V$ is:

$$\Delta V/V = \frac{4}{3}\pi \int_0^\infty R^3 N(R)\mathrm{d}R. \tag{4.35}$$

When analysing the merits of the fuel elements for fast reactors (LMFBR), semiempirical formulae are currently used in order to estimate the degree of swelling of the metallic alloy used for cladding, as a function of temperature, fluency and metallurgical state. For instance, for the stainless steel type 316, the change in volume is given by [35]:

$$\Delta V/V(\%)=(\tilde{\Phi}.10^{-22})^{(2.05-27/\theta+78/\theta^2)}.[(T-40).10^{-10}].\exp(32.6-5100/T-0.015T) \tag{4.36}$$

where $\theta = (T-623)$ K and $\tilde{\Phi}$ is the fluency in n/cm$^2$. The equation is valid at temperatures 623 K < $T$ < 900 K.

## 4.5. FERRITIC STEELS

### 4.5.1. Properties

For the construction of pressure vessels of LWRs ductile materials with high fracture strength are in order. The pressure vessels of all commercial reactors of this type built so far are made of carbon steel or lightly alloyed ferritic steels. Parts such as shells, hemispherical heads and rings obtained by cold machining (forging, rolling) of ingots are assembled by welding.

Figure 4.13 introduces, in its main features, the structure, in two variants, of a PWR reactor pressure vessel as it may be encountered in a nuclear power station; the left side presents a structure evolving from an assembling technology that relies on welding smaller parts, while the right side emphasizes the present assembling tendencies — by welding large parts [36].

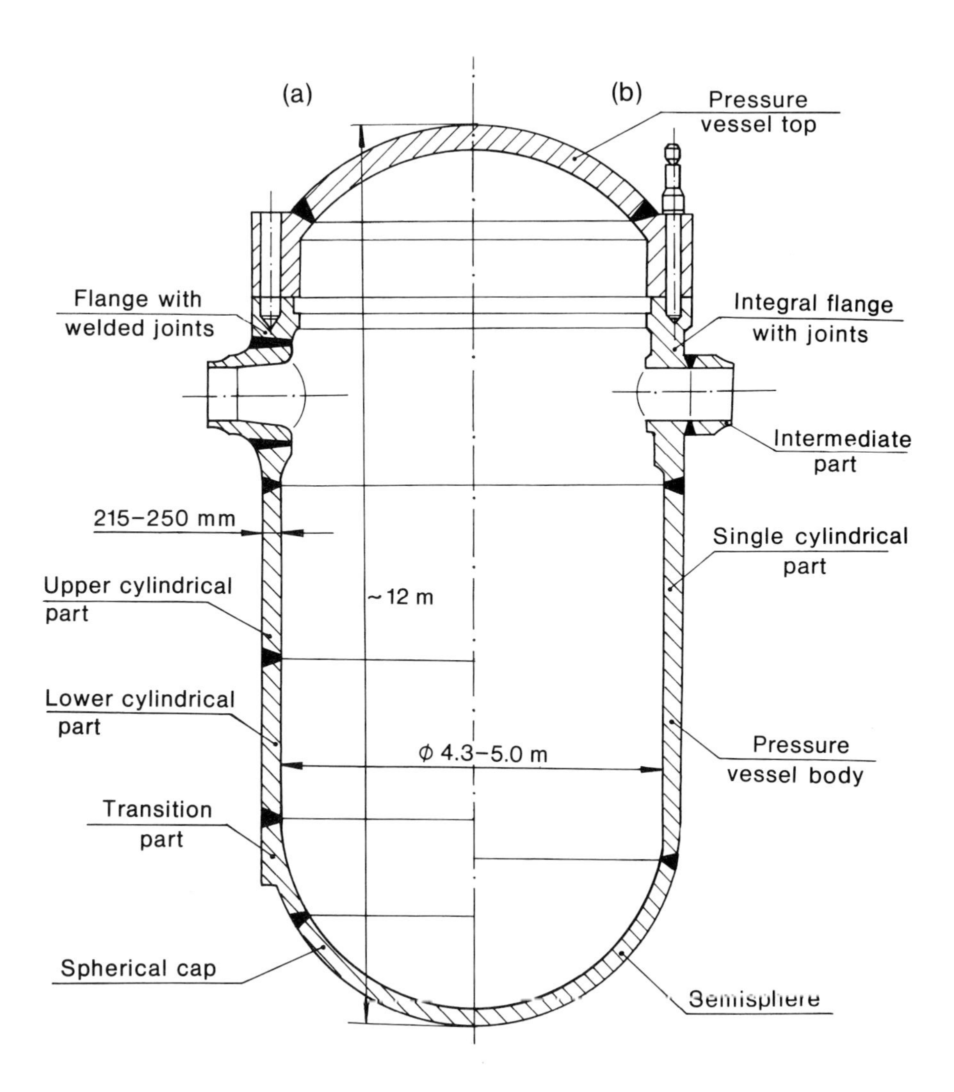


Fig. 4.13. Longitudinal section through a PWR pressure vessel, manufactured: (a) starting from small-dimension pieces obtained by forging and lamination (besides the radial welds indicated, the big cylindrical pieces contain also 3 – 6 longitudinal welds); (b) starting from large pieces obtained by forging.

Although, in principle, a large variety of steels are eligible for manufacturing nuclear pressure vessels, in practice only a few of them are actually used. The choice of the most suited materials is determined by a number of factors, such as their physical and mechanical characteristics, the way they respond to welding and forging, full control over technology warranting confidence in a constant quality of materials, and the global cost of manufacturing the pressure vessel.

Following the common wisdom in the petrochemical industry, the first PWR pressure vessels were made of only lightly alloyed steels. Thus, at Saxton and Indian Point-1 nuclear power plants in the United States steels with perlitic structure, alloyed with manganese (C: 0.3%, Mn: 0.8 — 1.2%) were used in the manufacture of the respective pressure vessels. As the geometric dimensions of the pressure vessels, and consequently their wall thickness grew ever larger, these varieties of steel proved inadequate — owing to their limited mechanical strength and tenacity — and were replaced accordingly with Mn-Mo-type steels. These steels are used in normalized and recovered or in quenched and recovered state, featuring a bainitic, or a bainitic — martensitic microstructure.

Further designed growth in the wall thickness called for increased tenacity. This has been achieved by adding Ni in a proportion of 0.40 — 1.00%, and Cr. Such continued developments, required in order to improve the physical, mechanical and technological merits, have led, in the manufacture of modern pressure vessels, to a choice of steels with chemical compositions given in Table 4.13 [36], and mechanical characteristics given in Table 4.14 [36]. Their description, except for the Soviet variety, observes the US terminology.

The requirements regarding the composition, fabrication procedures and mechanical characteristics of the materials for the pressure vessel components are specified in standards. In the USA, for instance, they are gathered in the ASTM (American Society for Testing Materials) standards, and the ASME (American Society for Mechanical Engineers) codes. The ASME codes and ASTM standards establish the general framework for the calculation, designing, manufacturing and testing of pressure vessels, as well as the materials specifications. The requirements to be met as regards the proper integrity and reliability of the nuclear pressure vessels during their entire lifetime (of maximum 40 years) under neutron irradiation are much more severe than the ones provided by the above-mentioned codes and standards. These requirements are periodically reviewed and revised, with the aim of ensuring the safety and reliability of the pressure vessels' operation; they also provide norms and conditions for the fabrication of the appropriate steels, and for procedures of fabrication of pressure vessels. Products manufactured according to these standards are certified as being of "nuclear quality". A fairly large number of countries interested in the development and implementation of technologies for nuclear materials and equipment have adopted the said system of codes and standards.

Ferritic steels are used in pressure tubes and fittings for the present commercial reactors of the CANDU-PHW type. Figure 4.14 displays the structure of a fuel channel [37], while Fig. 4.15 gives details regarding the junction pressure tube-end fitting, both featuring the CANDU-PHWR concept [38]. The material used in the fabrication of end-fittings according to the USA standards is the steel AISI 403, whose composition is given in Table 4.15 [38]. In the same table, a comparative presentation of the equivalent steels as per the AECL — Canada norm versus the one practised in Romania is offered,. A comparative view on the mechanical characteristics specified by the ASME and the AECL codes, respectively, is given in Table 4.16.

### 4.5.2. Irradiation Behaviour

A constant, flawless integrity of the pressure vessel is of essence. Consequently, the basic material of the vessel and the welded joints (welds, chuck) must ensure an adequate mechanical resistance and tenacity during the entire lifetime of the reactor, both in normal operation conditions and in case of accident.

TABLE 4.13 Typical Composition of Steels Used in the Manufacture of Modern Pressure Vessels

average chemical composition (% in weight)

| Constituents | A 533 B-I USA | A 533 B-II Japan | A 508-II USA | A 508-II Japan | A 508-III forgeable Japan | A 508-III forgeable France |
|---|---|---|---|---|---|---|
| | *Material type* | | | | | |
| C | 0.218 | 0.201 | 0.218 | 0.206 | 0.200 | 0.16 |
| Mn | 1.367 | 1.371 | 0.682 | 0.803 | 1.398 | 1.338 |
| Ni | 0.547 | 0.616 | 0.600 | 0.844 | 0.753 | 0.722 |
| Mo | 0.547 | 0.520 | 0.596 | 0.585 | 0.505 | 0.503 |
| Si | 0.236 | 0.243 | 0.284 | 0.231 | 0.243 | 0.235 |
| Cr | 0.074 | 0.136 | 0.374 | 0.374 | 0.097 | 0.195 |
| P | 0.009 | 0.007 | 0.006 | 0.006 | 0.007 | 0.010 |
| S | 0.014 | 0.006 | 0.0011 | 0.006 | 0.007 | 0.010 |
| Cu | 0.117 | 0.049 | 0.040 | 0.048 | 0.054 | 0.065 |
| Al | — | 0.025 | — | 0.029 | 0.028 | — |
| As | — | 0.007 | — | max.0.008 | max.0.007 | 0.016 |
| Sn | — | 0.008 | — | max.0.009 | max.0.008 | 0.011 |
| Sb | — | 0.002 | — | max.0.002 | max.0.002 | — |
| Co | — | — | — | max.0.010 | max.0.009 | 0.017 |

| Constituents | 20 MnMoNi 55 (type A A 533B) FR of Germany | 22 NiMoCr 37 (type A 508-II) forgeable FR of Germany | 22 K (type A 212B) USSR |
|---|---|---|---|
| C | 0.17 - 0.23 | 0.17 - 0.25 | 0.19 - 0.26 |
| Mn | 1.20 - 1.50 | 0.50 - 1.00 | 0.75 - 1.00 |
| Ni | 0.50 - 0.80 | 0.60 - 1.00 | max.0.30 |
| Mo | 0.45 - 0.60 | 0.50 - 0.80 | — |
| Si | 0.15 - 0.30 | max.0.35 | 0.20 - 0.40 |
| Cr | max.0.20 | 0.30 - 0.50 | max.0.40 |
| P | max.0.010 | max.0.025 | max.0.025 |
| S | max.0.015 | max.0.025 | max.0.025 |
| Cu | max.0.10 | max.0.20 | max.0.030 |
| Al | 0.010 - 0.040 | max.0.05 | — |
| As | — | — | — |
| Sn | — | — | — |
| Sb | — | — | — |
| Co | max.0.030 | — | — |

Phenomena that may affect adversely the vessel integrity are:

(a) fatigue;
(b) radiative embrittlement;
(c) embrittlement due to excessive exposure to high temperatures;
(d) tenacity at the level of the higher plateau of the ductile — fragile curve.

The (a), (b) and (d) processes are determined by the behaviour of the material under reactor operation conditions (pressure, temperature, number of operation cycles, water chemistry — most especially pH level and oxygen content, generation of defects by irradiation and their migration).

TABLE 4.14 Metallurgical State and Mechanical Properties of Pressure Vessel Steels

| *Material denomination* | *Type* | *Thermal treatment* | *Conventional yield limit* (MPa) | *Rupture limit* (MPa) | *Minimum relative elongation (%) (testing length: 50 mm)* |
|---|---|---|---|---|---|
| *Plate* | | | | | |
| A 533 B | Mn-Mo-Ni | Q/A | 345 | 620-793 | 18 |
| I class 22 K | | | 220 | | max. 20 |
| *Forged* | | | | | |
| A 508 | Mn-Mo-Ni | Q/A | 345 | 550-725 | 18 |
| II class | | | | | |
| A 508 | Mn-Mo-Ni | Q/A | 345 | 550-725 | 18 |
| III class 22 K | | | 220 | | max. 20 |

Q = quenched; A = annealed.

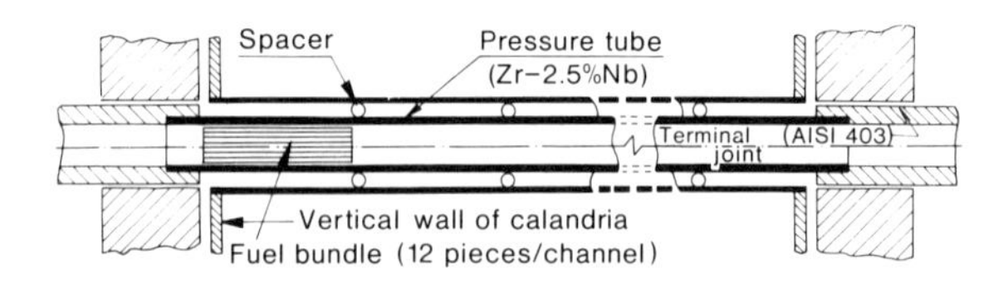


Fig. 4.14. Longitudinal section through a fuel channel of a CANDU-PHW-600 MWe.

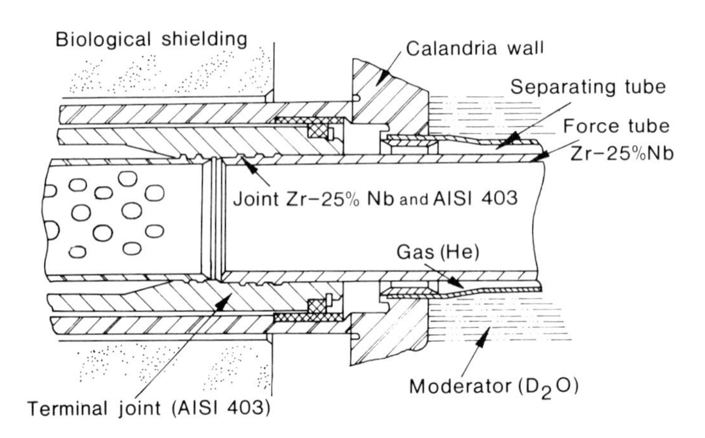


Fig. 4.15. Detail of a joint — force tube/terminal fitting, in the fuel channel of a 600 MWe CANDU-PHWR.

TABLE 4.15 AISI 403 Steel[1] Standard Composition, as in use in the USA, Canada and Romania

| | average chemical composition (% in weight) | | |
|---|---|---|---|
| *Constituents* | AISI 403 USA | AISI 403 Canada | AISI 403 m[2] Romania |
| C | 0.10 - 0.40 | 0.06 - 0.13 | 0.12 - 0.13 |
| Mn | 0.40 - 0.70 | 0.25 - 0.80 | 0.56 - 0.60 |
| Ni | 0.10 - 0.40 | max.0.50 | 0.13 |
| Si | 0.10 - 0.40 | max.0.50 | 0.26 - 0.29 |
| Cr | 11.70 - 12.50 | 11.50 - 13.00 | 11.80 - 11.85 |
| P | max.0.030 | max.0.030 | 0.01 - 0.013 |
| S | max.0.030 | max.0.030 | 0.008 |
| Cu | max.0.20 | — | 0.17 - 0.18 |
| Co | max.0.025 | — | max.0.025 |
| Mo | — | — | 0.02 |
| V | — | — | 0.05 |
| B | — | — | max.0.022 |

[1]Quenched, annealed; [2]m — modified.

TABLE 4.16 AISI 403 Mechanical Properties, as per the ASME Code, Compared with those Indicated by AECL-Canada

| | *Country* | *Conventional yield limit* (MPa) | *Rupture strength* (MPa) | *Elongation* (%) |
|---|---|---|---|---|
| AISI 403 | USA | 621 | 756 | 16 |
| AISI 403 | Canada | 586 | 724 | 12 |

Radiative embrittlement appears as a major potential nuisance especially for the cylinder shell of the pressure vessel near the reactor core. The material in this area is subject to temperatures of about 570 K and, during the entire lifetime, to a fluency of about $3.5 \times 10^{19}$ n . $cm^{-2}$ ($E > 1$ MeV). Such levels are characteristic for the area that is at a distance of 1/4 of the vessel wall thickness. Neutron irradiation may influence the mechanical properties of a lightly alloyed steel by increasing the yield strength ($\sigma_{0.2}$) and the fracture strength. In addition, irradiation may lead to an increase of the ductile — fragile transition temperature and to an energy reduction at the level of the higher plateau of the ductile — fragile transition curve.

Such phenomena have been the subject of very serious investigations. The chemical composition of steels constitutes a decisive factor as to the irradiation behaviour; this, in turn, depends on the irradiation parameters, such as the energy spectrum of the neutrons, the fluency and the irradiation temperature.

The effects of alloying with certain elements such as Ni, Cu, V, P, B and N have been studied by Steele [39]. The nitrogen content influences radiative embrittlement at low temperatures, while Cu and P play an important part in radiative embrittlement at 570 K. At present, the norms for PWR pressure vessel steels limit the Cu content to 0.010% in weight, and the P content to 0.015% in weight. These limits are permanently reviewed on the basis of the results from programmes of monitoring the irradiation effects.

The mechanism of steel embrittlement is not entirely understood and the effects noticed may be due to a concurrence of radiative and thermal embrittlement effects.

In current practice the effects of radiative embrittlement can be removed by the annealing of the reactor pressure vessel at a temperature of 670 K.

### 4.6. MAGNESIUM ALLOYS

Lightly alloyed magnesium has been used in the manufacture of clads for metallic uranium fuel elements designed for graphite-moderated, $CO_2$-cooled power reactors of the Magnox (UK) and EDF (France) types.

Owing to its very small absorption cross-section — several times smaller than that of aluminium or zirconium — magnesium was used as cladding material for the natural uranium graphite-moderated power reactors, where the economy of neutrons is of great importance.

The main characteristics of pure magnesium are presented in Table 4.1. Magnesium has a hexagonal close-packed structure, with the lattice parameters $a$ = 3.203 Å and $c$ = 5.200 Å. Density is very low: 1.74 x $10^3$ kg/$m^3$. Melting point is very close to that of aluminium; the mechanical resistance of magnesium at about 670 K — which is the clad temperature in natural uranium graphite-moderated power reactors — looks much better than that of aluminium. At this temperature, that is relatively close to the melting point, magnesium presents a strong tendency to crystal growth, that is partially compensated by irradiation.

Magnesium has a thermal conductivity lower than aluminium's. The expansion coefficient between 290 and 370 K is 26.1 x $10^{-6}$ $K^{-1}$.

Magnesium exhibits a good compatibility with metallic uranium. In a $CO_2$ environment is shows a good chemical resistance up to 670 K, owing to the protection provided by the MgO + $MgCO_3$ film that forms on the surface, which leads to magnesium's logarithmic oxidation kynetics. Above this temperature the oxidation kynetics become linear, and the oxidation rate is high. Small beryllium admixtures (~ 0.01%) can lift this temperature threshold up to almost 870 K.

Corrosion of magnesium alloys in dry steam is a complex process. At temperatures lower than 390 K, corrosion rate increases with temperature and the corrosion product is the magnesium hydroxide. This one stays as the predominant corrosion product up to 497 K, but the process rate decreases after reaching a maximum. Above 512 K magnesium oxide is generated, and the rate of the process increases again with temperature.

At temperatures higher than 440 K among the corrosion products outstanding is the magnesium hydride, its concentration increasing with temperature. At even higher temperatures, only the magnesium oxide is present.

In the sequel, certain experimental data are presented in connection with the magnesium corrosion processes [40]. At temperatures of 410 K the weight increment over approx. 16 hours was reported to be 89.2 x $10^{-3}$ kg/$m^2$, the resulting product being $Mg(OH)_2$. At 550 K the presence of both $Mg(OH)_2$ and $MgH_2$ was noticed, the change in weight over approx. 16 hours being 31.4 x $10^{-3}$ kg/$m^2$. At $T \simeq 570$ K, the weight increment is relatively small, i.e. 0.741 x $10^{-3}$ kg/$m^2$ over 48 hours, the reaction product being MgO.

Table 4.17 gives the composition of the main magnesium alloys used as cladding materials [40 — 43]. Magnox Al 80 is used in the Magnox reactors while magnox Zr 55 is used in the EDF reactors.

TABLE 4.17 Composition of some Major Magnesium Alloys (% in weight)

| *Constituent elements* | *Alloy type* | | | | |
|---|---|---|---|---|---|
| | *Mg — Al* | *Mg — Al — Ca* | *Mg — Al — Sn* | *Magnox Al 80* | *Magnox Zr 55* |
| Al | 7.7 | 5.2 | 5.1 | 0.8 | 0.003 |
| Be | - | - | - | 0.008 | - |
| Ca | < 0.01 | 0.20 | < 0.01 | 0.005 | < 0.003 |
| Ce | - | - | - | - | - |
| Cu | < 0.001 | < 0.001 | < 0.001 | 0.006 | 0.0006 |
| Fe | 0.004 | 0.003 | < 0.001 | 0.014 | < 0.003 |
| Mg | The remainder | The remainder | The remainder | The remainder | The remainder |
| Mn | < 0.01 | < 0.01 | < 0.01 | 0.004 | 0.003 |
| Ni | 0.001 | < 0.001 | 0.001 | 0.003 | 0.0005 |
| Pb | < 0.01 | < 0.01 | < 0.01 | - | - |
| Si | < 0.01 | < 0.01 | < 0.01 | < 0.005 | 0.0045 |
| Sn | < 0.01 | < 0.01 | 1.6 | - | - |
| Ti | - | - | - | - | - |
| V | - | - | - | - | - |
| Zn | < 0.02 | < 0.02 | < 0.02 | - | 0.002 |
| Zr | - | - | - | 0.008 | 0.55 |

## 4.7. OTHER STRUCTURAL MATERIALS

Besides the materials discussed so far, that are used on a large scale, other types of structural materials can also be employed. Among them, *beryllium* and *graphite* (with a small neutron absorption effective cross-section) can be used for fuel cladding in thermal neutron reactors (see Chapter 5). Metallic beryllium is not employed at industrial scale, because it is difficult to process, is not stable at high temperatures, and is subject to embrittlement caused by the helium atoms that result from the metal's reaction with neutrons. With the exception of the fuel elements for high-temperature reactors, graphite does not have major applications as structural material.

On the criterion of high temperature and corrosion resistance, the attention has been directed to nickel, as well as to the *refractory metals* such as niobium, vanadium, molybdenum, tantalum, etc. Of all these, only *nickel* was actually used, since it is relatively cheaper and features a very good corrosion resistance.

There are several nickel alloys (hastelloy, inconel) notable for their corrosion resistance in molten salts. In comparison with steels, but especially with Al, Mg, Zr, nickel alloys exhibit a high mechanical strength at high temperature. Although the behaviour of nickel and its alloys is satisfactory, their rather large capture cross-section for neutrons constitutes an obstacle to using them in thermal neutron reactors. The nickel alloys are the most suited for the molten salt-cooled, pressurized water reactors, for the enriched uranium reactors, and for the fast neutron reactors.

Among refractory metals, of special interest is *niobium*, due to its smaller capture cross-section, high melting temperature and good mechanical characteristics, although over long periods of operation embrittlement occurs.

In the reactor structure, especially of those with very high operating temperatures, *ceramic materials* such as oxides and carbides as well as the *cermet*-type materials may be used along with the metallic materials.

## 4.8. NEUTRON IRRADIATION-INDUCED CHANGES IN THE MECHANICAL PROPERTIES OF STRUCTURAL MATERIALS

The chief result of neutron irradiation on structural materials is their hardening. One notes also:

(a) a decrease in ductility, materialized in an increase in yield and fracture strength;

(b) a change in the ductile — fragile transition temperature, in the sense of material embrittlement at temperatures close to the operating ones (radiative embrittlement).

Hardening and embrittlement of materials under irradiation are due to interactions of the dislocations with the defects induced by irradiation. At irradiation temperatures below 0.35 $T_m$ ( $T_m$ is the material's melting temperature) the vacancies and interstitial atoms produced by irradiation agglomerate in Seeger areas and in small dislocation loops (see Chapter 2).

In order to assess the global effects of irradiation on the mechanical properties of materials, conventional tests are generally applied, namely:

(a) tensile strength tests,
(b) impact bending tests,
(c) tests to determine resilience under a plane strain state.

The fundamentals of these testing methods are extensively discussed in the specialist literature [44,45].

For the practical methodology of these tests, and in order to provide for comparability of the experimental results, ASTM standards are generally employed (Table 4.18).

TABLE 4.18 ANSI/ASTM Standards for Mechanical Testing

| *Standard* | *Method denomination* |
|---|---|
| ANSI/ASTM.E.6-81 | Standard Definitions of Terms Relating to *Methods of Mechanical Testing* |
| ANSI/ASTM.E.8-81 | Standard Methods of *Tension Testing of Metallic Materials* |
| ANSI/ASTM.E.21-79 | Standard Recommended Practice for *Elevated Temperature Tension Tests of Metallic Materials* |
| ANSI/ASTM.E.23-81 | Standard Methods for *Notched Bar Impact Testing of Metallic Materials* |
| ANSI/ASTM.E.399-81 | Standard Test Method for *Plane-Strain Fracture Toughness of Metallic Materials* |

The characterization, by means of optical and electronic microscopy before and after irradiation, of the microstructure of structural materials provides for developing models of radiative degradation of the mechanical characteristics. The correlation of the macroscopic features with the microscopic details opens the way to calculating the changes in the mechanical characteristics of a material, taking into account the microstructural alterations induced by irradiation. An important microstructural determination is the quantitative description, by means of scanning electron microscopy, of the material fracture before and after irradiation. At present, besides scanning electron microscopy, combined use is made of the Auger electron spectrometry and mass spectrometry. Table 4.19 gives an overview on the potential of the methods used in microstructural determinations for the assessment of neutron irradiation effects on structural materials. In order to carry out comparative studies regarding the mechanical characteristics and the microstructure of materials, the pertinent techniques and methods should be applied to both irradiated and non-irradiated samples.

TABLE 4.19 Techniques Employed in Microstructural Determinations for Testing Structural Material in Irradiation Conditions

| *Method* | *Kind of information* | *Relevance* |
|---|---|---|
| 1. Optical microscopy (metallography) | (a) Material microstructure (metallurgical state) | Major factors of radiative downgrading of mechanical properties |
| 2. Scanning electron microscopy (SEM) | (a) Irradiated surface imaging | Study of the degree of downgrading by measurements on the distribution, density and size of defects |
| | (b) Secondary electrons imaging of a ruptured or deformed zone | Detection of embrittlement degree from rupture's or deformation's aspect |
| 3. Electron probing | (a) Secondary electrons imaging of a ruptured or deformed zone | Same as 2(b) |
| | (b) X rays imaging of ruptured surface | Evidence on the segregation and precipitation effects produced by irradiation |
| 4. Auger electron spectroscopy | (a) Grain boundary composition | Evidence on the segregation and precipitation effects produced by irradiation |
| | (b) Composition near defects | |
| | (c) Composition near rupture | |

A correct assessment of the irradiation effects on structural materials requires both the determination of the mechanical characteristics altered by irradiation and a good knowledge of the reactor operation conditions. Studies regarding the role of the neutrons of various energies have proved that atomic displacements, that are chief causes of changes in the metal characteristics, are mostly produced by the high-energy neutrons.

If the neutron flux and fluency, the energy spectrum of the neutrons and the defect rate (the number of defects corresponding to various energy levels) are known, then the sensitivity of the mechanical characteristics to the irradiation conditions can be assessed.

The factors conditionning the accuracy in the determination of the fast neutron flux and fluency are:

(a) a good knowledge of the neutron spectrum;
(b) a proper selection of the neutron dosimeters;
(c) the knowledge of the irradiation "history".

Detection of neutrons using various metallic materials is based on their capacity to interact (metal activation), and on the accompanying emission of characteristic radiations. The main reactions employed to this effect are:

(a) (n,p), with gamma decay; e.g. $^{54}Fe(n,p)^{54}Mn$;
(b) (n,n'), with gamma decay and X emission; e.g. $^{115}In$ (n,n')$^{115m}In$;
(c) (n,f), with gamma decay; e.g. $^{235}U(n,f)^{99}Mo$;
(d) (n,α), with gamma decay; e.g. $^{63}Cu(n,\alpha)^{60}Co$.

The dosimeters based on such reactions are also known as threshold activation or fission detectors. Table 4.20 gives the characteristics of the main detectors that are employed in fast neutron dosimetry. The detectors come as sheets, wires or pellets.

TABLE 4.20 Elements used as Detectors for Fast Neutrons

| *Threshold energy* (MeV) | *Element* | *Nuclear reaction* | *Chemical species* | *Purity* (%) |
|---|---|---|---|---|
| 0.5 | Neptunium | $^{237}Np$ (n,f) $^{140}Ba$ | $Np_2O_3$ | 99.9 |
| 1.0 | Indium | $^{115}In$ (n,n') $^{115}In$ | In, In-Al | 99.99 |
| 1.45 | Uranium | $^{238}U$ (n,f) $^{140}Ba$ | U, U-Al, $UO_2$ | 99.9 |
| 1.75 | Thorium | $^{232}Th$ (n,f) $^{140}Ba$ | Th, $ThO_2$ | 99.95 |
| 2.1 | Titanium | $^{47}Ti$ (n,p) $^{47}Sc$ | Ti | 99.7 |
| 2.2 | Iron | $^{54}Fe$ (n,p) $^{54}Mn$ | Fe | 99.8 |
| 2.7 | Nickel | $^{60}Ni$ (n,p) $^{60}Co$ | Ni, Ni-Al | 99.9 |
| 2.9 | Nickel | $^{58}Ni$ (n,p) $^{58}Co$ | Ni, Ni-Al | 99.9 |
| 6.1 | Copper | $^{63}Cu$ (n,α) $^{60}Co$ | Cu, Cu-Al | 99.99 |
| 6.3 | Magnesium | $^{24}Mg$ (n,p) $^{24}Na$ | Mg, Mg-O | 99.7 |
| 6.8 | Titanium | $^{48}Ti$ (n,p) $^{48}Sc$ | Ti | 99.7 |
| 7.5 | Iron | $^{56}Fe$ (n,p) $^{56}Mn$ | Fe | 99.8 |
| 8.7 | Aluminium | $^{27}Al$ (n,α) $^{24}Na$ | Al, $Al_2O_3$ | 99.999 |
| 11.4 | Copper | $^{63}Cu$ (n,2n) $^{62}Cu$ | Cu, Cu-Al | 99.99 |
| 13.2 | Nickel | $^{58}Ni$ (n,2n) $^{57}Ni$ | Ni, Ni-Al | 99.9 |

In order to determine the spectrum, both calculations and experimental measurements are in order; to yield the neutron spectrum the results of these are fed into computer programs, such as, for instance, SAND-II [46].

The integrated flux, or fluency, is computed from activation data. A detector containing a known quantity of nuclei to be activated is subjected to the neutron flux; after removing it from the reactor, and following a certain interval of de-activation, the induced activity is determined by measuring the disintegration rate.

The simplest procedure for the flux calculation from the measured activity $A$, after an exposure $t$ of the dosimeter to a known spectrum, starts from the equation:

$$A = N_0\tilde{\sigma}\Phi(1-\exp(-\lambda t)), \qquad (4.37)$$

where $N_0\tilde{\sigma}\Phi$ is the saturation activity of the detector, $N_0$ is the total number of atoms contained in the detector before irradiation, $\tilde{\sigma}$ is the neutron activation effective cross-section spectrally weighed, $\lambda$ is the half-life of the resulting isotope, and $t$ is the duration of irradiation (corrected to account for reactor power variations).

The calculation of the flux and fluency from experimental data can be performed by means of the aforementioned deconvolution code SAND-II.

Table 4.21 gives the ASTM standards that establish the neutron flux measuring methodology as used in the USA [47].

TABLE 4.21 ANSI/ASTM Neutron Dosimetry Standards

| | |
|---|---|
| ASTM E.170-76 | Standard Definitions of Terms Relating to *Dosimetry* |
| ASTM E.181-76 | Standard Test Methods for *Analysis of Radioisotopes* |
| ASTM E.261-77 | Standard Practice for *Measuring Neutron Flux, Fluency and Spectra by Radio-activation Techniques* |
| ASTM E.262-77 | Standard Test Method for *Measuring Thermal Neutron Flux by Radioactivation Techniques* |
| ASTM E.263-77 | Standard Test Method for *Measuring Fast-Neutron Flux by Radioactivation of Iron* |
| ASTM E.264-77 | Standard Test Method for *Measuring Fast-Neutron Flux by Radioactivation of Nickel* |
| ASTM E.419-73 (rev. 1979) | Standard Guide for *Selection of Neutron Activation Detector Materials* |

REFERENCES


1. Tipton, Jr., C. R. (1960). In *Reactor Handbook*, vol. 1, Materials, Interscience Publ., New York; Dobrescu, L. (1960). *Tehnica reactoarelor nucleare*, Edit. Academiei, Bucharest.
2. Smith, C. O. (1967). *Nuclear Reactor Materials*, Addison Wesley Publ. Co., Reading (Mass.).
3. Mezer, R. M., and Morris, E. T. (1977). *J.Nuclear Materials*, 71, 36.
4. Kolkim, O. P. (1955). Sbornik Nauchnyh Trudov Moskovskovo Inst. *Tsvetnih Metal. y Zolota*, 26, 195.
5. Dalimarsky, J., and Kolatti, A. A. (1953). *Ukrain.Him, Jurnal*, 19, 372.
6. Boltax, A. (1964). In *Nuclear Engineering Fundamentals*, vol. 4, *Nuclear Materials*. Mc Graw-Hill Book Co., New York.
7. Fromont, J. P. and others (1979). *J.Nuclear Materials*, 80, 267.
8. Northwood, D. O., and Rosinger, H. E. (1980). *J.Nuclear Materials*, 89, 147.
9. Korobkov, L. L., and Ignatov, V. D. (1958). *Proc.Second.Conf.Geneva*, 5, 60.
10. Sarkisov, E. S., and Tchebotarev, N. T. (1958). *Atomnaya Energhya*, 5, 550.
11. Stehle, H. (1975). *Nuclear Eng. and Design*, 33, 155.
12. Hobson, D. O. (1972). Report ORNL-4758, USAEC, Washington.
13. Ballinger, R. G., and Dobson, W. G. (1976). *J.Nuclear Materials*, 62, 213.
14. Baker, L., and Just, L. C. (1962). Report ANL-6548, USAEC, Washington.
15. Rosinger, H. E., and Northwood, D. O. (1979). *J.Nuclear Materials*, 79, 170.
16. Maxim, I. V. (1969). *Materiale nucleare*. Edit. Academiei, Bucharest; Maxim, I. V., and Biscoveanu, I. (1974). *Materiale nucleare*. Edit. didactica si pedagogica, Bucharest.
17. Himusin, F. F. (1967). *Nerzhaveyustchie staly*. Metallurgya, Moskow.
18. Williams, C. D., and Gilbert, R. W. (1969). In *Proc.Symp.on Radiation Damage in Reactor Materials*, vol. 1, IAEA, Vienna.
19. Howe, L. W. (1960). *J.Nuclear Materials*, 2, 248.
20. Wood, D. S. (1966). *J.Electrochem.Tech.*, 4, 250.
21. Ross-Ross, P. A. (1968). *J.Nuclear Materials*, 26, 2.
22. Bement, A. L. (1965). In *Proc.AIME Symp.on Irradiation Effects, Asheville*. Gordon and Breach Science Publ., New York.
23. Makin, M. J. (1960). *Acta Metallurgica*, 8, 691.
24. Fleischer, R. L. (1962). *Acta Metallurgica*, 10, 835.
25. Debraza, A. J. (1980). *J.Nuclear Materials*, 88, 236.
26. Knorrand, D. B., and Pelloux, R. M. (1977). *J.Nuclear Materials*, 71, 1.
27. Hesketh, R. V. (1969). In *Proc.Symp.on Radiation Damage in Reactor Materials*, vol. 1, IAEA, Vienna.
28. Sawatzky, A. (1963). *J.Nuclear Materials*, 9, 364.
29. Sawatzky, A. (1960). *J.Nuclear Materials*, 2, 62.
30. Sawatzky, A., and Vogt, E. (1961). *Trans.Met.Soc.AIME*, 221, 819.
31. Sagues, A. A., and others (1978). *J.Nuclear Materials*, 78, 289.
32. Cawthorne, C. (1967). *Nature*, 216, 567.
33. Olander, D. R. (1976). *Fundamental Aspects of Nuclear Reactor Fuel Elements*. Nat.Techn.Inf.Service, US Dept.of Commerce, Springfield, TID-26711-P1.
34. Clement, C. F., and Wood, M. H. (1980). *J.Nuclear Materials*, 89, 1.
35. Murray, P. (1972). *Reactor Technol.*, 15, 16.
36. Druce, S. G. (1980). *Nuclear Technology*, 19, 347.
37. Cheadle, B. A. (1981). Report AECL-7564.
38. Hosbons, R. R. (1976). Report AECL-5325.
39. Steele, L. E. (1975). *Neutron Irradiation Embrittlement of Reactor Pressure Vessel Steels*. Technical Reports Series No.163, IAEA, Vienna.
40. Frisknev, C. A. (1981). *J.Nuclear Materials*, 99, 165.
41. Strucken, E. F. (1979). *J.Nuclear Materials*, 82, 39.
42. IAEA (1960). *Fuel Element Fabrication*, Proc.Symp.Vienna.
43. Gerasimov, V. V., and Monahov, A. S. (1982). *Materialy iadernoi tehniki*. Energoizdat, Moskow.
44. Dieter, Jr., G. E. (1970). *Metalurgie Mecanica*. Edit. Tehnica, Bucharest.
45. Cioclov, D.(1977). *Mecanica ruperii materialelor* . Edit.Academiei, Bucharest.

46. McElroy, W. N. and others (1967). *A Computer Automated Iterative Method for Neutron Flux Spectra Determination by Foil Activation*, AFWL-TR-67-41, vol. I-IV.
47. ANSI (1980). American Society for Testing Materials Nuclear Standards Part 45.

ADDENDUM

The study of the influence of neutron irradiation on the mechanical properties of materials requires both the analysis of the changes in properties and the establishment of certain metodologies for correlating these with the irradiation defects.

Irradiation-induced defects are generated by collisions of neutrons with lattice atoms. Such collisions could lead to the displacement of certain atoms from their lattice sites, and these, in turn, can generate so-called higher-order knock-outs (secondary, tertiary, etc.).

The irradiation-induced defects depend on the chemical and structural composition of the material, on temperature and on the energy of the primary neutrons.

During irradiation experiments only two integral quantities can be measured: the global level of damage $D$, correlated with the change in certain physical characteristics, and the activation $A$ of the flux detectors. The quantities $D$ and $A$ depend on the product of two energy functions. Ignoring their dependence on other parameters (material composition, irradiation time, irradiation temperature and physical characteristics) the functions $A$ and $D$ read:

$$A \sim \int_0^\infty \varphi(E)\sigma_{act}(E)dE \text{ , (disintegrations/s)} \quad \text{(A.1)}$$

$$D \sim \int_0^\infty \varphi(E)\sigma_d(E)dE \text{ , (displacements/s)} \quad \text{(A.2)}$$

where $\sigma_d(E)$ is the displacement cross-section, $\sigma_{act}(E)$ is the activation cross-section, $\varphi(E)$ is the neutron spectrum.

In eq. (A.1), only $\sigma_{act}(E)$, that corresponds to the fluency detectors, is relatively well known, and can be tabulated using, for instance, the nuclear data library ENDF/B [1,2].

The methodology for the experimental determination of the neutron flux, the fluency and the neutron spectrum, by activation techniques and the measurement of the quantities $A$ is described in detail in Ref. 3.

Eq. (A.2) represents the displacement rate of the lattice atoms and it introduces a new variable in the physics of materials, namely the number of displacements per atom (dpa), defined as:

$$\text{dpa/s} = \int_0^\infty \sigma_d(E)\Psi(E,t)dE. \quad \text{(A.2')}$$

The new quantity, dpa, is an exposure indicator, and its magnitude can be calculated by time integration of the displacement rate:

$$\text{dpa} = \int_0^{t_{irad}} \Phi_{tot}(t) \int_0^\infty \sigma_d(E)\Psi(E,t)dEdt, \quad \text{(A.3)}$$

where $\Phi_{tot}(t)$ is the time-dependent flux intensity, and $\Psi(E,t)$ is the normalized neutron spectrum.

This quantity is a material constant. The neutron spectrum in the reactor depends on the reactor type, on the irradiation position and, to a lesser extent, on the reactor loading. Consequently, the comparison of properties of some materials exposed to the same fluency becomes a difficult problem if the neutron spectrum and its time variation are not taken into account.

Assuming that the flux intensity and neutron spectrum are constant during the irradiation time $t_{irad}$, eq. (A.3) becomes:

$$\text{dpa} = \Phi_{tot} \cdot t_{irad} \int_0^\infty \sigma_d(E)\Psi(E)dE = \Phi_{tot} \cdot t_{irad} \cdot \overline{\sigma}_d , \qquad (A.4)$$

where $\overline{\sigma}_d$ represents the displacement cross-section averaged over the neutron spectrum. On this simplifying assumption, implying that $\Phi_{tot} \cdot t_{irad}$ and the $E$ spectrum can be measured, the determination of dpa depends only on $\overline{\sigma}_d$.

The displacement cross-sections can be determined in experiments using charged particles, accelerated at various energies [4].

The methodology to characterizing via the dpa the neutron exposure of ferritic steels employed in the manufacture of pressure vessels is described in Ref. 5.

Generally, the establishement for a given material of a correspondence between dpa and a certain physical or mechanical property is a difficult problem, now under thorough theoretical and experimental investigation. The establishment of correlations between changes in certain properties of the same material irradiated with various neutron spectra is possible, by using the corresponding dpa values of these spectra.

REFERENCES

1. Magurno, B. A. (1975). ENDF/B-IV, Dosimetry Files, BNN-NCS-50446 (ENDF-216), Brookhaven National Laboratory.
2. Magurno, B. A., and Ozer, O. (1975). ENDF/B, File for Dosimetry Applications, *Nuclear Technology*, vol. 25, pp. 376-380.
3. ANSI/ASTM, E 261-77 (1980). Standard Method for Determining Neutron Flux, Fluence and Spectra by Radioactivation Techniques, *Annual Book of ASTM Standards*, Part 45.
4. ANSI/ASTM, E 521-79 (1980). Recommended Practice for Neutron Irradiation Damage Simulation by Charged-Particle Irradiation, *Annual Book of ASTM Standards*, Part 45.
5. ANSI/ASTM, E 693-79 (1980). Standard Practice for Characterizing Neutron Exposures in Ferritic Steels in Terms of Displacements per Atom (DPA), *Annual Book of ASTM Standards*, Part 45.

CHAPTER 5

# Moderator Materials

## 5.1. GENERAL

The process of downscaling the energy of the fission-generated neutrons, from an average 2 MeV to values of (0.025 — 1) eV — which correspond to the thermodynamic equilibrium with the environment in the reactor and favours the $^{235}U$ fission in thermal reactors, is known as *neutron moderation*, and the materials in the reactor core that produce this effect are, accordingly, called *moderator materials*.

Moderation is performed by repeated elastic collisions of neutrons with the nuclei of the moderator material. At each collision a neutron transfers part of its energy to the moderator nucleus. The smaller the mass of the target nucleus, the larger the transferred energy, that reaches its maximum when the moderator nucleus has the same mass as the neutron's, which is only the case with hydrogen.

Consequently, a good moderator must have a small atomic mass, $A$, a large microscopic scattering cross-section $\sigma_s$, and a small microscopic absorption cross-section $\sigma$ for the neutrons resulting from fission. From a practical point of view, it is more advantageous to characterize the moderators by means of their macroscopical scattering and absorption cross-sections $\Sigma_s$ and $\Sigma_a$, as defined in Chapter 1. As has been shown, these cross-sections represent the product between the corresponding microscopic cross-sections and the number $N$ of atoms in the unit-volume:

$$\Sigma = \sigma N, \tag{5.1}$$

where

$$N = (N_A \rho)/A, \tag{5.2}$$

$N_A$ is the Avogadro number, and $\rho$ is the material's density. Table 5.1 displays magnitudes of the cross-section, for several elements and compounds [1-4].

TABLE 5.1 Neutron Cross-Sections

| *Element (compound)* | *Atomic (molecular) mass* | $\sigma_a \cdot 10^{28}$ ($m^2$) | $\sigma_s \cdot 10^{28}$ ($m^2$) | $\Sigma_a$ ($m^{-1}$) | $\Sigma_s$ ($m^{-1}$) |
|---|---|---|---|---|---|
| H | 1.008 | 0.330 | 38 | 0.18* | 11* |
| D | 2.015 | 0.00046 | 7.0 | 0.00025* | 1.8* |
| Be | 9.012 | 0.0095 | 7.0 | 0.1174 | 86.52 |
| C | 12.011 | 0.0034 | 4.8 | 0.02728 | 38.51 |
| O | 16.000 | 0.00028 | 4.2 | 0.00001* | 2.1* |
| Na | 22.990 | 0.505 | 4.0 | 1.2 | 7.6 |
| $H_2O$ | 18.016 | 0.664 | 103 | 2.220 | 344.3 |
| $D_2O$ | 20.030 | 0.003 | 13.6 | 0.0085 | 45.0 |
| $UO_2$ | 270.0 | 7.7 | 16.7 | 17.0 | 37.2 |

*For the gases $H_2$, $D_2$ and $O_2$ an arbitrary density 100 times the density at NTP is assumed [16].

To characterize the moderating capacity of various materials, certain specific quantities were introduced:

(a) The *mean logarithmic decrement*, $\xi$ , which describes the mean energy loss of a neutron, following its collision with a nucleus of the moderator, is defined by the expression:

$$\xi = \overline{\ln(E_1/E_2)}, \tag{5.3}$$

$E_1$ and $E_2$ are the neutron energies before and after collision, respectively. One may prove that the energy loss at collision is independent of the neutron's initial energy, but depends upon the moderator's atomic mass, $A$ , and the scattering angle $\vartheta$ [4]:

$$E_2/E_1 = \frac{1}{2}\,[(1+R) + (1-R)\cos\vartheta], \tag{5.4}$$

where

$$R = [(A-1)/(A+1)]^2 \tag{5.5}$$

By averaging $\ln(E_1/E_2)$ — where $E_1/E_2$ is given by eq. (5.4) — over all possible values of the angle $\vartheta$ , one obtains the magnitude of the mean logarithmic decrement, as a function of only the moderator's atomic mass:

$$\xi = 1 + R\ln R/(1-R). \tag{5.6}$$

For materials with $A > 10$, the above equation may be approximated by the expression:

$$\xi = 2/(A+2/3), \tag{5.7}$$

the error thus introduced being smaller than 1%.

For a substance consisting of several varieties of nuclei (1, 2,..., $n$ ), the mean value of $\xi$ is obtained with the equation:

$$\bar{\xi} = (\xi_1\sigma_{s1} + \xi_2\sigma_{s2} + \ldots + \xi_n\sigma_{sn})/(\sigma_{s1} + \sigma_{s2} + \ldots + \sigma_{sn}). \tag{5.8}$$

The mean logarithmic decrement may be used to calculate the mean collision number (MCN), that is required for a neutron resulting from fission ($E_1 = 2 \times 10^6$ eV) to attain the energy level of the thermal neutrons ($E_2 = 0.025$ eV):

$$\text{MCN} = \ln(E_1/E_2)/\xi = 18.2/\xi \tag{5.9}$$

(b) The *moderating power*, MP; this quantity not only describes the material capacity to dissipate the energy of fast neutrons, but also takes into account the scattering cross-section of its nuclei. Moderating power is defined by the expression:

$$\text{MP} = \xi\Sigma_s. \tag{5.10}$$

(c) The *moderating ratio*, MR, enhances the information given by the moderating power, also taking into account the absorption cross-section of the moderator:

$$\text{MR} = \xi\Sigma_s/\Sigma_a. \tag{5.11}$$

(d) The *moderating density*, MD, represents the number of neutrons per unit-volume and time, of the same energy $E$.

Were there no neutron absorption and consequently no losses, then MD would be equal with the number of neutrons resulting from fission, and, consequently, independent on their energy. In effect, the moderation process involves two distinct stages — the neutron slow-down, where the neutron distribution is inversely proportional to their energy, and the thermalization, where the energy distribution of the neutrons is a Maxwell — Boltzmann one. While in the first stage the moderating process can be described by using the classical neutron — nucleus interaction, in the second stage one has to use an interaction model based on quantum mechanics [5-9].

TABLE 5.2 Characteristics of some Moderators

| *Moderator* | $\xi$ | *MP* | *MR* | *MCN* |
|---|---|---|---|---|
| Light water | 0.95 | 1.28 | 58 | 19 |
| Heavy water | 0.51 | 0.18 | 5000 | 36 |
| Graphite | 0.1589 | 0.065 | 200 | 114 |
| Beryllium | 0.2078 | 0.16 | 130 | 86 |

Table 5.2 shows some of the characteristic parameters of the chief moderator materials. One may notice that heavy water has the best moderating ratio, whereas light water, that has the greatest moderating power, provides for smaller sizes of the reactors that use it as a moderator [4].

Besides the essential nuclear criteria reviewed above, in selecting a moderator other material characteristics should also be ensured, such as: a good thermal and irradiation stability; chemical compatibility with the structural materials and cooling agents; a high thermal conductivity; convenient manufacturing at low costs.

In thermal reactors the moderator materials that have a high scattering cross-section may also be used as neutron reflectors. Both light and heavy water fulfil this condition. In fast reactors the role of reflector may only be played by heavy-weight materials, such as iron, lead, bismuth, thorium, uranium, etc.

The following paragraphs give a more detailed description of the main moderators used in nuclear reactors.

## 5.2. NUCLEAR GRAPHITE

Graphite was the moderator in the first nuclear reactor, built by E. Fermi in 1942. A family of nuclear reactors, moderated with graphite and cooled with light water, has been developed in the USSR; the first nuclear power station in the world, Obninsk, of 5 MWe, was of this type. Graphite is still in use with the GCR and AGR reactors, and will probably have a part to perform with the forthcoming HTGR reactors. For the GCR reactors, the fuel is in the form of metallic natural uranium, and for the AGR and HTGR ones it comes as enriched uranium. For HTGR reactors, a variety of nuclear fuel in the form of small, graphite-coated grains of fissionable material is being developed.

As shown in Table 5.2, the graphite has a low moderating power, and, therefore, the average number of collisions that the high-energy neutrons must undergo to get thermalized is high. This is why the graphite-moderated reactors are large, require large quantities of moderator and, besides, in order to reach criticality, they cannot use natural uranium in the form of $UO_2$, as a fuel.

### 5.2.1. Properties

Table 5.3 shows some of the physical properties of the nuclear graphite [10-12]. At ambient pressure, graphite sublimates at 3923 K, a liquid phase being only possible at pressures higher than 10 MPa.

TABLE 5.3 Physical Properties of Nuclear Graphite

|  |  |
|---|---|
| Theoretical density (kg/m$^3$) | $2.27 \cdot 10^3$ |
| Crystalline structure | hexagonal<br>$a_0 = 2.456 \cdot 10^{-10}$ m<br>$c_0 = 6.697 \cdot 10^{-10}$ m |
| Sublimation temperature (K) | $3923 \pm 25$ |
| Linear expansion coefficient (K$^{-1}$) | $1.5 - 1.7 \cdot 10^{-6}$ |
| Thermal conductivity (W/(m . K)) | 146.5 |
| Electrical resistivity ($\Omega$ . m) | $8 \cdot 10^{-6}$ |

Owing to its structure, which is markedly anisotropic, the mechanical properties of the graphite are largely sensitive to the fabrication procedures. Table 5.4 exhibits some features of the graphite which was used for Magnox reactors, AGR and HTGR [13,14]. The values of these quantities are given in the table, for the parallel and perpendicular direction to the extrusion or moulding directions. Superior characteristics are obtained in the extrusion direction.

TABLE 5.4 Thermal and Mechanical Properties of the Graphite Used in Various Reactor Types

| *Reactor type* | *Magnox* | *AGR* | *HTGR* |
|---|---|---|---|
| Density $10^3$ . (kg/$m^3$) | 1.7 | 1.82 | 1.78 |
| Elasticity modulus $10^4$ (MPa) | | | |
| parallel | 1.35 | 1.0 | 1.6 |
| transversal | 0.74 | 1.0 | 0.7 |
| Tensile strength (MPa) | | | |
| parallel | 17.0 | 17.5 | 14.2 |
| transversal | 7.5 | 17.5 | - |
| Compression strength (MPa) | | | |
| parallel | 32.0 | 70 | 28.6 |
| transversal | - | 70 | 27.7 |
| Bending strength (MPa) | | | |
| parallel | 16.5 | 23 | 17.8 |
| transversal | 11.5 | 23 | 10.9 |
| Expansion coefficient ($10^{-6}$/K) | | | |
| parallel | 0.90 | 4.6 | 0.50 |
| transversal | 2.90 | 4.6 | 2.40 |
| Thermal conductivity (W/(m . K)) | | | |
| parallel | 209 | 134 | 189 |
| transversal | 126 | 134 | - |
| Resistivity $10^{-6}$ (Ω . m) | 6.0 | 9.0 | 6.7 |

One specific property of graphite is that the mechanical strength increases with temperature up to ~ 2800 K. Figure 5.1 shows its variation with temperature for a graphite of density of 1.67 x $10^3$ kg/$m^3$, up to the sublimation temperature. The compressive strength varies similarly.

Carbon used as moderator should be very pure. It should not contain elements with high neutron absorption cross-section such as boron (which is commonly found as traces in natural carbon) and volatile substances (such as hydrogen and hydrocarbons), which may cause ignition or explosion of the material at higher temperatures. Table 5.5 shows the neutron capture cross-section for graphite, when the impurity content is $10^{-4}$% in weight [15].

The impurity content of the graphite of reactor quality is limited at $10^{-2}$%. These can be only elements with small neutron absorption cross-section, such as Si, Ca, etc., so that the increment in the graphite absorption cross-section would not exceed 10%. Table 5.6 lists the main impurities that may be found in graphite of nuclear quality [15].

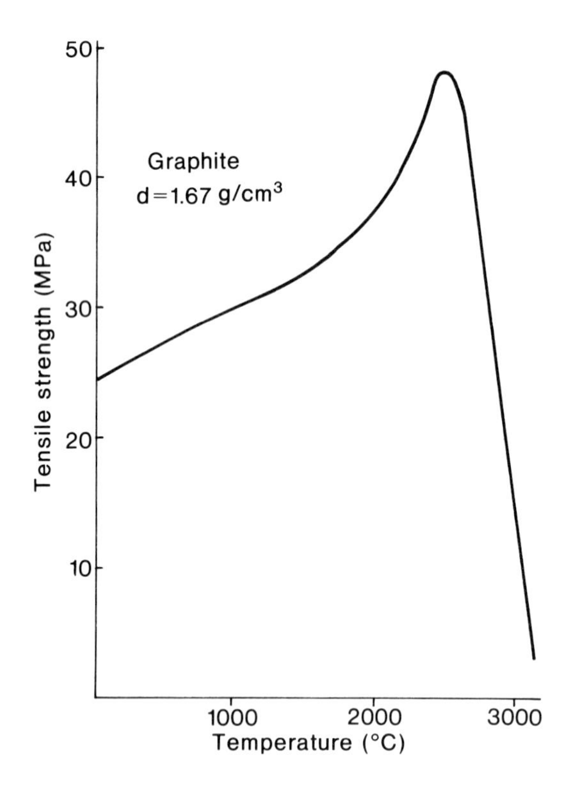


Fig. 5.1. Temperature effect on graphite tensile strength.

TABLE 5.5 Modification of the Neutron Absorption Cross-Sections of Graphite due to Impurities

| *Natural impurity* | *Thermal neutrons absorption cross-section of impurities* ($10^{-28}$ $m^2$) | *Percentage increment of* $\sigma_a$ *of the graphite, due to an impurity content of* $10^{-4}$% |
|---|---|---|
| $_{64}Gd$ | 46,000 | 103.3 |
| $_{5}B$ | 755 | 24.7 |
| $_{52}Sm$ | 5,600 | 13.16 |
| $_{63}Eu$ | 4,300 | 9.99 |
| $_{48}Cd$ | 2,450 | 7.70 |
| $_{3}Li$ | 71 | 3.61 |
| $_{66}Dy$ | 950 | 2.06 |
| $_{77}Ir$ | 440 | 0.81 |
| $_{80}Hg$ | 380 | 0.67 |
| $_{49}In$ | 191 | 0.59 |
| $_{17}Cl$ | 33.8 | 0.34 |
| $^{1}_{1}H$ | 0.332 | 0.116 |

TABLE 5.6 Chief Impurities in Nuclear Graphite

| *Element* | *Concentration* ($10^{-4}$% in weight) | *Increment in the nuclear graphite absorption cross-section* ($10^{-31}$ $m^2$) |
|---|---|---|
| B | 0.2 – 1.0 | 0.17 – 0.84 |
| V | 50 – 200 | 0.05 – 0.2 |
| Cl | 3 – 30 | 0.03 – 0.3 |
| Ti | 20 – 100 | 0.03 – 0.13 |
| H | 10 – 40 | 0.03 – 0.14 |
| Ca | 50 – 500 | 0.01 – 0.07 |
| Si | 50 – 200 | 0.01 – 0.05 |
| Rare earths | 0.05 – 0.5 | 0.01 – 0.02 |

The quality control of nuclear graphite requires that the total porosity should be about 25%, and the permeability to gases of (1-3) . $10^{-4} m^2$ . $s^{-1}$. The graphite surface (measured by gas adsorption) is of several square meters per gram [15].

TABLE 5.7 Contribution of the Air Included in Pores to the Increment in the Graphite's Capture Cross-Section

| *Graphite apparent density* ($10^3$ kg/ $m^3$) | *1.5* | *1.6* | *1.7* | *1.8* |
|---|---|---|---|---|
| Open porosity (%) | 33.0 | 29.0 | 22.0 | 20.0 |
| Increment in capture cross-section caused by air ($10^{-31}$ $m^2$) | 0.3 | 0.24 | 0.19 | 0.14 |

Owing to its porosity and large specific surface the graphite may store large quantities of gas, either inside the pores, or adsorbed on the surface, which causes an increase in the neutron capture cross-section. Table 5.7 gives the relation between the apparent density of graphite and the increment in the neutron capture cross-section caused by air [15].

### 5.2.2. Corrosion Resistance

Although graphite does not react with most of the substances, it may, however, oxidize and form compounds at higher temperatures, in certain gaseous, liquid or solid media. Table 5.8 gives an overview on how graphite reacts to some corrosion agents [16].

Graphite-moderated power reactors are usually cooled with pressurized carbon dioxide. The following reaction may take place in the reactor:

$$CO_2 + C \rightarrow 2CO \qquad (5.12)$$

TABLE 5.8 Graphite Reactions with Various Corrosion Agents

| *Agent* | *Temperature* (K) | *Effects or reaction products* |
|---|---|---|
| Gases | | |
| oxygen | > 720 | Carbon oxides |
| hydrogen | 1170-1270 | Methane (in the presence of catalysts) |
| | > 2200 | Traces of acetylene |
| fluorine | 300 | Interlamellate compounds |
| | 1170-2170 | Carbon fluorides |
| nitrogen | ≦ 3200 | No reaction products |
| water vapours | ≦ 1100 | Negligible reaction |
| | ≧ 1100 | Reaction rate raises steeply |
| carbon dioxide | ≦ 1100 | Negligible reaction |
| | ≧ 1100 | Carbon oxide |
| Aqueous solutions | | |
| dilute acids or bases | < boiling point | No reaction products |
| concentrated acids | > 570 | Graphite oxidation |
| KOH (50%) | > 620 | Graphite dissolution |
| Solids | | |
| alkali hydroxides | = melting point | No reaction products |
| metals | ≧ 1800 | Metallic carbides |
| metallic oxides | ≧ 1800 | Metallic carbides and carbon oxides |

A graphite consumption of 1% per year results from this reaction. Investigations with $^{14}C$ on graphite have pointed out that the corrosion process takes place in the pores of the graphite and that the corrosion rate increases with the dilation of pores [17]. The corrosion rate decreases significantly when methane gas is added; a concentration of 0.1% $CH_4$ in $CO_2$ diminishes the corrosion rate by about 40 times.

### 5.2.3. Irradiation Effects

The main effects of irradiation on graphite are: changes in geometric dimensions; energy storage; enhanced corrosion; changes in electric resistivity and in thermal conductivity.

*Variation of geometric dimensions* following irradiation is explained by the displacement of the carbon atoms from the crystal lattice sites, caused by the fast and intermediate neutrons. The amount of carbon atoms displaced at the moderation of a 2 MeV high-energy neutron is estimated at about 20,000 [18]. This determines structure modifications (distortions of the elementary cell), modification of macroscopic dimensions, of porosity and of other physical properties.

By irradiation, graphite atoms in the basal planes are displaced to form clusters of interstitial atoms in between the planes that are perpendicular to the $c$-axis. This determines an increase of the lattice parameter in the $c$ direction, and, accordingly, its shrinking in the $a$ direction. Experimental results (see Fig. 5.2) support this interpretation [19]. The values of the $c$ parameter grow proportional to the neutron fluency. The relative variation $\Delta c/c$ also depends on temperature. The relative modification of the $a$ parameter, $\Delta a/a$, is about 1/10th of the absolute value of $\Delta c/c$. The magnitude of $\Delta a/a$ does not seem too much sensitive to the irradiation temperature.

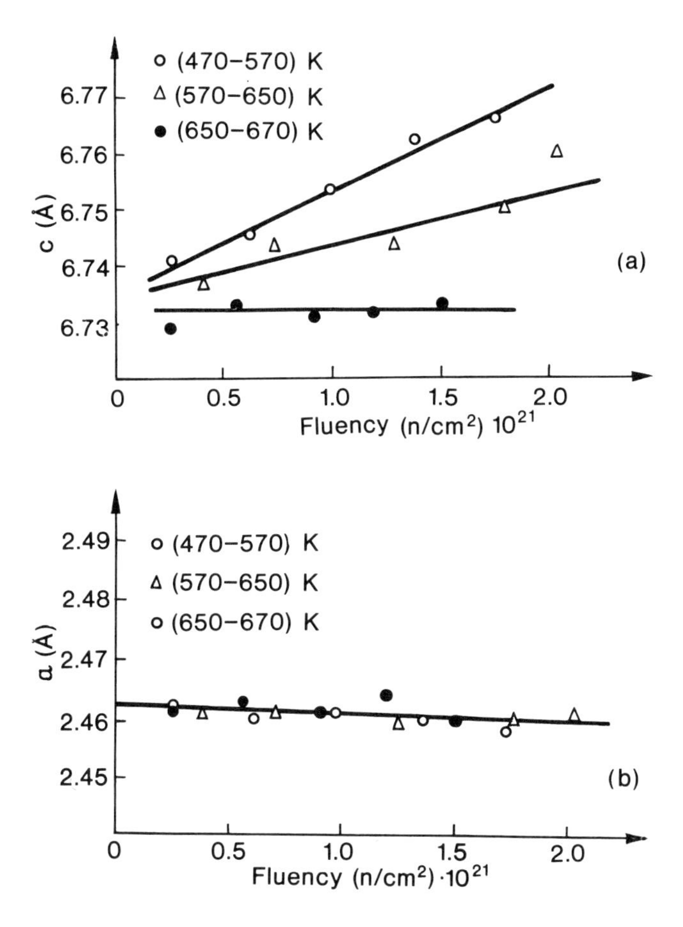


Figure 5.2. Irradiation effects on graphite lattice parameters: (a) on the *c*-axis; (b) on the *a*-axis.

Graphite thermal expansion coefficients at $T < 670$ K are positive for the *c* parameter and negative for the *a* parameter. Experimental results indicate a linear relationship between neutron irradiation dose and the expansion coefficient in the *c* and *a* directions, respectively. As a result, the dimensional variation after irradiation may also be determined by a measurement of the thermal expansion coefficient. As a rule, irradiation effects upon graphite are estimated through the modifications in its electric resistivity.

Figure 5.3 displays relative length modifications of various graphite samples that are used as moderators in the Calder Hall Reactor (England), as a function of temperature, fluency and distance from the bottom of the reactor core. At high fluencies, that correspond to the F-2 curve, the material expands up to 0.03%, and then, with the irradiation going on and the temperature raising, it shrinks. At low fluencies, as described by the F-1 curve, the relative length change, $\Delta l/l$, does not depend on either temperature or fluency [19].

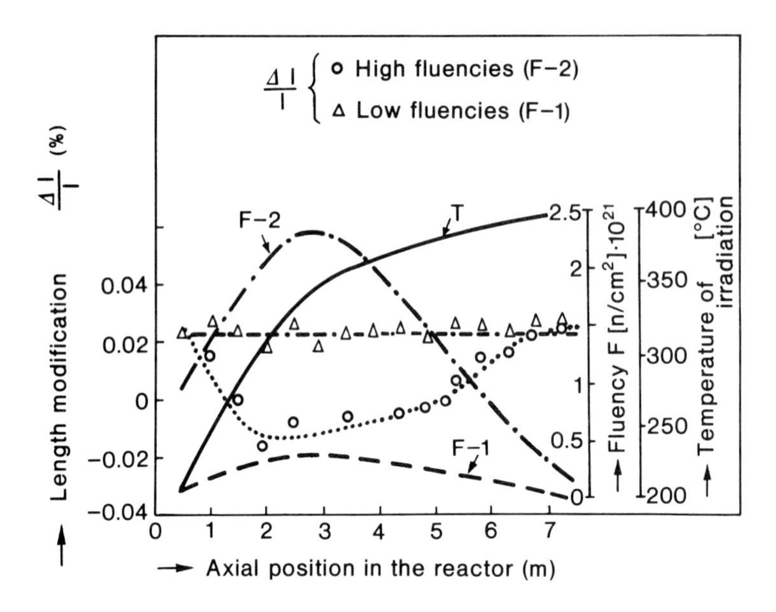


Fig. 5.3. Length modification of some graphite samples as function of irradiation fluency and temperature.

Change in dimensions of the KC, CSF and TSGBT graphites in the directions perpendicular and parallel to extrusion axis, as a function of the burn-up (at an irradiation temperature of 303 K) points out that the TSGBF variety is superior in behaviour, showing only reduced dimensional changes [18,20].

Most of the graphite varieties that have been selected for employment in the high-temperature reactors (HTR) are of gilso-carbon type, prepared from high-density natural bitumen (1030 – 1100 kg/m$^3$). A spherical arrangement of the graphite layers leads to a better dimensional stability at irradiation due to the material's isotropic behaviour. The irradiation behaviour of this particular variety of graphite differs from the behaviour of the previously introduced graphite types, as can be seen in Fig. 5.4. At 600 $^0$C the shrinkage of the graphite in question depends only weakly upon the neutron flux, whereas at 1150 – 1225 $^0$C a strong dependence upon the neutron flux and the initial elasticity modulus is recorded. As shown in Fig. 5.4, for an operation temperature of about 1500 K a graphite having an initial elasticity modulus of (7.4-9.0)x10$^9$ N . m$^{-2}$ should be preferred [21].

*Energy storage* is one of graphite's interesting properties. Stored energy, also called Wigner energy, is due to a rise in the internal energy of the crystalline lattice as a result of atoms' displacement in metastable interstitial positions under the action of the neutron flux. When the atoms return to their normal positions, their potential energy may be released as heat. At normal temperatures the recovery is slow. The process is accelerated by heating the graphite.

The energy stored in graphite depends upon the irradiation temperature and the flux of fast neutrons. Wigner energy is insignificant over 750 K, and hence one may conclude that the release of this energy represents a problem only in reactors operating at relatively low temperatures (~ 500 K).

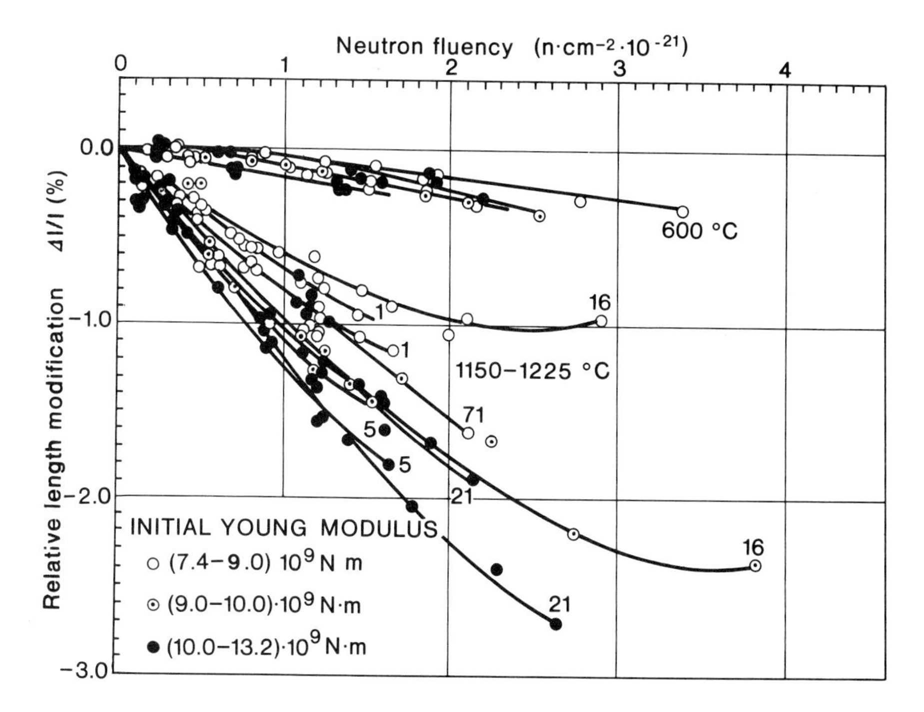


Fig. 5.4. Dimensional modifications of graphite used in high-temperature reactors, at 600 °C and 1150 – 1225 °C temperatures, respectively.

Figure 5.5 shows the temperature dependence of the rate of release of stored energy for a graphite irradiation at 300 K, and various irradiation durations (expressed by the fuel burn-up). By heating the sample, the stored energy is released all over a certain temperature interval; the longer the duration of irradiation, the larger this interval. For low fluencies, a recombination of defects takes place at relatively low temperatures ($T$ < 500 K), where the energy release rate is higher. Short time irradiations at low temperatures generate defects that may easily recombine. Thus, when the moderator temperature is low, frequent heating is in order, with the aim of releasing the stored energy. The accumulation, in time, of quite a large amount of energy may lead to a rise in the moderator and fuel elements temperature; the latter may even melt down. Such an accident occurred in 1957 at the Windscale reactor (UK) [22]. Techniques of controlled release of the energy stored in graphite, as applied to the French reactors that use this moderator, are described by Savineau [23].

### 5.2.4. Preparation of Nuclear Graphite

The raw materials for the preparation of nuclear graphite are oil coke, the tar resulting from brown coal processing, and carbon black. Special care is exercised in the selection of raw materials, so that these should not contain impurities with high neutron absorption cross-section (see Tables 5.5 and 5.6). The processing aims at the removal of volatile substances ($H_2$, hydrocarbons) and at the complete conversion to graphite of the entire sample.

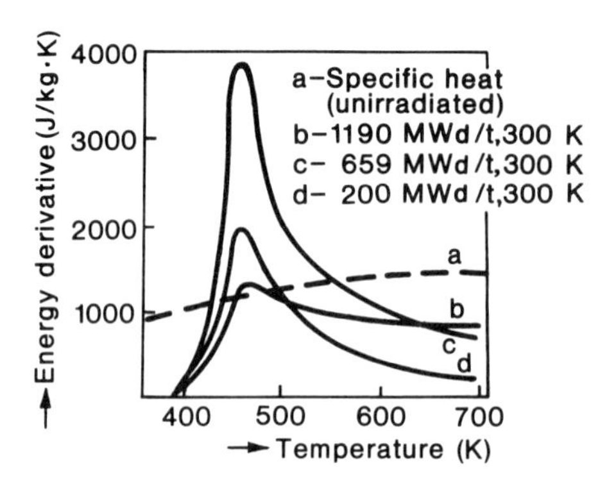


Fig. 5.5. Stored energy derivative with respect to temperature.

The main steps in the graphite preparation process are [1,18,24]:
(a) Paste preparation. Oil coke is calcinated at 1600 K to remove the volatile products and to form the crystalline lattice. After grinding and sorting out a granulation of $10^{-3}$ — $10^{-6}$ m, coke is mixed with tar and lubricating oil at a temperature of 438 K, with an admixture of 10% carbon black. The density of the mixture is thus raised to $1.67 \times 10^3$ kg/m$^3$ [18]. The paste is then extruded at 375 K, in pieces of various shapes, then heated up at 1100 — 1300 K in a coal powder bed. To avoid the occurrence of internal stresses, the heating and cooling are performed slowly, the cycle duration lasting up to 3 weeks [24].

(b) Impregnation. The calcinated material is very porous and, to increase its density, an impregnation operation is in order, at 525 K and 0.7 MPa; the result is subjected to a new annealing. The impregnation — annealing cycle is repeated until the desired density is ultimately obtained. The bigger the density, the better the quality of the graphite.

(c) Graphitation. To obtain the graphite's specific structure, the impregnated material is heated up to temperatures of 2900 — 3300 K, in special electric ovens. The density of the material is thus brought to $1.8 \times 10^3$ kg/m$^3$.

(d) Purification. This operation aims at reducing the amount of impurities down to values of the total absorption cross-section that are close to that of the graphite's. In the final stage this operation may be performed thermically or chemically. Thermal purification consists in heating the graphite up to 3100 — 3300 K, for 2 days. As a result of the high temperatures most of the impurities evaporate and are thus removed from the graphite. However, this operation cannot remove boron and vanadium carbide. The chemical purification takes place at 2800 K. Gases like $Cl_2$, freon 12, $SF_6$, $CCl_4$ are then employed, which react with the impurities that are removed as volatile halogenides.

For the graphite used in the HTGR reactors, the required density is higher — namely $(2.0\text{-}2.1) \times 10^3$ kg/m$^3$. To obtain such a higher density, raw materials with superior properties are used. The proportion of carbon black grows to 30% and the number of impregnations for filling the pores is larger. To increase the corrosion resistance, a good impregnation is necessary, in order to fill the pores and to break the links between them. It is considered that, at each annealing of the material, pores volume diminish by ~ 50%.

The graphite preparation procedure is laborious and requires a high consumption of electric energy. The graphitation electric oven, for example, employs currents of the order of an ampere/cm$^2$ and the heating lasts for 3 — 4 days. Considering the low level of impurities admitted in the final product (see Table 5.6) the raw materials for the preparation of nuclear graphite must be extremly pure, and their handling raises special problems.

## 5.3. NATURAL (LIGHT) WATER

The convenient hydrodynamic and thermodynamic properties of natural water have prompted its utilization in nuclear power generation as a moderator, cooling agent and shielding material. Water has a great moderating power (see Table 5.2) and, consequently, the core of reactors that use light water (LWR) is smaller. The water is also a cheap cooling agent, with good thermal properties both in the pressurized water reactors (BWRs) and in the boiling water reactors (PWRs). The use of water as a nuclear material is, however, limited by its comparatively low critical temperature (647 K) and also by a high absorption cross-section for thermal neutrons. The first of these constraints makes it impossible to use water in high-temperature reactors, whereas the latter makes essential the use of enriched uranium in order to reach criticality in the LWRs.

In the following sections the chief physical and technological aspects related to the use of water in nuclear reactors are reviewed.

### 5.3.1. Water Activation

Water is a subtle alliance of hydrogen and oxygen isotopes. Nuclear reactions leading to radioisotope formation take place under the influence of high energy neutrons, photons, and $\gamma$-radiation. The main activation products are due to the oxygen isotopes. Data concerning the nuclear reactions and the activation products are found in Table 5.9. Among these reactions the most important are: $^{16}O(n,p)^{16}N$ ($^{16}O$ is found in significant amounts, and the $\beta^-$, $\gamma$-radiations emitted by $^{16}N$ are relatively strong) and $^{18}O(n,\gamma)^{19}O$ (the neutron capture cross-section is important, and the $^{18}O$ proportion is by no means negligible). In both cases the half-life is short and, consequently, the saturation activity is quickly attained: its magnitude is 3.8 Ci/m$^3$ for $^{19}O$ and 0.4 Ci/m$^3$ for $^{16}N$, at a flux of $10^{13}$ neutrons/cm$^2$ . s. The resulting activity is relatively high and, consequently, it is necessary for water to follow long enough routes before coming into contact with the operating personnel. Metallic joints are in order throughout, since the plastic ones are destroyed by the active oxygen.

### 5.3.2. Activation of Impurities

Impurities that are either ordinarily in water, or occur as a result of corrosion processes (such as iron, aluminium, sodium, etc.) are strongly activated by neutrons. When found in water even in very small amounts, the product of the aluminium activation — an isotope with a short half-life, of 120 s — contributes strongly to the moderator's activation. However, a shut-down of only a few hours is enough for the water activity to disappear completely. As for iron, the situation is a bit more difficult, for the irradiation induces the occurrence of the radioactive isotopes $^{59}Fe$, with $T_{1/2}$ = 4x10$^6$ s, and $^{60}Co$, with $T_{1/2}$ = 1.67x10$^8$ s, which allows the water activity to decrease sensibly only over some weeks after irradiation has ceased. In pressurized water reactors, where special steels are used, the impurities that are important as far as activation is concerned are tantalum, cobalt, sodium and manganese.

TABLE 5.9 Nuclear Reactions in Light and Heavy Water during Irradiation

| *Isotope* | *Natural occurrence* (%) | *Nuclear reaction* | $T_{1/2}$ (s) | *Radioactivity* (MeV) | | |
|---|---|---|---|---|---|---|
| | | | | β *radiation* | γ *radiation* | *Neutrons* |
| $^{16}O$ | 99.76 | $^{16}O(n,p)^{16}N$ | 7.43 | 3.32;4.39;10.4 | 6.13;7.10 | - |
| $^{16}O$ | 99.76 | $^{16}O(n,\alpha)^{13}N$ | 600.0 | 1.2 ($\beta^+$) | - | - |
| $^{17}O$ | 0.0374 | $^{17}O(n,p)^{17}N$ | 4.14 | 3.7 | - | 0.92 |
| $^{17}O$ | 0.0374 | $^{17}O(n,\alpha)^{14}C$ | $18 \times 10^{10}$ | 0.155 | - | - |
| $^{18}O$ | 0.204 | $^{18}O(p,n)^{18}F$ | 6840.0 | 0.65 | - | - |
| $^{18}O$ | 0.204 | $^{18}O(n,\gamma)^{19}O$ | 29.43 | 4.5 | 0.20; 1.36 | - |
| $^{1}H$ | 99.985 | $^{1}H(n,\gamma)\ ^{2}H$ | - | - | - | - |
| $^{2}H$ | 0.015 | $^{2}H(n,\gamma)\ ^{3}H$ | $39 \times 10^{6}$ | 0.018 | - | - |
| $^{2}H$ | 0.015 | $^{2}H(\gamma,n)\ ^{1}H$ | - | - | - | - |

Apart from radioactivity, the impurities contribute to the consumption of thermal neutrons in the reactor core, and to corrosion acceleration.

### 5.3.3. Water Radiolysis

As a result of water irradiation with fast neutrons and γ radiation, $H_2$ and $H_2O_2$ molecules occur. This phenomenon, called water radiolysis, has been a subject of enduring investigations aiming at elucidating the reaction mechanism [25-28].

Water radiolysis involves three distinct stages:

(a) In the *physical stage* (of a duration inferior to $10^{-15}$ s) primary excitations and ionizations of water molecules are produced through Frank – Condon transitions induced by the incident particles.

(b) In the *physicochemical stage* (that lasts $10^{-13}$ s or more), the energy absorbed is redistributed over the other molecules, either stable or excited, and ions, by sheer interaction, breaking up of polyatomic molecules, or ion — molecule reactions. It is worth noting that ion — molecule reactions do not necessarily imply displacement of the ionized molecule, since the interactions may take place in the liquid at distances of an order of magnitude of several intermolecular spacings [29].

(c) The *chemical stage* (of a duration over $10^{-10}$ s) is the one where reactions occur between the species that are generated in the preceding stages.

During the first stage excited $H_2O^*$ molecules, ionized $H_2O^+$ molecules and also secondary electrons of high kinetic energy are produced:

$$H_2O \rightarrow H_2O^*, \quad (5.13)$$

$$H_2O \rightarrow H_2O^+ + e^-. \quad (5.14)$$

The secondary electrons, as well as the Compton or photoelectric electrons, are quickly slowed-down and thermalized, and then they are promptly picked up by water molecules:

$$H_2O + e^- \rightarrow \dot{H} + OH^-. \quad (5.15)$$

Since $H_2O^*$ and $H_2O^+$ are not stable, the following stage of molecular radiolysis will entail the reactions:

$$H_2O^* \rightarrow \dot{H} + \dot{O}H, \quad (5.16)$$

$$H_2O^+ + H_2O \rightarrow H_3O^+ + OH. \quad (5.17)$$

The $H_3O^+$ ionized molecule may be neutralized by an electron:

$$H_3O^+ + e^- \rightarrow H_3O, \quad (5.18)$$

and dissociates

$$H_3O \rightarrow H_2O + \dot{H}, \quad (5.19)$$

$$H_3O \rightarrow e^- + H_3O^+. \quad (5.20)$$

Resulting radicals will combine:

$$\dot{H} + \dot{H} \rightarrow H_2 \quad (5.21)$$

$$\dot{O}H + \dot{O}H \rightarrow H_2O_2 \quad (5.22)$$

$$\dot{O}H + \dot{H} \rightarrow H_2O. \quad (5.23)$$

During the chemical stage, recombinations between molecules, radicals and free electrons take place [25]:

$$\dot{O}H + H_2 = H_2O + \dot{H}, \quad (5.24)$$

$$\dot{H} + H_2O_2 = H_2O + OH, \quad (5.25)$$

$$e^- + H_2O_2 = \dot{O}H + OH^-. \quad (5.26)$$

A reaction providing for the interpretation of the excess concentration in $H_2O_2$ is:

$$\dot{O}H + H_2O_2 = H_2O + \dot{H}O_2, \quad (5.27)$$

$\dot{H}O_2$ vanishing as a result of the reaction:

$$\dot{H}O_2 + \dot{H} = H_2O_2. \quad (5.28)$$

Due to $\dot{H}O_2$, molecular oxygen is also produced:

$$\dot{H}O_2 + \dot{H}O_2 = H_2O_2 + O_2. \quad (5.29)$$

In the presence of dissolved oxygen, the following reaction takes place:

$$\dot{H} + O_2 = \dot{H}O_2, \quad (5.30)$$

Reaction (5.24) explains why dissolved hydrogen inhibits water radiolytic decomposition. The steep acceleration of the radiolysis under the effect of radical collectors can be explained by the capture of $\dot{H}$ and $\dot{O}H$, that prevent the recomination reactions (5.24) and (5.25) taking place.

The solvated electron has reducing properties:

$$e^- + H_2O = \dot{H} + OH^-, \quad (5.31)$$

and also the properties of a base :

$$e^- + H^+ = \dot{H}. \quad (5.32)$$

Overall, radiolysis is an undesirable process because it generates explosive gases that enter the water circuit, and also yields oxygenated water and free radicals that enhance the water corrosive action upon reactor structural materials. To limit radiolysis effects, water should be maintained at a high purity (electric resistivity over 1 MΩ cm). Water purification is achieved with ion-exchange resins. Continual monitoring of water purity is carried out by measurements of its electric conductivity.

## 5.4. HEAVY WATER

The characteristics of heavy water as a moderator are shown in Table 5.2. The high moderating ratio enables the heavy water-moderated nuclear reactors to use the natural uranium in the form of $UO_2$ as fuel. It is this noteworthy advantage that is ordinarily taken into consideration against the relative inconvenience of heavy water's high cost — in excess of $250 per kg [30], and the large quantity of moderator required; thus, each MWe installed in the most widespread heavy water reactors — the CANDU PHWRs — requires 850 kg of moderator [31].

### 5.4.1. Irradiation Properties and Effects

Table 5.10 compares physical properties of $D_2O$ and $H_2O$. Their specific weight differs by about 11%, whereas the other properties are closely alike [32].

TABLE 5.10 $D_2$ and $H_2O$ Physical Properties

| *Property* | | $D_2O$ | $H_2O$ |
|---|---|---|---|
| Molecular mass | (a.m.u.) | 20.03 | 18.016 |
| Density | (kg/m$^3$) | 1107 | 1000 |
| Boiling point | (K) | 377.5 | 373.1 |
| Freezing point | (K) | 276.96 | 273.15 |
| Critical temperature | (K) | 644.6 | 647.3 |
| Temperature corresponding to the maximum density | (K) | 284.35 | 277.13 |
| Vaporization heat | (kJ/kg) | 2073.4 | 2232.9 |
| Thermal conductivity | (W/(m . K)) | 1.96 | 2.12 |
| Viscosity | (kg/(m . s) | $1.2514 \cdot 10^{-3}$ | $1.005 \cdot 10^{-3}$ |
| Refractive index | - | 1.3283 | 1.3326 |

One specific problem when using heavy water is the so-called *isotopic pollution*. The main source is atmospheric humidity. Consequently, in the reactor a slightly pressurized dry inert gas should be introduced above the level of heavy water. Also, the heavy water should be allowed access in the ducts only after carefully drying these. The purification installations with ion-exchange resins can only be used subsequent to their deuteration. As a consequence of possible light water leaks into heavy water due to corroded heat-exchangers, the moderator concentration in the nuclear reactor should be continually monitored [30] by means of a special instrumentation.

The development of *heavy water radiolysis* follows the same pattern as light water's. Yet, for heavy water, the deuterium in gaseous state has to be recovered and $D_2O$ regeneration is in order, by a reforming process. As for light water, the radiolysis rate may be lowered by an initial purification of water and subsequent, systematic removal of corrosion products.

Another important effect of irradiation is the formation of radioactive products (Table 5.9). Unlike light water, heavy water gives a nuclear reaction from which tritium emerges; tritium is a $\beta$ radioactive nucleus emitting a radiation of 0.018 MeV and having a long half-life. This leads to heavy water *tritiation*, causing a radioactivity that may reach 40 Ci/dm$^3$. So, the moderator and the cooling agent should be submitted to a tritium-scrubbing process. Another characteristic reaction is $^2H(\gamma,n)^1H$ that leads to a permanent *depletion in deuterium* of heavy water, by hydrogen formation. Elimination of hydrogen that had entered the moderator in various ways is carried out in heavy water reconcentration facilities, based on the vacuum distillation process.

### 5.4.2. Heavy Water Production

In principle, deuterium is present in all hydrogen compounds, but the natural deuterium to hydrogen ratio is very low — around 1/7000. The deuterium found in the surface water is in the form HDO; its concentration is 0.0132 – 0.0156% D/(D+H).

A great deal of processes have been proposed for deuterium extraction and enriching, but only few of them proved technically reliable and economically sound [33-35]. Processes currently in practical use are:

1. The chemical exchange between
   (a) hydrogen sulphide and water,
   (b) hydrogen and water,
   (c) hydrogen and ammonia.
2. Water distillation,
3. Hydrogen distillation,
4. Water electrolysis,

Also, under development and with promising results is laser separation.

1. Chemical exchange

This is a process based upon a reaction that reads, generically

$$mBA_{p-i}A_i^* + nCA_{q-j}A_j^* \rightleftarrows mBA_{p-i-n}A_{i+n}^* + nCA_{q-j+m}A_{j-m}^* \tag{5.33}$$

where $A^*$ and $A$ are isotopes of the same element, and $m$ and $n$ are the stoichiometric coefficients. The isotope distribution at equilibrium, as described in eq. (5.33), may be characterized by the equilibrium constant $K$ and the separation coefficient, $\alpha$.

The *equilibrium constant* is defined by the expression [36]:

$$K = \frac{[BA_{p-i-n}A_{i+n}^*]^m \ [CA_{q-j+m}A_{j-m}^*]^n}{[BA_{p-i}A_i^*]^m \ [CA_{q-j}A_j^*]^n} \tag{5.34}$$

The magnitude of the equilibrium constant is temperature-dependent.

The *separation coefficient* $\alpha$ is the ratio of the relative concentrations of the isotopes of the element $A$, in the two substances that react with each other i.e. $BA$ and $CA$:

$$\alpha = \frac{X_{CA}}{Y_{BA}} = \frac{x/(1-x)}{y/(1-y)}, \tag{5.35}$$

where $x$ and $y$ are the molar fractions of the $A^*$ isotope in liquid state and in gaseous state, respectively, and $X$ and $Y$ the corresponding relative concentrations.

The following relation subsists between the equilibrium constant $K$ and the separation coefficient $\alpha$:

$$\alpha = (K/K_\infty)^{1/mn}, \tag{5.36}$$

where $K_\infty$ is the equilibrium constant at high temperatures, when the probabilities of an even distribution of the isotope of interest between the substances that react are equal. $K_\infty$ may be calculated with the equation [36]:

$$K_\infty = \left(\frac{x_{i+n}}{x_i}\right)^m \cdot \left(\frac{x_{j-m}}{x_j}\right)^n. \quad (5.37)$$

The difference between $K$ and $K_\infty$ is very small, for most isotopes. The hydrogen isotopes are, however, an exception, as may be seen from Tables 5.11 and 5.12 [36].

TABLE 5.11 Isotopic Exchange Equilibrium Constants of Hydrogen, Deuterium and Water

| *Reaction* | *Temperature* (K) | | | | | | |
|---|---|---|---|---|---|---|---|
| | *273* | *298* | *323* | *348* | *373* | *398* | $\infty$ |
| 1. $H_2O + HD \rightleftarrows HDO + H_2$ | 4.19 | 3.62 | 3.20 | 2.88 | 2.62 | 2.43 | 1 |
| 2. $H_2 + D_2 \rightleftarrows 2HD$ | 3.17 | 3.26 | 3.33 | 3.39 | 3.44 | 3.48 | 4 |
| 3. $H_2O + D_2O \rightleftarrows 2HDO$ | 3.76 | 3.80 | 3.83 | 3.85 | 3.87 | 3.89 | 4 |
| 4. $HDO + D_2 \rightleftarrows D_2O + HD$ | 3.54 | 3.11 | 2.78 | 2.53 | 2.33 | 2.17 | 1 |
| 5. $H_2O + D_2 \rightleftarrows D_2O + H_2$ | 14.82 | 11.25 | 8.90 | 7.29 | 6.13 | 5.27 | 1 |

Note: Separation coefficient for reactions 1, 4 and 5 is $\alpha$ = K.

TABLE 5.12 Isotopic Exchange Equilibrium Constants for Ammonia, Water and Hydrogen

| *Reaction* | *Temperature* (K) | | | | | | | |
|---|---|---|---|---|---|---|---|---|
| | *273* | *298* | *373* | *473* | *573* | *673* | *773* | $\infty$ |
| 1. $NH_3 + HDO \rightleftarrows NH_2D + H_2O$ | 1.64 | 1.61 | 1.56 | 1.52 | 1.50 | 1.49 | 1.49 | 3/2 |
| 2. $NH_3 + HD \rightleftarrows NH_2D + H_2$ | 6.85 | 5.83 | 4.10 | 3.03 | 2.53 | 2.24 | 2.07 | 3/2 |
| 3. $NH_3 + NHD_2 \rightleftarrows 2NH_2D$ | 2.92 | 2.94 | 2.97 | 2.99 | 3.00 | 3.00 | 3.00 | 3 |
| 4. $NH_2D + ND_3 \rightleftarrows 2NHD_2$ | 2.88 | 2.90 | 2.94 | 2.96 | 2.98 | 2.99 | 2.99 | 3 |

Note: Separation coefficient for reactions 1 and 2 is $\alpha$ = (2/3)K.

1(a) *Hydrogen sulphide and water.* Among the processes to separate deuterium by chemical exchange, the most employed is the exchange between hydrogen sulphide and water, which ensures 90% of the world production. The process is based on the reversible reaction:

$$(H_2O)_l + (HDS)_g \underset{2}{\overset{1}{\rightleftarrows}} (HDO)_l + (H_2S)_g. \quad (5.38)$$

The equilibrium constant of this reaction:

$$K_{gl} = ((HDO)_l/(H_2O)_l)/((HDS)_g/(H_2S)_g) \quad (5.39)$$

varies inversely with temperature. When temperature decreases, $K_{gl}$ increases and the equilibrium is shifted to the right (arrow 1).

The temperature effect upon equilibrium is used to separate deuterium by circulating a water flow against a gaseous $H_2S$ flow through two cascading columns at different temperatures, which constitute a separation stage. Thus, the $H_2O$ — $H_2S$ exchange process is a "bithermal" one. The multiplication of the elementary separation effect evolves a profile of deuterium concentration throughout each stage; maximum concentrations are reached at the basis of the cold column and on top of the hot one. Figure 5.6 shows schematically a primary separation stage. Deuterium enrichment obtained in one stage does not suffice, so that the installation is made up of two or three cascading stages. The final deuterium concentration that can be obtained by this process is 15 — 20% D/(D+H). Further isotopic enrichment of heavy water up to 99.80% D/(D+H) is achieved by vacuum distillation.

The bithermal separation procedure involves mass transfer columns that work at different temperatures. The installation must then include heat exchangers that would convey the heat from the warm fluids that leave the column, to cooler fluids that enter the column. There is a de-humidifier at the base of the cold column, and a humidifier at the base of the hot column, so named owing to the effect of temperature upon the water vapour content in $H_2S$.

Hydrogen sulphide dissolves in water; its solubility is directly proportional to pressure and decreases with temperature. The gaseous $H_2S$ is saturated with water vapours; its humidity is inversely proportional to pressure and increases with temperature. Each phase in the separation columns contains the molecular species $H_2S$, HDS, $H_2O$, HDO. As a result, the deuterium distribution among different phases is intricate, the magnitude of the global deuterium separation factor (for D/(D+H) ratios up to 1%), being given by the equation [37-40]:

$$\beta = \frac{(S + (\alpha_{H_2S}/\alpha_{H_2O})K_g)\ (1+H)}{\alpha_{H_2S}(1+K_g\ .\ H)\ (1+S)} \tag{5.40}$$

$S$ designates the $H_2S$ solubility in water, defined through (5.41); $H$ is the $H_2S$ humidity, as defined by (5.42); $\alpha_{H_2O}$, $\alpha_{H_2S}$ are the relative volatilities for water and $H_2S$, defined in (5.43) and (5.44), respectively; $K_g$ is the equilibrium constant in gaseous phase, given by (5.45); one has:

$$S = \text{dissolved } H_2S\text{, moles/liquid } H_2O\text{, mole;} \tag{5.41}$$

$$H = H_2O \text{ vapours, moles/gaseous } H_2S\text{, mole;} \tag{5.42}$$

$$\alpha_{H_2O} = \frac{[HDO]_g}{[H_2O]_g} \cdot \frac{[H_2O]_l}{[HDO]_l} = 1.1596\ \exp(-65.43/T); \tag{5.43}$$

$$\alpha_{H_2S} = \frac{[HDS]_g}{[H_2S]_g} \cdot \frac{[H_2S]_l}{[HDS]_l} = 1.034\ \exp(-8.037/T); \tag{5.44}$$

$$K_g = \frac{[HDO]_g}{[H_2O]_g} \cdot \frac{[H_2S]_g}{[HDS]_g} = 1.01\ \exp(233/T). \tag{5.45}$$

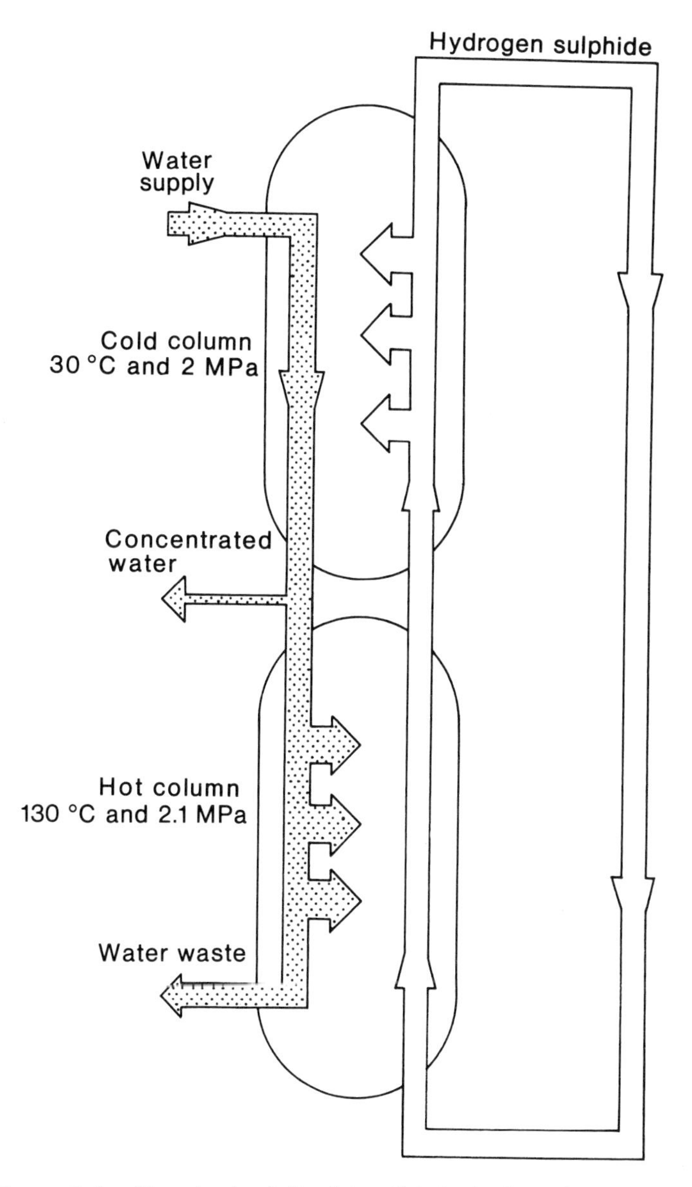


Figure 5.6. Flow-chart of the $H_2O$ — $H_2S$ isotopic exchange process.

A simplified expression, yielding results close to those given by eq. (5.40) may be obtained admitting that:

$$\alpha_{H_2O} \simeq \alpha_{H_2S} = \alpha \tag{5.46}$$

The resulting equation, that enjoys frequent quotation in the literature on the $H_2O - H_2S$ exchange, is [41]:

$$\beta = \frac{(1+H)(K_g+S)}{\alpha(1+HK_g)(1+S)} . \qquad (5.47)$$

The calculation of the global separation factor on the basis of concentrations is rather cumbersome and requires mastering of appropriate computer programs [39,40].

Figure 5.7 presents the variation with temperature and pressure of the global deuterium separation factor in the cold column, $\beta_R$, and in the hot column, $\beta_c$, plotted according to eq. (5.47).

The operating parameters of the columns are chosen considering the physico-chemical properties of the reactants and also criteria of economic optimization [41-43]. The cold column temperatures and pressures are determined, first, by the constraint following from the formation of a crystal-hydrate that solidifies at 303 K; another constraint is the liquefaction of the hydrogen sulphide at pressures over 2.2 MPa. Table 5.13 indicates the temperature and pressure, when, at equilibrium, in the $H_2O$ — $H_2S$ system, the third, liquid, phase of $H_2S$; and the hydrate appear, respectively [40]. As for the hot columns, the higher the temperature, the better, because in this way the deuterium extraction, as defined by $100\times(1-\beta_c/\beta_R)$ [44] will be enhanced. Economic considerations that seek to limit the heat and electricity consumption, put a limit of 400 – 410 K to the operating temperature.

Ordinarily the operation range for the cold columns is 305 — 308 K. For the hot ones it is 403 — 408 K at an average pressure of 1.96 — 2.06 MPa. The pressure is lower in the cold column, by about 0.08 MPa. Under these circumstances the global deuterium separation factor in the cold column is $\beta_R = 2.22$, and in the hot column $\beta_c = 1.76$, which leads to a global factor per stage of $\beta_R/\beta_c = 1.26$ and to a ratio of 20% for the deuterium extraction from the feed water.

The contact elements of the two phases coupled in countercurrent are the *plates*. The most employed are the *bubble plates* and the *sieve plates*.Figure 5.8 introduces the bubble plate, where the intimate contact between the liquid and gaseous phases is achieved by means of the so-called bubble cap. A more efficient device for the $H_2O$ — $H_2S$ exchange is the sieve plate (Fig. 5.9), that allows for operating at bigger loading factors and entails a lower pressure drop. Modern facilities for heavy water separation use big, multi-overfall plates (that exceed 8 m in diameter) [45,46]. On the sieve plate, the bubbling occurs as the gas lifts up the plate openings and through the liquid mass that flows along the plate; this process is shown in Fig. 5.9.

The main limitations in operating the plates are determined by foaming, and deposits of corrosion products that occur mainly in the cold column. In both cases the column's loading factor decreases and pressure drop increases, which induces a corresponding diminishing in output and a raise in the specific energy consumption.

The foaming is a consequence of the quality of water and hydrogen sulphide and of an accumulation of foaming-inducing substances in extremely low quantities. Foaming is avoided either by adding anti-foam substances or — more appropriately — by making use of only high-quality raw materials, as well as of plates equipped with foam-breaking devices. In order to reduce corrosion product deposits, a technology is being developed — to form stable layers of pyrites on the inside surface of the columns and pipes; a reduction in the

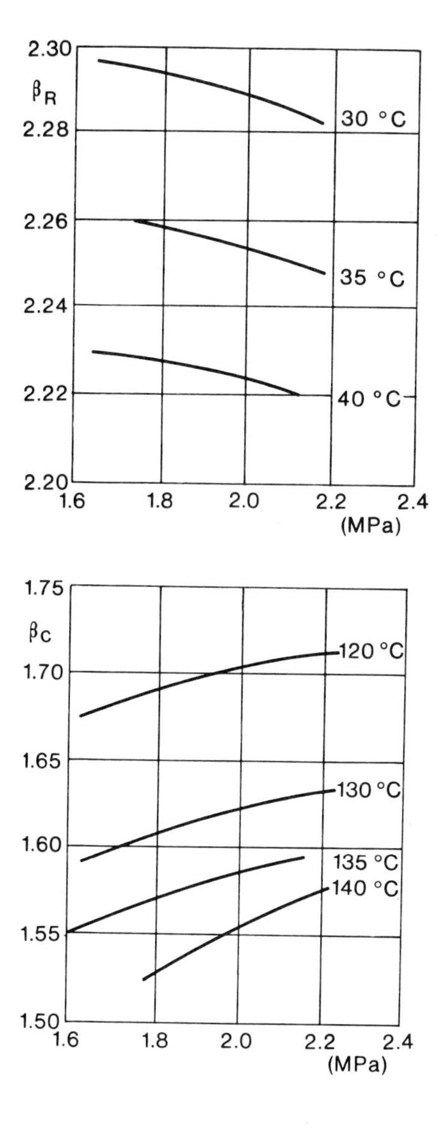


Fig. 5.7. Variation of separation factors for the $H_2O - H_2S$ isotopic exchange process, as functions of temperature and pressure.

formation rate of the atomic hydrogen whose effects on the metal internal structure is very pronounced is also sought. To monitor the plates' operation and the height of the foam strata, special devices using $\gamma$ sources have been created, that may be installed on modern separation columns (diameter 8.6 m, 100 m high) to permanently beam information to the installation's control panel [47].

TABLE 5.13 Three-phase Equilibrium Conditions in the $H_2O - H_2S$ System

| *Pressure* (MPa) | *Temperature* (K) | *Third phase* |
|---|---|---|
| 1.0 | 295.4 | Hydrate |
| 1.2 | 297.1 | Hydrate |
| 1.4 | 298.6 | Hydrate |
| 1.6 | 299.8 | Hydrate |
| 1.8 | 300.9 | Hydrate |
| 2.0 | 301.7 | Hydrate |
| 2.239 | 302.55 | Hydrate + liquid $H_2S$ |
| 2.3 | 303.9 | Liquid $H_2S$ |
| 2.4 | 305.7 | Liquid $H_2S$ |
| 2.5 | 307.6 | Liquid $H_2S$ |

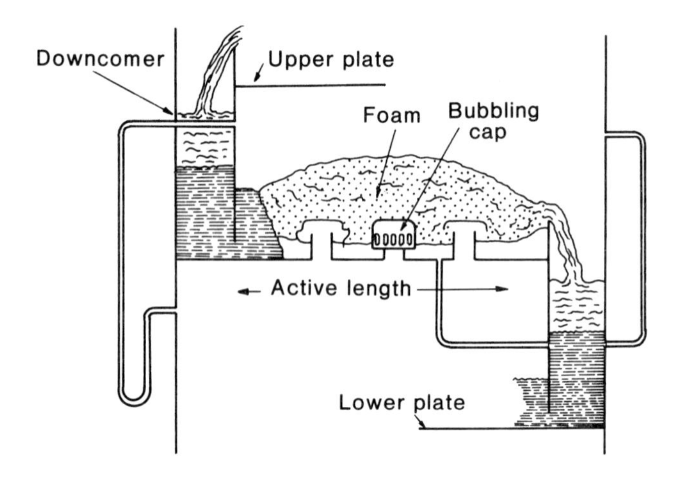


Fig. 5.8. Bubble plate.

In order to establish the required height of the isotopic separation columns, it is necessary to calculate the amount of *theoretical plates* and to experiment on the *isotopic effectiveness* . Several calculation methods have been proposed to determine the amount of theoretical plates; among them are the McCable — Thiele method, the Colburn method and the method based on a Riccati-type equation [48]. The latter has the advantage of being an analytical method, applicable for any isotopic concentration. Various algorithms and computer programs have been evolved, based on these methods [49].

The "theoretical plate" is an idealized plate, on which one supposes the liquid and gaseous phases to reach the isotopic equilibrium at the moment when they leave the plate. This conjecture is, however, never true on a real plate. In order to take this into account, a quantity called "Murphree isotopic effectiveness" was introduced. This is defined as the ratio of the actual concentration growth to the concentration growth in case all the phases were at isotopic equilibrium when leaving the plate (theoretical plate). The

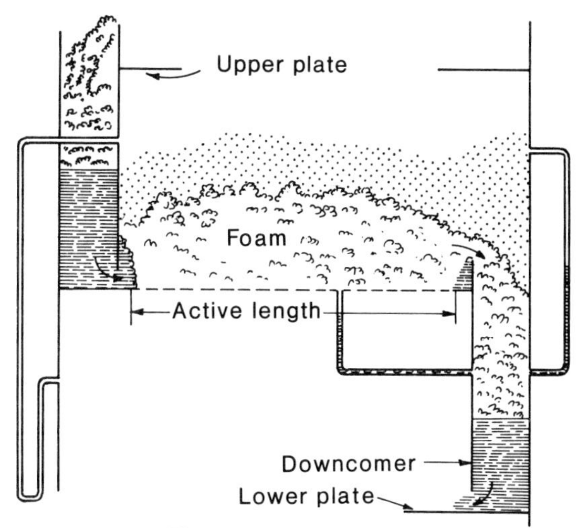


Fig. 5.9. Sieve plate.

expression of the effectiveness, as defined for the liquid phase, on which measurements are usually performed, is:

$$E_M = (x_i - x_{i+1})/(x_i^* - x_{i+1}). \tag{5.48}$$

The meaning of notation is given in Fig. 5.10 and $x_i^*$ represents the concentration of the liquid phase at equilibrium whith the concentration of the $y_i$ gaseous phase that leaves the plate. The equilibrium equation for the bithermal process $H_2O$ — $H_2S$ reads:

$$x_i^* = \frac{\beta y_i}{1 + (\beta-1)y_i} \tag{5.49}$$

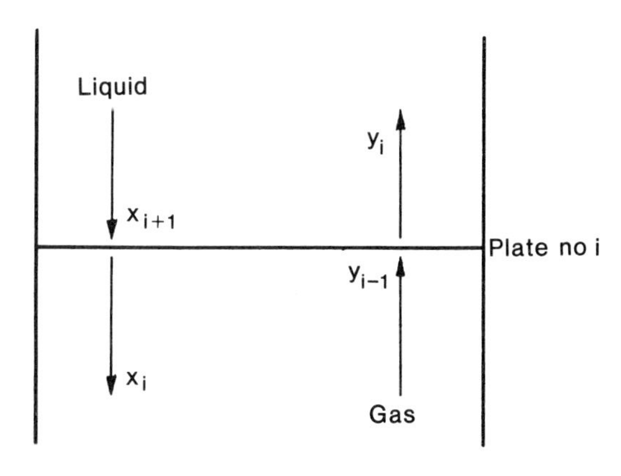


Fig. 5.10 Variation of isotopic gas concentration ($y$), and of isotopic liquid concentration ($x$), on the $i$th plate.

The experimental determination of a plate effectiveness is difficult, and is usually accompanied by severe errors. That is why the usual method adopted is to determine an overall, or mean, effectiveness, over a great number of plates of one column. To this purpose, several determination methods have been proposed, for the stationary [50,51] and non-stationary [52] operating states. Ordinarily, isotopic analyses in gaseous phase are avoided, because of too many error sources that are mainly due to the sample extraction procedures [53].

An installation for deuterium extraction and enrichment through $H_2O$ — $H_2S$ exchange is composed of 2 — 3 stages; the first stage has the largest volume, and also the highest energy consumption. As a rule this first stage achieves a separation below 10, which means a raise in concentration from the natural level, i.e. 0.0150% D/(D+H), to 0.150% D/(D+H).

Although the separation thus achieved is low, about 90% of the heavy water cost goes with this stage. It is noteworthy that the volume of the first stage and that of the next are in a 6 to 1 ratio, whereas the number of real plates in these stages is in a 3 to 5 ratio. Coupling between stages may be discharged on the liquid, gaseous, or both phases. The deuterium transfer between stages in liquid phase has the merit of being very stable, because the flows, when passing from one stage to the other, are 2.5 times lower, as compared to those that feature the transfer between stages in gaseous phase; however, the thermal energy consumption is 10% higher [54]. A connection between stages both at the gas contact line and at the liquid contact line combines the advantages offered by the two systems. In this latter case the operating temperature and pressure of the stages are set by the parameters of the first stage. Figure 5.11 gives a schematic representation of a three-stage separation facility, cascaded on the gas contact line.

One of the problems specific to the isotopic exchange between water and hydrogen sulphide is the high sensitivity of the operation, especially at variations of the ratio between the liquid and gas flow. A 1.5% deviation from the optimum of the gas — liquid ratio causes a 1% decrease in output, while bigger deviations are accompanied by even more important losses. Since the phase flows cannot be determined accurately enough, the liquid phase isotopic concentrations should be measured at the mid-level of the separation columns, on the knowledge that between the concentration ratio and the flow ratio there are well established relations [55,56]. Concentrations in the first stage are determined by mass spectrometry, a method allowing the control of production at a ±0.3% accuracy. In order to determine experimentally the plates' effectiveness, methods to measure the deuterium content with the highest accuracy are also in order [57].

One asset of great consequence in controlling a heavy water production facility is to know the duration of the transitory regime till a stationary state is reached; it is a fact that this duration is unusually long, as compared to what happens ordinarily in other chemical facilities. This has required the development of complex mathematical models, calling for numerical solutions on large computers for appropriate differential equations [58-60].

Due to the corrosiveness of the $H_2S$ aqueous solution and of the humid $H_2S$, and also due to the very large quantities of toxic gas that are found practically in a pure state in a heavy water facility, high-quality materials and special precautions for the personnel and environmental protection are required. Particular care is devoted to recovering $H_2S$ from the waste water before its dumping into rivers, so that the residual concentrations should be around $10^{-5}$%.

Material and energy demands in a $H_2O$ — $H_2S$ heavy water plant are rather high, as may be seen from Table 5.14 [31]. In order to extract 800 tons of heavy water per year, the amount of feed water must exceed $3 \times 10^7$ tons (the cooling water not included). One should also mention that the thermal

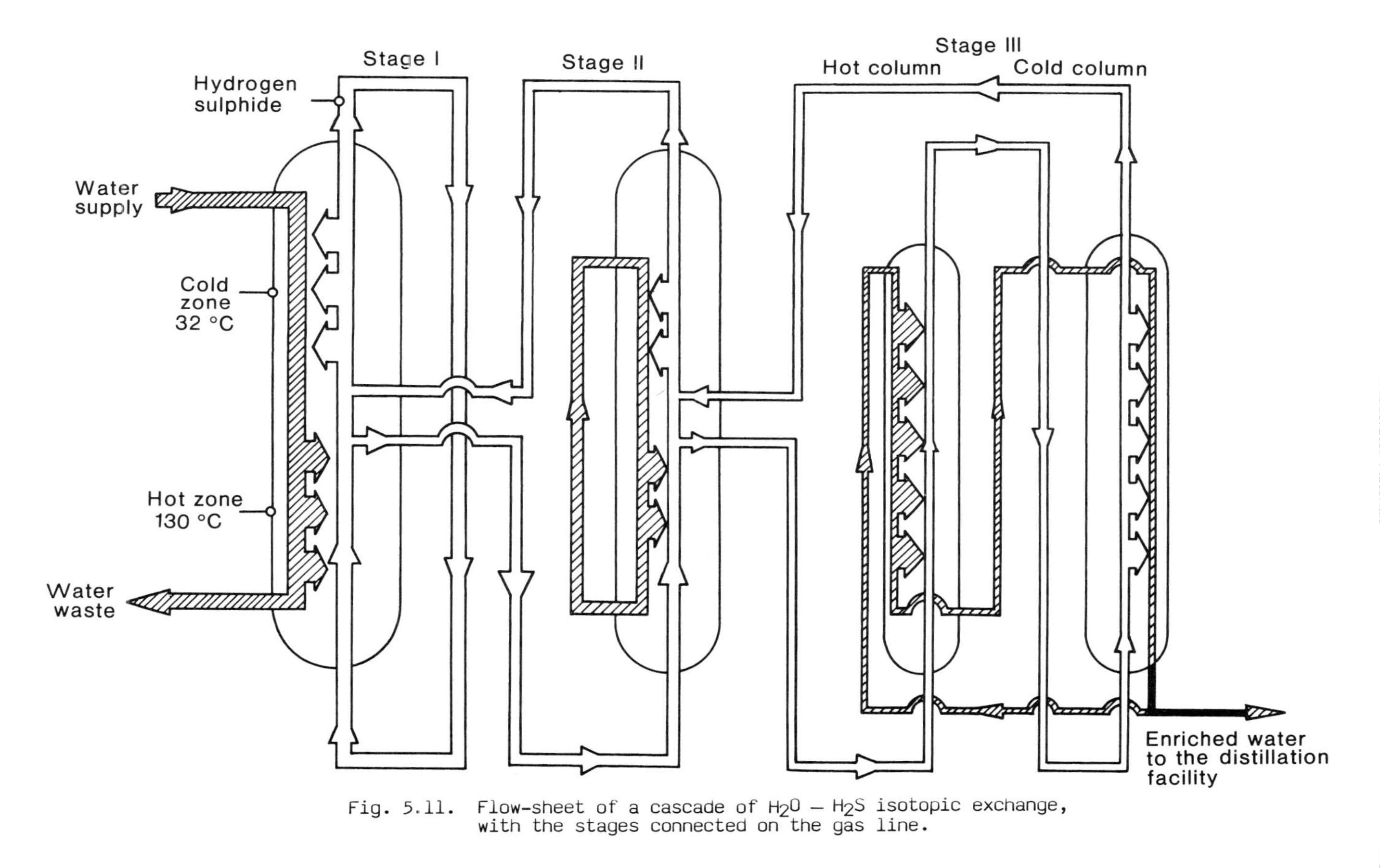


Fig. 5.11. Flow-sheet of a cascade of $H_2O$ – $H_2S$ isotopic exchange, with the stages connected on the gas line.

TABLE 5.14 Material and Energy Requirements for an 800 t $D_2O$/year Plant

| | |
|---|---|
| Process water | 35 t/kg $D_2O$ |
| Cooling water | 10 t/s |
| Thermal energy | 600 MWt (23 GJ/kg $D_2O$) |
| Electric energy | 67 MWe (700 kWh/kg $D_2O$) |

energy consumption, of 23 GJ/kg $D_2O$, can be maintained at this level only if a high-performance recovery system, working at about 70% efficiency, is in permanent operation [61]. However, it is believed that $H_2O$ — $H_2S$ chemical exchange will continue to be the chief heavy water production process, on two essential grounds: first, the deuterium source is cheap and readily available in large quantities; second, the reaction does not need a catalyst, which allows employment of simple and large-size contact elements.

1(b) *Hydrogen and water*. Developed in several variants, the chemical exchange between water and hydrogen was adopted for its cheap deuterium source, and the non-corrosive character of the process. The main drawback of this avenue is a low exchange rate, which calls for the use of nickel- or platinum-based catalysts. Among the processes that use this kind of exchange are:

(a) The hydrogen — water vapours exchange reaction, at nearly ambient pressure and at about 343 K, using nickel or chromium trioxide and platinum on coal, as catalysts. This process has been used since 1944 at the Rjukan and Glonifjord Plants in Norway, at a 20 t/year rated capacity, and at the Trail Plant, Canada, at a 6 t/year capacity.

(b) The hydrogen — liquid water reaction, in the presence of a catalyst such as colloidal platinum spread on charcoal, or on a homogeneous NaOH catalyst. Experiments conducted in the Federal Republic of Germany at very high pressures, of 20 MPa, were not satisfactory.

(c) The hydrogen — liquid water reaction in the presence of a nickel or platinum catalyst spread on a porous support coated with a fine resin layer, impermeable to water, but permitting the diffusion of hydrogen. Research carried out in Canada seems promising, but the utilization of this system is likely to be more adequate for the detritiation of heavy water than for deuterium separation [62]. Although a full economic assessment of this process is not yet possible, the chances that it would eventually become competitive with the $H_2O$ — $H_2S$ process look remote.

1(c) *Hydrogen and ammonia*. The hydrogen — ammonia exchange uses as deuterium source the hydrogen employed in the ammonia synthesis, and as catalyst the potassium amide $KNH_2$, disolved in liquid phase in a 1 — 2% ratio. The presence of the amide is a complication, since:

(a) it is expensive, thus having to be be recovered and recycled;
(b) it reacts violently with the oxygen found as impurity in the synthesis gas.

The reaction on which the process relies is:

$$NH_3(l) + HD(g) \rightleftarrows NH_2D(l) + H_2(g) \quad (5.50)$$

The process may be monothermal or bithermal. The monothermal process has been employed at the Mazingarbe Plant, France [63]. Figure 5.12 shows the simplified flow-chart of the monothermal process; the production of the plant was of about 26 t per year, and the deuterium recovery 85% of the amount identified in the feed water (0.0132% D/(D+H)).

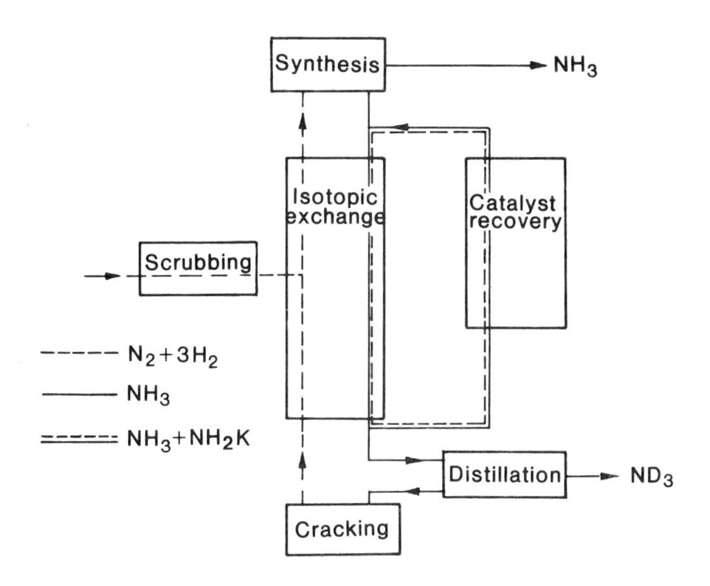


FIG. 5.12 Flow-sheet of the process of monotherm ammonia — hydrogen isotopic exchange.

The main drawbacks of this process stay with the limitations in the deuterium source (a production of 70 t per year requiring that the heavy water facility is coupled to a chemical plant rated at 1000 t of ammonia per day), and with the sophistication of the exchange plate, which must be provided with special ejectors to increase the reaction interface.

Few details are known as yet on the bithermal process. In India, at Talcher, the Uhde company has built a plant based on this process. Table 5.15 shows some data in comparison to the monothermal process. The bithermal process looks disadvantageous for two reasons: it requires more plates than the monothermal process, that is 72.9 theoretical plates against 5.7; and the energy recovery as well as the humidifying systems are cumbersome.

The chemical exchange methylamine — hydrogen is in many ways similar to the $NH_3 - H_2$ exchange. It starts from the reaction:

$$CH_3NH_2(l) + HD(g) \rightleftarrows CH_3NHD(l) + H_2(g), \qquad (5.51)$$

which is catalysed with potasium methylamide, $CH_3NHK$. Merits of this process are the low vapour pressure of the liquid phase; a high hydrogen solubility in the liquid phase; and a superior reaction rate. No validation at an industrial scale of this process is known of.

TABLE 5.15 A Comparison of the Monothermal and Bithermal $NH_3 - H_2$ Exchange Process

| | *Procedure* | |
|---|---|---|
| | *Monothermal* | *Bithermal* |
| Flow rate (kmol/hr) | | |
| Ammonia | 439 | 467 |
| Synthesis gas | | |
| Stripping | 2642[1] | 4957[1] |
| Enrichment | 886[1] | 3202[1] |
| Number of theoretical plates | - | 3202[2] |
| Stripping | 2.6[1] | 19.9[1] |
| Enrichment | 3.1[1] | 13.0[1] |
| | | 40.0[2] |
| Total | 5.7 | 72.9 |

[1] In the cold column; [2] In the hot column.

## 2. Water distillation

Water includes three molecular species, namely $H_2O$, HDO and $D_2O$, that are in equilibrium according to the reaction:

$$H_2O(l) + D_2O(l) \rightleftharpoons 2HDO(l). \tag{5.52}$$

Starting from the equation (5.35) above, the separation factor for deuterium between the liquid and gaseous phase may be written as:

$$\alpha = \left[\frac{x_{HDO} + 2x_{D_2O}}{2x_{H_2O} + x_{HDO}}\right] \cdot \left[\frac{2y_{H_2O} + y_{HDO}}{y_{HDO} + 2y_{D_2O}}\right]. \tag{5.53}$$

On the conjecture that:

(a) liquid and gaseous phases are ideal;
(b) HDO vapour pressure is the geometric mean value of the $H_2O$ and $D_2O$ vapour pressure;
(c) the equilibrium constant of the reaction (5.52) is 4;
(d) the liquid stands always in equilibrium;

one may show [64] that the equation (5.53) becomes:

$$\alpha = \sqrt{\frac{p_{H_2O}}{p_{D_2O}}} \tag{5.54}$$

The lower the pressure, the higher the separation factor, and, consequently, the operation of a water distillation facility is conducted at pressures ranging from 15 to 30 kPa. For all that, the magnitude of $\alpha$ remains rather small ($\sim$ 1.05) [65].

Water distillation for the primary concentration of deuterium — a process employed years ago in three large US plants — is not seen any longer as economically sound, owing to the immense amounts of water to be vehiculated, and to exceedingly high heat demands.

However, due to the rather straightforward nature of the process and good safety record, vacuum distillation is successfully applied to upgrade heavy water to 99.8% D/(D+H), using as feed the primary concentrated product, at (5 — 30)% D/(D+H). Such facilities employ columns equipped with fillings of corrugated phosphorus bronze wire nets, folded in packs.[65]

To cascade a distillation facility in order to provide for a higher extraction factor and a lower energy consumption, the modelling of the process is needed that would also help establishing the number and size of columns [66,67]. A satisfactory operation of the installation also requires a good acquaintance with the effects on production of the perturbations in the liquid and vapour flows in the columns [68,69].

## 3. Hydrogen distillation

Hydrogen is distilled at low temperatures (22 K), with a high separation coefficient (1.7), and a high extraction factor (85%). Work pressure is low, and the environment is neither toxic nor corrosive. The main problem with the implementation of this process is the high-tech, highly priced cryogenic equipment required.

Production capacity is limited by that of the hydrogen source. The hydrogen should be carefully purified prior to liquefaction, in order to eliminate paramagnetic impurities such as oxygen (bringing them below $10^{-4}$%), in order to avoid the ortho- to parahydrogen conversion. A high consumption of work is required to evacuate the heat released during this transformation (superior to the hydrogen's latent vaporization heat). The heat release is however strictly necessary in order to avoid a quick loss of liquid from the storage vessels.

One feature of deuterium production through hydrogen distillation is the exclusive enrichement in the HD species, since the equilibration between the $H_2$, HD and $D_2$ molecules takes place only in the presence of a specific catalyst. As a result, the maximum molar deuterium concentration that may be reached is 50%. Further deuterium concentration is, preferably, achieved by vacuum distillation or by the electrolysis of water that has resulted from the burn-up of the hydrogen as produced in the distillation installation. The overall specific consumption of energy in this process is high, reaching up to 5500 KWh/kg $D_2O$ [70].

## 4. Water electrolysis

This is the earliest technique of deuterium separation. In comparison with the other processes described, this process features a high separation factor. Table 5.16 shows the magnitudes of the separation factors for various cathodes [71]. They depend not only on the nature of the cathode, but also on other factors such as the quality of the electrode surfaces, temperature, electrolyte purity, etc. The separation factor increases significantly with the lowering of electrolytic bath temperature. The usual operating temperature is 250 K, and the separation factor obtained stays between 6 and 8.

An electrolysis plant for heavy water enrichment consists of cells arranged in a cascade. The hydrogen depleted in deuterium leaves the cell and is directed to the preceding one, whereas the remaining water that is richer in deuterium passes to the next cell. The power consumption for the production of 1 kg of $D_2O$ is very high, about 125 MWh, and, consequently, the costs are high as compared to other processes [33,34].

TABLE 5.16 Separation Coefficient, at Electrolysis

| *Cathode* | *Separation coefficient* |
|---|---|
| Graphite | 6.7 |
| Iron | 9.9 |
| Nickel | 5.1 |
| Platinum | 6 – 14 |

Water electrolysis is mainly used for the final upgrading of heavy water that has resulted from other chemical exchange processes, but also for pre-concentration, followed by liquid hydrogen distillation [37].

*Laser separation.* The use of lasers for deuterium separation is quite a recent avenue, and a promising one. Ever since 1972, separation factors equal to 6 have been obtained by using the selective decomposition of formaldehyde ($H_2CO$), while in 1975 a 14 times enrichment in the deuterium content has been achieved by the method of the selective excitation of HDCO. Research in this field is still in progress, based on complex programmes that envisage the development of this process at an industrial scale [72].

Selecting a specific isotope separation process to implement heavy water production requires a thorough technical and economic assessment, including comparative evaluation of alternative blueprints and proper consideration of the local and international economic environment. This highlights the importance of finding and mastering comprehensive criteria that would also enable comparisons of different variants of the same process, to allow for optimal decisions ever since the initial stages of the project [73]. As a rough illustration only, Table 5.17 compares merits and drawbacks of the heavy water production processes, now commercial, that have been discussed.

TABLE 5.17 Several Characteristics of Industrial Processes for Heavy Water Production

| $H_2O - H_2S$ *exchange* | $H_2O - H_2$ *exchange* | $NH_3 - H_2$ *exchange* | $H_2O$ *distillation* | $H_2$ *distillation* |
|---|---|---|---|---|
| Merits | | | | |
| Independent | Independent | - | Independent | - |
| Non-catalysed | - | - | Non-catalysed | Non-catalysed |
| Conventional equipment | - | - | Conventional equipment | - |
| - | Non-corrosive | Non-corrosive | Non-corrosive | Non-corrosive |
| - | Non-toxic | - | Non-toxic | Non-toxic |
| Good α | Very good α | Very good α | - | Good α |
| - | Good extraction factor | Good extraction factor | - | Very good extrac tion factor |
| Drawbacks | | | | |
| - | - | Dependent | - | Dependent |
| - | Catalysed | Catalysed | - | - |
| - | Demanding on equipment | Demanding on equipment | - | Demanding on equipment |
| | | | - | |
| Corrosive | - | - | - | - |
| Toxic | - | Toxic | - | - |
| - | - | - | Very low α | - |
| Low extraction factor | - | - | Low extraction factor | - |

Regardless of the production process, one major problem with the operation, quality assurance and control in a heavy water production facility is the acquisition of a fully satisfactory system of isotope analyses, proper equipment included, that would match reliability with sensitivity [74-77].

The deuterium isotopic analysis had to cover a large range of concentrations, from $10^{-2}$% up to 99.9% D/(D+H) and be adequate to the specifics of the separation technologies for deuterium oxide. Of particular importance are the isotopic analyses in the areas of extreme concentrations, where a maximum measurement accuracy is required. Thus, in a heavy water plant the rate of deuterium extraction from the raw material is made known via analyses at the level of natural concentrations, while the analyses in the range over 99% D/(D+H) should bring evidence on the isotopic merits of the final product of nuclear quality [57,78-81].

Besides assurance of a 99.80% D/(D+H) isotopic concentration (at a $\pm$0.01% D/(D+H) accuracy), the quality conditions required for the heavy water to qualify for a valid moderator specify a low value of the admissible content in impurities.

It is believed that Appendix 8 may stand for a useful complement, on the line of quality inspection analyses in the heavy water production process. It gives details of the measurement methods employed in various concentration ranges and on the quality conditions imposed upon heavy water, as well as upon the measuring devices to determine the $H_2/N_2$ ratio in vapour phase, and the flow rates of several liquids and gases that are ordinarily encountered in the heavy water enterprise.

## 5.5. BERYLLIUM

Beryllium is a light metal (1850 kg/m$^3$) with a relatively high melting point. In nature it is found in small quantities as beryl, $Be_3Al_2Si_6O_{18}$ and as crysoberyl, $BeAl_2O_4$. Beryllium is an excellent material for moderators and reflectors, owing to its small neutron absorption cross-section and high neutron scattering cross-section, respectively. The material also features the remarkable property of multiplying the number of neutrons by ($\gamma$,n) and (n,2n) reactions. However, such merits are counter—balanced by the metal's low ductility (which translates into difficult mechanical processing), its relatively low corrosion resistance, and high cost.

Along with metallic beryllium, magnesium — beryllium alloys are also employed as moderators. Alloys with a high content of beryllium have special thermal properties, being able to resist high temperatures for longer durations [7,82]. Thus, in air at 1120 K, magnesium — beryllium alloys with a 2 — 5% content of beryllium resist oxidation for 12,000 to 14,000 hours.

Beryllium oxide, a refractory material, was also considered for use as a reflector. However, the mechanical merits of sintered beryllium oxide are strongly downgraded at a fluency of $2 \times 10^{21}$ n/cm$^2$, which limits the usefulness of this material [83].

### 5.5.1. Properties

Beryllium's physical properties are presented in Table 5.18. The mechanical behaviour of this metal depends on its anisotropy and purity. When obtained by extrusion, beryllium is anisotropic. Anisotropy is still present even after annealing at temperatures of ~ 1270 K. The material obtained by hot pressing of beryllium powders does not exhibit anisotropic mechanical properties. Beryllium presents a high tensile strength, a high elasticity modulus and a

TABLE 5.18 Physical Properties of Beryllium

| | |
|---|---|
| Density ($kg/m^3$) | 1850 |
| Allotropic transformation | $T$ = 1250 °C |
| Crystalline structure hexagonal close-packed | $a_o$ = 2.2858 Å; $c_o$ = 3.5842 Å |
| Melting point (°C) | 1283 ± 3 |
| Boiling point (°C) | 2970 |
| Linear expansion coefficient ($(°C)^{-1}$) | |
| 20 — 200 °C | $13.3 \cdot 10^{-6}$ |
| 20 — 700 °C | $17.8 \cdot 10^{-6}$ |
| Thermal conductivity at 20 °C (W/m . K) | 349.5 |
| Electric resistivity (μΩ . m) | 3.9 - 4.3 |
| Neutron absorption cross-section (b) | $9 \cdot 10^{-3}$ |

small ductility. Table 5.19 shows the strength and elongation of this metal at ambient temperature, for various processing methods. The data render evident the important differences between properties in the extrusion direction, as compared to those in a direction perpendicular to it.

TABLE 5.19 Mechanical Properties of Various Varieties of Beryllium

| | *Direction parallel to extrusion axis* | | *Direction perpendicular on the extrusion axis* | |
|---|---|---|---|---|
| | *Flow limit* (MPa) | *Elongation* (%) | *Flow limit* (MPa) | *Elongation* (%) |
| Obtained through casting | 467 | 0.36 | 277 | 0.3 |
| Obtained through extrusion | 571 | 1.82 | 237 | 0.18 |
| Same + annealing at 800 °C | 665 | 0.55 | 415 | 0.3 |

### 5.5.2. Corrosion Resistance

Beryllium is not particularly corroded in air, at temperatures below 670 K. At higher temperatures ( ~ 1088 K) a white, adherent layer of beryllium oxide appears on the surface of the metal. In the 970 — 1070 K range the oxidation resistance of beryllium cast in vacuum, then extruded, is superior to the oxidation resistance of the beryllium obtained by powder metallurgy techniques.

In terms of water corrosion resistance, beryllium rates lower than zirconium. The corrosion rate decreases when pH $> 6.5$, and is influenced by the presence of such impurities as copper and chlorine in the aqueous medium.

### 5.5.3. Irradiation Effects

Beryllium is stable under irradiation. Its exposure to neutron fluencies up to $2.4 \times 10^{21}$ n/cm$^2$ does not entail any sizeable dimensional changes, nor does it change its electric resistivity and corrosion resistance. As for the other properties — the thermal conductivity slightly decreases, tensile strength increases from 500 MPa to about 714 MPa, whereas the elongation decreases from 1.4 to 0.3%. Notably, the hardness increases by about 30%. Also the generation of helium as a result of irradiation causes beryllium to become brittle.

## 5.6. METAL HYDRIDES

As argued before, one element with good moderating properties is hydrogen. Accordingly, substances rich in hydrogen are expected to be good moderators.

Several elements in the fourth group of the periodic system such as zirconium, yttrium, and titanium, show a high affinity for hydrogen. When heated in a hydrogen atmosphere these metals may absorb considerable amounts of this gas ($7.25 \times 10^{22}$ atoms/cm$^3$ for zirconium, and $3.8 \times 10^{23}$ atoms/cm$^3$ for titanium).

In small quantities, hydrogen absorption leads to the formation of metal — hydrogen solid solutions, without essentially modifying the metal's original properties; as the content of hydrogen rises, metallic hydrides are evolved [84]; their physical and mechanical properties are different from those of the pure metals.

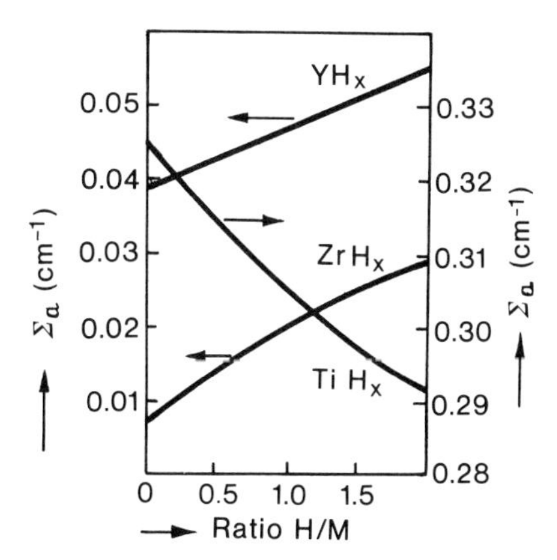


Fig. 5.13 Neutron macroscopic absorption cross-sections for yttrium, zirconium, and titanium hydrides.

Zirconium and yttrium hydrides are mainly used as moderator materials, as shown in Fig. 5.13; they have small neutron absorption cross-sections.

The physical and mechanical properties of zirconium hydride depend upon the hydrogen content and temperature: The specific heat of $ZrH_{1.76}$

is ~ 570 J/kg . K at 470 K and rises to 800 J/kg . K at approximately 800 K [85]. Zirconium hydride's thermal conductivity is to a lesser extent affected by either temperature or hydrogen content. Its typical value is 20 W/m . K [85].

Mechanical properties of zirconium hydride depend on the H/Zr ratio (see Table 5.20). The higher the temperature, the brittler the material becomes [86].

TABLE 5.20 Mechanical Properties of Zirconium Hydride

| *Compound* | $x_H \cdot 10^{22}$ (atoms/cm$^3$) | *Temperature* (K) | *Elasticity modulus* (MPa . $10^5$) | *Tensile strength* (MPa) | *Elongation* (%) | *Section reduction* (%) |
|---|---|---|---|---|---|---|
| $ZrH_{0.72}$ | 2.7 | 300 | 0.85 | 157 | 0 | 0 |
| | | 700 | 0.77 | 155 | 1.5 | 1 |
| | | 1000 | 0.40 | 19.3 | 49 | 93 |
| | | 1200 | 0.30 | 11 | 72 | 100 |
| $ZrH_{1.01}$ | 4.0 | 300 | 0.81 | 91 | 0 | 0 |
| | | 700 | 0.61 | 131 | 1 | 1 |
| | | 1000 | 0.44 | 19.3 | 53 | 100 |
| | | 1200 | 0.17 | 22.7 | 22 | 61 |

At 270 K the yttrium hydride with $4.8 \times 10^{23}$ hydrogen atoms/cm$^3$ presents a specific heat of 360 J/kg . K, that rises up to 900 J/kg . K at 1150 K. The $YH_{1.60}$ thermal conductivity is 28 W/m . K at 350 K, and 12 W/m . K at 1150 K. The $YH_{1.62}$ elasticity modulus decreases from $1.1 \times 10^5$ MPa at 300 K down to $0.51 \times 10^5$ MPa at 1150 K. Tensile strength varies in the same way as with zirconium hydride, both with the hydrogen content, and temperature. For $YH_{1.62}$ it decreases from ~ 50 MPa at 300 K, to ~ 23 MPa at 1500 K. In the case of the $YH_{1.83}$ hydride, the tensile strength varies between 24 MPa at 300 K and 35 MPa at 1500 K.

Apparently, the physical properties of the metallic hydrides are not affected too much by irradiation. When yttrium hydride was irradiated at a fluency of $1.3 \times 10^{20}$ n/cm$^2$ and a temperature of 1200 — 1300 K, a good dimensional stability and weight conservation have been demonstrated [87]. A similar behaviour has been noticed in the case of zirconium hydride ($ZrH_{1.66}$), irradiated at 320 K, at a $2.5 \times 10^{18}$ n/cm$^2$ fluency.

To perform as moderators, metallic hydrides are either mixed with the fuel, or are clad in stainless steel tubes that provide protection against corrosion from the cooling agents with which the moderators come into contact.

## REFERENCES

1. Hughes, D. J., and Harvey, J. A. (1955). Neutron Cross Sections, BNL 325. USAEC Report, Upton, Brookhaven National Labs.
2. Argonne National Labs.(1963). Reactor Physics Constants,ANL 5800.USAEC Rep.
3. Maxim, I. V. (1969). *Materiale nucleare*. Edit. Academiei, Bucharest.
4. Ram, K.S. (1977). *Basic Nuclear Engineering*. Wiley Eastern Ltd., New Delhi.
5. Egelstaff, P. A. (1962). *Nuclear Science Eng.*, 12, 250.
6. Egelstaff, P. A., and Schofield, P. (1962). *Nuclear Science Eng.*, 12, 260.
7. Cadilhac, M. (1964). Report CEA, R-2368.
8. Williams, M. R. (1966). *The Slowing Down and Thermalization of Neutrons*. North-Holland Publ. Co., Amsterdam.

9. Purica, I. I. (1967). *Rev.Roum.Phys.*, 12, 87; (1970). *St.cerc.fiz.*, 22, 4.
10. Boltax, A. (1964). *Nuclear Engineering Fundamentals*. Vol.4, McGraw-Hill Book Co., New York.
11. Veatkin, S. E., and Deev, A. N. (1967). *Iadernyi grafit*. Atomizdat, Moscow.
12. Matsuo, H. (1980). *J.Nuclear Materials*, 89, 9.
13. Smith, C. O. (1967). *Nuclear Reactor Materials*. Addison Wesley Publ. Co., Reading (Mass.).
14. Akai, T., and Oku, T. (1979). *J.Nuclear Materials*, 79, 227.
15. Roth, E. (1968). *Chimie Nucléaire Appliquée*. Masson et Co., Paris.
16. Tipton, C. P. (1960). *Reactor Handbook*. Vol.1, Materials, New York, p.893.
17. Lind, R. and Wright, J. (1964). Communication 566 a la 3-ieme Conf.Int. Util.Pacif.En.Atom., Geneve.
18. Nightingale, R. E. (1962). *Nuclear Graphite*. Academic Press, New York, London.
19. Nagasaki, R. (1971). *Proc.Int.Conf.Peaceful Uses of Atomic Energy*. Vol.10, Geneva.
20. National Carbon Company (1969). *The Industrial Graphite Handbook*.
21. Blackstone, R. and others (1969). *Symposium on Radiation Damage in Reactor Materials*. Paper SM-120/H-8, Vienna.
22. Weissman, I. (1977). *Elements of Nuclear Reactor Design*. Elsevier, p.33.
23. Savineau, M. (1966). *Bull.Inf.Scient.et Techn. CEA*, 105, 17.
24. Dawson, J. K., and Sowden, R. G. (1963). *Chemical Aspects of Nuclear Reactors*. Vol.I, Butterworths, London.
25. Allen, A. O. (1961). *The Radiation Chemistry of Water and of Aqueous Solutions*. Van Nostrand.
26. Dyne, P., and Kennedy, J. (1958). *Can.J.Chem.*, 36, 1518.
27. Hayon, E. (1964). *Trans.Farad.Soc.*, 60, 1059.
28. Am.Chem.Soc.Publ. (1965). *Advances in Chemistry*, Washington.
29. Lampe, F. W., Franklin, I. L., and Field, F. H. (1961). *Kinetics of the Reactions of Ions with Molecules*. Pergamon Press.
30. Barringer, A. (1978). *Can.Chem.Proc.*, 5, 26.
31. Haywood, L. R., and Lumb, P. B. (1975). *Chemistry in Canada*, 27, 15.
32. Kirillin, V. A. and others (1963). *Tyazhelaya Voda, Teplofizitcheskye svoystva*. Gosenergoizdat, Moskow.
33. Becker, E. W. (1962). *Heavy Water Production*, Review Series No.21, IAEA, Vienna.
34. Levins, D. M. (1970). *Heavy Water Production, A Review of Processes*, AAEC Report - TM 562.
35. Vasaru, G., Ursu, D., Mihaila, A., and Szentgyorgyi, P. (1975). *Deuterium and Heavy Water, A Selected Bibliography*. Elsevier Sci.Publ. Co.
36. Rozen, A. M. (1960). *Teoryia razdelenyia izotopov v kolonnah*. Atomizdat, Moskow.
37. London, H. (1961). *Separation of Isotopes*, George Newnes Ltd., London.
38. Galey, M. R. and others (1973). AECL Report 4255.
39. Neuburg, H. J., Atherley, J. F., and Walker, L. G. (1977). AECL Report 5702.
40. Pavelescu, M., and Peculea, M. (1975). Report ICEFIZ.
41. Bebbington, W. P. and others (1959). USAEC Report DP-400.
42. CNEN (1970). Symposium on the Technical and Economic Aspects of Heavy Water Production, Torino.
43. Proctor, J. F., and Thayer, V. R. (1962). *Chem.Eng.Prog.*, 4, 53.
44. Pratt, H. R. C. (1967). *Countercurrent Separation Progresses*. Elsevier Publishing Co., London.
45. Perryman, E. C. W. (1974). Proceedings CNA-74-305, Canadian Nuclear Association, Ottawa.
46. Chang, S. D., and Davies, S. M. (1975). *Chemistry in Canada*, 27, 25.
47. Smith, B. D. (1963). *Design of Equilibrium Stage Processes*. McGraw-Hill Book Co.
48. Mickley, H. S., and Sherwood, T. K. (1957). *Applied Mathematics in Chemical Engineering*. McGraw-Hill Book Co.
49. Constantinescu, D. M., Dumitrescu, M., and Peculea, M. (1977). *St.cerc.fiz.* 29, 323.

50. Proctor, J. F. (1963). *Chem.Eng.Prog.*, 58, 3.
51. Peculea, M., and Constantinescu, D. M. (1976). *St.cerc.fiz.*, 29, 557.
52. Hodor, I, and Boian, M. (1979). *Revista de Chimie*, 30, 668.
53. Pavelescu, M. (1977). Report ICEFIZ.
54. Pavelescu, M., and Olariu, A. S. (1981). Report ICEFIZ.
55. Morris, J. W., and Scotten, W. C. (1962). *Chem.Eng.Progr.Sym.Series*, 58,39.
56. Pavelescu, M., and Olariu, A. S. (1981). Report ICEFIZ.
57. Steflea, D., and Pavelescu, M. (1977). Report ICEFIZ.
58. Constantinescu, D. M., Dumitrescu, M., Peculea, M., and Ursu, I. (1974). Report ICEFIZ.
59. Constantinescu, D. M., Isbasescu, G., and Peculea, M. (1977).Report ICEFIZ.
60. Peculea, M. (1977). *Revista de Chimie*, 28, 962.
61. Vasaru, G. (1968). *Izotopi stabili*. Edit. Tehnica, Bucharest.
62. Hammerli, M., Stevens, W. H., Bradley, W. J., and Butler, J. P. (1976). AECL Report 5512.
63. Lefrancois, B. (1970). Proc.of Conf.on Techn.and Ec.of Prod.of Heavy Water, Torino.
64. Benedict, M., Pigford, T. H., and Levi, H. W. (1981). *Nuclear Chemical Engineering*. Second Edition, McGraw-Hill Book Co.
65. Alexandrov, A. A., and Ershova, Z. A. (1981). *Urovnenya krivoy nasystchenya dlya obychnoy i tyasheloy vodyi.* Inzhinerno-fizitcheskyi zhurnal, 40, 894.
66. Constantinescu, D. M., and Peculea, M. (1976). Report ICEFIZ.
67. Constantinescu, D. M., Dimulescu, A., Isbasescu, G., Peculea, M., and Stefanescu, I. (1980). Report ICEFIZ.
68. Constantinescu, D. M., Peculea, M., and Ursu, I. (1974). Report ICEFIZ.
69. Constantinescu, D. M., Dimulescu, A., Peculea, M., and Ursu, I. (1979). *St.cerc.fiz.*, 31, 119.
70. Malkov, M. P. (1961). *Vydelenie deyteria iz vodoroda metodam glubokogo ohlajdenyia*. Gosatomizdat, Moskow.
71. Chemla, M., and Perie, J. (1974). *La séparation des isotopes*. Presses Univ.de France.
72. Los Alamos National Laboratory (1982). Molecular Laser Isotope Separation Report.
73. Peculea, M. (1981). *Critica proceselor de separare a apei grele*. Simpozionul Separarea si utilizarea izotopilor, Cluj-Napoca.
74. Pavelescu, M., Dogaru, E., Steflea, D., and Stanciu, V. (1980). Report ICEFIZ.
75. Mercea, V. (1966). *Isotopen Praxis*, 8.
76. Mercea, V., and others (1978). *Introducere in spectrometrie de masa*, Edit. Tehnica, Bucharest.
77. Mercea, V. (Ed.) (1981). *Separarea si utilizarea izotopilor*. A Symposium in Cluj-Napoca.
78. Ursu, I., Peculea, M., Pavelescu, M., Demco, D., Hirean, I., Steflea, D., and Stanciu, V. (1982). Measurements of the D/H Isotopic Ratio in Water, Ammonia and Hydrogen, IAEA, Vienna, Contract 3036/RB.
79. Ursu, I., Peculea, M., Simplaceanu, V., and Demco, D. (1978). Romanian Patent, No 67839.
80. ASTM (1976). *Annual Book of ASTM Standards*, Part 31, Water. Chap.Deuterium Oxide.
81. Pavelescu, M. (1978). Report ICEFIZ.
82. Ivanov, V. E. (1963). *New Nuclear Materials including Non-Metallic Fuels*. IAEA, Vienna, p.183.
83. Beauge, R. and others (1980). *New Nuclear Materials including Non-Metallic Fuels*. IAEA, Vienna, p.497.
84. Mueller, W., Blackledge, H. P., and Libowitz, G. G. (1980). *Metal Hydrides*. Academic Press, New York, London.
85. Beck, R. L. (1962). *Trans. ASM*, 52, 556.
86. Jaeger, R. G. (1975). *Engineering Compendium of Radiation Shielding*. Vol.II, *Materials*, New York, p.327.
87. Marshall, I. C. (1964). *Trans.Amer.Nucl.Soc.*, 7, 123.

CHAPTER 6

# Materials for Reactor Reactivity Control

## 6.1. GENERALS

Operation and control of the reactor, i.e. the steering of the chain fission reaction, are performed by monitoring and adjusting the effective multiplication factor $k_{eff}$. To this effect several alternative methods are available, among which are those based on neutron absorption, variation induced upon certain characteristics of the moderator, modification of neutron leakage from the system, and change of fuel geometry or quantity [1]. The method most frequently used consists in changing the quantity of neutron absorbent materials (control materials) in the reaction zone. Control materials and systems are used for:

(a) operating the reactor, that is performing start-up, shut-down and setting the power rate;

(b) reactor scram (fast reactor shut-down) in abnormal conditions, as an emergency procedure to secure nuclear safety;

(c) keeping the neutron flux at a constant, pre-set value;

(d) smoothing the spatial distribution of the neutron flux over the volume of the active zone;

(e) compensation of reactivity variations in the long run, that are due to fuel consumption and to the accumulation of fission products that absorb neutrons ("poisons"); compensation of reactivity variations following a short shut-down, when the content in $^{135}Xe$ increases as a result of its continuous generation from the existing $^{135}I$ and of discontinuation of its consumption by neutron capture (see Chapter 1).

The materials for reactivity control perform in the reactor as *control rods* whose position can be manually or automatically adjusted, providing for reactor operation steering, ensuring safety and compensating reactivity variations. They may also be introduced as *neutron absorbing substances* dispersed into the fuel, moderator or coolant, to ensure compensation of reactivity variations, a spatially-even neutron flux and nuclear safety.

In order to evaluate the reactivity correction induced by the control rods one starts with the multiplication coefficient for a reactor of large sizes:

$$k = k_\infty/(1 + (L^2 + \tau)B^2) = k_\infty/(1 + M^2B^2), \tag{6.1}$$

where $k_\infty$ is the multiplication coefficient in an infinite medium, $L$ is the diffusion length, $\tau$ is the Fermi age of the neutrons, and $B$ is a parameter characterizing system's geometry, its shape and sizes (see Chapter 1).

The control rod acts only upon the geometry parameter $B^2$, so that the variation of the multiplication coefficient is:

$$\Delta k = -\frac{k_\infty M^2}{(1 + M^2B^2)^2}\Delta(B^2) = -\frac{k^2}{k_\infty}M^2\Delta(B^2). \tag{6.2}$$

The reactivity that is induced by the control rod is then defined as:

$$\frac{\Delta k}{k} = -\frac{kM^2}{k_\infty}\Delta(B^2) \simeq -\frac{M^2}{k_\infty}\Delta(B^2) \tag{6.3}$$

The reactivity that is compensated by the rod (antireactivity) should be equal and of opposite sign to the expression (6.3) of $\Delta k/k$:

$$\rho = [M^2/k_\infty]\Delta(B^2). \tag{6.4}$$

In a thermal reactor, the average life-time ($\tau$) of the prompt neutrons is roughly equal to their diffusion time. For a slightly supercritical reactor, of reactivity $k \simeq 1$, for each neutron of the first generation there will be $(1+\Delta k)$ neutrons in the second generation. Likewise, in the third generation there will be $(1+\Delta k)^2$ neutrons. Thus, after $(m+1)$ generations, there will be $(1+\Delta k)^m$ neutrons in the reactor. The neutron flux $(nv)$ is directly proportional to the number of neutrons:

$$nv = (nv)_o \cdot (1 + \Delta k)^m. \tag{6.5}$$

Taking the logarithm of equation (6.5) one obtains:

$$\ln[(nv)/(nv)_o] = m\ln(1+\Delta k). \tag{6.6}$$

For very small $\Delta k$ values, $\ln(1+\Delta k) \simeq \Delta k$, so that the equation (6.6) reads:

$$nv = (nv)_o\exp(m\Delta k). \tag{6.7}$$

The number $m$ of neutron generations over a time $t$ is $m = t/\tau$, and thus:

$$nv = (nv)_o\exp((t/\tau)\Delta k). \tag{6.8}$$

The reactor period ($T_r$) is defined as the time required for the neutron flux to increase by $e$ ($\simeq 2.71$) times:

$$T_r = \tau/\Delta k. \tag{6.9}$$

As an example, let us consider a reactor rated at a power of 1 W, where the neutrons' lifetime is $\tau = 5.10^{-4}$ s. Let us determine to what extent the reactor power increases after 0.1 s, then after 1 s, assuming a quick lift of the control rods inducing a variation of $k$, $\Delta k = 0.005$. Note that:

$$T_r = \tau/\Delta k = 5 \cdot 10^{-4}/0.005 = 0.1 \text{ s};$$
$$nv = 2.7(nv)_o, \text{ after } 0.1 \text{ s};$$
$$nv = 22{,}000(nv)_o, \text{ after } 1 \text{ s}.$$

The generated power being directly proportional to the neutron flux, it follows that even for a very small change in reactivity ($\Delta k = 0.005$) the power will increase to 2.7 W in 0.1 s, and will soar to 22,000 W in just 1 s. It is practically impossible to keep such a reactor under control.

In this analysis the fact that some neutrons do occur from fission fragments which, after a $\beta^-$-disintegration, have an excess of neutrons has not been accounted for. For example, $^{87}_{35}Br$, losing the excess neutron decays into $^{87}_{36}Kr$. Owing to these neutrons, called delayed neutrons (see Chapter 1), the average lifetime $\tau$ of the neutrons is longer than $\tau_o$ — the lifetime of the prompt neutrons. By grouping the nuclides which emit delayed neutrons into six groups resulting from "forerunners" with comparable decay constants, the expression for $\tau$ is:

$$\tau = (1-\beta)\tau_o + \sum_{i=1}^{6} \beta_i(\tau_i + \tau_o), \tag{6.10}$$

where $\beta_i$ and $\tau_i = 1/\lambda_i$ are the fraction of delayed neutrons and the inverse of the mean decay constant of group $i$, respectively, and

$$\beta = \sum_{i=1}^{6} \beta_i$$

is the total fraction of delayed neutrons. Table 6.1 presents the properties of the groups of delayed neutrons that result from the fission of $^{235}U$ [2], in which case $\tau \simeq 0.1$ s. For $\Delta k = 0.005$, the reactor period $T_r = \tau/\Delta k = 20$ s is long enough to allow the operation of the control systems.

If the delayed neutrons are taken into account, an equation like (6.8) is not exactly satisfactory when it comes to inferring flux evolution. In this case, the kinetic equation of the reactor reads (for $t > 0$):

$$\tau \frac{d(nv)}{dt} = \Delta k(nv) - \int_o^t \left(\frac{d(nv)}{dt}\right)_{t_1} f(t-t_1)dt_1 , \tag{6.11}$$

where $\Delta k = \Delta k(t)$ is the increment in reactivity and $f(t)$ is the yield function of the delayed neutrons:

$$f(t) = \sum_{i=1}^{6} \beta_i \exp(-\lambda_i t). \tag{6.12}$$

TABLE 6.1 Properties of the Groups of Delayed Neutrons at the Fission of $^{235}U$ with Thermal Neutrons

| *Average life-time of the group (s)* ($\tau_i$) | *Fraction of the total number of neutrons* ($\beta_i$) | $\beta_i \tau_i$ *(s)* |
|---|---|---|
| 0.071 | 0.00025 | 0.0000178 |
| 0.62 | 0.00084 | 0.000521 |
| 2.19 | 0.0024 | 0.00526 |
| 6.50 | 0.0021 | 0.01365 |
| 31.7 | 0.0017 | 0.0539 |
| 80.2 | 0.00026 | 0.0208 |
| | $\Sigma\beta_i$ = 0.00755 | $\Sigma\beta_i\tau_i$ = 0.0941488 |

For a very small $t$, one has $f(t) = \beta$ and $\tau_0[d(nv)/dt] = (\Delta k - \beta)(nv) + \beta(nv)_0$, which, in the particular case of $\Delta k = \beta$, leads to $nv = (nv)_0(1 + \beta t/\tau_0)$, and therefore to a linear flux growth. For $\Delta k > \beta$ one obtains an exponential asymptotic variation:

$$nv = (nv)_0 \exp\left[\frac{1}{\tau_0}\int_0^t (\Delta k-\beta)dt'\right] \cdot \left\{1 + \frac{\beta}{\tau_0}\int_0^t \exp\left[-\frac{1}{\tau_0}\int_0^{t''}(\Delta k-\beta)dt'\right]dt''\right\} \quad (6.13)$$

It follows that $\beta$ is an important parameter, which determines the evolution of the flux; as it increases, the flux growth rate decreases. Parameter $\beta$ differs considerably from one fissionable nuclide to another. Thus,

for $^{233}U$, $\beta$ = 0.248%
for $^{235}U$, $\beta$ = 0.75%
for $^{239}Pu$, $\beta$ = 0.364%.

A similar effect is produced by $D(\gamma,n)H$ reactions, that are of consequence in the heavy water reactors, leading to the so-called delayed photoneutrons ($\beta^*$ = 0.1%).

In order to perform their functions, control materials have to meet the following requirements [2]:

(a) To have a total capacity of absorbing neutrons as big as possible, over the entire energy spectrum. *The total absorption capacity*, $C_a$, is defined as the total mass of neutrons that can be absorbed in the unit-volume. No sizeable alteration in the total absorption capacity of the control rod should occur, over its entire time of operation in the reactor.

(b) To exhibit fair mechanical, structural and dimensional stability. In manufacturing control rods there is need that materials free of dimensional changes (swelling) and changes in the mechanical properties be used.

(c) To be corrosion- and wear-resistant in their strongly aggressive environments (high temperatures, pressurized water, water vapours, liquid metals). Corrosion and mass transfer may lead to damage of the control rods and, consequently, to the contamination of the cooling agents with strongly radioactive nuclides.

(d) To be stable under steep and cyclic heating.

(e) To be as low-weight as possible, in order to allow quick displacement of the rods, while not overstraining the driving mechanisms.

(f) To exhibit high melting points (for solid materials), or high boiling points (for liquids), thus enabling their use at high temperatures.

(g) To exhibit a high thermal conductivity, for a fast evacuation of heat that is generated on neutron absorption.

Table 6.2 presents the absorption cross-sections for thermal and resonance neutrons, as well as the total absorption capacity of the chemical elements which meet, more or less, the conditions above [2].

Figure 6.1 displays the dependence, on the energy of the neutrons, of the macroscopic cross-section of absorption, for some highly neutron-absorbant materials [3].

Originally, control rods were made of metallic cadmium. Since this material exhibits a low melting point (312 $^{0}C$), other metals and alloys with higher melting temperatures, and ceramic materials, are currently employed.

## 6.2. REACTOR CONTROL AND SHUT-DOWN SYSTEMS

In keeping with the functions it has to perform, the reactor control system includes units in charge with shut-down, adjustment, compensation of antireactivity that occurs in transient conditions, and reactor power zonal control.

The *shut-down rods* assembly allows the reactor shut-down in normal conditions, and in case this is required by abnormal operation. The antireactivity injected in the reactor is very high (usually in excess of 50 m*k*). The mechanisms of vertical displacement must ensure a high speed of rod-loading into the reactor. The total duration of the assembly shift-in is of 1 — 2 s, but most of the antireactivity is generated in the first few tenths of a second.

For a shut-down that is forced by an abnormal operation of the reactor (reactor scram) a second shut-down system may also be used, with soluble poisons injected in the moderator.

The *control rods* assembly allows reactor power control through their positioning in the active zone at different heights. This system has a lower potential to generate antireactivity as compared with the shut-down system, and also a lower operating speed in the active zone; during reactor shut-down, it ensures an excess of antireactivity over the one introduced by the control rods.

The *adjusting* and *overreactivity rods* assemblies allow the compensation of reactivity variations that accompany changes in the reactor power yield, or at its start-up after a shut-down of short duration. The adjusting rods perform antireactivity compensation when withdrawn from the active zone, while the overreactivity rods, normally outside, compensate it when being moved into the active zone. As a result of their permanent location in the active zone, the adjusting rods absorb neutrons continually, thus affecting adversely, in principle, the overall economy of neutrons. To diminish this shortcoming, cobalt is used as an absorption material in some PHWR reactors; it may subsequently be turned to good account as a radiation source. The active

TABLE 6.2 Nuclear Properties of several Neutron-absorbent Materials

| *Element* | *Atomic mass* | *Isotopic fraction* | *Absorption cross-section for thermal neutrons* | | *Cross-section for resonance neutrons* | | *Neutron capture reac-tions* | $C_a$ |
|---|---|---|---|---|---|---|---|---|
| | | | $\sigma_a$ (b) | $\Sigma_{a_1}$ ($cm^{-1}$) | *Energy* (eV) | $\sigma_a$ (b) | | ($10^3 kg/m^3$) |
| Silver | 107.868 | - | 62 | 3.64 | - | - | (n,γ) | 0.096 |
| | 107 | 0.519 | 30 | 1.77 | 16.6;42; 45;52 | 630 | - | - |
| | 109 | 0.481 | 84 | 4.87 | 5.12;31; 41;57;72;86 | 12500 | - | - |
| Gold | 186.97 | - | 98 | 6.1 | 4.9 | 30000 | (n,γ) | 0.098 |
| Boron | 10.811 | - | 755 | 106.7 | - | - | (n,α) | 0.044 |
| | 10 | 0.188 | 4010 | 596.46 | - | - | (n,α) | - |
| Cadmium | 112.41 | - | 2550 | 118 | 0.178 | 7200 | (n,γ) | 0.009 |
| Dys-prosium | 162.70 | - | 1100 | 35 | 5.5 | 7500 | (n,γ) | 0.12 |
| | 164 | 0.282 | 2600 | 81.43 | - | - | (n,γ) | - |
| Europium | 151.96 | - | 4600 | 95 | - | - | (n,γ) | 0.136 |
| | 151 | 0.478 | 9000 | 188.4 | 0.46;1;3.4; 7.2 | 11000 | (n,γ) | - |
| | 153 | 0.522 | 420 | 8.67 | 2.46;3.8;9 | 3000 | (n,γ) | - |
| Gado-linium | 157.25 | - | 46000 | 1400 | 2.58 | 1000 | (n,γ) | 0.015 |
| | 155 | 0.148 | 70000 | 2145.8 | - | - | (n,γ) | - |
| | 157 | 0.157 | 160000 | 4842.2 | - | - | (n,γ) | - |
| Hafnium | 178.49 | - | 105 | 4.71 | - | - | (n,γ) | 0.090 |
| | 177 | 0.184 | 380 | 17.1 | 2.36 | 6000 | (n,γ) | - |
| | 178 | 0.271 | 75 | 3.34 | 7.8 | 10000 | (n,γ) | - |
| | 179 | 0.138 | 65 | 2.88 | 5.89 | 1100 | (n,γ) | - |
| | 180 | 0.354 | 13 | 0.57 | 73.9 | 130 | (n,γ) | - |
| Indium | 114.82 | - | 190 | 7.26 | - | - | (n,γ) | 0.063 |
| | 113 | 0.042 | 58 | 2.25 | 1.46;3.86; 9.10 | 30000 | (n,γ) | - |
| | 115 | 0.958 | 197 | 7.52 | - | - | (n,γ) | - |
| Iridium | 192.22 | - | 430 | 30.4 | 1.3 | 5600 | (n,γ) | 0.116 |
| Rhodium | 102.905 | - | 150 | 10.9 | 1.26 | 4700 | (n,γ) | 0.122 |
| Samarium | 150.36 | - | 5500 | 152 | - | - | - | 0.006 |
| | 149 | 0.138 | 50000 | 1523.68 | 0.096;0.86;5 | 16000 | (n,γ) | - |
| | 152 | 0.266 | 140 | 4.18 | 8.2 | 15000 | (n,γ) | - |

material for overreactivity rods is $^{235}U$ or Pu. Such rods are used in some PHWR reactors instead of the adjusting rods.

The system relying on added liquid "poisons" injects liquid neutron absorbants (solutions of Gd nitrates, boric acid) into the moderator. In time, Gd

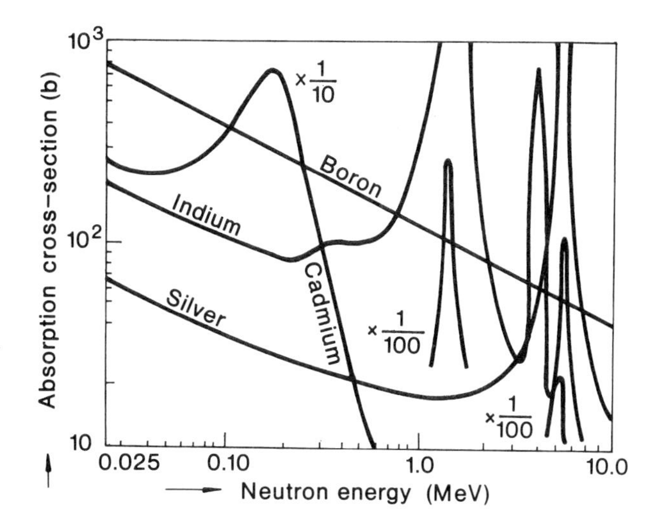


Fig. 6.1. Absorption cross-sections dependence on neutron energy, for certain control materials.

compensates the reactivity changes that appear in the operation after a shut-down of long duration (compensation of the deficit in $^{135}Xe$). Boron compensates the lack of fission products featuring the beginning of a new run, on fresh fuel.

The *zonal control* system allows the adjustment of the neutron flux in different portions of the active zone, which constitutes a special way of changing the curve of power distribution in the reactor. In a PWR reactor this can be done by control rods independently acted upon, whereas in PHWRs, by a special system of cellular rods filled with light water, that allow setting of the local reactivity by changing the amount of water in the cells.

There is a variety of types of control rods [4]; Table 6.3 presents some characteristics of the control rods used in GCR, BWR and PWR reactors.

As an illustration, Fig. 6.2 features the reactivity control assembly in a PHWR-CANDU reactor [5]. It consists of five systems:

(a) *The adjusting rods system*, composed of 21 units. During operation they are in the reactor. By their total withdrawal from the reactor, reactivity increases by 15 *mk*. The excess reactivity is used to compensate the effect of the growth in concentration of $^{135}Xe$, that appears following a reactor shut-down (of less than 30 min), or its operation at low power.

(b) *The light water control system* is employed in the zonal control of the neutron flux. It is composed of 6 rods, each having 2-3 compartments (cells). The amount of light water in the 14 sections can be controlled, both on the whole and in each cell. This system can cut reactivity down by at most 7.5 *mk*, at a maximum speed of 0.14 *mk*/s.

(c) *The solid-absorbent control rods system* consists of 4 rods that can move vertically with variable speed. During operation these rods are normally placed outside the active zone. They are used, together with the light water control system, to set the level of the neutron flux. The rods may

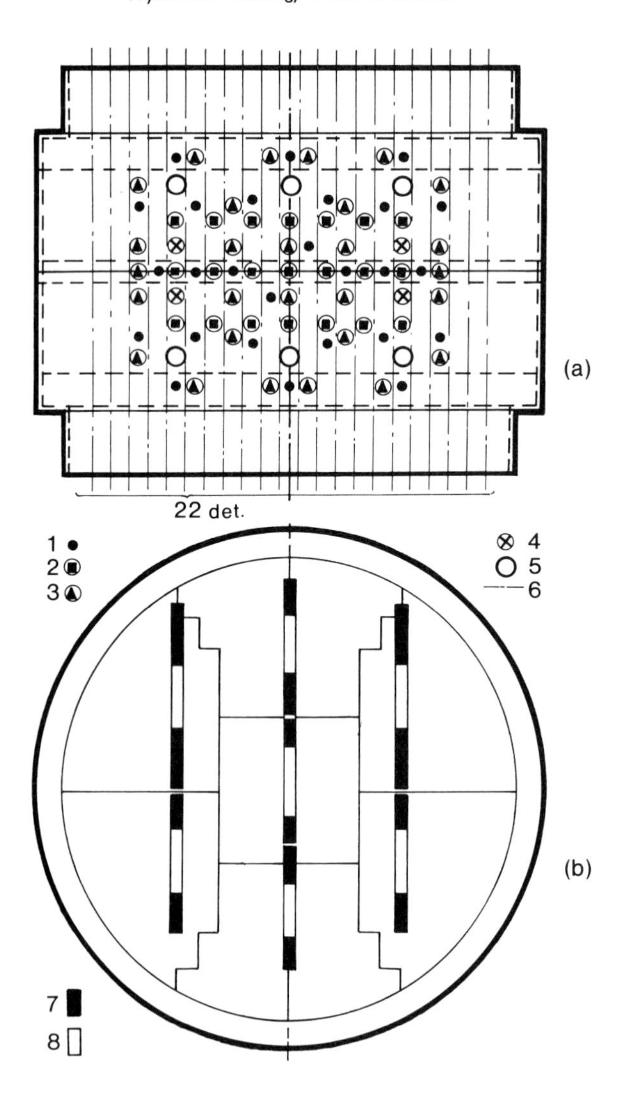


Fig. 6.2. The reactivity control system of a CANDU-PHWR. (a) top view, (b) vertical section showing the location of absorbent light water elements for zonal control: 1 - vertical flux detector (26 detectors), 2 - adjustment rods (21 rods), 3 - shut-down rods (28 rods), 4 - control rods with solid absorbents (4 rods), 5 - control rods with light water (6 rods), 6 - horizontal flux detectors (22 detectors), 7 - flux detectors for the light water zonal control rods, 8 - light water compartments (14 cells) in the zonal control rod.

TABLE 6.3 Characteristics of the Control Rods

| *Characteristics* | *Reactor type* | | | | | |
|---|---|---|---|---|---|---|
| | *GCR* | | *BWR* | | *PWR* | |
| | *EDF 1* | *Latina* | *Kahl* | *Würgassen* | *BR-3* | *Chooz* |
| Reactor power (MWe) | 70 | 200 | 15 | 640 | 10.5 | 266 |
| Number of control rods | 60 | 100 | 21 | 109 | 12 | 30 |
| Absorbent material | $B_4C$ | Boron steel | $B_4C$ | $B_4C$ | Ag-In-Cd (80/15/5) | Ag-In-Cd (80/15/5) |
| Rods dimensions: | | | | | | |
| active length (m) | 6.44 | 7.01 | 1.455 | 3.670 | 1.40 | 2.97 |
| diameter (mm) | 62 | 69.2 | 225.4[1] | 247[1] | 131[2] | 190[2] |
| lattice spacing(mm) | 600 | -[3] | 273.3 | 305 | - | 284 |
| Rod weight (kg) | - | 118 | - | 93 | - | 204 |
| Cladding material | Stainless steel | - | Stainless steel | Stainless steel | - | - |
| Shifting speed (cm/s) | 15 | - | 2.5 | - | 0.32 | 0.64 |
| Setting manner | Continuous | Continuous | Continuous | Continuous | Discontinuous | Discontinuous |

[1]Cross arms length
[2]Lateral cross arms length
[3]Unavailable

introduce a maximum antireactivity of 10.3 m*k*. Also, they are used to prevent initiation of the chain reaction in the event of an accidental lifting of the shut-down rods during intervals when the reactor is not in operation.

(d) *The shut-down rods system*, composed of 28 units, plays the main part in the reactor shut-down procedure.

(e) *The liquid "poisons" control system* allows injection of Gd nitrate or boric acid in the moderator. The system compensates the deficit of $^{135}Xe$ on the reactor start-up following a long shut-down period, or the effect of low concentration of fission products following a fresh fuel reloading. It can also be used for reactor shut-down.

## 6.3. BORON-BASED CONTROL MATERIALS

Boron is one of the best neutron absorbents (Table 6.2). The estimated concentration of boron in the earth's crust is about 0.001%. It is found as borides or boric acid (in the volcanic areas). Pure boron is usually amorphous, and comes as a fine black powder. By distilling this powder crystalline boron is obtained. The isotopic composition of natural boron is 18.8% $^{10}B$ and 81.2% $^{11}B$.

Absorption of thermal neutrons observes the reactions:

$$^{10}_{5}B + n \rightarrow ^{7}_{3}Li + ^{4}_{2}He; \qquad \sigma_a = 4010 \text{ b}$$

$$^{10}_{5}B + n \rightarrow ^{10}_{4}Be + ^{1}_{1}H; \qquad \sigma_a < 0.2 \text{ b}$$

$$^{11}_{5}B + n \rightarrow ^{12}_{5}B + \gamma; \qquad \sigma_a = 0.05 \text{ b.}$$

The high absorption capacity for thermal neutrons is to be ascribed to the isotope $^{10}B$, whose absorption cross-section is 4010 b. The isotope $^{11}B$ has a much smaller absorption cross-section, namely 0.05 b. By neutron capture, boron releases low-energy gamma radiations. The absorption cross-section of the neutrons in the 0.01 — 100 eV range varies inversely proportional to the square root of the energy (the $1/v$ law).

Table 6.4 presents some of the physical properties of boron [2,6].

TABLE 6.4 Physical Properties of Boron

|  |  |
|---|---|
| Density ($kg/m^3$) | |
| amorphous | $2.35 \cdot 10^3$ |
| crystalline | $2.48 \cdot 10^3$ |
| Melting temperature (K) | 2273 — 2348 |
| Boiling temperature (K) | 2823 |
| Knoop hardness | 3300 |
| Specific heat (J/kg . K) | |
| $T$ = 273 K | 600 |
| $T$ = 573 K | 651 |
| $T$ = 1073 K | 1760 |
| Thermal expansion coefficient in the temperature range 293 — 1023 K ($10^{-6}$/K) | 8.3 |

These are usually sensitive to the fabrication process. Also, the irradiation changes the physical properties of boron.

In the manufacture of control rods boron comes as boron carbide, boron-alloyed steels, alloys of boron and aluminium (boral) etc.

*Boron carbide* ($B_4C$) is a black crystalline powder. It is obtained through the reaction of boron oxide with high-purity graphite at 2450 °C [7]. The resulting product is powdered by grinding. To manufacture control rods, the powder is compacted in stainless steel claddings, by vibration. Another way of making rods consists in sintering $B_4C$ pellets at ~ 2200 °C (if binders are not used) or at 600 — 1000 °C, if iron or aluminium binders are used [8]. $B_4C$ may also be dispersed in a host metallic phase. Both pellets and dispersion are then clad in stainless steel tubes.

Some physical properties of $B_4C$ are displayed in Table 6.5 [2,9].

$B_4C$ has a high thermal conductivity. As a result, the temperature distribution over the control rod is relatively even. Occurrence of "hot points" on the surface of the control rod is thus avoided, although the neutron absorption takes place especially at the rod surface.

TABLE 6.5 Physical Properties of Boron Carbide

|  |  |
|---|---|
| Density (kg/m$^3$) | 2.51 . 10$^3$ |
| Melting temperature (K) | 2723 |
| Boiling temperature (K) | > 3773 |
| Knoop hardness | 2800 |
| Tensile strength (MPa) | 6900 |
| Specific heat (J/kg . K) | |
| 300 – 400 K | 1200 |
| 400 – 700 K | 1600 |
| Thermal conductivity (W/m . K) | |
| 373 K | 121 |
| 573 K | 92 |
| 873 K | 75 |
| 1073 K | 65 |
| Thermal expansion coefficient in the temperature range 300 – 1000 K (10$^{-6}$/K) | 4.5 |
| Absorption cross-section for thermal neutrons (b) | 755 |

Boron carbide behaves well in oxidizing environments up to ~ 800 K and, in He, up to 1000 K [7,10]. In the presence of air or carbon dioxide, $B_2O_3$ is generated, which has a relatively low melting point. Thus, it is necessary to avoid a direct contact of $B_4C$ with the air, $CO_2$ or $H_2O$, especially at the relatively high temperatures encountered in the reactor.

For the manufacturing of control rods based on $B_4C$, one has to consider the fact that boron reacts easily with nickel and, to a smaller extent, with stainless steel, but it does not react with copper.

Post-irradiation examination of $B_4C$ control rods using $^{10}B$, irradiated for almost 3000 hours (at a 3.4% burn-up), has confirmed a good irradiation behaviour [11].

Boron carbide is used in the manufacture of control rods for research and material testing reactors, and for power reactors as well (PWR, BWR, HTGR, FBR).

In BWRs, the control rods on $B_4C$ have a cross-shaped transversal cut — Fig. 6.3,a. They are placed between the fuel assemblies — Fig. 6.3,b, their number being about a quarter of the number of fuel assemblies. In this case the stainless steel tubes are filled up with $B_4C$ powder compacted at a

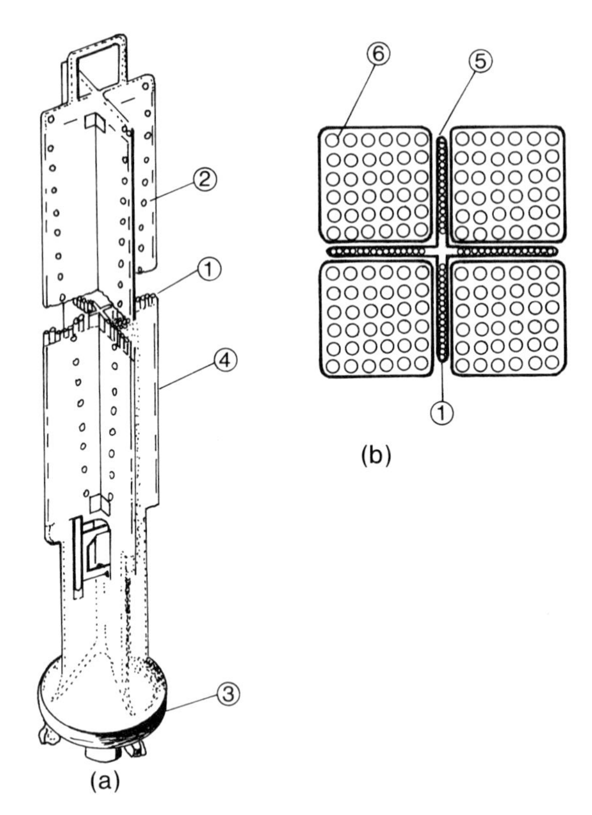


Fig. 6.3. The assembly of reactivity control rods of boron carbide, for a BWR reactor (a); structural pattern of the control system (b). 1 - clad control rod; 2 - roller; 3 - limiter; 4 - flap; 5 - stainless-steel clad; 6 - fuel element.

density of about 65% TD and sealed up by welding with end plugs. On each side of the control rod there are slots where $B_4C$ rods are inserted.

The control rods are fitted on their entire length with rolls to ensure easy sliding between the fuel assemblies.

$B_4C$ is used as a reactivity adjusting material in PWR reactors as well [12]. In this case $B_4C$ pellets, having a density of ~73% of the theoretic density, are clad in stainless steel tubes. These tubes are provided with enough room to store the He that is generated in the $^{10}B$ (n,$\alpha$) reaction and also the water vapours that emerge from $B_4C$. The rods have a circular transversal cut and are fastened on a grid in groups of four elements. The grid is moved across the active zone by a carrying mechanism. Eventually the grid is provided also with a central rod that carries temperature sensors and the detectors of flux measuring devices. Usually, the length of the rods is equal to that of the active zone, but shorter rods may also be used. Part of the fuel assemblies (almost a third) have 4 or 5 fuel rods replaced by guiding tubes, in which rods of the reactivity control assembly operate.

For HTGR reactors current designs stipulate control rods with a length of almost 6 m, and an outer diameter of ~ 0.08 m. Ring-shaped pellets of boron carbide, obtained by hot pressing, are clad in coaxial stainless steel tubes. The surfaces of the tubes in direct contact with $B_4C$ are copper-coated [13]. Heat generated by neutron and photon absorption is evacuated by appropriately chosen cooling agents, a prime candidate being He. The helium generated by the irradiation of $B_4C$ is caught in the circulation of the cooling agent. To avoid damage of the cladding as a result of $B_4C$ swelling, spacings are provided between the absorbent elements and the stainless steel claddings.

The heat density generated per unit-length of a control rod based on $B_4C$ is about 6800 W/m. Figure 6.4 presents the axial distribution of temperatures in a control rod [10,13]. The figure also provides a comparative reading of the distribution of both the cladding surface temperature and that of the cooling agent (helium).

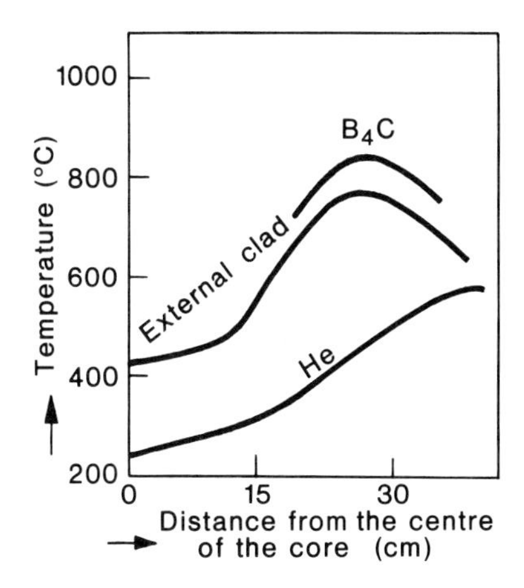


Fig. 6.4. Axial temperature distribution in a boron-carbide control rod (EGCA reactor).

The $B_4C$ rods are also used in fast neutron reactors. To this purpose, boron enriched up to 90% $^{10}B$ is required [14].

*Steels with boron* may also be used as control materials. Figure 6.5 presents part of the Fe — B phase diagram [15]. It shows that the proportion of boron that is still soluble in γ-iron falls far below the concentration that would enable the use of boron as an ingredient in the manufacture of stainless steel control rods (0.2 – 2%).

When in excess to the solubility limit, boron precipitates as borides. In Fig. 6.6 the dependence on boron concentration of elongation and tensile strength, $\sigma_r$ and $\sigma_{0.2}$ respectively, of a stainless steel are represented [16].

The elongation strongly decreases for a concentration of 0.6 — 0.8 % B in weight. The values of $\sigma_{0.2}$ reach a maximum at 0.7% B in weight, whereas $\sigma_r$ increases continually with the boron content.

A decline in ductility and a swelling, as a result of He formation, are recorded during irradiation. The swelling process is not isotropic, as (n,α)

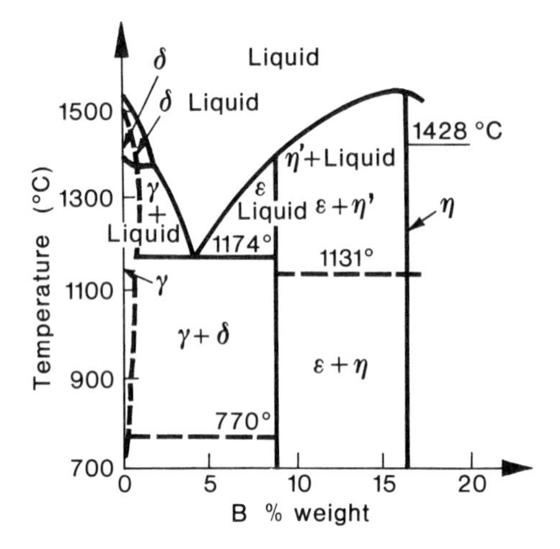


Fig. 6.5. Phase diagram of the Fe — B system in the range of 0 — 20 % boron in weight.

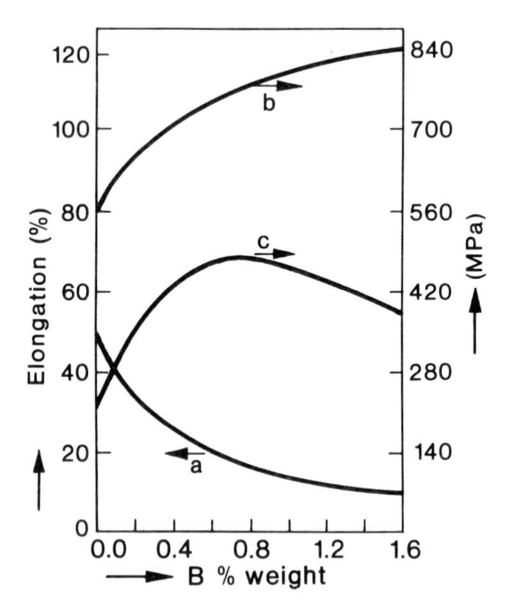


Fig. 6.6. Mechanical properties of a boron stainless-steel vs. the boron content: (a) elongation, (b) rupture strength $\sigma_r$, (c) $\sigma_{0.2}$.

reactions take place at the surface of the material. Swelling becomes sizeable at burn-ups of ~ 5%; at about 16% the surface is destroyed. To avoid damage of the control rod, a certain non-even distribution of boron across the rod must be obtained, the boron concentration increasing with the distance from the material surface [17].

The *Boral* compound is obtained by mixing boron in melted aluminium, or by pressing together $B_4C$ and aluminium powders, the product being then sintered at temperatures below 650 °C. The material is delivered as plates ~ 6 mm thick, boron content being of about 50% in weight. The plates have a density of ~ $2.5 \times 10^3$ kg/m$^2$. Thermal conductivity at 500 K is 330 W/m . K and the specific heat is about 730 J/kg . K.

## 6.4. CADMIUM-BASED ALLOYS

Cadmium is a good neutron absorbent (Section 6.2). Its uses as a control material are, however, limited due to its relatively low melting temperature, and to the penetrating gamma radiations that are generated during neutron capture.

In the earth's crust there is cadmium in proportion of 0.0005% in weight. Fresh cadmium is a shiny metal, very ductile and, consequently, easy to process. Table 6.6 presents certain physical and mechanical properties of cadmium [18].

TABLE 6.6 Physical Properties of Cadmium

|  |  |
|---|---|
| Density (kg/m$^3$) | $8.64 \cdot 10^3$ |
| Melting point (K) | 594 |
| Boiling point (K) | 1038 |
| Rupture strength (MPa) | 72 |
| Elongation (%) | 50 |
| Elasticity modulus (MPa) | $0.5 \cdot 10^5$ |
| Poisson ratio | 0.30 |
| Brinell hardness | 21 – 23 |
| Specific heat at 293 K (J/kg . K) | 230 |
| Thermal conductivity (W/m . K) at temperatures of: | |
| 293 K | 96 |
| 373 K | 94 |
| 473 K | 92 |
| Thermal expansion coefficient in the temperature range 293 – 373 K ($10^{-6}$/K) | 31.8 |

Cadmium oxidizes in air and is commonly covered with a grey oxide layer. Over ~ 573 K its colour changes to brown. At temperatures below 378 K cadmium does not react with water; the corrosion rate increases very much with temperature.

Cadmium exhibits a high cross-section for thermal neutrons, namely 2550 b. As the neutron energy gets higher than 0.178 eV, the cross-section starts to decrease rapidly. Therefore, this element has a low absorption capacity for epithermal neutrons.

Pure cadmium cannot be used as a control material. For the cadmium-based control rods manufacture silver-cadmium alloy, silver-indium-cadmium alloy or cadmium tantalate are used.

*Cadmium- and silver-based alloys* are used as reactivity control materials for research reactors, where the temperatures reached by the control rods are relatively low. Ag-Cd alloys combine silver's large absorption cross-section for resonance neutrons with cadmium's large absorption cross-section for thermal neutrons. They have a good mechanical resistance only at relatively low temperatures. Ag-Cd alloys used as control materials are α-solid solutions, with a fcc structure [2]. The cadmium content is 20 — 30 % in weight, as shown in Fig. 6.7. Liquidus curve is at ~800 °C. The alloy re-crystallizes at about 250 °C.

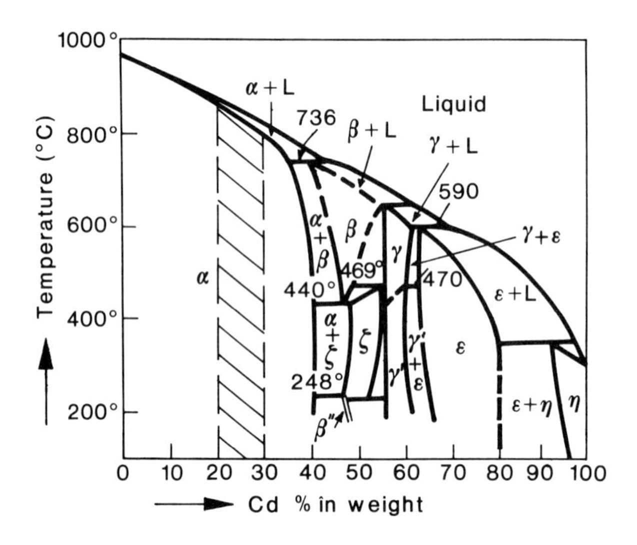


Fig. 6.7. Phase diagram for the indium — cadmium system.

Indium is a good absorbent for resonance neutrons. Thereby its presence in Ag-Cd alloys improves absorption properties. It also improves the mechanical and corrosion resistance at high temperatures of Ag-Cd alloys. This allows the use of Ag-Cd-In control rods in LWRs. The composition designed for control rods is Ag, 80% in weight; In, 15% in weight; and Cd, 5% in weight, and they are obtained by moulding and hot extrusion. Finally, the material is cased in a stainless steel cladding, with which the alloy makes a metallurgical bond.

The Ag-In-Cd alloy as described is widely used for reactivity control in PWR reactors. The rods, of a circular section, are clad in stainless steel tubes. They slide in guiding tubes (about 20) scattered across the net of fuel rod assemblies.

About one-third of the assemblies contain guiding tubes. The control rods bundle is fastened at the bottom by a grid driven by the control rods mechanism. One pattern of distribution of the guiding tubes inside a fuel assembly is rendered in Fig. 6.8. (viz. also Chapter 9).

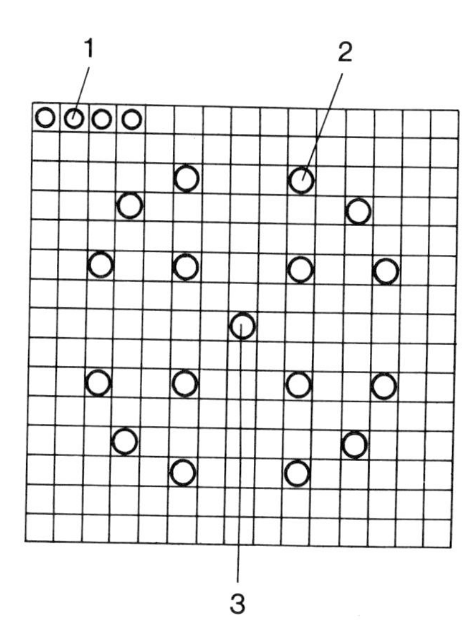


Fig. 6.8. Pattern of guide tubes inside a fuel element holder: 1 - fuel element, 2 - guide tube for Ag — Cd or Ag — Cd — In control rods, 3 - available for measuring devices.

*Cadmium tantalate* has better absorbent properties than pure cadmium. By neutron absorption, the isotope $^{181}Ta$ turns into $^{182}Ta$. This isotope has a very large (17,000 b) neutron absorption cross-section. The accumulation of $^{182}Ta$ preserves the absorption capacity of the material, thus compensating Cd "burn-up".

Cadmium tantalate is obtained by heating together $Ta_2O_5$ and CdO at 1223 K. The powder, of a density of $8.21 \times 10^3$ kg/m$^3$ [19] consists of irregularly shaped particles, with sizes of $4 \times 10^{-6}$ to $27 \times 10^{-6}$ m.

Compacts resistant to thermal shocks are obtained by pressing and sintering the material at 1223 K for 15 hours. Cadmium tantalate is compatible with nickel (at $T < 1073$ K) and with copper (at $T < 1273$ K) [19]. Thus, cadmium tantalate can be dispersed in nickel or copper matrices, that are then clad in stainless steel tubes.

## 6.5. HAFNIUM

Hafnium may be used in the manufacture of reactivity control rods for PWR reactors. It has a high corrosion resistance in water, at high temperatures [2]. Hafnium isotopes have considerable absorption cross-sections for thermal neutrons, and display outstanding resonances for epithermal neutrons — Table 6.2. As a result of neutron absorption, hafnium isotopes generate nuclides that are, in their turn, efficient neutron absorbents. Thereby the total absorption capacity of hafnium can be secured over a long time.

However, as a result of the fact that natural hafnium is available only in minute quantities, its use in the manufacture of control rods is limited.

## 6.6. RARE EARTHS-BASED CONTROL MATERIALS

Table 6.7 presents some of the physical properties of the elements known as "rare earths"; the fact that gadolinium, samarium, europium and dysprosium can be used as control materials owing to their high neutron absorption cross-section is outstanding among these [20].

TABLE 6.7 Physical Properties of the Rare Earths

| *Element* | *Charge number* | *Atomic mass* | *Ionic radius* (Å) | *Crystalline structure* | *Density* $(kg/m^3) \cdot 10^3$ | *Melting point* (K) | *Absorption cross-section* (b) |
|---|---|---|---|---|---|---|---|
| Gd | 64 | 157.25 | 1.1 | - | 7.95 | 1473 | 46000 |
| Sm | 62 | 150.36 | 1.13 | - | 6.93 | 1573 | 5500 |
| Eu | 63 | 151.96 | 2.04 | bcc | 5.22 | 1373-1473 | 4600 |
| Dy | 66 | 162.50 | 1.77 | hcp | 8.56 | 1773 | 1100 |
| Er | 68 | 167.20 | 1.75 | hcp | 9.15 | 1713 | 173 |
| Tm | 69 | 169.40 | 1.74 | hcp | 9.35 | 1823-1923 | 137 |
| Lu | 71 | 174.99 | 1.73 | hcp | 9.74 | 1923-2023 | 113 |
| Ho | 67 | 164.94 | 1.76 | hcp | 8.76 | 1773 | 65 |
| Nd | 60 | 144.27 | 1.82 | hcp | 6.98 | 1093 | 46 |
| Tb | 65 | 159.20 | 1.77 | hcp | 8.33 | 1673 | 46 |
| Yb | 70 | 173.04 | 1.93 | fcc | 7.01 | 2073 | 37 |
| Pr | 59 | 140.92 | 1.83 | hcp | 6.78 | 1223 | 11.3 |
| La | 57 | 138.92 | 1.87 | hcp | 6.19 | 1139 | 8.9 |

Rare earths have a weak resistance to corrosion and they easily oxidize. To overcome this shortcoming, in the manufacture of control rods rare earth oxides and dispersions in metallic matrices are employed instead. Thus, in high-temperature reactors oxides of the rare earths are in order, while in PWRs dispersions of the rare earth oxides in stainless steel, clad in stainless steel tubes, are preferred. The advantage of this solution is that, on an accidental rupture of the cladding, only the surface layer of oxides will hydrate.

The total high absorption capacity of europium is explained by the fact that along the chain $^{153}Eu \rightarrow {}^{154}Eu \rightarrow {}^{155}Eu$ capture cross-sections are comparable. Thus $^{153}Eu$, found in natural Eu in a proportion of 52%, is replaced over its consumption by equally efficient neutron absorbing nuclides. But the amount of $Eu^{151}$ and $Eu^{155}$ nuclides is dwindling in time; the latter, as a consequence of the short half-life of $Eu^{156}$, practically breaks the chain [21] (Fig. 6.9).

Europium, as $Eu_2O_3$, can be dispersed in stainless steel [22]. The steel must have a low content in silicon to prevent the formation of silicates, that make the material brittle. These metallo-ceramic dispersions may contain $Eu_2O_3$ up to 50% in weight. Control rods are obtained by cladding the dispersion in stainless steel claddings, sometimes followed by rolling or extrudation. As noted already, the stainless steel cladding is meant to avoid the contact of the dispersion with water, with which $Eu_2O_3$ might react.

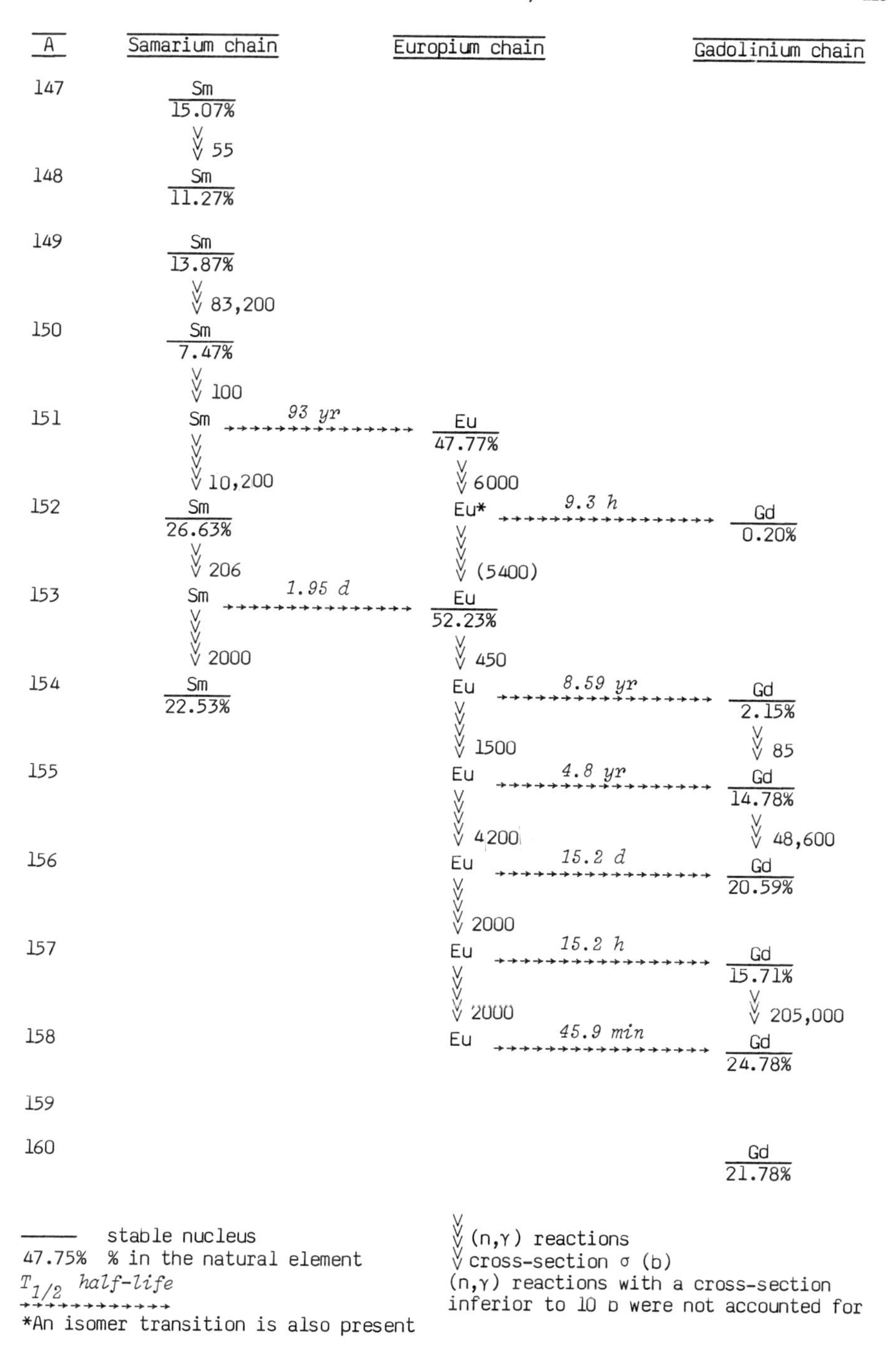


Fig. 6.9. The decay pattern of Sm, Eu and Gd nuclides.

Control rods like the ones described are used in special reactors, that require a small core.

Recent designs of fast reactors also contemplate the use of $Eu_2O_3$-based dispersions as control materials [14].

Figure 6.10 shows the temperature dependence of $Eu_2O_3$ thermal conductivity. Compared with stainless steel, $Eu_2O_3$ has a lower thermal conductivity. Moreover, the linear coefficient of thermal expansion $\Delta l/l_0$ is lower than that of stainless steel [22]. Thus, the heating of clad rods does not raise any particular problem.

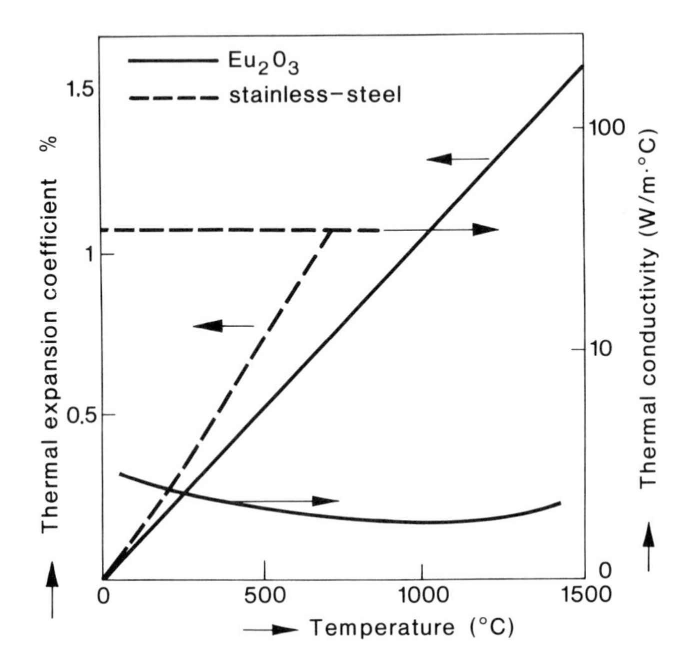


Fig. 6.10. $Eu_2O_3$ and stainless steel thermal conductivity and thermal expansion coefficient, as functions of temperature.

Hydration of $Eu_2O_3$ is inhibited in the presence of certain oxides — of titanium, aluminium, wolfram etc.

Samarium and gadolinium have their nuclides chains shown in Fig. 6.9. As control materials their properties differ from those of europium. Thus, for samarium, the natural nuclide $Sm^{149}$ has the greatest contribution to the capture cross-section; when this isotope turns into $Sm^{150}$, the cross-section drops by a factor of $10^2$ — $10^3$, which makes the element a consumable absorbent.

Like samarium, gadolinium is a consumable absorbent.

To enhance corrosion resistance, gadolinium is usually alloyed with a stainless steel, having a low content in carbon ($< 0.02\%$ in weight), at a level of 5 — 30 % gadolinium in weight.

Table 6.8 gives the mechanical properties of some alloys of gadolinium with stainless steels in the AISI-304 L class [23] (see also [22]).

TABLE 6.8 Properties of certain Alloys: Stainless Steel[1] – Gadolinium

| *Gadolinium concentration (% in weight)* | *Mechanical properties* | | | |
|---|---|---|---|---|
| | *Yield strength* (MPa) | *Rupture strength* (MPa) | *Elongation* (%) | *Cross-section reduction* (%) |
| 0 | 388.8 | 639.8 | 66 | 80.5 |
| 0.1 | 314.2 | 596.7 | 72 | 80.6 |
| 0.2 | 307.2 | 608.2 | 57 | 76.4 |
| 0.5 | 386.6 | 650.4 | 44 | 63.5 |
| 0.75 | 311.5 | 632.8 | 56 | 59.1 |
| 1.00 | 315.0 | 629.3 | 56 | 56.4 |
| 2.00 | 292.5 | 376.2 | 6 | 14.8 |
| 3.00 | 289.7 | 457.0 | 7 | 4.68 |

[1]AISI-304 L stainless steel.

The flow and fracture strength, as well as the elongation, decrease with the growth in gadolinium content.

Samarium and gadolinium oxides might be used as steel-clad dispersions in the manufacture of consumable control rods, instead of boron rods. Gadolinium oxide ($Gd_2O_3$) is used as a consumable ingredient in the fuel. To this effect, pellets of $UO_2 - Gd_2O_3$ are prepared.

## REFERENCES


1. Galanin, A. D. (1957). *Teorya iadernyh reaktorov na teplovyh neitronah.* Atomizdat, Moskow; Ganev, I. H. (1981). *Fizika i rastchet reaktora.* Energoizdat, Moskow.
2. Tipton, C. R. (1964). *Reactor Handbook.* Vol. I, *Materials* , New York; USAEC (1960). USAEC Reactor Handbook. Vol. I, *Physics.* Interscience Publ., New York.
3. Weissman, J. (1977). *Elements of Nuclear Reactor Design.* New York.
4. Burducea, C., Motoiu, C., Vasilescu, N., Mingiuc, C., and Wlezek, C. (1974). *Centrale nuclearoelectrice de putere mare.* Edit. Tehnica, Bucharest; Semenov, V. N., and others (1963). Opredelenye effectivnosty reguliruyuschyh steryney v sborkah reaktora VVER. In *Proc.Symp.on Physics and Material Problems of Reactor Control Rods*, Vienna, p. 153.
5. Cooper, W. R. (1981). *Aparate de control şi instrumentatie CANDU*, Seminar tehnic, Bucharest; AECL (1977). *Data Base for CANDU-PHW Operating on a Once-Through Natural Uranium Cycle*, AECL 6593.
6. Cooper, H. S. (1961). *Rare Metals Handbook*, New York.
7. Jaeger, R. G. (1975). *Engineering Compendium on Radiation Shielding.* Vol. II, *Materials.* Springer-Verlag, Berlin-Heidelberg-New York.
8. Suri, A. K., and Gupta, C. K. (1978). *J.Nuclear Materials*, 74, 114.
9. Deem, H. W. (1951). Report BNL-713.

10. Michel, W. (1964). *Design and Thermal Analysis of EGCR Control Rods.* USAEC Report, ORNL 3503.
11. Brammer, H. A. (1963). Proc.Symp. Physics and Material Problems of Reactor Control Rods, Vienna.
12. Pederson, E. S. (1978). *Nuclear Power.* Vol. I, Ann Arbor Science Publ.Inc., Ann Arbor, p.135
13. Michel, W. (1963). Proc.Symp. Physics and Material Problems of Reactor Control Rods, Vienna, p.415
14. Cuculeanu, V. (1982). *Fizica şi calculul reactorilor nucleari cu neutroni rapizi*, Edit. Tehnica, Bucharest.
15. Hansen, P. (1958). *Constitution of Binary Alloys.* McGraw-Hill Book Comp.
16. Anderson, W. K. (1960). In *Boron Reactor Handbook*, Vol. II, 2nd ed., *Materials*, Interscience Publishers Inc., New York.
17. Beaver, R. J. (1972). *Nucl.Technology*, 16, 187.
18. Ageron, P. (1963). *Technologie des Réacteurs Nucléaires.* Vol. I, *Matériaux*, Edit. Eyroles.
19. Priesler, E. (1963). Proc.Symp.Physics and Material Problems of Reactor Control Rods, Vienna.
20. Hughes, D. J. (1960). Raport BNL-325.
21. Walker, W. H. (1975). FISSPROD-2 AECL 5105.
22. Gelchrist, K. E. (1975). Thermophysical Property Measurements of RFL Springfield, *Atom*, 257, 240; Savelev, E. G.(1963). Materialy dispersionogo typa dlya pogloshayuschyh steryney iadernyh reaktorov. In *Proc.Conf.on New Nuclear Materials including Non-Metallic Fuels, Prague 1963*, Vol. II, p. 237; Portnoy, K. I. (1963). Zakonomernost' izmenenya svoystvosty pogloshayuschyh materyalov v zavysimost' ot kontsentratsyi provoditelya. In *Proc.Symp.on Physics and Material Problems of Reactor Control Rods*, Vienna, p. 283.
23. Copeland, M. (1963). Proc.Symp.Physics and Material Problems of Reactor Control Rods, Vienna.

CHAPTER 7

# Coolant Materials

## 7.1. GENERAL

The thermal energy generated in the fuel elements as a result of nuclear reactions is evacuated from the reactor core by means of selected fluids known as *coolants*. The heat at the surface of the fuel claddings is transfered to the coolant, which transports it to the steam generators.

Considering the manner in which the coolant stores the thermal energy when passing through the reactor one distinguishes:

1. coolants that keep their physical state, storing the heat by the growth of their internal energy with no phase transition;

2. coolants that change their physical state, storing the heat as vaporization heat;

3. coolants that change their chemical state, storing the heat as reaction heat.

The first two types are currently in use. The third, illustrated by gaseous $N_2O_4$ which, passing through the reactor, dissociates (into $NO_2$ and $NO+O_2$), is presently studied, in view of a design of an advanced single-circuit fast reactor [1].

As a rule, a nuclear reactor has two cooling circuits known as the primary and the secondary circuit respectively, the heat being transferred from the reactor core to the primary, then to the secondary circuit and then to a heat exchanger, which is, in effect, a steam generator.

To give an idea of how the cooling circuits work, their operation schemes in heavy water and pressurized light water reactors are briefly commented on in the following sections.

Figure 7.1 presents the cooling system of a PHWR reactor [2]. It consists of two loops, (1) and (2); each loop ensures the cooling of half of the 380 pressure tubes in which fuel bundles (380x12) are introduced. The coolant (pressurized heavy water) in loop (1) passes through the core in the direction

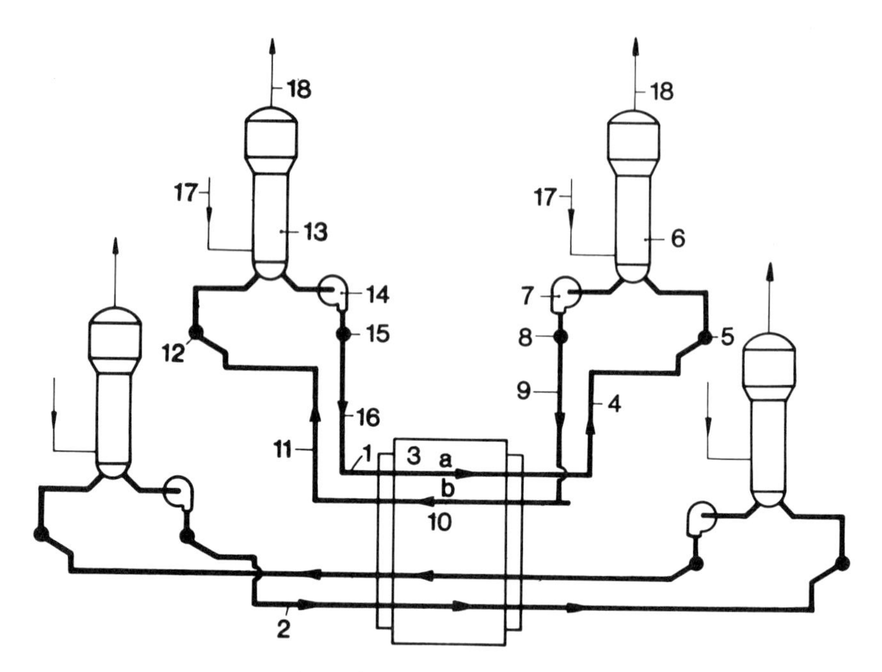


Fig. 7.1. PHWR (CANDU) primary cooling circuit: 1,2 - cooling loop; 3,10 - pressure tube; 4,9,11,16 - connecting pipe, between pressure tube and collector; 5,8,12,15 - collector; 6,13 - steam generator; 7,14 - pump; 17 - steam generator water supply (secondary circuit); 18 - steam outlet.

indicated by the arrow (a) through 95 pressure tubes, and then, through the connecting pipes (4) reaches the collector (5) and the steam generator (6). The heavy water is then sucked by pump (7) and sent to the collector (8), wherefrom, through the connecting pipes (9) it overpasses the core, in the direction of the arrow (b) through the second group of 95 pressure tubes (10) of the loop (1). Then, the heavy water reaches the collector (12) and the steam generator (13) through the connecting tubes (11). From here it is sucked by pump (14), that brings it to the collector (15); then, through the connecting pipes (16) the water re-enters the first group of pressure tubes. The operation of loop (2) is identical to that of loop (1), as described. The temperature of the primary agent (heavy water) at the inlet of the pressure tubes is 540 K, and the pressure is 11.05 MPa. At the outlet the temperature of the heavy water is 585 K, and the pressure 10.3 MPa. Generators (6) and (13) produce steam at 4.55 MPa and 531 K.

Figure 7.2. shows the scheme of the coolant circuit of a pressurized light water reactor, type VVER-440 [3]. The primary circuit consists of six loops that circulate a total volume of 39,000 $m^3$ per hour. The water enters the reactor pressure vessel (1) through six pipes and cools the core by a bottom-to-top circulation. The water is heated from an inlet temperature of 541 K to the outlet temperature of 574 K. Owing to the pressurizer (3) the water pressure remains constant at a value of ~12.7 MPa. Then the water enters the six steam generators (4), where its temperature returns to 541 K; then it is pumped back in the pressure vessel. The steam at 532 K and 4.8 MPa, as

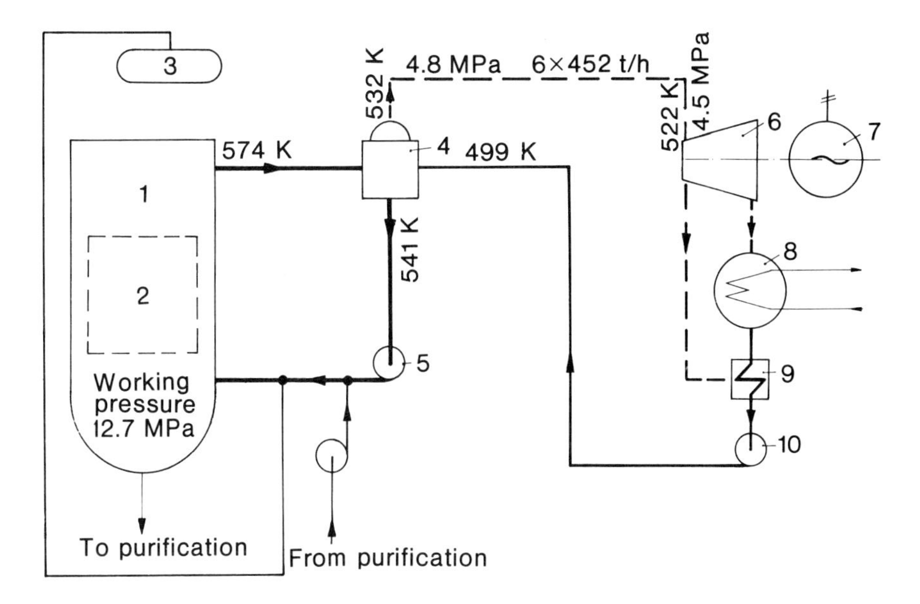


Fig. 7.2. PWR (VVER-440 Novovoronezh 3-4) primary cooling circuit: 1 - reactor; 2 - core; 3 - pressurizer (38 m$^3$, 2.4 m internal diameter); 4 - steam generator (6 units, 2440 m$^2$ each); 5 - pump (6 units, 6500 m$^3$/h each); 6 - turbine (2 units); 7 - electric generator (2 units, 220 MWe, 6.3 kV each); 8 - condenser (4 units, 18275 t/h at 330 Pa each); 9 - pre-heater; 10 - pump.

obtained in the generators, drives two turbines (6), that in turn drive the generator (7). After passing through the turbines, the steam goes to the condensers (8) and the preheater (9); then, through the loading pump (10) the secondary cooling circuit closes.

Primary coolants that change their physical or chemical state are especially considered for one-circuit-cooling systems. In this case, the coolant is injected directly into the turbine.

Considering the important part they play and the special working conditions they are subject to, the cooling materials must meet certain special requirements. For the gas coolants they list as follows:

1. a good chemical and radiochemical stability;

2. low levels of induced radioactivity; short half-life of the nuclei generated by neutron-induced reactions;

3. high density, specific heat and thermal conductivity;

4. a good and lasting compatibility with the structural and moderator materials with which they come into contact; at least a short-lived compatibility with the nuclear fuel in case of accidental damage to the claddings.

For the liquid cooling materials, the following requirements are further in order:

1. a low melting temperature, a critical temperature as high as possible, and a critical pressure as low as possible;

2. low viscosity;

3. small neutron capture cross-section (this general request is much more important for liquids than for gases, where the absorption macroscopic section $\Sigma_a = \sigma N$ is small, due to the low values of $N$).

The liquids used for cooling fast reactors must also have a moderating power as weak as possible, to avoid the growth of the fraction of low-energy neutrons, which would reduce the efficiency in the generation of fissionable nuclides.

Table 7.1 presents the physical properties of some coolants [4]. Among the gaseous coolants mention must be made of the air, used in the first-generation reactors designed for plutonium production; and carbon dioxide — for the power reactors with natural uranium moderated with graphite, and for advanced reactors on enriched uranium. For high and over-high temperature reactors the use of He as coolant is envisaged. The cooling gas most frequently used in power reactors, is, at present, carbon dioxide.

TABLE 7.1 Some Characteristics of the Coolants

| *Material* | *Melting point* (°C) | *Boiling point* (°C) | *Temperature*[4] (°C) | *Density* $\rho$ ($kg/m^3$) | *Specific heat* $c_p$ ($kJ/kg\cdot K$) | *Viscosity* $\mu$ ($N\cdot s/m^2$) | *Absorption cross-section for thermal neutrons* (b) |
|---|---|---|---|---|---|---|---|
| $H_2$[1] | - | - | 100 | 0.066 | 14.356 | $1\times10^{-5}$ | 0.66 |
| He[1] | - | - | 100 | 0.13 | 5.193 | $2.2\times10^{-5}$ | $8\times10^{-3}$ |
| Air[1] | - | - | 100 | 0.95 | 1.004 | $2.1\times10^{-5}$ | 1.5 |
| $CO_2$[1] | - | - | 100 | 1.44 | 0.932 | $1.83\times10^{-5}$ | $5\times10^{-3}$ |
| $H_2O$[2] | 0 | 100 | 20 | 998.2 | 4.183 | $1004\times10^{-6}$ | 0.6016 |
| $H_2O$[2] | 0 | 100 | 100 | 958.4 | 4.220 | $282\times10^{-6}$ | 0.6016 |
| $D_2O$[2] | 3.82 | 101.43 | 20 | 1105 | 4.208 | $1253\times10^{-6}$ | $9.2\times10^{-4}$ |
| Na | 97.3 | 878 | 100 | 928 | 1.386 | $714\times10^{-6}$ | 0.45 |
| K | 63.7 | 760 | 100 | 818 | 0.816 | $458\times10^{-6}$ | 2.5 |
| NaK[3] | -19 | 825 | 100 | 852 | 0.955 | $517\times10^{-6}$ | 2.0 |
| NaOH | 323 | - | 350 | 1790 | 2.318 | $1\times10^{-3}$ | 0.78 |

[1] At 1 atm.
[2] Liquid.
[3] With 25% Na.
[4] Reference temperature for $\rho$, $c_p$ and $\mu$.

For the power reactors on thermal neutrons (PWR, BWR and PHWR) the coolant is light water (either pressurized or boiling) and pressurized heavy water, respectively. Liquid Na or the liquid alloy Na — K are used as cooling agents for LMFBRs. Designs were prepared of gas-cooled fast breeder reactors (GCFBR) using carbon dioxide, helium or nitrogen oxide as coolants, the latter having superior nuclear qualities.

Finally, a certain class of thermal reactors rely on the use of organic liquids as coolants.

## 7.2. REMOVAL OF HEAT FROM NUCLEAR REACTORS

The thermal energy produced in the nuclear fuel comes mainly from the kinetic energy of fission fragments [4], as shown in Chapter 1. For this reason, the heat generated in the unit-volume is proportional to the fissioned fuel fraction in the unit of time:

$$Q = a\Phi N_o \sigma_f , \tag{7.1}$$

where $N_o$ represents the number of atoms in a cubic meter of fuel, $a$ is the energy released in the fuel at each fission act (~ 181 MeV), $\Phi$ is the neutron flux and $\sigma_f$ is the fission cross-section.

The total heat (MeV/s) produced in the nuclear reactor is:

$$Q_t = a\bar{\Phi} N_o \sigma_f V, \tag{7.2}$$

where $\bar{\Phi}$ is the average neutron flux and $V$ is the fuel volume ($m^3$).

The heat recovery is performed by the cooling fluids whose temperature at the inlet channel of the reactor, $T_{f_o}$, gradually increases with the distance covered in the channel. The thermal balance of the channel is expressed by the equation:

$$Gc_p dT_f = q(z) n\pi r_o^2 dz, \tag{7.3}$$

where $G$ is the flow of the cooling agent, $c_p$ is the specific heat at constant pressure, $dT_f$ represents the increment in the fluid temperature as it covers a distance $dz$ in the channel, $q(z)$ is the heat generated in the unit-volume of fuel, $r_o$ is the radius of the fuel element and $n$ is the number of rods in the channel.

Thus, the temperature of the cooling agent at level $z$ is obtained by integrating the equation (7.3):

$$T_f(z) = T_{f_o} + n(\pi r_o^2)(Gc_p)^{-1} \int_o^z q(z) dz. \tag{7.4}$$

The difference in temperature between the outer surface of the cladding and the fluid is obtained from the equation:

$$\varphi = h(T_s - T_f), \tag{7.5}$$

where $\varphi$ is the local heat flow through the unit-area of the clad/coolant contact interface, and $h$ is the clad-to-coolant heat transfer coefficient.

The evacuation of heat from *PWR and PHWR reactors* is done by the "forced convection" of the pressurized water. The general equation to determine the transfer coefficient is the Dittus-Boelter equation:

$$\mathcal{N}u = a\,\mathcal{R}e^{0.8}\,\mathcal{P}r^{0.4}\;, \tag{7.6}$$

where $\mathcal{N}u$ is the Nusselt number ($\mathcal{N}u = [hd_n/K]$, $h$ is the heat transfer coefficient, $d_n$ is the equivalent diameter, $K$ is the fluid thermal conductivity); $a$ is a constant value ($a$ = 0.023); $\mathcal{R}e$ is the Reynolds number ($\mathcal{R}e = Vd_n/(\mu/\rho)$, $V$ is the average speed of the fluid in the considered section, $\rho$ is the density of the fluid and $\mu$ is its dynamic viscosity; $\mathcal{P}r$ is the Prandtl number ($\mathcal{P}r = \mu c_p/K$).

If the fluid flows under a big difference between its temperature and the contact surface, the transfer coefficient is determined from the equation [1]:

$$hd_n/K = 0.023\,\mathcal{R}e^{0.8}\,\mathcal{P}r^{1/3}\,(\mu/\mu_s)^{0.14}, \tag{7.7}$$

where $\mu_s$ is the dynamic viscosity of the cooling agent at the temperature of the fluid film adherent to the cladding surface. The relation (7.7) is valid in the case $L/d_n > 60$, where $L$ is the length of the channel. When the tubes are cooled by natural convection the transfer coefficient is obtained from

$$\mathcal{N}u = 0.5(\,\mathcal{G}r\,.\,\mathcal{P}r\,)^{0.25}(\mu/\mu_s)^{0.25}, \tag{7.8}$$

where $\mathcal{G}r$ is Grashof number given by the equation:

$$\mathcal{G}r = \beta\Delta T\,\frac{gd_n^3\rho^2}{\mu^2}\;.$$

Here $\beta$ denotes the volume expansion coefficient of the fluid, $g$ is the gravity acceleration and $\Delta T$ is the difference between the average temperatures of the cladding wall and the coolant.

In the reactors cooled with boiling water (BWR) and, to some extent, in those cooled with pressurized water (PWR and PHWR) the heat transfer is accompanied by the occurrence of vapours in the cooling agent. For that reason this type of heat transfer is called a "two-phase heat transfer". This allows the obtaining of higher transfer coefficients than in the case of a one-phase heat transfer as described by Dittus – Boelter equation above.

Growth in thermal flow, decline in coolant flow and pressure drops may lead to an increase in the temperature of the surface to be cooled. If the temperature of the fluid in the considered section of the channel stays below the boiling temperature at local pressure, the occurrence of vapours in the immediate neighbourhood of the surface is limited, and this type of boiling is called "subcooled boiling". In this case there is no proportionality between the heat flow and the difference in temperature between the surface and the coolant, to allow obtaining of a heat transfer coefficient similar to the one-phase case. In such a situation the Jens and Lottes equation [5] can be used; it links the difference $\Delta T_s$ between the temperatures of the surface and the boiling temperature of the coolant, at local pressure $P$, under the thermal flow $\varphi$,

$$\Delta T_s = a\varphi^{1/4}\exp(-P/P_o), \tag{7.9}$$

where $a = 7.9\ ^{o}C\ cm^{0.5}\ W^{-0.25}$ and $P_o$ = 6.43 MPa.

If the temperature of the fluid in the considered section of the channel is slightly above the boiling temperature at local pressure, the heat transfer is achieved by the so-called "nucleate boiling", during which vapour bubbles are formed, stimulated by the cooling environment, which thus turns two-phase over its entire volume. However, since the vapour content is rather low, the liquid phase will remain the continuous one. The content in vapours (steam title) in a PHW-CANDU reactor is about 0.03 — 0.04 kg steam/kg agent, thus increasing by more than 10% the amount of heat transported by a unit of mass of the agent.

If the temperature of the cooled surface exceeds considerably the boiling temperature of the cooling agent in the considered section of the channel (the case of BWR reactors that operate at a lower pressure and flow of the cooling agent than the PWR reactors) the vapour content of the cooling environment considerably increases, the vapour phase becoming the continuous one, and the liquid phase a suspension in it. The cooled surface will still be covered with a film of liquid, that further provides for a very high heat transfer coefficient: ~ $6\times10^4 W/m^2 \cdot K$ in BWR, compared with ~ $3\times10^4 W/m^2 \cdot K$ in PWR. The film of liquid is continually supplied with drops of the suspension in the environment.

Any further growth in the surface temperature will lead to a temporary fragmentation of the liquid film adherent to the cooled surface. However, the sprinkling of the surface with drops of liquid in suspension will continue as long as the heat flow remains below a value, called "critical flow", that depends on the local conditions. Beyond the critical flow there is the "heat transfer crisis", featured by a sudden drop in the transfer coefficient, on account of the fact that heat transfer is restricted only to the one-phase mechanism.

The thermal transfer coefficient immediately prior to the crisis can be determined with the equation [5]:

$$\mathcal{Nu}/(\mathcal{Re}^{0.8}\,\mathcal{Pr}^{0.4}) = k_1 B_o + k_2(1/\chi)^n, \tag{7.10}$$

where

$$\chi = ((\Delta p)_l/(\Delta p)_g)^{1/2}.$$

Here $\Delta p$ denotes the pressure losses in the two phases (water and vapours), $B_o = \varphi/d_n H_{gl}$ ($\varphi$ = thermal flow, $H_{gl}$ = enthalpy of the two-phase mixture).

The thermal transfer coefficient during the crisis $H_{cr}$ is related to the critical heat flow $\varphi_{cr}(W/m^2)$ by a linear relation of the type of the equation (7.5):

$$\varphi_{cr} = h_{cr}(T_s - T_{st}) \quad , \tag{7.11}$$

where $T_s$ is the surface temperature under heat transfer crisis, and $T_{st}$ is the temperature of the vapours at saturation.

The critical flow is obtained from the Kutateladze formulae [6,7]

$$\varphi_{cr} = 0.14 r\rho_v^{1/2}[g\sigma(\rho_L - \rho_v)]^{1/4}, \tag{7.12}$$

where $r$ (J/kg) is the vaporizing latent heat, $\rho_L$ and $\rho_V$ are the densities of the liquid and vapours at saturation respectively (kg/m$^3$), $\sigma$ is the superficial tension in N/m and $g$ is the gravitational acceleration.

*In gas-cooled reactors* the heat transfer is done by *forced convection*. For a gaseous thermal agent the heat transfer coefficient can be derived with a relation similar to (7.6), taking for the respective quantities values corresponding to the average temperature of the fluid film marked by m:

$$\mathcal{Nu}_m = 0.0205\, \mathcal{Re}_m^{0.8}\, \mathcal{Pr}_m^{0.4}, \qquad (7.13)$$

which differs, if water is used, by a somewhat lower value of the coefficient $a$.

The equations that were established for the forced flow of fluids are not applicable to *liquid metals*. The thermal transfer coefficient for circular pipes with constant heat flow, where heat removal is performed by *turbulent flow of molten metals* can be estimated by an equation of the type:

$$\mathcal{Nu} = 0.625\, \mathcal{Pe}^{0.4}, \qquad (7.14)$$

where $\mathcal{Pe}$ is the Peclet number ($\mathcal{Pe} = Vd_n \rho c_p / K$).

As an example, Table 7.2 presents the values of certain hydrodynamic parameters of several cooling agents.

TABLE 7.2 Hydrodynamic Parameters of certain Types of Reactors

| *Reactor* | *Coolant* | *Pressure* (MPa) | *Fluid speed* (m/s) | $\mathcal{Re}$ | $\mathcal{Pr}$ |
|---|---|---|---|---|---|
| GI[1] | Air | 0.1 | 20 | 26,500 | 0.73 |
| EL-4[1] | $CO_2$ | 5.9 | 50 | 820,000 | 0.7 |
| MTR | $H_2O$ | 0.1 | 1 | 9,000 | 4.57 |
| VVER-440[2] | $H_2O$ | 12.5 | 3.7 | 330,000 | 0.90 |
| FBR | Na | - | 5 | 93,000 | 0.36 |

[1] Power reactors in France.
[2] Power reactors in USSR.

## 7.3. GASEOUS COOLANTS

As mentioned above, for removing heat from the reactor core gaseous cooling agents may also be used. Hydrogen seems to be an ideal cooling medium — Table 7.1. However, its merits as described in this Table are diminished by the problems with handling this highly diffusive and hazardous gas in reactor operation conditions, and by the brittling effect upon the materials with which hydrogen is bound to come into contact. Requirements of safety, easy re-circulation etc., prompted rather the use of carbon dioxide, helium, and, lately, nitrogen oxide as coolants.

7.3.1. Carbon Dioxide

The thermal and thermodynamic properties that are pertinent to the use of carbon dioxide as a cooling medium are relatively modest. As a result of the low specific heat of this gas high flows are necessary to effectively evacuate heat from the reactor core.

The temperature dependence of the specific heat at constant volume/pressure, of the thermal conductivity and of the dynamic viscosity, respectively, of the carbon dioxide are shown in Fig. 7.3 [8]. They grow with temperature. Some values of these parameters, at a pressure of 3 MPa, are given in Table 7.3 [9].

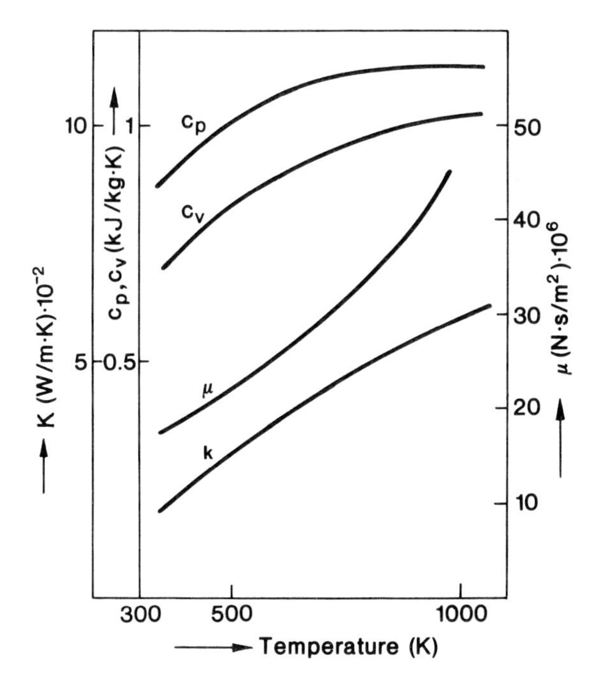


Fig. 7.3 Carbon dioxide specific heat at constant volume ($c_v$) and at constant pressure ($c_p$), its thermal conductivity ($K$) and dynamic viscosity ($\mu$), vs. temperature.

TABLE 7.3 Temperature Variation of $CO_2$ Thermal and Hydrodynamic Properties (3MPa)

| *Temperature* (°C) | *Density* ρ (kg/m$^3$) | $c_v$ (kJ/kg . K) | $c_p$ (kJ/kg . K) | μ (N . s/m$^2$) | $K$ (W/m . K) |
|---|---|---|---|---|---|
| 20 | 66 | 0.695 | 1.128 | 1.65x10$^{-5}$ | 1.98x10$^{-5}$ |
| 200 | 34 | 0.809 | 1.044 | 2.29x10$^{-5}$ | 3.71x10$^{-5}$ |
| 400 | 24 | 0.923 | 1.135 | 2.99x10$^{-5}$ | 4.96x10$^{-5}$ |
| 600 | 18 | 1.003 | 1.208 | 3.60x10$^{-5}$ | 6.96x10$^{-5}$ |

As the temperature decreases and the critical temperature (304 K) is approached, the carbon dioxide density becomes greater than that corresponding to an ideal gas.

Carbon dioxide has a good stability at high temperature. The effect of radiations on $CO_2$ depends on pressure [10]. At normal pressure, as a consequence of the recombination reactions, no decomposition processes are recorded. At a pressure of 1 MPa the reaction:

$$CO_2 \rightarrow CO + O \qquad (7.15)$$

can be noticed, and, to a lesser extent,

$$CO_2 \rightarrow C + O_2.$$

The presence of carbon monoxide contributes to the growth of the recombination rate (the reaction reverse to that indicated by (7.15)). Because of that, small quantities of CO [10] are introduced in the composition of the cooling gas, as shown in Table 7.4. This also contributes, as we shall see later on, to securing a better compatibility of the cooling agent with the graphite.

TABLE 7.4 The Composition of the $CO_2$-based Coolant

| *Constituents* | *Percentage in volume* |
|---|---|
| $CO_2$ | 99.5 |
| $CO$ | 0.38 |
| $H_2O$ | 0.001 (in weight) |
| Ar | 0.0003 |
| $H_2$ | 0.0016 |
| $O_2$ | 0.0017 |
| $N_2$ | 0.0094 |
| $C_2H_6$ | 0.0001 |
| $CH_4$ | 0.0013 |

An important problem in the use of carbon dioxide as coolant is its compatibility with graphite, used as moderator and structural material of the fuel bundle, respectively. Carbon dioxide reacts with graphite, according to the equilibrium reaction:

$$CO_2 + C \rightleftarrows 2CO. \qquad (7.16)$$

The reaction equilibrium is shifted to the left at low temperatures and high pressures, and to the right at high temperatures and low pressures, as it appears from Table 7.5 [11]. The reaction (7.16) reaches the equilibrium at temperatures $T \geq 820$ K. This value sets a threshold for the contact temperature of $CO_2$ and graphite.

Study of the behaviour of $CO_2$-based coolants [12] highlights the role of CO and $CH_4$ in determining the reaction rate of the process described by (7.16). Thus, Fig. 7.4 shows the reaction rate, $k$ vs the CO content. One can see this

TABLE 7.5 Constants featuring the Reaction $CO_2 + C \rightleftarrows 2CO$
$p = p_{CO_2} + p_{CO} = 3$ MPa

| *Temperature* (K) | $K = \frac{(p_{CO})^2}{p_{CO_2}}$ | $p_{CO}/p$ (%) |
|---|---|---|
| 500 | $1.77\times10^{-9}$ | 0.0008 |
| 600 | $1.87\times10^{-6}$ | 0.025 |
| 700 | $2.67\times10^{-4}$ | 0.30 |
| 800 | $1.10\times10^{-2}$ | 1.9 |
| 900 | 0.193 | 7.1 |

rate declining markedly at concentrations up to 0.4% CO. As shown in Fig. 7.5, the process is also inhibited by admixtures of $CH_4$. A decline by two orders of magnitude in the reaction rate is recorded when $CH_4$ is brought into the gas mixtures at a level of 0.15%.

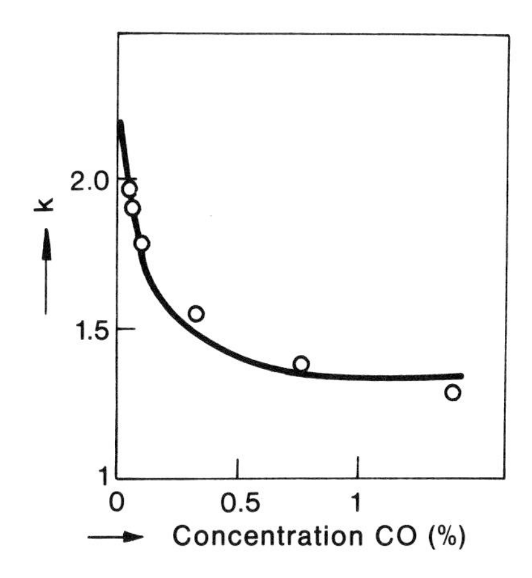


Fig. 7.4 Radiolytic oxidation rate constant of graphite, as function of the CO concentration.

Sequels of the reaction (7.16) are the fact that the carbon is carried to the cooler parts of the primary circuit, and that the graphite deteriorates. At high temperatures the equilibrium of the reaction (7.16) is shifted to the right and thus the CO concentration increases. Once the cooling gas reaches the cooler portions of the circuit the reaction equilibrium is deviated to the left and a carbon deposition can take place [13].

The study of graphite oxidation in $CO_2$ or $CO_2/CO$ type gaseous media brings into focus the fact that, if graphite weight is reduced by 2%, its open porosity increases by approximately 40%. More open pores contribute to the expansion of the contact surface of graphite and coolant, which intensifies the oxidation process [14]. Corrosion reduces the mechanical strength of graphite. The $\sigma_r$ values are diminished by ~40% at a weight reduction of 10%, as a consequence of oxidation [15].

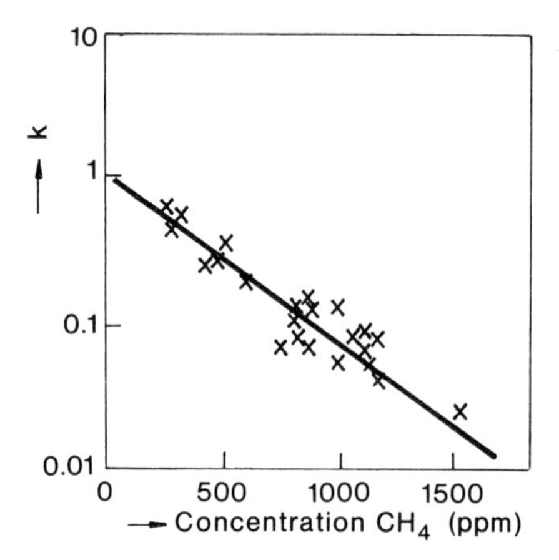


Fig. 7.5. Radiolytic oxidation rate constant of graphite as function of the content in $CH_4$

The compatibility of carbon dioxide with structural materials like magnox alloys, zircaloy and stainless steels was previously discussed. Carbon dioxide is compatible with magnesium alloys in fuel element claddings, at relatively low temperatures ($T \leq 700$ K). In the EL 4 reactors, in order to increase the temperature of the fuel elements when $CO_2$ is used as coolant, Zr-Cu alloys are used. Pure zirconium reacts with $CO_2$, resulting in zirconium oxide and carbon oxide. By alloying it with up to 4% copper the corrosion rate at temperatures of ~ 850 K, in a $CO_2$ atmosphere (6 MPa pressure) [16] is considerably reduced – Fig. 7.6.

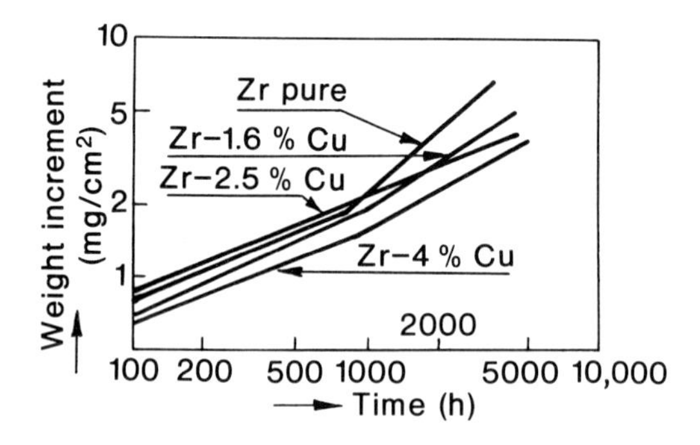


Fig. 7.6 Time-variation of Zr-Cu alloys weight in a $CO_2$ atmosphere, at $T = 850$ K and $p = 6$ MPa

Carbon dioxide has small activation cross-sections for neutrons with the energy spectrum specific to thermal and fast reactors. The reactions with neutrons of the $^{12}C$ (98.9%), $^{13}C$ (1.1%), $^{16}O$ (99.757%), $^{17}O$ (0.039%) and $^{18}O$ (0.204%) constituent nuclides, in the $(n, \gamma)$ case, lead to the following chains:

$^{12}C(n,\gamma)\ ^{13}C(n,\gamma)\ ^{14}C(\beta;\ T_{1/2} = 5570$ years)

$^{16}O(n,\gamma)\ ^{17}O(n,\gamma)\ ^{18}O(n,\gamma)\ ^{19}O(\beta;\ T_{1/2} = 27$ s)

Considering that the absorption cross-sections of thermal neutrons are very small (4.5 mb for $^{12}C$, 1 mb for $^{13}C$, < 0.2 mb for natural oxygen), the above reactions are not of consequence for the radioactivation; of particular interest are, instead, the reactions:

$^{18}O\ (n,\alpha)\ ^{14}C$

$^{16}O\ (n,p)\ ^{16}N\ (\beta,\gamma;\ T_{1/2} = 7.4$ s)

The last reaction with fast neutrons (a threshold at 4.5 MeV, cross-section in the fission spectrum 0.014 mb) leads to the $^{16}N$ radionuclide which emits (with an 80% yield) very strong gamma rays (6.2 — 7 MeV). These reactions make compelling the sealing and shielding of the primary circuit components. The gamma radioactivity declines rapidly, however, as soon as the reactor is shutdown, which allows access to the cooling circuit.

Argon and nitrogen impurities lead to the formation of $^{41}Ar$ and $^{14}C$.

Table 7.6 gives a list of $CO_2$-cooled, graphite moderated Magnox-type nuclear reactors (with natural uranium, at a maximum $CO_2$ temperature of 670 K), and of AGR type reactors (with enriched uranium, at a maximum $CO_2$ temperature of 870 K) [17].

TABLE 7.6 $CO_2$-cooled, Graphite-moderated, Magnox and AGR-type Reactors

| *Reactor* | *Type* | *Power* (MWe) | *Pressure* (MPa) |
|---|---|---|---|
| $G_2$ and $G_3$ | Magnox | 2 x 30 | 1.5 |
| EDF 3 | Magnox | 430 | 2.8 |
| Oldbury | Magnox | 2 x 300 | 2.5 |
| St.Laurent I | Magnox | 480 | 2.8 |
| Wylfa | Magnox | 2 x 590 | 2.8 |
| St.Laurent II | Magnox | 545 | 2.8 |
| Bugey I | Magnox | 540 | 4.3 |
| Vandelos | Magnox | 500 | 2.8 |
| Dungeness B | Magnox | 2 x 600 | 3.2 |
| Hinkley Point B | AGR | 2 x 625 | 4.2 |
| Hunterston B | AGR | 2 x 625 | 4.2 |
| Hartlepool B | AGR | 2 x 625 | 4.2 |

Carbon dioxide has built its reputation as a coolant especially on its low cost. Thus, as far as numbers, $CO_2$-cooled reactors rank immediately after water-cooled reactors.

### 7.3.2. Helium

Helium is a very good gaseous cooling medium, with a high specific heat (0.5 KJ/kg . K at 323 K) and a very small neutron absorption cross-section ($7 \times 10^{-3}$ b). The thermal conductivity of helium is also high. Figure 7.7. presents the temperature and pressure dependence of helium density, as well as its main thermal and thermodynamic properties [2].

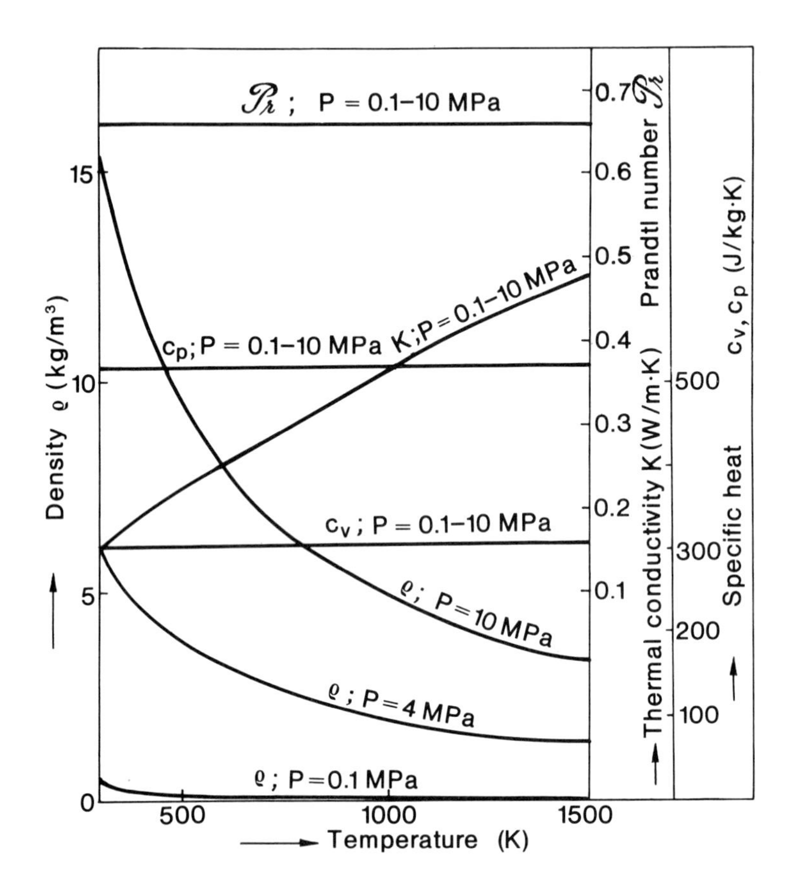


Fig. 7.7 Physical properties of helium, vs. temperature and pressure.

Being an inert gas, helium is stable from a chemical point of view. It does not create problems with instability under irradiation. With regard to radioactivation by thermal neutrons, only $^{3}He$, which is found in a concentration of 0.00013% in natural helium, is important. Due to the (n,p) reactions with a 5400 b cross-section, tritium generation results; tritium is β-active, with a half-life of $T_{1/2}$ = 12.1 years. Thus the neutron absorption cross-section of natural helium is $7 \times 10^{-3}$ b.

The helium used as coolant contains gaseous impurities like argon and nitrogen, that are activated by the neutron-capture reactions.

The compatibility of helium with all materials with which it may come in contact in the reactor is good, even at high temperatures (~ 1300 K). This is why helium is used in high-temperature graphite-moderated reactors (HTGR). Table 7.7 introduces certain reactors that use helium as a coolant [17,18].

TABLE 7.7 Helium-cooled Reactors

| *Reactor* | *Power* (MWe) | *Outlet temperature of helium* (K) | *Helium pressure* (MPa) | *Remarks* |
|---|---|---|---|---|
| Peach Bottom | 40 | 810 | 10 | |
| Fort St.Vrain | 330 | | | |
| AVR | 15 | 1110 | 1 | |
| THTR | 300 | 1020 | 4 | In the design stage |
| HTGR | 1100 | 1120 | 4.8 | In the design stage |

The 4 — 5 MPa pressure of the helium in the primary circuit is lower than the water pressure in the secondary circuit, 4 — 24 MPa. Thereby water can leak into the primary circuit, if the heat exchanger loses tightness. Such a prospect necessitates the purification of helium, in order to avoid graphite corrosion by water contact.

High costs limit the use of helium in reactor cooling. Also, the high diffusiveness of helium makes a tight sealing of the primary circuit mandatory, for a large variety of materials.

### 7.3.3. Nitrogen Oxide $N_2O_4$

Nitrogen oxide $N_2O_4$ is being considered in the USSR as an alternative to cooling fast reactors. Its use would provide for a higher breeding coefficient in comparison with the liquid natrium cooling system.

The use of nitrogen oxide implies only one cooling circuit, the heated gas being directly blown into the turbine. Cooling with nitrogen oxide considerably reduces the weight of turbine per unit of energy output, as well as the requirement of pumping power [10].

At temperatures of 300 – 1470 K and 10 MPa one encounters the dissociation reactions:

$$N_2O_4 \rightleftarrows 2NO_2 \rightleftarrows 2NO + O_2 \qquad (7.17)$$

As the temperature rises, the reaction equilibrium shifts from the left to the right, 57.3 J/mol being stored on the generation of $2NO_2$, and 112.9 J/mol on the generation of $2NO + O_2$ respectively. The process is reversible. Overall, such a reaction heat gives the coolant an apparent specific heat that ranks very high.

The compatibility with stainless steel of the gaseous fractions that result from the reactions (7.17) is satisfactory. Corrosion rate at 770 K and 5 MPa is $5 \times 10^{-3}$ g/m$^2$ hour. A prolongation of the exposure time from $10^2$ to $10^4$ hours results in a decline by a factor of $10^2$ of the corrosion rate, before it settles down at a constant level [10].

## 7.4. LIQUID COOLANTS

Liquid coolants now in use or showing prospects of future utilization are light and heavy water, liquid metals, organic liquids, and molten salts.

### 7.4.1. Light Water and Heavy Water

The moderating properties of natural water and heavy water and some considerations on their stability in the core of nuclear reactors have been presented in Chapter 5. Table 7.8 presents some thermodynamic properties of light and heavy water, from the angle of their heat transfer performance [4].

TABLE 7.8 Properties of Light and Heavy Water

| *Property* | *Coolant agent* | |
|---|---|---|
| | $H_2O$ | $D_2O$ |
| Molecular weight | 18.016 | 20.029 |
| Density (kg/m$^3$) | 1000 (at 4 $^oC$) | 1106 (at 11.2 $^oC$) |
| Melting point ($^oC$) | 0 | + 3.8 |
| Critical pressure (MPa) | 22.13 | 36.43 |
| Critical temperature ($^oC$) | 374.2 | 371.5 |
| Thermal transfer coefficient (W/m$^2$ . K) | $22.55 \times 10^{3*}$<br>$13.95 \times 10^{3**}$ | |
| Reaction with fuel | slow | slow |

* v = 6 m/s, channel diameter 12.5 mm, temperature 25 $^oC$.
** v = 1.25 m/s, channel diameter 20 mm, temperature 25 $^oC$.

Figure 7.8 presents the temperature dependence of some physical properties of light water [8]. One notes that from a thermodynamic point of view, the water (whether light or heavy) has indeed properties that recommend its use as a coolant. As the boiling point is low and vapour pressure soars at high temperatures, the reactor must operate at high pressure. This is an obvious source of engineering problems.

The temperatures are limited to 560 K in BWR reactors and to 600 K in PWR reactors, in both cases the water playing the parts of both coolant and moderator. In heavy water reactors (PHWR) the temperature of the coolant-heavy water is lower than 585 K, whereas that of the moderator-heavy water is maintained at about 340 K. In Beloyarsk reactors (graphite moderated) one section of the core is cooled with boiling light water, while another one is cooled by the secondary circuit steam that is thus superheated ("nuclear superheating").

### 7.4.2. Organic Liquids

At 670 K the organic liquids in the polyphenyl class with a high boiling point have a relatively low vapour pressure (< 2 MPa), thus constituting a good alternative, especially for cooling heavy water reactors.

Up to 670 K the liquids in the polyphenyl class have a good chemical stability. Under nuclear irradiation modifications similar to those caused by a rise in temperature are induced. These consist of sediments that result from

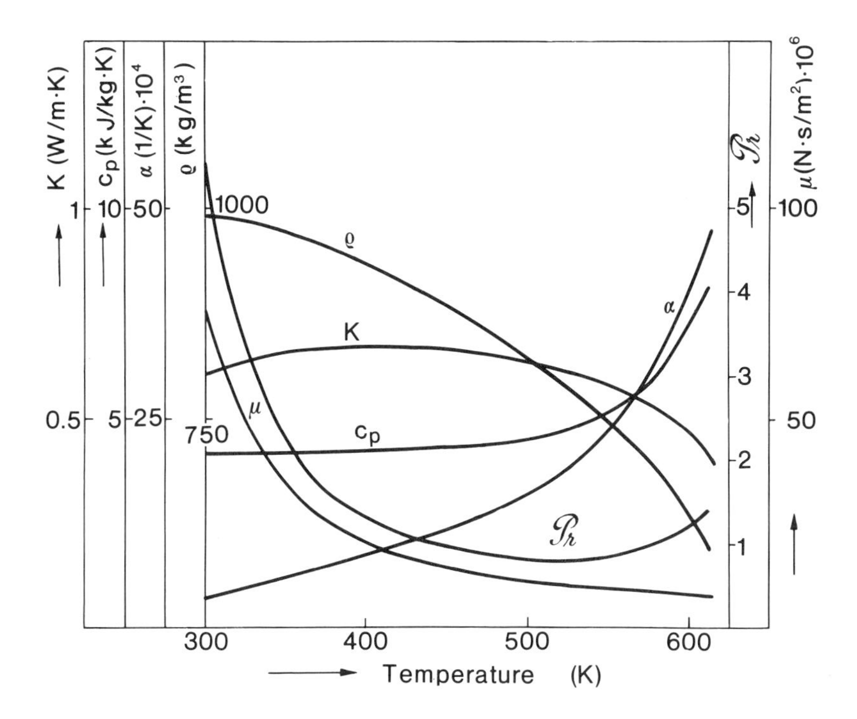


Fig. 7.8. Temperature dependence of several physical properties of the natural water: thermal conductivity ($K$), specific heat ($c_p$), expansion coefficient ($\alpha$), density ($\rho$), dynamic viscosity ($\mu$), and Prandtl number ($\mathcal{P}r$).

thermal and radiolytic decomposition in the hot sections of the cooling circuit, such as the fuel element clads.

Radiolysis is less pronounced with polyphenyls than with water. The lack of oxygen prevents the making of $\gamma$-active nitrogen ($^{16}N$). Specific heat and thermal conductivity of polyphenyls are lower than those of water.

Organic coolants are usually a mixture of polyphenyls — diphenyl, terphenyl etc. As physical properties of these substances, Fig. 7.9 shows the temperature dependence of diphenyl density, specific heat, thermal conductivity and viscosity [8]; thermal conductivity, varies a little with temperature; dynamic viscosity quickly declines at temperatures between 300 and 400 K; specific heat at constant pressure increases.

The boiling temperature of diphenyl (527 K) is higher than that of water. Vapour pressure is lower: 1.4 MPa at 700 K and 2.1 MPa at 730 K, respectively.

Compared with water, polyphenyls have a better compatibility with structural materials and with those of which the cooling system is manufactured. Because the organic liquids are good lubricants, the wear of piping is lower than with water as a coolant.

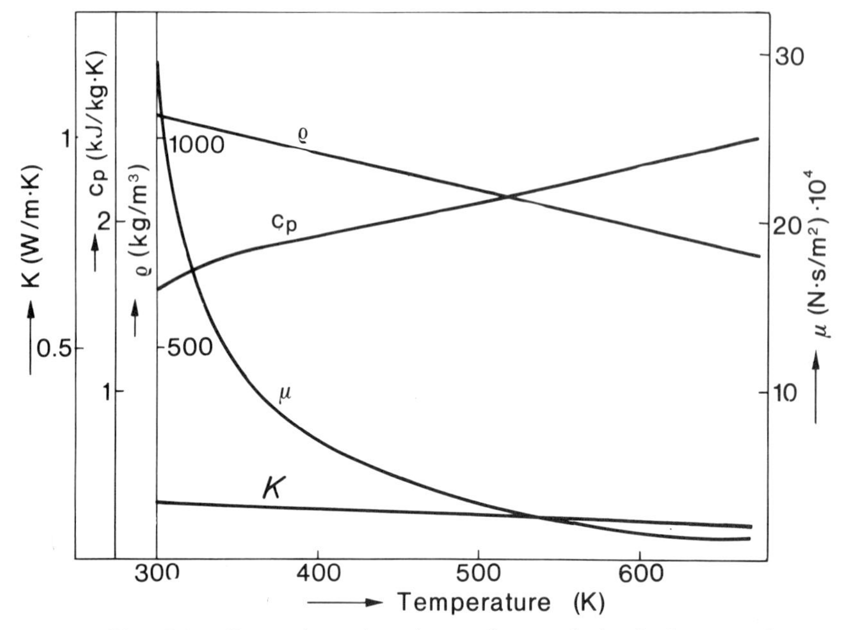


Fig. 7.9. Temperature dependence of several physical properties of the diphenyl: thermal conductivity ($K$), specific heat ($c_p$), density ($\rho$), and dynamic viscosity ($\mu$).

Due to the high hydrogen content ($0.23 \times 10^{29}$ atoms/$m^3$), diphenyl is a good moderator. The macroscopic cross-section for neutron absorption is smaller than that of water.

Diphenyl has a high melting point ( ~ 340 K), so that to maintain it in a liquid state it is necessary to keep it warm, especially when the reactor is shutdown. To avoid this inconvenience, the eutectic diphenyl-terphenyl (which melts at ~ 290 K), or the partially saturated diphenyl (which stays liquid at room temperature) may be used. However, these particular organic liquids are more unstable than diphenyl under irradiation.

Specific heat and thermal conductivity of polyphenyl are lower than those of water, whereas its viscosity is higher. As a consequence, the thermal transfer coefficient at $T \simeq 540$ K is only 22% of that of water, and does not exceed 50% at 720 K.

When irradiated, diphenyl splits into compounds of the terphenyl type, and hydrogen is released [19]. Hydrogen that results from radiolysis contributes to the embrittlement of the fuel cladding by hydriding.

Due to high operating temperatures, the presence of small quantities of water, oxygen or chloride in the coolant lead to a strong corrosion of parts exposed to their action. On the other hand, iron may be transferred from the low-temperature sections of the circuit on the surface of the fuel cladding. The layers thus deposited cause deterioration in the heat transfer at the clad — fluid interface [20]. The permanent purification of the coolant prevents such inconveniences. Polyphenyls are compatible with $UO_2$ and UC, so that clad deterioration is not a particular problem.

The critical thermal flux, $\varphi_c$ [W . $m^{-2}$] has been experimentally determined for various organic liquids [9]. Thus,

1. for diphenyl:

$$\varphi_c = 3.154\ (454\ \Delta T\ v^{0.63} + 111{,}600) \tag{7.18}$$

2. for a mix of 1% in weight diphenyl and 99% in weight terphenyl (10 orto-, 53 meta- and 36 para-) known as Santowax-R

$$\varphi_c = 3.154\ (552\ \Delta T\ v^{0.67} + 152{,}000) \tag{7.19}$$

$\Delta T$ ($^{o}C$) denotes the subcooling and $v$ is the velocity of the liquid, in cm/s.

Organic liquids of the polyphenyl type are used as cooling agents in the ORME, Idaho (SUA) and WNR (Whiteshell — Canada) reactors.

### 7.4.3. Liquid Metals

Liquid metals are used as coolants in fast reactors. In principle, any metal that is normally in liquid state at room temperature, or at temperatures slightly higher, can be employed to this effect. Among these are mercury, sodium, potassium, lead, lithium, tin, bismuth, etc. Mercury has a large absorption cross-section for thermal neutrons, and is also very toxic. Lead, bismuth and tin alloys have large inelastic scattering cross-sections for neutrons, so that their use would entail undesirable moderation of fast neutrons. Lithium, in natural state, has an absorption cross-section for thermal neutrons of 70 b. The isotope $^{7}Li$, which comprises 80% of natural lithium, has an absorption cross-section for thermal neutrons of 0.033 b. Yet the high cost of $^{7}Li$ makes its use as a coolant in fast reactors prohibitive.

All these considered, it is liquid sodium that constitutes the coolant most widely used at present in fast reactors. This choice has been largely determined by its low moderating power and vapour pressure. The thermal and hydrodynamic parameters of sodium allow the cooling of compact active zones, that are typical for fast reactors. Such moderator-free cores consist, in fact, of fuel assemblies with fissile and fertile materials.

Figure 7.10 shows the variation with temperature of sodium thermal conductivity, viscosity and specific heat [21]. These values decrease within the temperature range 370 — 870 K. Vapour tension is low, that is 1.86 Pa at 600 K and 0.114 MPa at 1100 K [21]. Therefore, sodium may be used at high temperatures and low operation pressures.

Sodium compatibility with structural materials is determined by its purity, temperature and circulation rate. Sodium compatibility with stainless steel used at present to manufacture clads and fuel assemblies, and also with oxidic fuel materials ($UO_2$ and $UO_2$ — $PuO_2$) is very good. The presence of oxygen as an impurity strongly affects the corrosion rate, so that the content in oxygen must be limited. Table 7.9 shows the maximum concentration in impurities as admitted in molten sodium [22]. This limitation requires continuous purification of liquid sodium.

For an oxygen content of 10±5 ppm and temperatures up to 920 K, the corrosion in stainless steels tested in a liquid sodium flow of $v$ = 6 m/s is weak, i.e. 0.04 mm/year [23]. According to Zebroski [24], the corrosion rate of

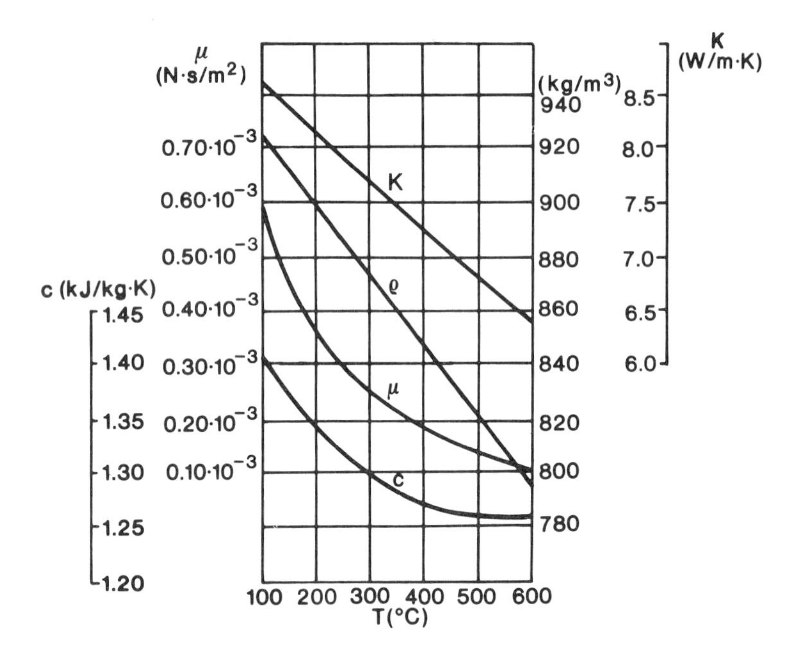


Fig. 7.10. Thermal and hydraulic properties of molten sodium, as function of temperature: $\rho$ - density; $K$ - thermal conductivity; $\mu$ - viscosity; $c$ - specific heat.

TABLE 7.9 Permissible Impurity Concentration, in Molten Sodium

| *Element* | *Concentration (ppm in weight)* | *Element* | *Concentration (ppm in weight)* |
|---|---|---|---|
| Al | < 10 | K | < 200 |
| B | < 2 | Li | < 5 |
| Ba | < 10 | Mg | < 10 |
| C | < 15 | Mn | < 5 |
| Ca | < 10 | N | < 5 |
| Cd | < 5 | Ni | < 10 |
| Co | < 5 | O | 10±5 |
| Cr | < 5 | Pb | < 10 |
| Fe | < 25 | Rb | < 50 |
| H | < 10 | Si | < 10 |
| Halogens | < 20 | Sn | < 10 |

austenitic stainless steels is given by the equation:

$$R = v^{0.834} C_o^{1.156} \exp\left[12.845 - \frac{13,300}{T + 273} - 0.00676\,\frac{L}{D} + \frac{2.26}{t + 1}\right]. \tag{7.20}$$

where $R$ is the corrosion rate (mg/dm$^2$ x month); $v$ is the fluid circulation speed, $C_o$ is the primary oxygen concentration (ppm), $t$ is the time (in months); $L$ is the distance from the inlet to the area where speed and temperature may be taken as constant, and $D$ is the hydraulic diameter.

Another process that affects sodium compatibility with stainless steel is carburation and decarburation of parts of the circuit, respectively. Carbon transportation by sodium mainly appears where materials with various concentration of carbon are used within the same cooling circuit, or where the sodium is accidentally impurified by carbon. The source that causes the cooling fluid impurification is the stainless steels that have a higher carbon content, and are usually used to manufacture the cold parts of the circuit.

The decarburation of some steels and the carburation of the low carbon content ones may have unfavourable effects upon the mechanical properties of structural materials. Carbon transfer can be reduced by alloying high-carbon steels with Ni, V and Ti [25].

Some of the disadvantages of sodium utilization as a coolant are its high chemical reactivity and intense activation.

Molten sodium "ignites" in air, and sodium hydride is formed in the presence of hydrogen at ~ 500 K. Sodium reacts strongly with water, releasing hydrogen. Consequently, such a high reactivity requires that the design of the cooling systems provides for prevention of any direct contact of the liquid sodium with air, water or hydrogen, especially in the steam generators.

Natural sodium (100% $^{23}Na$) is activated by the reaction $(n,\gamma)$, resulting in the formation of $^{24}Na$ which is $(\beta,\gamma)$ active, with $T_{1/2}$ = 15 hours. This necessitates a secondary cooling circuit with liquid sodium, that is placed between the intensely radioactive primary circuit and the steam generator.

*Sodium activation*, as the molten metal passes through the reactor core, can be estimated as follows. Let us consider a unit-volume of liquid metal that passes through the core in a time $t_1$, and through the recirculation circuit, placed outside the reactor core, in a time $t_2$. The variation in time of the concentration of radioactive atoms during the passage through the reactor core is:

$$\frac{dC}{dt} = N\sigma\Phi - \lambda C \tag{7.21}$$

where $N$ is the number of Na atoms per unit volume, $\Phi$ is the neutron flux, and $\lambda$ is the decay constant of the newly formed radioactive element.

The relation (7.21) may also be written as:

$$\frac{d}{dt}[C(t)\exp(\lambda t)] = N\sigma\Phi\exp(\lambda t) \tag{7.22}$$

then, by integration, one obtains:

$$C_2\exp(\lambda t_1) - C_1 = [N\sigma\Phi/\lambda][\exp(\lambda t_1) - 1], \tag{7.23}$$

where $C_1$ and $C_2$ are the concentrations at the inlet and outlet of the reactor, respectively.

After a certain equilibrium state is reached, the concentration at the reactor inlet equals that at the outlet, corrected by a coefficient that takes into account the radioactive decay and the deactivation that is achieved by retaining the active products on filters.

Upon equilibrium, one obtains the following:

$$C_1 = C_2 \alpha \exp(-\lambda t_2) \tag{7.24}$$

where $\alpha$ is the fraction of the recirculated agent.

By combining eqs. (7.23) and (7.24) one obtains, in the case $t_2 >> t_1$:

$$C_2 = (N\sigma\Phi/\lambda) \cdot (1-\exp(-\lambda t_1))/(1-\alpha\exp(-\lambda t_2)). \tag{7.25}$$

By expanding in series the exponential at the numerator and retaining only the leading terms, one gets:

$$C_2 = (N\sigma\Phi \cdot t_1/(1-\alpha\exp(-\lambda t_2)) \tag{7.26}$$

and if the decay constant is small as compared with the reciprocal of a cycle duration $\lambda << (t_1+t_2)^{-1}$:

$$C_2 = N\sigma\Phi t_1/(1-\alpha) \tag{7.27}$$

Knowing that sodium has a 0.45 b neutron absorption cross-section, that this element is monoisotopic and has a half-life of about 15 hours, one may calculate, by using eq. (7.26) or (7.27), the activity that results from its passing through the reactor core, as a function of the reactor characteristics ($\Phi$,$\alpha$).

As far as structure, there are two concepts of sodium-cooled fast reactors: the loop, and the pool — Fig. 7.11 (see e.g. [2]). With the loop-type, the Na active/Na non-active heat exchanger is situated outside the reactor vessel (3), whereas with the pool-type the heat exchanger is placed inside the vessel, immersed in the active sodium, which reduces the risk of losing the primary agent in case the pipes or the intermediate heat exchanger get damaged. Both systems are in current use, as indicated in Table 7.10 [24].

### 7.4.4. Molten Salts

Molten salts can also be used as coolants in nuclear reactors. They must have a very small capture cross-section for thermal neutrons. Such a requirement is met by fluorides, because fluorine has $\sigma_c$= 0.0096.

$^7Li$ fluoride mixed with beryllium fluoride is considered for use as thermal agent in the primary circuit. Because the thorium and uranium fluorides are soluble in this mixture, a variety of reactors with circulating liquid fuel has been envisaged.In the intermediary circuit salts of the type $NaBF_4$— NaF, with a composition corresponding to the eutectic, are considered for use.

The coolants discussed have high melting points. Thus, the salts corresponding to the eutectic composition have melting temperatures above 600 K. Therefore

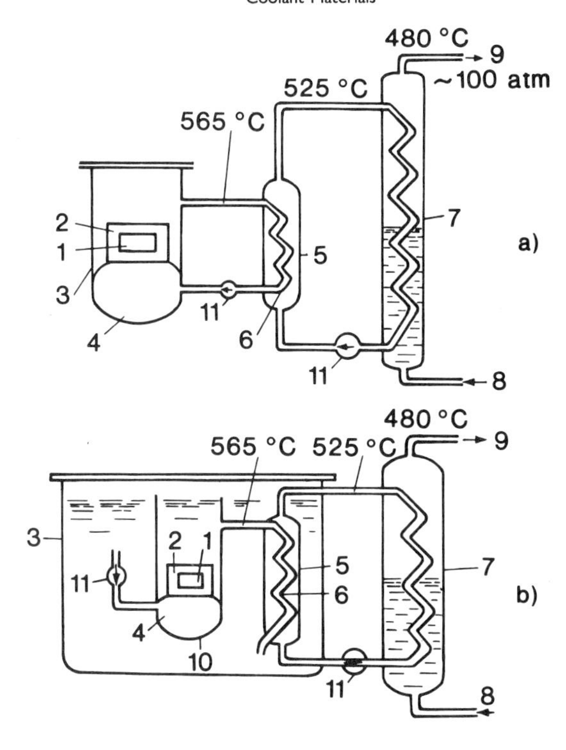


Fig. 7.11. Cooling systems for sodium-cooled fast reactors:
a - loop-type; b - pool-type.
1 - core; 2 - fertile reflector; 3 - reactor vessel; 4 - primary cooling circuit, with active sodium; 5 - Na — Na intermediate heat exchanger; 6 - intermediate cooling circuit, with inactive sodium; 7 - steam generator; 8 - water supply of secondary circuit; 9 - vapours secondary circuit; 10 - vessel to conduct the flow of active sodium; 11 - sodium pumps.

TABLE 7.10 Fast Power Reactors Currently in Operation

| *Reactor* | *Cooling system* | *Rated power* (MWe) |
|---|---|---|
| BN 350 (USSR) | Loop | 350 |
| BN 600 (USSR) | Loop | 600 |
| PFR (UK) | Pool | 250 |
| Phoenix (France) | Pool | 250 |

special measures are in order, to avoid solidification. This disadvantage is compensated by the low vapour pressure, a fact which makes it possible to handle the cooling fluid at atmospheric pressure, at temperatures in excess of 1000 K.

Molten fluorides considered for coolants have a low viscosity, comparable to that of kerosene, which allows high specific powers per core unit-volume.

The compatibility at high temperature of fluorides with the graphite used as moderator is good. Fluorides also have a good compatibility with nikel-alloyed steels of the hastelloy type. Consequently, these alloys are used in the fabrication of both circuits and heat exchangers.

A problem that still lacks adequate study is the behaviour of oxides and hydroxides formed in meltings as a consequence of the penetration through leaking seals of water vapours at ~24 MPa from the secondary circuit.

Molten fluorides are considered for coolants in graphite-moderated MSBRs (Molten Salt Breeder Reactors) [25], that operate on the Th/$U^{233}$ cycle with continuous fuel reprocessing. Fuel reprocessing allows keeping the $Xe^{135}$ and $Pa^{233}$ concentration, as well as the concentration of non-saturable fission products, at levels sufficiently low to enable the superbreeding of fissile material. In this way this reactor type can ensure a doubling time of the fissile material of 19 years, which compares fairly with the doubling time specific to fast reactors. Moreover, the fissile material inventory is much smaller than the one in fast reactors.

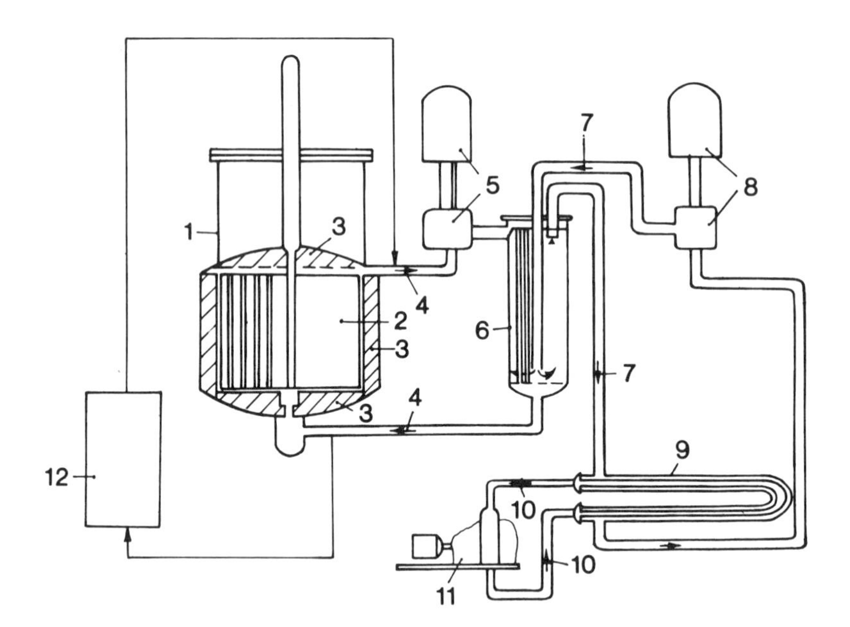


Fig. 7.12. Cooling circuits of a MSBR: 1 - reactor vessel; 2 - core, of graphite blocks; 3 - graphite reflector; 4 - primary coolant (molten fluorides); 5 - primary agent pump; 6 - heat exchanger; 7 - intermediate coolant (molten fluorides); 8 - intermediate circuit pump; 9 - steam generator; 10 - secondary circuit (steam); 11 - turbo-generator; 12 - facility for the primary agent re-treatment.

The primary agent containing the fuel has the following composition (% mols): $^7LiF$ (71.7%), $BeF_2$ (16%), $ThF_4$ (12%) and $UF_4$ (0.3%). The temperature corresponding to the liquidus curve is 772 K. The inlet temperature is 839 K and the outlet temperature is 977 K.

The coolant in the intermediate circuit ($NaBF_4$ — NaF eutectic) has the following composition (% mols): $NaBF_4$ (92%) and NaF (8%). The temperature corresponding to the liquidus curve is 658 K. At the heat exchanger inlet it has $T$ = 727 K and at the outlet the temperature is 894 K. This allows production in the generator of high-grade steam (811 K, 24 MPa) which brings the net efficiency of the reactor up to 44%.

Figure 7.12 presents the cooling scheme of a MSBR. The reactor is placed in vessel (1), the core (2) being surrounded by the graphite reflector (3). The core consists of prismatic rods coated in a pyrocarbon layer; these are cooled by the melting consisting of $^7LiF$ — $BeF_2$ — $ThF_4$ — $UF_4$. After passing through the reactor the melting is directed by the pump (5) in the intermediate heat exchanger (6), wherefrom it returns into the reactor. Part of the melting passes through the reprocessing installation (12) and is then reintroduced in the primary circuit. The cooling liquid is driven by pump (8) through the intermediate circuit, in the heat exchanger (6) thereafter going to the steam generator (9). The steam circuit (10) feeds the generator (11).

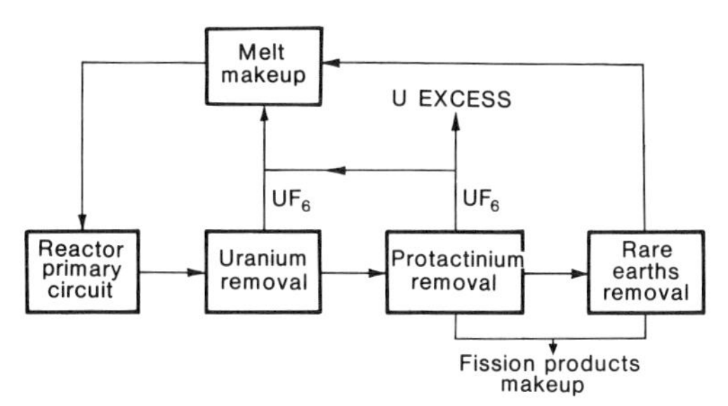


Fig. 7.13. Flow-chart of re-treatment, for the molten salts in the primary circuit of a MSBR.

Figure 7.13 presents the flow-chart in a reprocessing installation (12). In an initial phase the recovery of uranium from the melting by fluoration (as $UF_6$) is performed; in the second phase one removes the protactinium that, after a deactivation interval, yields $U^{233}$ which is then recovered as $UF_6$ again. The following step has in view the fission products removal (rare earths). To restore the initial composition, $UF_4$ is introduced in the melting that mixes lithium, beryllium and thorium.

REFERENCES

1. Krasin, A. K and others (1977). In *Proc.Conf.Nuclear Power and its Fuel Cycle, Salzburg*, Vol. 1, p. 569.
2. Ingolfsrud, L. G. (1981). *Proiectarea centralelor CANDU*, Seminar tehnic, Bucharest; AECL (1973). *Data Base for a CANDU-PHW Operating on a Once-Through Natural Uranium Cycle*. AECL 6593; Weissman, J. (1977). *Elements of Nuclear Reactor Design*. Elsevier Sci. Publ. Comp., New York.
3. IAEA (1976). *Directory of Nuclear Reactors*. Vol. X, *Power and Research Reactors*, Vienna.
4. Leca, A., Pop, M., Stan, N., Badea, A., and Lucia, L. (1980). *Procese si instalatii termice in centrale nuclearoelectrice*. Edit. didactica si pedagogica, Bucharest.
5. Jens, W. H., and Lottes, P. A. (1952). Raport ANL-4915.

6. Kutateladze, S. S. (1962). *Teploperedacha pri kondensatsyi i kipenyi.* Masghiz, Moskow; Polyanin, L. N., Ibragymov, M. K., amd Sebelev, G.I. (1982). *Teploobmen v iadernyh reaktora*, 5th ed. Energoizdat, Moskow.
7. Kutateladze, S. S., and Leontiev, A. J. (1972). *Teplomasoobmen i trenye v turbulentnom pogranychnom sloye.* Energya, Moskow.
8. Raznjevic, K. (1978). *Tabele si diagrame termodinamice.* Edit. Tehnica, Bucharest.
9. Vugalovici, M. P., and Altunin, V. V. (1965). *Teplofizicheskye svoystva dvuokysy ugleroda.* Atomizdat, Moskow.
10. Gerasimov, V. V., and Monahov, A. S. (1962). *Materialy iadernoy tehniky.* Energoizdat, Moskow.
11. Rossini, F. D., and others (1945). *J.Research Natl.Bur.Standards*, 34, 143.
12. Wright, I. (1971). *Proc.Conf.on Peaceful Uses of Atomic Energy.* Vol. 10, *Effects of Irradiation on Fuels and Materials*, Geneva.
13. Davidge, P. C., Edwards, H. S., and Leons, R. N. (1960). Gas Coolant. In *Reactor Handbook I.*
14. Standring, I. (1965). *Carbon*, 3, 157.
15. Board, I. A. (1966). *Proc.Second Conf.on Carbon and Graphite*, London, p. 355.
16. Baque, P. (1963). Raport CEA-2393.
17. Pedersen, E. S. (1978). *Nuclear Power.* Vol. I, *Nuclear Power Plant Design*, Ann Arbor Science Publishers Inc., Ann Arbor.
18. IAEA (1976). *Directory of Nuclear Reactors.* Noex, IAEA, Vienna.
19. De Halas (1958). *Proc.Int.Conference, Geneva*, paper 611.
20. IAEA (1967). *Organic Liquids as Reactor Coolants and Moderators.* IAEA Technical Report 70.
21. Ageron, P. (1959). *Technologie des reactéurs nucléaires*, Vol. I, *Matériaux*, Eyrolles, p.356.
22. McKisson, R. L. (1967). Report ANL-7380, p.134.
23. Rowland, M. C. (1965). Report GEAP-4831.
24. Zebroski, E. L. (1966). *Proc.Conf.Alkali Metal Coolants*, Vienna, p.198; Ryneyskyi, A. A. (1970). K voprosu vybora gydravlitscheskoy reaktory na bystryh neytronah. In *Proc.Symp.on Sodium-cooled Fast Reactors Engineering, Monaco 1970*, p. 107; Ryneyskyi, A. A. (1970). Nasosy dlya perekatchyvanya zhydkyh metallov. In *Proc.Symp.on Sodium-cooled Fast Reactors Engineering, Monaco 1970*, p. 299.
25. Ebrasco-Services Inc. (1972). *1000-MW(e) Molten Salt Breeder Reactor Conceptual Design Study*, Final Report — Task 1.

CHAPTER 8

# Shielding Materials

## 8.1. GENERAL

Shielding materials are meant to reduce the radiation and heat flows generated in the core of a nuclear reactor, down to levels corresponding to irradiation doses and temperatures low enough to ensure biological safety to the personnel, as well as the complete integrity of the equipment and installations.

In order to illustrate the distribution of the radiation field typically associated to a PWR reactor, Fig. 8.1 indicates the gamma exposure rates as well as the equivalent (biological) doses, for the various components [1].

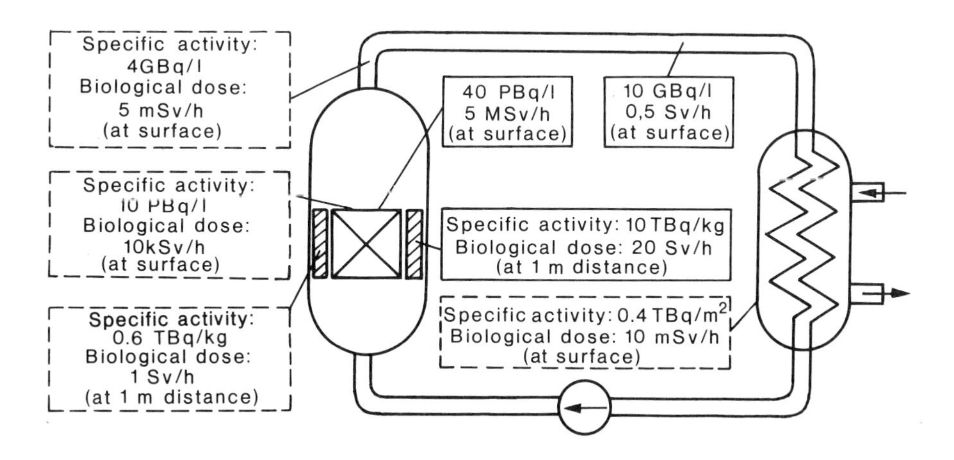


Fig. 8.1. Specific activities and characteristic equivalent doses for certain components of a PWR. Values during operation are given in full-contour rectangles, whereas shutdown values are in broken-line rectangles.

At the surface of the core the equivalent dose is around 1400 Sv/s. Since this dose is much higher than the acceptable limit values of $0.7 \times 10^{-8}$ Sv/s ($2.5 \times 10^{-3}$ R/h) a considerable attenuation of the radiation intensity is necessary — by a factor of $10^{11}$. On the other hand, shielding the equipment calls for meeting certain requirements, depending upon the type of reactor. For reactors with steel pressure vessels, such as, for instance, BWR or PWR, in order to avoid radiative embrittlement it is necessary to limit the fast neutron irradiation (at $E \geq 1$ MeV) at fluencies lower than $10^{19}$ n/cm$^2$, over an operation duration of 40 years. Similar conditions must be fulfilled for the HTR and AGR reactors with the primary circuits integrated in prestressed concrete pressure vessels. Limiting the effects of irradiation must also be considered in view of protecting the control instrumentation.

Shielding materials are assembled in various configurations, thus achieving what is generically called a *shielding screen*. Shielding screens are designed in such a way that, by attenuating the radiations and heat flux, they limit especially the activation of certain reactor components that require repair or inspection during reactor shutdown.

A good shielding screen is bound to meet a number of basic technical and economic requirements, such as:

1. to provide a large moderating capacity for fast neutrons, and a large absorption cross-section for the thermal neutrons evolved in the shielding screen itself following the moderation of the fast neutrons;
2. to ensure an adequate attenuation of the gamma radiations emitted in the reactor core or produced by neutron capture in the shielding screen;
3. to warrant confidence in a satisfactory mechanical stability at the temperatures and under the mechanical stresses they are subjected to, as well as under irradiation;
4. lowest possible costs per unit-volume.

The slow-down of the fast neutrons is achieved by elastic collisions with the nuclei of the elements constituting the screen.

In an elastic collision, the smaller the mass of the target-nucleus, the larger the energy transfer to it. From this angle, the most efficient moderating element is hydrogen and, consequently, the shielding materials should contain substances rich in hydrogen. On the other hand, the use of materials made of heavier nuclei ensures cuts in the energy of the fast neutrons by processes of inelastic collisions. In short, in order to reduce the energy of the neutrons, materials with a high $Z$ as well as materials rich in hydrogen are equally commendable in the manufacture of shielding screens. Table 8.1 gives the hydrogen content of several materials that are used in the fabrication of shielding screens.

TABLE 8.1 Hydrogen Content of Certain Materials

| *Material* | *Hidrogen content* (atoms/m$^3$ . $10^{28}$) | *Material* | *Hidrogen content* (atoms/m$^3$ . $10^{28}$) |
|---|---|---|---|
| Water | 6.69 | Concrete | 0.4–0.7 |
| Octane | 6.81 | High-density concrete: | |
| Lithium hydride | 5.85 | barytine | 0.4 |
| Calcium hydride | 5.45 | magnetite | 0.3 |
| Zirconium hydride | 7.25 | limonite | 0.3 |
| Titanium hydride | 9.08 | Boron concrete | 0.2 |
| Paraffin | 7.9 | High-temperature concrete | 1 |

The thermal neutrons originating either directly from the reactor or from the damping of the fast neutrons in the shielding screen, are absorbed in the screen through nuclear reactions of the types (n,$\gamma$), (n,$\alpha$), (n,p), etc. The shielding screens shall therefore contain elements with a large effective cross-section. On the other hand, it is desirable that the absorption of neutrons be achieved by means of reactions that do not generate high energy photons. Consequently (n,$\alpha$) reactions are preferred. The $\alpha$ particles that result, as well as the low-energy gamma radiations, do not raise any additional shielding problem. In this class of materials boron is outstanding. Nevertheless, most elements absorb neutrons by (n,$\gamma$) reactions that generate a penetrant gamma (secondary) radiation. Its shielding requires complex systems of protection.

The attenuation of gamma radiations generated either in the reactor or through secondary reactions takes place via the mechanisms already described in Chapter 2. As pointed out, the efficiency of the attenuation depends on the density of the shielding material.

The attenuation in the radiation intensity or in the neutron flux that penetrate an absorbent material of thickness $x$ and macroscopic cross-section $\Sigma$, is given by:

$$I = I_0 \exp(-\Sigma x) = I_0 \exp(-N\sigma x) = I_0 \exp(-x/\lambda), \qquad (8.1)$$

where $I_0$ is the intensity of the incoming radiation flux, $I$ is the intensity of the radiations outgoing from the absorbing screen and $\lambda$ represents the mean free path of the radiation quantum, or particle in the absorbing material (the relaxation length), $\Sigma$ is the macroscopic cross-section, $\sigma$ is the microscopic cross-section of interaction, and $N$ is the density of nuclei.

Figure 8.2 shows the correlation of the screen thickness with the material density and its dependence on the radiation energy, in order to reduce 10 times the intensity of the gamma radiations [2].

Generally, the composition and dimensions of the shielding screens attenuating the neutron flux ensure also the attenuation of the gamma radiations. If the screen thickness must have a predetermined value, the attenuation of the radiations is achieved by blending with materials of large atomic mass.

Table 8.2 shows the relaxation lengths of photons and neutrons (fast, intermediary and thermal) for certain materials used in the manufacture of shielding screens [3]. It also gives the global relaxation length, $\lambda_t$, i.e. the highest among the relaxation lengths that characterize the material under consideration. Screens that combine the above-mentioned materials provide superior attenuation characteristics, as compared to those made of a single component.

A material is warranted as having good $\lambda$ and neutron-radiation attenuation characteristics if the relaxation lengths of the two types of radiation are approximately equal to each other ($\lambda_\gamma \simeq \lambda_n$).

An equation of the type (8.1) describes the attenuation of a collimated beam of gamma radiation. In fact, the problem of screening the gamma radiation is much more complicated by the presence of diffuse radiation. Thus, in case of wide-angle gamma radiation beams, and also in very thick absorbents, "secondary" gamma radiations appear as a result of the Compton effect within the energy span 0.1 — 10 MeV. A photon undergoes 5 — 10 scattering events, during which its energy decreases, before the particle is absorbed. Thus the initial narrow-angle radiation beam turns into an all-angle diffusion background.

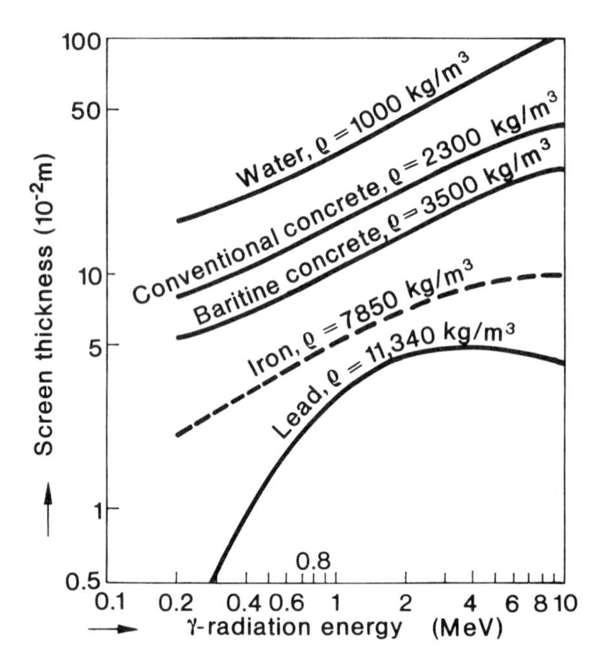


Fig. 8.2. Screen thickness required to reduce 10 times the intensity of the gamma radiations.

TABLE 8.2 Relaxation Lengths

| *Material* | *Density* ($kg/m^3$) | *Relaxation length* (cm) *Photons* ($\lambda_\gamma$) | *Neutrons* *Fast* ($\lambda_{fn}$) | *Inter-mediate* ($\lambda_{in}$) | *Thermal* ($\lambda_{tn}$) | $\lambda_t$ |
|---|---|---|---|---|---|---|
| Iron | 7,800 | 4.6 | 5.9 | 17 | 1.1 | 17 |
| Lead | 11,400 | 2.2 | 10 | 8 | 14 | 14 |
| Concrete (conventional) | 2,300 | 17 | 12 | 6 | 7-13 | 17 |
| Barytine concrete | 3,500-4,000 | 7.7-12 | 8-11 | 6 | 2.1-7 | 8-12 |
| Water | 1,000 | 39 | 10 | 2.6 | 2.8 | 39 |
| Polyethene | 900 | 25 | 8.5 | 2.5 | 3 | 25 |
| Iron — water | | 6.8 | 7 | 7 | 2 | 7 |
| Lead — water | | 3.4 | 9.6 | 9.6 | 4 | 9.7 |
| Iron — polyethene | | 6.6 | 6.6 | 6.6 | 2 | 6.6 |
| Lead — polyethene | | 8.7 | 8.7 | 2.9 | 4 | 8.7 |

Figure 8.3 shows the attenuation of the gamma radiation in a channelled (a) and unchannelled (b) geometry, respectively [4]. As a result of the diffusion processes, the radiation attenuation depends on the wall thickness as well, and it is slower than the one described by eq. (8.1). Considering Fig. 8.3, eq. (8.1) can be improved by introducing a *correction coefficient B*, so that

$I = BI_0 \exp(-x/\lambda)$. The coefficient $B$ can be linked to $\lambda$, in a way that $B = x\lambda^{-1}$. This approximation in known as the *linear correction*.

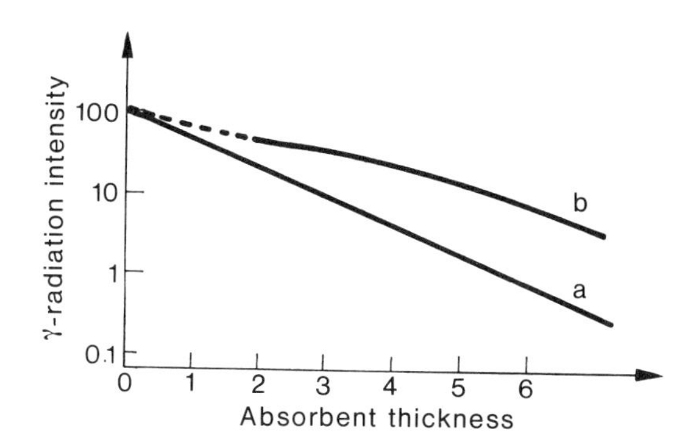


Fig. 8.3. The decrease of the radiation flux intensity for a channelled (a), and unchannelled (b), flux, respectively, vs. the absorbent thickness.

Other methods of analysis of how the gamma radiation intensity changes in screens have also been developed; generally, these imply cumbersome calculations [4-6]. Mathematical treatment distinguishes between deep and average penetration of the material, depending on whether the screen thickness is higher or smaller than (10 — 20) $\lambda$. The latter case is to be encountered more often is practice. In this situation the *method of moments* [7,8] is used. The calculations are in fair agreement with the experimental data, with differences below 5 — 20 %.

A similar situation is recorded with the attenuation of neutron beams. An equation of the type (8.1) is valid only when capture processes and nuclear reactions are at the origin of the attenuation. As a result of elastic or inelastic scattering, a correction factor $B(E_0,x)$ is introduced in eq. (8.1), that depends on the initial energy $E_0$ and on the neutron penetration depth $x$. This factor can be determined, in principle, in a way similar to the case of the gamma radiations [7,8]. Since the cross-sections for neutrons at $E > 5$ — 6 MeV are not known to a satisfactory degree of accuracy, two other semiempirical methods of approximation are used. The first method is convenient when accuracy is not particularly required, as in the case of preliminary designs, or of reactors where weight is of little consequence. In such a case, an exponential attenuation equation of the type (8.1) is used, where *an estimated average relaxation length* $\lambda_m$ is introduced. Values of $\lambda_m$ are tabulated (see e.g. [4]). The second method works on the assumption that, from the standpoint of protection, of primary consequence is the attenuation of neutrons with $E = 8$ MeV. Starting from this assumption, it was shown that it is possible to account for the additional attenuation introduced by lead, iron or concrete slabs, when these join a water screen. In this case the attenuation is described by a variant of equation (8.1), where a cross-section appears in the exponent. This parameter that is independent of the thickness $x$ is known as the *effective absorption cross-section*, or *displacement cross-section*, $\sigma_{disp}$. Such an effective cross-section does not have a physical meaning. However, it may be correlated with the total cross-section of the neutrons with the energy $E = 8$ MeV, namely: $\sigma_{disp} \simeq 0.6\ \sigma_{tot}(8\ \text{MeV})$. This relation is valid only for screens made up of alternating layers of materials and substances that contain hydrogen (generally water). The $\sigma_{disp}$ values are to be found in tables, and they can be satisfactorily represented by the relation $\sigma_{disp}/\rho = 0.085\ A^{-1/3}$, where $\rho$ is the density and $A$ is the atomic mass [4].

In practice shielding screens for gamma radiation and neutrons comprise several components [4]. Consequently, the approximations above have been extended in order to take into account the attenuation in screens made up of more than one type of material [4-6]. Since screen thickness is relevant to radiation attenuation, quite often the relaxation lengths are given on certain thickness ranges.

## 8.2. RADIATION EFFECTS ON SHIELDING MATERIALS

The cooling agents can not take up the entire amount of energy generated in the nuclear reactor, since the conversion into heat of the energy of radiations does not take place exclusively in the reactor core. The "leaks" of energy will be "stopped" by the reactor shielding and turned into heat in the materials that are part of it. The dissipation of the energy of primary gamma radiations, neutrons, and secondary gamma radiations in the shielding materials entail heating of these materials as well as changes in their mechanical characteristics and structural stability (direct interaction).

At the origin of the consequences of the direct interaction (also analysed in the previous chapters) are the defects induced by irradiation. Among them are: dimensional instability; variation of mechanical characteristics (tensile strength, wear strength, etc.) variation of other physical properties (diminishing of thermal conductivity) etc.

The nuclear heating of shielding materials is due to absorption of the gamma radiation $\gamma$ and the neutron slow-down. The generation of heat in the reactor shielding material can lead to undesirable effects, such as the establishment of thermal gradients that induce important stresses (especially in concrete); a reduction in the humidity content of concrete, etc. Regular concrete has shielding qualities as long as it contains 8 — 10% water. The water content can be preserved under normal atmospheric conditions for a period of 20 — 50 years. As temperature increases, a de-hydration begins at ~ 70 $^{\circ}$C.

*Gamma radiations* (from fission and capture) interact with the materials in three modes, that are:

1. the Compton scattering, a phenomenon practically encountered over the entire energy range of the $\gamma$-photons (0.5 — 10 MeV);
2. the photoelectric effect, which prevails in case of low-energy photons;
3. the pair (electron — positron) generation, a phenomenon occurring at $E > 1.02$ MeV.

*The neutrons* lose energy by elastic collisions with the nuclei. In an electric collision the mean energy loss (in the centre-of-mass system) is given by

$$g = \frac{E - E'}{E} = \frac{2A}{(A + 1)^2}\ , \tag{8.2}$$

where $E$ is the initial energy, $E'$ is the energy after collision and $A$ is the atomic mass.

If the energy of the neutron excedes $\sim 10\ A^{2/3}$ MeV [9] the scattering is no longer isotropic and the energy loss is:

$$g = \frac{E - E'}{E} = \frac{2A}{(A + 1)^2} \cos \vartheta , \tag{8.3}$$

where $\vartheta$ is the scattering angle in the centre-of-mass system.

If the energy spectrum of the neutrons, $\varphi(E,\vec{r})$, is known, then the heat generated per unit-volume and time in a point of position vector $\vec{r}$ is:

$$Q(\vec{r}) = K \int_{E_0}^{E_1} \Sigma_{el}(E) \cdot \varphi(E,\vec{r})\, Eg\, dE, \tag{8.4}$$

where $K$ is a constant ($K = 1.6 \times 10^{-13}$ W . s/MeV if energy is expressed in MeV) and $\Sigma_{el}$ is the macroscopic elastic scattering cross-section.

A neutron diffusion analysis [10], conducted in the two-group approximation and on the assumption that the energy transported by the neutrons in the thermal group is clearly inferior to the one transported by the fast neutrons, leads, for the thermal energy generated, to the expression

$$Q(\vec{r}) = K\Phi_f \overline{\Sigma_{el}E}\, g, \tag{8.5}$$

where

$$\overline{\Sigma_{el}E} = \int_{0}^{E_1} \Sigma_{el} E f(E)\, dE,$$

$f(E)$ is the spectrum of the fission neutrons and $\Phi_f$ is the flux of fast neutrons.

If the material has a density $\rho$ and is a homogeneous mix of $i$ constituents, of atomic masses $A_i$ and concentrations $\delta_i$, the thermal energy generated reads:

$$Q = K\Phi_f (\rho/100)\, \overline{\sum_i (\Sigma_{el})_i E}\, \delta_i g_i. \tag{8.6}$$

Another source of heat are the (n,$\alpha$) reactions that may come with certain constituents of the shielding materials (lithium and boron). In order to estimate the heat generated by such processes one skips the (n,$\alpha$) reactions that are induced by fast and intermediary neutrons, which have very small cross-sections as compared to the same type of interactions generated by thermal neutrons. In this way, the heat generated per unit of time in a material of a density $\rho$ containing a quantity   of natural boron or lithium is:

$$Q = K\Phi_{th} (\rho/100) \Sigma(n,\alpha) \delta E, \tag{8.7}$$

where $\Sigma(n,\alpha)$ has the following values [9]:

1. for natural boron (20% $^{10}B$): $\Sigma(n,\alpha) = 35.79$ cm$^{-1}$
2. for natural lithium (7.52% $^{6}Li$): $\Sigma(n,\alpha) = 2.89$ cm$^{-1}$.

## 8.3. SHIELDS

The attenuation, outside the reactor core, of gamma radiations and neutron leaks is achieved, as indicated above, by *shielding screens*, surrounding the core. As to their role, one can make a distinction between screens for *thermal shielding* and those for *biological shielding*. To avoid exposure of the entire shielding to the heat generated in the reactor and to attenuate radiations, nearest to the core are the *thermal shielding screens*. These are made of materials of high density, good thermal conductivity and high melting temperatures. Thermal screens absorb the high-energy gamma radiations and reduce the energy of the fast neutrons by inelastic collisions. Towards the exterior of the reactor, the thermal shielding screens are followed by *biological shielding screens*. These are made of water and/or concrete, and their main task is to attenuate the secondary gamma radiations, as well as to thermalize and absorb the neutrons. Biological shielding screens are so designed as to ensure that the equivalent (biological) doses specified for the exterior of the screen in the enforced regulations are not exceeded.

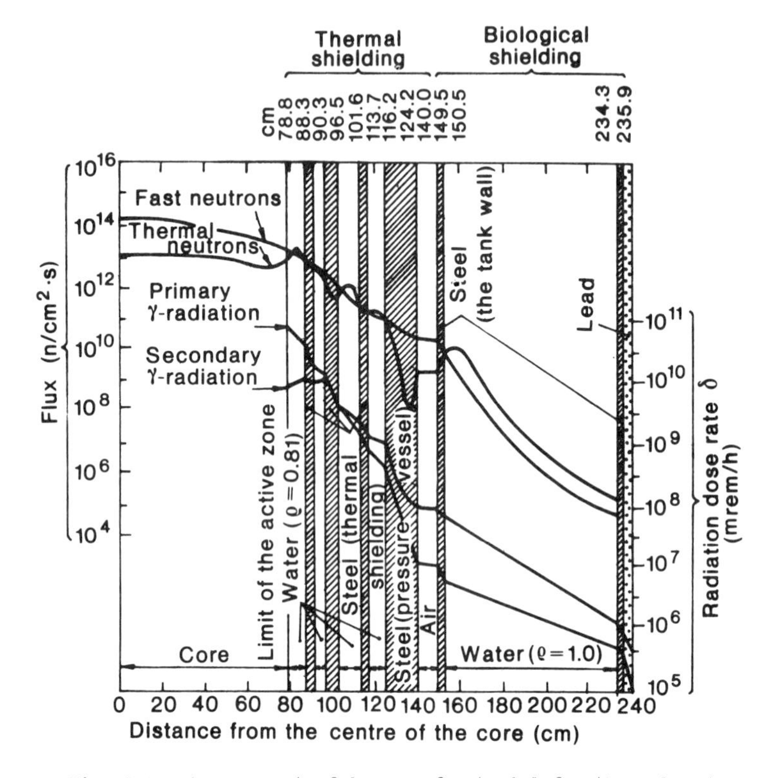


Fig. 8.4. Arrangement of layers of material for thermal and biological shielding; the corresponding reduction in the neutron flux and radiation dose rate (water density, ρ, in kg/dm$^3$).

In order to meet these requirements, the shielding screens are made of alternating layers of various materials. As an example, Fig. 8.4 presents schematically the structure of a shielding screen [11].

### 8.3.1. Thermal Shielding Materials

The very high neutron and photon fluxes at the periphery of the core determine powerful heating of the materials in direct contact with it, together with important temperature gradients. Considering these factors, the design of a thermal shielding also observes the type of reactor. The following sections describe the shielding screens employed in the leading types of reactors, as well as the materials they are made of.

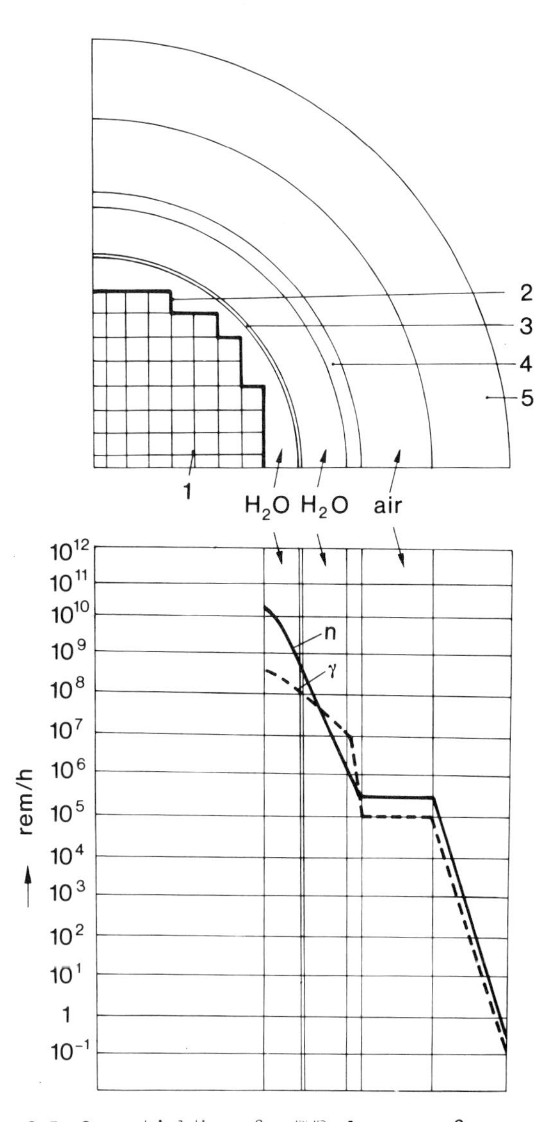


Fig. 8.5. Core shielding of a BWR: 1 - core; 2 - core shielding; 3 - cylinder supporting the core; 4 - pressure vessel; 5 - concrete shielding.

Figure 8.5 shows schematically the shielding system used for the BWRs and PWRs [12]. The lower part presents the gamma and neutron attenuation across the entire shielding. Figure 8.6 illustrates the configuration of a PWR thermal shielding [13]. The system is made of metallic cylinders that are concentric to the core; in between them the primary cooling agent is circulated. In this way one prevents direct thermal loading of the pressure

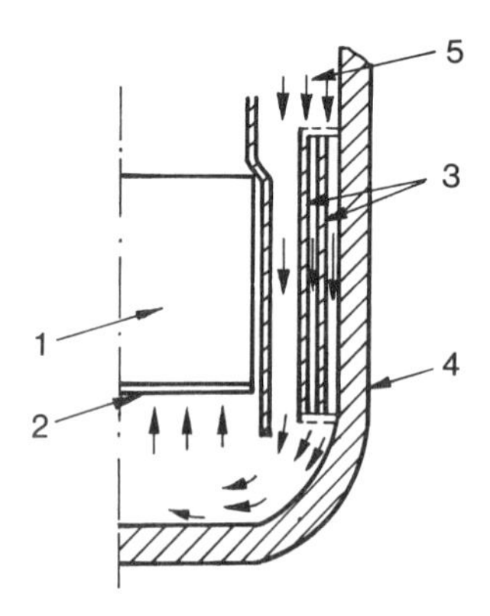


Fig. 8.6. The arrangement of thermal shielding screens of a PWR: 1 - core; 2 - core support; 3 - thermal shielding screens; 4 - pressure vessel; 5 - cooling agent.

vessel, also achieving a reduction of the fast neutron flux down to $10^8$–$10^9$ n/cm$^2$.s, at the level of the vessel walls, thus avoiding their radiative embrittlement. In the manufacture of thermal shielding screens for PWRs, iron, carbon steel or stainless steels are in order.

Figure 8.7 presents the arrangement of the outer shielding of a pressure vessel for a VVER-440 reactor [14]. The shielding consists of a ring-shaped vessel filled with water (2) that surrounds the core, and of the concrete shielding (3). In the radial direction the water screen thickness is 0.95 m and the total steel thickness (the vessel walls) is 0.025 m. The new generation of reactors, VVER-1000, employs as outer shielding of the pressure vessel a concrete screen with an admixture of serpentine. The high water content of the serpentine ensures the thermalization of fast neutrons.

*For the CANDU type reactors*, with pressure tubes, a radial shielding shown in Fig. 8.8a, as well as an axial shielding depicted in Fig. 8.8b, are employed. Components of the radial shielding are the $D_2O$ reflector (2) and the $H_2O$ thermal shielding (3) located between the calandria and the concrete vault (4). Figure 8.9 shows the variation in these shieldings of the dose rate, both for neutrons and gamma radiations. The thickness of the shielding elements is also indicated. The figure presents the change in the gamma radiation dose rate in the lateral concrete screen (5) of the reactor, that acts as a biological shielding. This wall is built on that side of the reactor where the rooms that allow the access of personnel are located. For the axial shielding, the thermal screen consists of the tubular plate (6) of the calandria vessel, the enclosure (7) between this plate and the outer tubular plate (8) which actually is the wall towards the room that hosts the fuelling machine, and the outer tubular plate (8). The enclosure is filled with steel beads and water, their volume ratio being 3/2, which corresponds to a close-packed arrangement of the spherical beads. Passing across this element of the screen the neutron flux decreases exponentially, from $10^{11}$ — $10^{12}$ n/cm$^2$ . s to approximately $10^6$ n/cm$^2$ . s. The pressure tubes are provided with screening plugs. The dose rate at the exterior of the tubular plate being $10^5$ R/h, the fuelling machine room is not accessible to the personnel.

In the case of *HTRs and AGRs* with a single cavity, that have the steam generators integrated in the prestressed concrete vessel of the reactor, an important part in the reduction of neutron and gamma radiation fluxes is

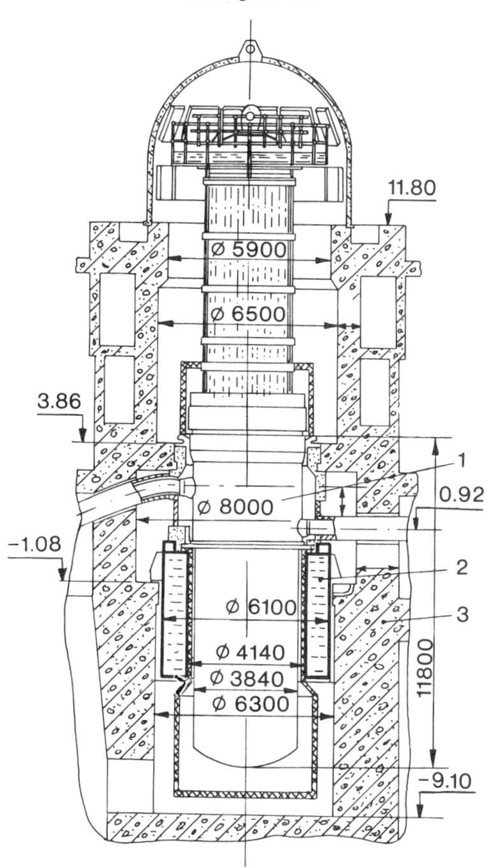


Fig. 8.7. The shielding system of a VVER-400 reactor:
1 - reactor body; 2 - water screen (water vessel);
3 - concrete shielding.

played by both the graphite reflector and the thermal shielding of the vessel; the latter consists of a steel blanket which separates the core from the steam generator.

*Thermal shielding of fast reactors* is mainly achieved by means of a complex screen made of stainless steel, sodium, a small-density screen made of steel alloyed with chromium and of sodium, with a ratio of the constituting volumes of 1/4 as well as a high-density shielding screen with a steel to sodium ratio of 2/1. The thermal shielding of the "Enrico Fermi" fast reactor [15] is 0.305 m thick and consists of stainless steel plates with liquid sodium circulating in between, to eliminate the heat generated by irradiation. For fast reactors one can use either iron and boral (boron carbide in aluminium) screens or lead and boral screens [16] arranged in an alternating manner.

*In research reactors*, shielding is generally provided by the water in which the core is immersed. For example, the thermal shielding screen of the ETR material testing reactor [15] is approximately 0.55 m thick and consists of a

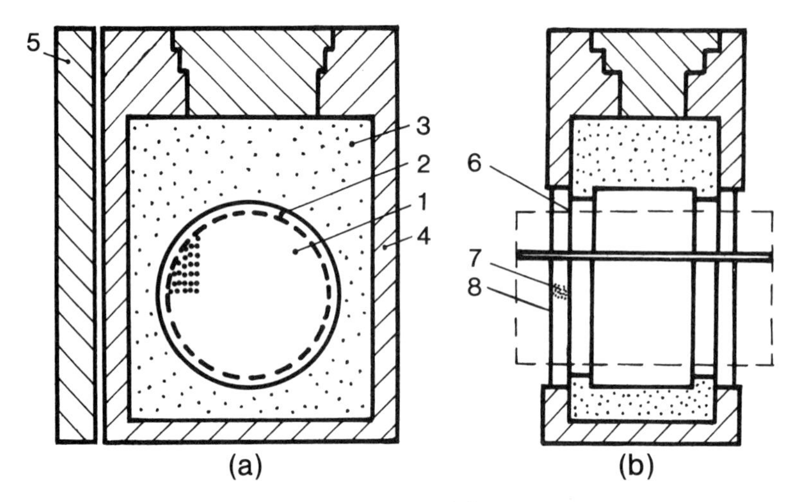


Fig. 8.8. Radial (a) and axial (b) shielding screens, for a PHWR: 1 - core; 2 - reflector; 3 - $H_2O$ axial thermal shielding; 4,5 - concrete screens; 6 - tubular plate of calandria; 7 - enclosure filled with stainless steel beads and water; 8 - outer tubular plate.

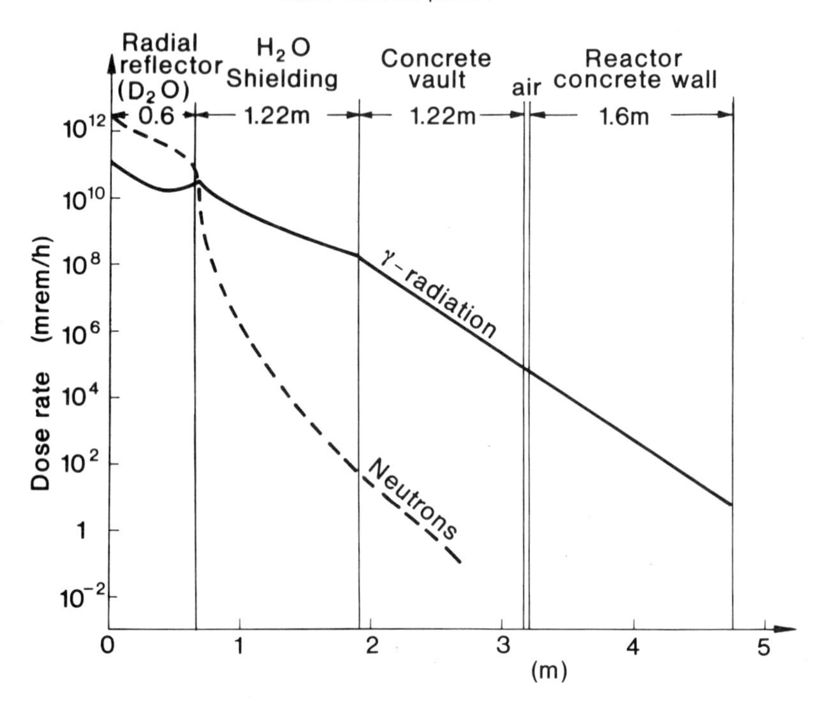


Fig. 8.9. Variation of the dose rate in the shielding screens of the PHWR-CANDU.

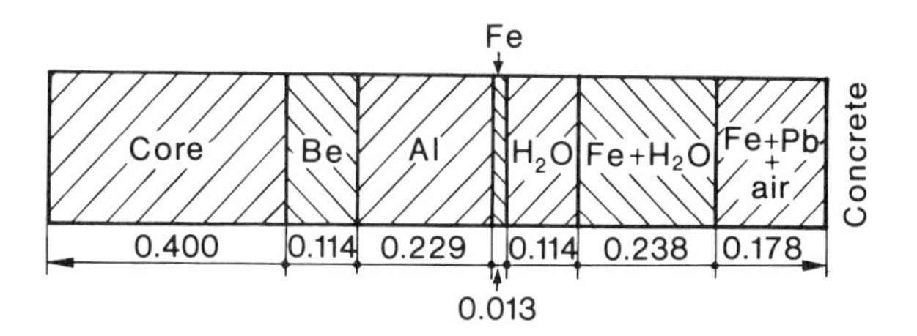


Fig. 8.10. Structure of the thermal shielding screen of the ETR.

sequence of four layers: iron, water, water + iron, and iron + lead + air, respectively (Fig. 8.10). The amount of heat generated in a thermal shielding screen is determined by the solution adopted for building the screen, by the intensity of the neutron and photon fluxes, as well as by physical characteristics of the materials that constitute the screen. As an example, Fig. 8.11 presents the total density of the energy generated in the constitutive layers of the thermal shielding screen of the ETR reactor referred to in Fig. 8.10, as well as the contributions of the neutrons and gamma radiations [15].

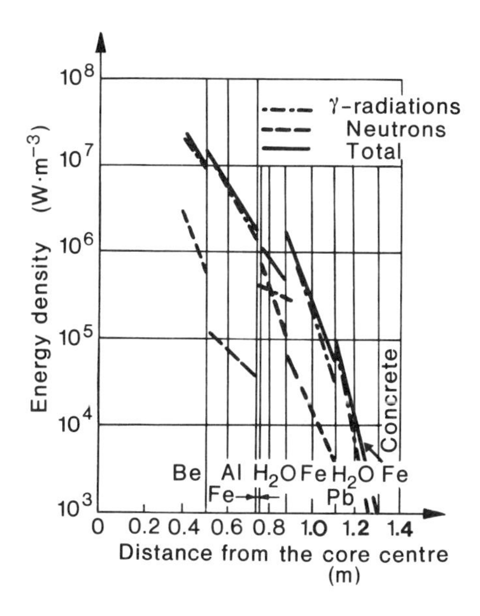


Fig. 8.11. Energy density as generated by gamma and neutron radiations in the thermal shielding of the ETR.

In all the cases described, the thermal energy is evacuated by recirculating the fluid that is an integral part of the shielding. If this fluid is distinct from the cooling fluid (v.PHWRs), the thermal shielding screen implies a special recirculation and purification system.

### 8.3.2. Biological Shielding Materials

Let us remember that the biological shielding screens are meant to attenuate the neutron and photon fluxes, to the extent that at the exterior surface of the reactor the equivalent doses should not exceed the values specified in the regulations. As noted before, in biological shielding the neutrons are slowed down and absorbed, and the secondary gamma radiations are further attenuated. The materials currently used for biological shielding screens are water and concrete, both regular and special.

8.3.2.1. Water shielding screens. Water shielding screens are mainly used with the reactors for testing materials and fuel elements, of the pool type. Water is not particularly efficient in the attenuation of gamma radiations ($\lambda_\gamma$ = 39 cm) — Table 8.2. For this reason the screens must be of a considerable thickness. This shortcoming is partially compensated by the easy access to the core in view of introducing the irradiation devices, even while the reactor is being operated. Among the reactors using water shielding systems mention can be made of TRIGA (USA, Romania), OSIRIS (France), BR-2 (Belgium) etc.

Other physical characteristics of water, featuring its merits as a shielding material, including stability under irradiation, were dealt with in Chapters 5 and 7.

8.3.2.2. Concrete shielding screens. The use of concrete as a shielding material calls for certain requirements that these materials have to meet, among which are: preservation of the initial properties for long periods of time; resistance to radiations and temperatures generated as a result of the interaction with the radiation fluxes; dimensional stability, etc. The design of concrete shielding screens must take into account the need to secure the initial water content of the material, and to limit the effects on the screen of the temperature, radiations and mechanical stresses. This, again, implies meeting certain requirements [17,18]. Thus:

1. the maximum temperature in the material should be limited to approximately 360 K for neutron screens and to ~ 450 K for gamma radiation screens, respectively;

2. the incident fluxes should not exceed ~ $5 \times 10^9$ n/cm$^2$ . s for neutrons, and ~ $4 \times 10^{10}$ $\gamma$-photons/cm$^2$ . s for gamma radiations with an energy of approximately 1 MeV;

3. the maximum incident energy flux should be limited to $4 \times 10^{10}$ MeV/cm$^2$ . s.

4. the heat generated in the screen per unit-volume should not exceed $10^6$ W/m$^3$;

5. the temperature increment as a result of screen heating should not exceed ~ 280 K;

6. the maximum permissible temperature gradient in the material is 100 K/m.

The varieties of concrete employed in the construction of shielding screens may be grouped into two classes — conventional (regular), and special.

*Conventional concrete* has a density of $(2.2 - 2.3)\times10^3$ kg/m$^3$ and a compression resistance of 25 — 35 MPa.

Among special concretes are heavy concrete, boron concrete and high temperature concrete.

*High-density heavy concrete* is obtained by admixing into conventional concrete certain ingredients based on iron, barium or lead. From this category mention is made of the concrete with baritine ($\rho = 3.5 \times 10^3$ kg/m$^3$); limonite ($\rho = 3 \times 10^3$ kg/m$^3$); hematite ($\rho = (3.5\text{-}4) \times 10^3$ kg/m$^3$); iron scraps ($\rho = (6\text{-}6.5) \times 10^3$ kg/m$^3$); and lead ($\rho = (9\text{-}9.5) \times 10^3$ kg/m$^3$). The increase in density contributes to the attenuation of gamma radiation and, to some extent, of neutrons. Studies on concrete with iron scraps, baritine and chromite, respectively, of densities between 2000 kg/m$^3$ and 5500 kg/m$^3$, evidence a linear increase of the attenuation coefficient $\mu$, from 4 m$^{-1}$ up to 16 m$^{-1}$ as the density increases [19]. A linear dependence on the iron content in the concrete is noticed for the macroscopic capture effective cross-section of thermal neutrons $\Sigma_c$ [19]. Consequently, the use of high-density concrete allows a reduction in the shielding screen thickness.

The use of *boron concrete* is in order when a high capacity of attenuation of thermal neutrons is required. Boron is introduced in the concrete through ingredients with a boron content (colemanite $Ca_2B_2O_{11} \cdot 5H_2O$), by adding pyrex glass powder to the sand that is part of the concrete content, or by mixing boric acid into the water used in the fabrication of the concrete. The macroscopic capture cross-section of the thermal neutrons depends linearily on the boron content in the concrete — Fig. 8.12 [19].

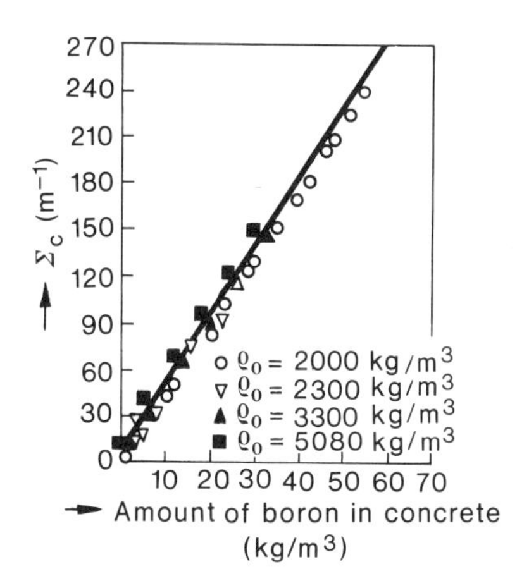


Fig. 8.12. Macroscopic effective cross-section of capture (in m$^{-1}$) vs. the boron content in concrete.

*High-temperature concrete* contains mineral ingredients where water remains chemically bound up to temperatures of 500 °C. Among the ingredients used, special mention is made of serpentine ($3MgO \cdot 2SiO_2 \cdot 2H_2O$). Figure 8.13 shows the temperature dependence of the quantity of water bound in the ingredients (aggregates). Serpentine concretes have a high water content, even at high temperatures. Although by adding serpentine the density of the concrete decreases, the hydrogen content increases approximately three times, as compared with the conventional concrete.

The physical characteristics that are determinant in selecting concretes for building shielding screens are density, compression resistance, thermal conductivity, hydrogen content, thermal stability of bound water, together with the nuclear characteristics of the entire collection of constituents.

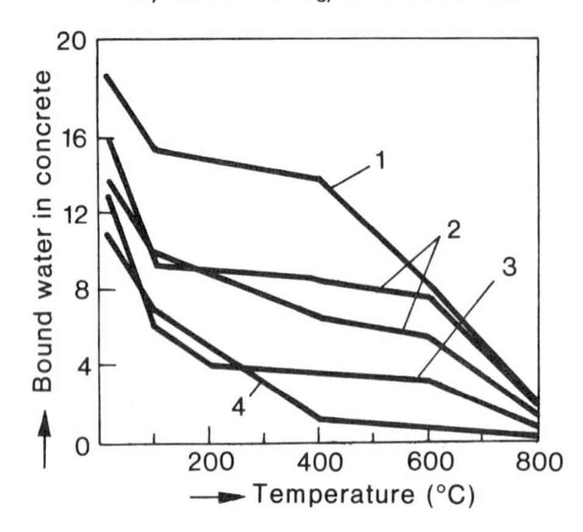


Fig. 8.13. Temperature dependence of the content of bound water in concrete: 1 - serpentine concrete; 2 - serpentine and hematite; 3 - chamotte; 4 - hematite chamotte.

Table 8.3 contains several physical properties of certain representative types of concrete, and Table 8.4 gives some of their nuclear properties [18].

TABLE 8.3 Mechanical and Physical Properties of Concretes

| *Concrete* | *Density* ($10^3$ kg/m$^3$) | *Composition (%)* | | | | *Specific heat* (J/kg.K) | *Thermal conductivity* (W/m.K) | *Thermal expansion coefficient* ($K^{-1}.10^{-6}$) | *Mechanical strength* | |
|---|---|---|---|---|---|---|---|---|---|---|
| | | *Cement* | *Additives* | *Water* | *Iron (metallic)* | | | | *Tensile* | *Compression* |
| Conventional | 2.3 | 8 | 85 | 7 | - | 650 | 0.87 | 14 | 2.1-3.2 | 21 |
| High density concrete: | | | | | | | | | | |
| barytine | 3.3 | 8 | 84 | 6 | - | 510 | 1.6 | | 5.18 | 24 |
| magnetite | 4.73 | 12 | 23 | 4 | 61 | | 2.4 | 9 | 5.6 | 20 |
| limonite | 4.54 | 13 | 21 | 5 | 61 | | 4.8 | 10.7 | 6.4 | 12.5 |
| Boron concrete (colemanite addition) | 5.36 | 7 | 9[1] | 3 | 81 | | | | 5.4 | 110 |
| High-temperature concrete (serpentine addition) | 2.06-2.2 | 11.1 | 75.4[2] | 13.5 | - | | | 18 | | 13 |

[1] 1.2% boron; [2] 52.5% serpentine.

TABLE 8.4 Nuclear Properties of several Varieties of Concrete

| Concrete | Neutron cross-sections | | | γ-attenuation coefficients ($cm^{-1}$) | | | | |
|---|---|---|---|---|---|---|---|---|
| | $\Sigma_R(cm^{-1})$ | $\Sigma_a(cm^{-1})$ | $\Sigma_t(cm^{-1})$ | 0.5 MeV | 1 MeV | 3 MeV | 5 MeV | 10 MeV |
| Conventional | 0.0849 | 0.0094 | 0.174 | 0.2017 | 0.1473 | 0.0841 | 0.0663 | 0.0520 |
| High-density | | | | | | | | |
| barytine | 0.0985 | 0.0224 | 0.199 | 0.2964 | 0.2030 | 0.1217 | 0.1047 | 0.0963 |
| magnetite | 0.127 | 0.0959 | 0.220 | 0.3989 | 0.2878 | 0.1719 | 0.1456 | 0.1315 |
| limonite | 0.121 | 0.0888 | 0.207 | 0.3827 | 0.2761 | 0.1650 | 0.1398 | 0.1263 |
| Ferro-phosphate | 0.125 | 0.0924 | 0.216 | 0.3946 | 0.2851 | 0.1697 | 0.1430 | 0.1280 |
| As in the shielding of the NRU reactor (Canada) | 0.0983 | 0.0766 | 0.187 | 0.2909 | 0.2106 | 0.1241 | 0.1028 | 0.0892 |

*Compression resistance* represents one important parameter in selecting the concrete for screens, since concrete also has structural functions in a nuclear reactor. By an adequate gradation of constituents a concrete structure with a high compression resistance can be achieved. The preparation of concrete must take into account simultaneously the dimension of the aggregates, the granulometric componence, the water/cement ratio, the mortar — aggregate adherence, the gradation of cement, water, and other ingredients. If a continuous granulometric distribution is secured, a plastic concrete can be obtained, that is easy to process, with no segregation risks. If the distribution of the granules is discontinuous, a very compact concrete is obtained, with a high mechanical resistance. For shielding screens, one seeks the best compromise between the compression resistance and the water content (hydrogen) that is necessary for the attenuation of neutron fluxes. For regular concrete the compression resistance is 20 — 40 MPa: by adding iron minerals-based incredients, it can increase to 38 – 46 MPa.

*The thermal conductivity* of regular concrete is 0.3 — 5 W/m . K, depending on the composition. Generally, the high density aggregates in the concrete also contribute to an increment in thermal conductivity. Thus, the thermal conductivity of baritine concrete is 1.8 W/m . K; it increases to 2 — 5 W/m . K for a concrete with iron minerals, and to 12 W/m . K for a concrete that includes iron scraps.

*The hydrogen content* in concrete is determined by the quantity of chemically bound water in constituents as well as by the quantity of organic materials in the concrete. Concrete used in the manufacture of shielding screens must contain hydrogen in a proportion of at least 0.5% in weight. Also, as shown before, the content of water in concrete must be maintained as long as possible, at a value as close as possible to the initial one. At ambient temperature, concrete dehydration is very slow. The initial water content diminishes to a half, over a period of 25 — 30 years. Concrete dehydration is accelerated by high temperatures, that are determined by the interaction of γ-radiations and neutrons with the constituent nuclei. The maximum permissible temperature for concrete shielding screens is limited to ~ 90 °C. Figure 8.13 points out the fact that thermal stability is conditioned by the aggregate used for the preparation of the concrete. In the case of regular concrete, at ~ 90 °C the content of bound water is of about~6%, increasing to ~16% in serpentine concrete.

The capacity of concrete to attenuate the neutron or photon fluxes is determined by their nuclear characteristics. In this sense, Table 8.4 presents certain nuclear constants, namely the scattering cross-sections (the reciprocal of the relaxation length), the absorption and the total cross-section. The same table gives the attenuation coefficients of the gamma radiation with energies of 0.5 – 10 MeV [20].

Figure 8.14 presents the dependence on the neutron energy of the total absorption cross-section [18].

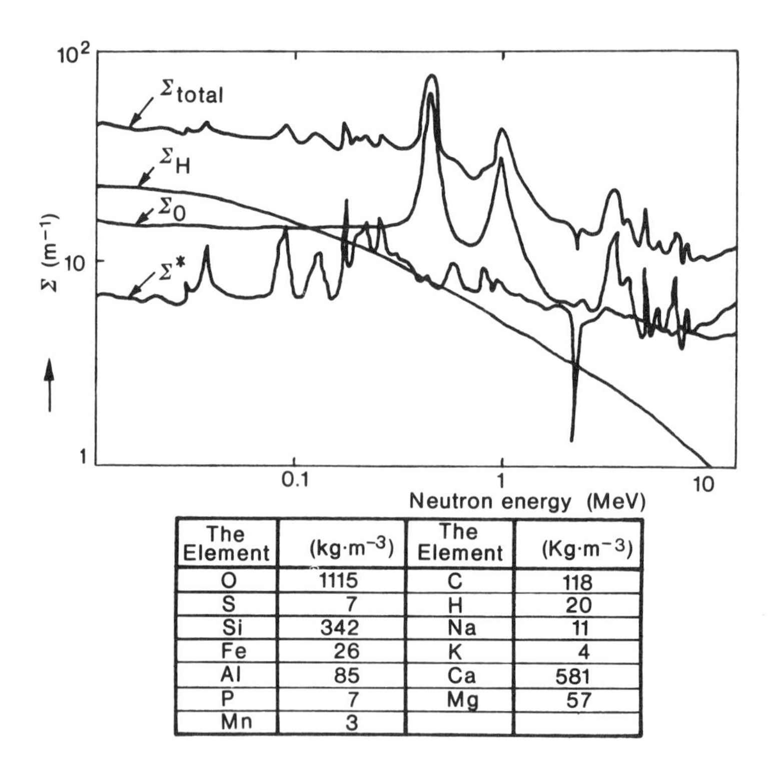


| The Element | $(kg \cdot m^{-3})$ | The Element | $(Kg \cdot m^{-3})$ |
|---|---|---|---|
| O | 1115 | C | 118 |
| S | 7 | H | 20 |
| Si | 342 | Na | 11 |
| Fe | 26 | K | 4 |
| Al | 85 | Ca | 581 |
| P | 7 | Mg | 57 |
| Mn | 3 | | |

Fig. 8.14. Neutron energy-dependence of the macroscopic effective cross-section, in concrete. Absorption cross-sections are $\Sigma_H$ in hydrogen; $\Sigma_O$ in oxygen and $\Sigma^*$ in other components, respectively. Total cross-section is denoted by $\Sigma_{total}$.

As shown before, owing to the difussion processes taking place in the shielding screens, the relaxation length depends on the thickness of the screen. In order to emphasize this, Table 8.5 gives the $\lambda$ values, as functions on the thickness and density of the concrete that makes the screen [19]. It should be noticed that the values of the relaxation length are approximately two times higher when the screen thickness increases by one order of magnitude.

TABLE 8.5 Relaxation Length (cm) as a Function of Concrete Thickness and Density

| *Type of concrete* | *Density* ($kg/m^3$) | *Shield thickness* (m) | | | | | | |
|---|---|---|---|---|---|---|---|---|
| | | *0.0–0.25* | *0.25–0.50* | *0.50–1.00* | *1.0–1.5* | *1.5–2.0* | *2.0–2.5* | *2.5–3.0* |
| Conventional | 2300 | 7.9 | 9.6 | 12.0 | 14.1 | 15.0 | 15.7 | 16.1 |
| With hematite | 3200 | 6.65 | 8.2 | 10.2 | 11.9 | 12.7 | 13.3 | 13.6 |
| With iron scrap | 4200 | 5.9 | 7.3 | 9.05 | 10.6 | 11.3 | 11.8 | 12.1 |

Of special interest are the *effects of irradiation on screens.* The main effect is the "nuclear heating" by radiation absorption. It entails higher local temperatures and, consequently, leads to expansion effects. The internal stresses caused by such expansion may generate cracks and inner cavities that contribute to a degradation of the original physical and nuclear characteristics of the material. Therefore, a limitation to $10^3$ W/m$^3$, of the quantity of heat generated per unit of time and volume is necessary. This corresponds to an incident energy flux of approximately $4 \times 10^{10}$ MeV/cm$^2$ . s.

Paragraph 8.2 referred to the thermal energy generated as a result of the interaction of radiations with the shielding screens. The equations developed in that context can be particularized for concrete shielding screens. The amount of heat generated per unit of volume and time $Q_n^o$ (W/cm$^3$) can be determined with the equation [18]:

$$Q_n^o \simeq 1.28 \;.\; 10^{-12}\; \Sigma_c(E)\Phi, \tag{8.8}$$

where $\Sigma_c(E)$ is the macroscopic capture cross-section for the neutrons of energy $E$ and $\Phi$ represents the neutron flux. The quantity $\Sigma_c$ is evaluated with the equation:

$$\frac{\Sigma_c}{\rho} = \sum_{i=1}^{N} \frac{N_o \sigma_{ci} \delta_i}{A_i}, \tag{8.9}$$

where $N_o$ is Avogadro's number, $\sigma_{ci}$ is the microscopic capture cross-section of the concrete constituent $i$, $\delta_i$ is the relative mass of the constituent $i$, $A_i$ is its atomic mass, $N$ is the maximum number of constituents and $\rho$ is the density of the material.

For the gamma radiation a relation similar to (8.8) is obtained:

$$Q_f^o(\text{W/cm}^3) \simeq 1.6 \;.\; 10^{-13}\; \mu_\gamma \Phi(E) E_\gamma, \tag{8.10}$$

where $\mu_\gamma$ represents the mass absorption coefficient and $\Phi(E)$ is the gamma radiation flux of energy $E_\gamma$ (in MeV).

The increment in the local temperature $\Delta T(x)$, resulting from the heating of the material, is given by:

$$\Delta T(x) = \frac{Q_o \lambda^2}{K} [1-\exp(-x/\lambda)], \tag{8.11}$$

where $x$ is the distance from the core surface, $Q_o$ the heat generated, $\lambda$ the local relaxation length for the considered radiation ($1/\Sigma_R$ for neutrons and $1/\mu$ for photons, respectively) and $K$ is the material's thermal conductivity.

The temperature gradient in a point $x$ is expressed as:

$$\frac{dT}{dx} = \frac{Q_o \lambda}{K} \exp(-x/\lambda). \tag{8.12}$$

Accordingly, the temperature gradient reaches its maximum value ($Q_o \lambda/K$) at $x = 0$, that is, on the screen surface situated on the side of the core.

The average mechanical stress, $\sigma_o$, at the surface $x = 0$ of the shielding screen — that results from the thermal expansion — can also be expressed as a function on the amount of heat generated, $Q_o$, as well as on the physical properties of the concrete:

$$\sigma_o = P_o - \frac{E_e Q_o \lambda^2 \alpha}{K(1 - \gamma_p)} \{1 - 2\frac{\lambda}{L} - (1 - \frac{2\lambda}{L})\exp(-L/\lambda)\}. \tag{8.13}$$

where $P_o$ is the average load in the concrete layer; $E_e$ is the elasticity modulus; $\alpha$ is the thermal expansion linear coefficient and $K$ is the thermal conductivity; $\gamma_p$ is the Poisson's ratio, which, for concrete, has a value of approximately 0.14, and $\lambda$ is the relaxation length.

Starting from the stress $\sigma_o$ the stress value $\sigma$ can be determined, for a point situated at a distance $x$ from the screen wall:

$$\sigma(x) = \sigma_o + \frac{\alpha E \Delta T(x)}{1 - \gamma_p}, \tag{8.14}$$

where $\Delta T(x)$ is given by eq. (8.11).

The numerical evaluation of the internal stresses points to the fact that the concrete resistance that is considered permissible can, under certain conditions, be lower than the value of the internal stresses — a fact that might cause cracks and cavities in the bulk of the material.

Table 8.6 assembles some data concerning the behaviour of concrete under irradiation [3].

TABLE 8.6 Irradiation Effects on Concrete

| *Irradiation duration* (months) | *Thermal neutron flux* $(n/cm^2 . s) . 10^{13}$ | *Fluency* $(n/cm^2) . 10^{19}$ | *Generated heat* $(W/cm^3)$ | *Weight loss* (%) | *Fracture strength* (MPa) |
|---|---|---|---|---|---|
| 2 | 1.1 | 0.5 | 0.011 | 2.1 | 7.4 |
| 6 | 1.2 | 1.6 | 0.012 | 2.6 | 6.3 |
| 12 | 1.3 | 3.0 | 0.013 | 2.2 | 5.6 |
| 24 | 1.4 | 7.0 | 0.014 | - | 4.3 |

The direct effects of irradiation, generated by the interaction with gamma radiations and neutrons, are relatively limited. To cause displacements in the material, with any sizeable effect on its mechanical characteristics, neutron fluencies higher by 2 — 3 orders of magnitude than those normally recorded in the shielding material ( ~ $10^{12}$ n/cm$^2$) are necessary. Consequently, the irradiation of concrete at a fluency of $10^{21}$ n/cm$^2$ (under normal conditions, for approximately 30 years) cuts to half the tensile strength of the material [3].

## REFERENCES


1. Kruger, F. W. (1982). *Proc. Int. Conference on Nuclear Power Experience*, IAEA-CN-42/84, Vienna.
2. Ageron, P. (1959). *Technologie des Réacteurs Nucléaires*, Vol. I, *Matériaux*. Eyrolles, Paris.
3. Price, B. T., Horton, C. C., and Spinney, K. T. (1957). *Radiation Shielding*. Pergamon Press, New York, London.
4. Gauzet, M., and Kahan, T. (1957). *Controle et protection des réacteurs nucléaires*. Dunod, Paris; Brodev, D. L. (1957). *Atomnaya Energya*, 3, 55; Kuktsevich, V. I., and Synytsyn, B. I. (1961). *Atomnaya Energya*, 10, 511.
5. Goldstein, H. (1959). *Fundamental Aspects of Reactor Shielding*. Reading, Mass. Addison-Wesley.
6. Harrison, I. R. (1959). *Nuclear Reactor Shielding*, Temple Press Limited, London.
7. Spencer, I. V., and Fano, U. (1951). *Phys.Rev.* 81, 464; *J.Res.Nat.Bur. Stand.*, 46, 446.
8. Fano, U. (1953). *Nucleo*, 11, 8, 55.
9. Davison, B. (1957). *Neutron Transport Theory*, Oxford.
10. CEA (1963). Report CEA-R-2253.
11. Maxim, I. V. (1969). *Materiale Nucleare*. Edit. Academiei, Bucharest.
12. Hahn, G., and Kicherer, C. (1980). *Nuclear Data and Benchmarks for Reactor Shielding*, Proc.of a Specialist Meeting, Paris, NEA, OECD, p.89.
13. Weissman, J. (1977). *Elements of Nuclear Reactor Design*, Elsevier Sci.Publ. Comp., New York.
14. Egorov, Iu. A. (1982). *Osnovy radiatsionnoy bezopasnosty atomnyh electrostantsyi*. Energoizdat, Moskow; Dubrovsky, V. B., and others (1979). *Stroitelstvo atomnyh electrostantsyh*. Energya, Moskow.
15. Jaeger, R. G. (1975). *Engineering Compendium on Radiation Shielding*. Vol. III, *Shield Design and Engineering*. Springer-Verlag, Berlin, Heidelberg, New York.
16. Davies, N. S. (1964). *Reactor Handbook*. Interscience Publ., N.Z.
17. Brodev, D. L., Zaytsev, L. N., Komokykov, M. V., Malkov, V. V., and Lyrev, P. S. (1967). *Beton v zastchyte iadernyh ustanovok*. Atomizdat, Moskow.
18. Jaeger, R. G. (1975). *Engineering Compendium on Radiation Shielding*. Vol. II, *Shielding Materials*. Springer-Verlag, Berlin, Heidelberg, New York.
19. Avram, C., and Bob, C. (1980). *Noi tipuri de betoane speciale*. Edit. Tehnica, Bucharest.
20. Walker, R. (1961). Report ANL-6443.

CHAPTER 9

# Nuclear Fuel Elements

A nuclear fuel element is a sophisticated assembly consisting of *solid fuel material*, either metallic or ceramic, contained in a non-fissile *cladding*, and *assembly pieces* . The fuel cladding gives the fuel the designed shape and spatial geometry, provides mechanical strength, prevents contamination of the cooling agent with fission products as well as fuel erosion and corrosion under the effect of the turbulent flow of the cooling agents.

Fuel elements are the very essence of the core of a nuclear reactor, triggering and craddling the self-sustained chain fission reaction. Also their design provides for convenient evacuation of the fission-released energy.

Design of fuel elements must warrant confidence that the following general requirements are observed:

1. an adequate mechanical resistance to the effects of extremely severe working environments (high temperatures and temperature gradients, diverse and concurrent mechanical stresses, unavoidable alteration of the original mechanical properties due to irradiation);

2. corrosion resistance to the attack of various agents (cooling fluids, fission products, etc.);

3. stability of components, both dimensional and as regards other physical properties;

4. appropriate composition and nuclear properties that would interfere only marginally with the neutron economy of the core.

Design and manufacturing of fuel elements must provide for the preservation of the cladding functions over the entire operation of the fuel inside the reactor, of the structure and optimal geometry of the fuel bundle, and to make sure that the chemical composition of the constituent materials remains, in time, satisfactorily compatible with the actual operation conditions (accounting for time-variation of power per fuel element), etc.

The selection of the materials employed in the manufacture of fuel elements should comply with the general criteria above, also considering the reactor type, costs, etc.

## 9.1. DESIGN OF FUEL ELEMENTS

The *design* of fuel elements has a meaning that goes beyond the ordinarily accepted notion, because of the complexity of the target-product whose components have to withstand the effects of a variety of uncommon phenomena with the potential of constraining service performance. Elaboration of the fuel element's design requires prior definition of the concept, according to the freedom and constraints following from the designed core, a detailed knowledge of the properties of materials for components (see Appendix 11), as well as convincing evidence on how these properties would vary when subject to real operation conditions — a kind of evidence that can be obtained either by effective testing under irradiation, or by appropriate simulation of the irradiation effects. Establishment of the geometrical shape and fuel element composition optimization are done based on the study of its behaviour under irradiation in experimental devices, and by testing experimental models made on the basis of preliminary designs. Irradiation behaviour analysis may be performed either by experimental simulation (out-of-pile testing simulating some in-pile effects), or theoretically — by mathematical modelling of fuel behaviour, for which computer programs have been made available.

### 9.1.1. Irradiation-induced Phenomena in Fuel Elements

The working conditions of different fuel element components translate the interaction of radiation with matter (neutrons, $\gamma$-rays, fission fragments, etc.).

Following the fission process, fission fragments transfer their entire energy to the bulk of the fuel, generating heat that must be transferred to the cooling agent, and also induce dimensional changes (fuel swelling), or changes in properties. As mentioned in Chapter 3, the sum-volume of the two atoms emergent from fission is much higher than the volume of the fissioned atom, which leads to an increment in the fuel volume. At sufficiently high temperatures and burn-ups, the fuel dimensions increase on the account of nucleation and growth of fission gas bubbles. In a short time following the reactor start-up a stationary thermal regime is reached (as far as temperatures and temperature gradients), which evolves a complex pattern of stresses and strains over the fuel element components (Fig. 9.1) [1].

The compound of these phenomena is also effected by the dynamics of fuel — cladding and cladding – coolant reactions (both adversely affecting cladding integrity); diffusion of gaseous fission products (that, leaving fuel, generate an internal pressure in the cladding); physical, mechanical and nuclear properties of the components, etc. Such "primary" phenomena generate, in turn, "secondary" effects inside the components' material, which have to be properly considered in the definition of the design concept, materials selection, manufacturing technology, quality control procedures, etc., in such a way as to ensure the requested operational parameters of the fuel element (linear power, burn-up, clad integrity, etc.).

Clad failure may occur esentially from permanent strains (either commonly plastic or creep) due to internal pressure of fission gases or to mechano-chemical interaction of cladding with fuel [2]. Mechanical behaviour of the cladding depends on the aforementioned "secondary" phenomena, from which, to date, only few are theoretically fully understood. Inside the fuel elements these processes develop simultaneously and interweave in an intricate manner. Figure 9.2 [3] is indicative of the complex relationship between such physical, chemical and mechanical processes (shown in the figure as rounded-corner rectangles) and their effects as observed in fuel elements (straight rectangles). In dotted rectangles the pellet – cladding interface phenomena are inserted.

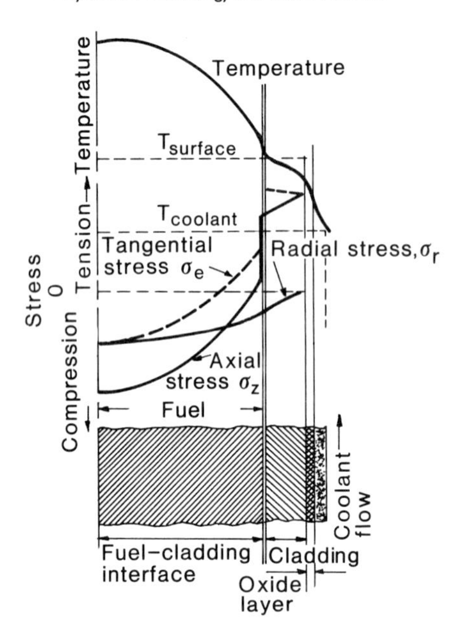


Fig. 9.1. Distribution of temperature, and radial, tangential and axial stresses in a fuel element.

Mathematical modelling of fuel behaviour is based on the recognition of the interconnections sketched in Fig. 9.2 and that translates into the computer programs employed in design.

In the following paragraphs some of these phenomena will be briefly discussed; a detailed account on them is given, for example, in Olander's book [3].

### 9.1.2. Temperature Distribution in Fuel Elements

Taking as reference a rod-shaped fuel element containing ceramic fuel, it has been found that, in a steady-state condition, a parabolic type temperature distribution sets across it. If the heat generation rate is $q$, then the equation describing the temperature distribution reads:

$$r^{-1} \frac{d}{dr} \left( rK \frac{dT}{dr} \right) + q = 0, \tag{9.1}$$

where $K$ is the thermal conductivity of the fuel and $r$ is the distance from rod axis to the point under examination. A very important parameter in this equation is thermal conductivity, closely related to fuel properties (porosity $p$, density $\rho$, O/U ratio etc.) and temperature. The theoretical functional structure of $K = f(p, \rho, O/U, \ldots)$ is, to date, in disagreement with the experimental data and, for this reason, only semiempirical formulae are in actual use. For example, to anticipate the $UO_2$ thermal conductivity in the temperature range of 800 — 2000 $^oC$ the following formula is in order [4]:

$$K = 0.0130 + [(0.038 + 0.45p)T]^{-1}. \qquad (9.2)$$

where $T$ is the temperature expressed in centigrades and $K$ is expressed in (W/cm . °C). In Appendix 11 more expressions obtained by processing of experimental data are provided, enabling determination of thermal conductivity.

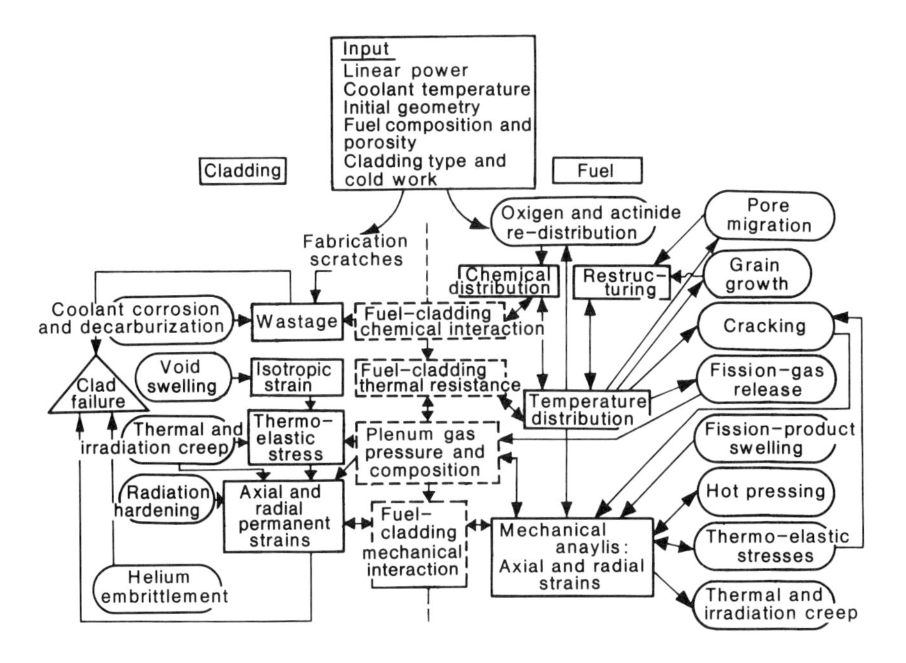


Fig. 9.2. Processes and phenomena that develop in the fuel element during in-pile operation (round cornered rectangles), their apparent effects (angular cornered rectangles), and clad/fuel interface phenomena (dotted rectangles).

A solution to eq. (9.1), when $K$ is given by (9.2) and considering the effect on the thermal conductivity of thermally activated changes in the fuel (grain growth, oxygen redistribution, etc.), can be found only by means of numerical methods.

Assuming $K$ and $q$ constant, the integration of eq. (9.1) with boundary conditions of the type: $T(R) = T_s$ and $(dT/dr)_{r=0} = 0$ results in:

$$T - T_s = \frac{1}{4}\frac{qR^2}{K}\left(1 - \frac{r^2}{R^2}\right). \qquad (9.3)$$

where $T_s$ is the temperature on the fuel surface.

The integration of eq. (9.1) when the dependence of $K$ on temperature is considered and the generation rate $q$ is taken as constant, gives:

$$rK(T)\frac{dT}{dr} = -\frac{1}{2}qr^2, \tag{9.4}$$

the integration constant being zero for a filled cylinder ($r \in (0,R)$). A new integration between $r = 0$ and $r = R$, where $T = T_c$ and $T = T_s$, respectively, gives:

$$\int_{T_s}^{T_c} K(T)dT = Q/(4\pi), \tag{9.5}$$

where $Q$ is the power generated on the unit-length of fuel rod:

$$Q = \pi R^2 q. \tag{9.6}$$

The conductivity integral:

$$\int_{T_s}^{T_c} K(T)dT \tag{9.7}$$

is of particular interest because:

1. it is directly related to the fuel linear power, which is easily measurable;
2. the maximum temperature of the fuel $T_c$ is independent of the rod diameter;
3. it is a fuel property. In many cases the conductivity integral is normalized to a certain surface temperature (for example 0 °C):

$$I(T) = \int_0^T K(T')dT'. \tag{9.8}$$

The conductivity integral for a $T_s$ that is specific to a given reactor can be obtained from the difference $I(T) - I(T_s)$.

Ordinarily, the recommended values for $I(T)$ are plotted [3]. For $UO_2$, $I = 93 \pm 4$ W/cm corresponds to a temperature associated with a central melting of the fuel.

Thermal transfer in fuel elements is a problem much more difficult to solve. Besides the aforementioned effects, the thermal transfer at pellet — cladding interface and the effect on thermal conductivity of the fuel cracking due to very high thermal gradients ($\sim 10^3$ K/cm for $UO_2$) are to be considered (cracks are barriers to the heat transfer). Thermal transfer in the gap between fuel and cladding is influenced by the roughness of surfaces in contact; by gaseous composition (the originally existent He is impurified during reactor operation with fission gases such as Xe and Kr that have thermal conductivities lower than helium's, and that, by accumulation, diminish the gap's thermal transmittance); by contact pressure between fuel and cladding, etc. The effect of structural changes (crystal growth, Fig. 9.1) on the temperature profile consists in diminishing the maximum temperature in the fuel — even by 300 K in some extreme cases [3]. This is explained by the fact that the columnar grains exhibit a much higher thermal conductivity than the original fuel material.

9.1.3. Structural Changes in Fuel

As mentioned in section 9.1.2, temperature and temperature gradients activate the grain growth in the fuel. Depending on fuel linear power, initial grain dimensions (as fabricated) and irradiation duration, the following sequence of restructuring zones can be observed across a fuel transversal cut: a peripheral zone having a structure practically identical to the original one, an equiaxial-grain zone, a columnar-grain zone and, for very large liniar powers, a central hole (Fig. 9.3) [3].

The equiaxial grain growth is determined by the natural tendency of the total free energy of the surface to decrease, that translates in a diminishing number of grains and, consequently, of the total grain area. In the process, intergranular surfaces shift in the direction of curvature, the large grains with concave surfaces growing on account of including the smaller grains with convex surfaces. The pores and impurities on intergranular surfaces diminish their movement rate so that the growth rate is related to material transfer across the pores (evaporation — condensation, surface diffusion, etc.).

Assuming that migration of the pores placed on grain boundaries takes place through a vapour transport mechanism, Nichols [5] has established the following relation between the initial cristallyte size $d_o$ and the final one $d$, after a certain time interval:

$$d^4 - d_o^4 = k_c t \exp(-Q_v/(RT)), \tag{9.9}$$

where $Q_V$ is the molar activation energy, $R$ is the universal gas constant, $T$ is the absolute temperature and $k_c$ is the growth constant.

Generally, the growth may be empirically described through a expression similar to (9.9), but involving an adjustment parameter $m$:

$$d^m - d_o^m = k_c t \exp(-Q_v/(RT)), \tag{9.10}$$

where $m > 2$. Thus, it was found [6] that for $UO_2$ the above equation may describe satisfactorily the experimental data if one takes $m = 2.5$ and $Q_v$ = 460 KJ/mol.

The columnar grain shape is explained by lenticular pore migration through temperature gradients, accompanied by a material transport in the opposite direction. The material transport is carried out, depending on the pore sizes, through either surface diffusion, volume diffusion or vapour transport, and is characterized by the masic flow $J_m$:

$$J_m = \frac{D_i}{RT} \cdot C_i \cdot \left(-\frac{Q_i^*}{T} \operatorname{grad} T\right), \tag{9.11}$$

where $D_i$ is the diffusion coefficient specific to a given migration mechanism, $C_i$ is the concentration of diffusive species (vacancies, vapours, etc.), $Q_i^*$ is the transport heat of the process, and $T$ is the absolute temperature. By using eq. (9.11) the material transport rate may be determined and, consequently, the growth rate [7,8,10].

During columnar grains growth the pores migrate in the direction of the thermal gradient and accumulate near the fuel centre, generating a central hole. Nichols [5] estimates the central hole size ($r_g$) by the equation:

Fig. 9.3. Structural alterations in a fuel element. Section enlargement: x45; burn-up: 20,000 MWday/t; linear power: 500 W/cm.

$$r_g = r_{g\cdot c}\, p^{1/2}, \tag{9.12}$$

where $r_{g.c}$ is the columnar grain zone radius and $p$ is the percentage porosity of the fuel.

Structural changes enhance fission gas release, irradiation creep, etc. An enhanced fission gas release, for example, has two chief effects: a considerable impurification of the gaseous environment between cladding and fuel (resulting in an increased fuel temperature) and an enhanced internal pressure of the fission gas, that results in clad stresses.

### 9.1.4. Fission Gas Release

At lower temperatures ($< 1300$ K), the mobility of gas atoms generated by fission is low, so that it is unlikely for them to leave the fuel material ($UO_2$ for example) by diffusion. In this temperature range, fission gas release is achieved only by direct recoil of fission fragments generated near the fuel surface, in a layer having a thickness lower than the fission fragment's free path in the material.

In the temperature range 1300 — 1900 K gas atom mobility is considerable, so that atoms can migrate across the intergranular and external fuel surfaces. On the assumption that fission gases are uniformly generated over the entire volume, that they do not vanish in time, and that the medium is homogeneous and isotropic, the diffusion of stable fission gases is described by the second Fick's law, for a spherical symmetry:

$$\frac{\partial c}{\partial t} = \frac{1}{r^2}\frac{\partial}{\partial r}\left(r^2 D\,\frac{\partial c}{\partial t}\right) + B \tag{9.13}$$

where $c(r,t)$ is the gas atom concentration, $B$ is their rate of generation by fission, and $D$ is the diffusion coefficient. By integration of eq. (9.13) over a sphere of radius $a$ (an idealized $UO_2$ grain), with boundary and initial conditions: $c = 0$ for $r = a$, $c = 0$ for $t = 0$ and defining the release fraction ($f$) as the ratio of the quantity that leaves the fuel by diffusion to the total quantity generated by fission, for $\pi D't < 1$ one obtains [9]:

$$f \simeq 4((D't)/\pi)^{1/2} - 3(D't)/2 \tag{9.14}$$

If $\pi D't \geq 1$ one has:

$$f \simeq 1 - \frac{1}{15\,D't} + \frac{6}{\pi^2 D't}\exp(-\pi^2 D't), \tag{9.15}$$

where $D' = D/a^2$ is the effective diffusion coefficient, and $t$ is the irradiation duration.

For temperatures higher than 1900 K, other release mechanisms are effective. Fission gas bubbles precipitated inside the fuel as well as the pores in the material have enough mobility to enable them to migrate over large distances across temperature gradients. Gas release from these cavities takes place when they come across cracks and channels in the fuel that are connected to the inner empty volume. A mathematical treatment of these phenomena is quite difficult. A detailed presentation of gas release mechanisms is given

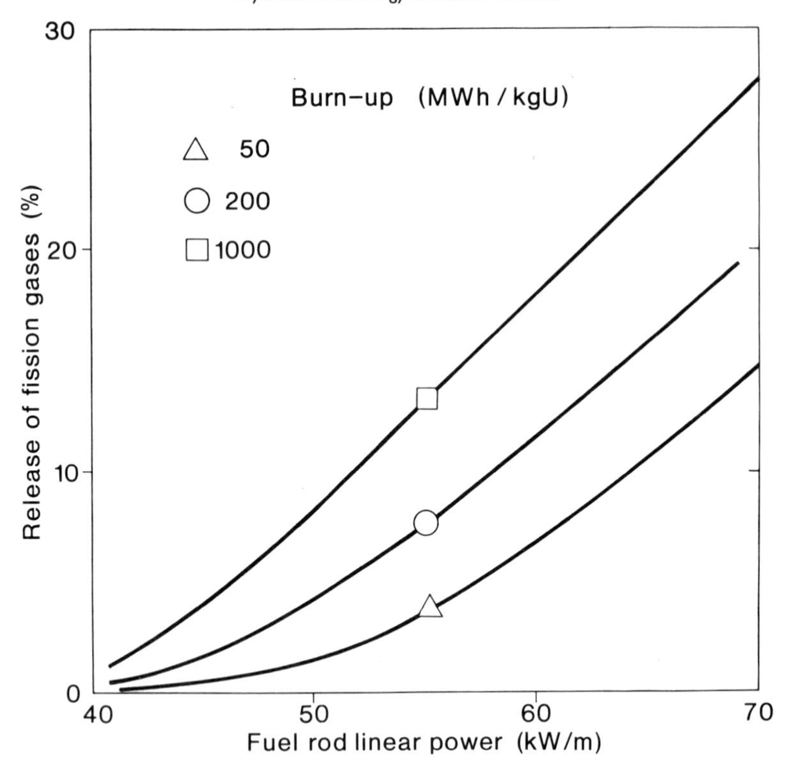


Fig. 9.4. Percentage release of fission gases vs. the linear power, at different burn-ups, in a PHWR fuel element.

elsewhere [3,10]. In practice, it is preferable to use semiempirical formulae, such as those given in Appendix 11.

Fission gas release is determined by material properties of the fuel (density, stoichiometry, grain size, etc.), by the evolution of these properties during irradiation, and by the thermal and neutron flux conditions of the irradiation, power burn-up, etc.) [11].

In Fig. 9.4 the fission gas release fraction as a function of irradiation parameters (power and burn-up) for a PHWR-type fuel element is shown [12].

### 9.1.5. Fission Gas Pressure

Internal pressure of fuel elements depends on the quantity of gas in the free volume available at a certain moment of irradiation, on the total free volume and on gas temperature. Gaseous phase consists of filling gases as provided in the design specifications, gases desorbed from the fuel at the current working temperatures (water vapours, hydrogen, etc.) and fission gases released from the fuel. Notley [13] has given a computational model that predicts the internal pressure evolution in a PHWR-type fuel element during irradiation. Figure 9.5 shows the internal pressure variation, experimentally measured, for the PHWR fuel containing $UO_2$ of different original densities, as a function of the irradiation duration [13,14].

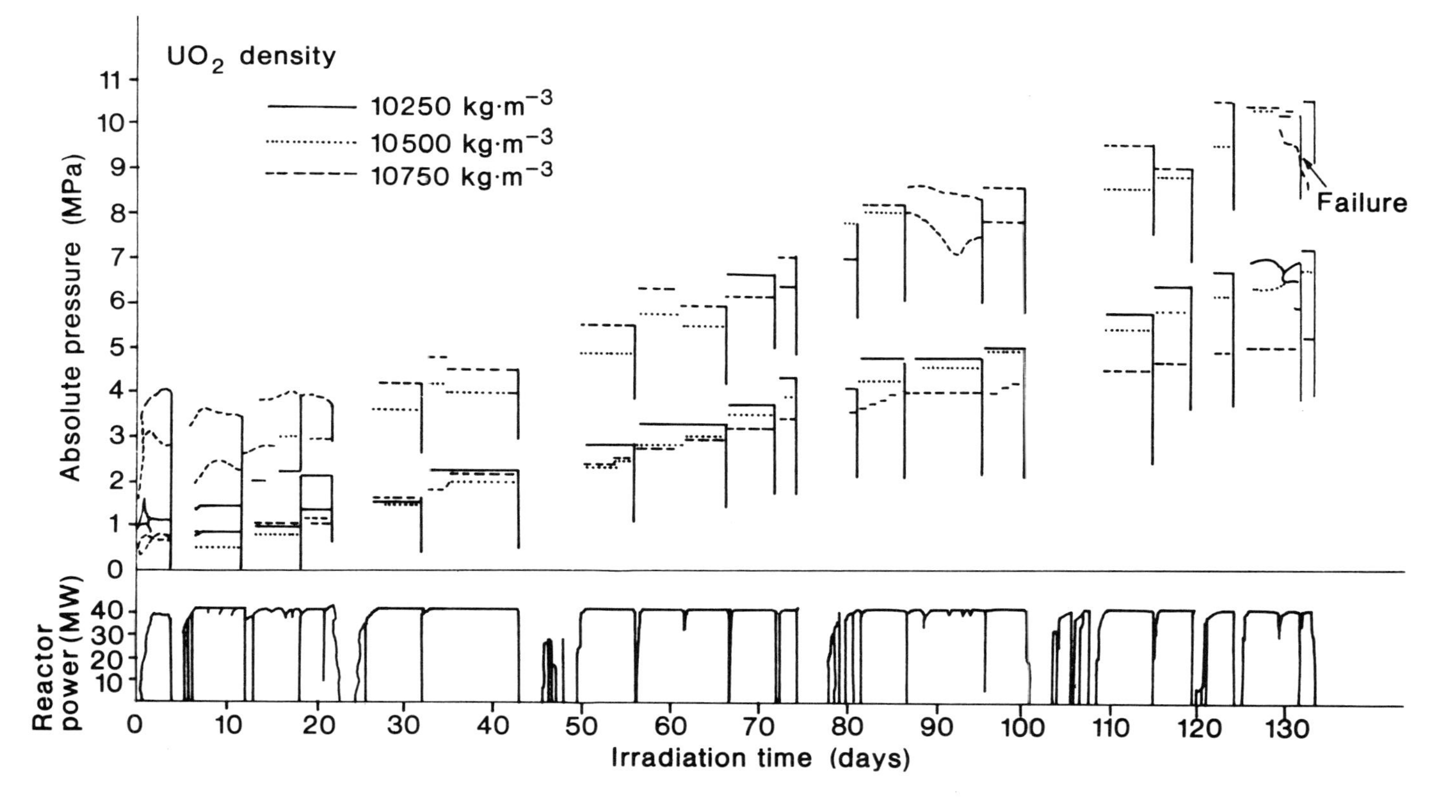


Fig. 9.5. Development in time of internal pressure, for PHW fuel elements of various $UO_2$ density.

### 9.1.6. Fuel Swelling

In Chapter 3 several problems concerning metallic fuel swelling were briefly tackled. The mathematical modelling of ceramic fuel swelling is an extremely complex issue [3,15]; with this in mind, we will confine ourselves here to only some qualitative considerations, supported by few experimental results [12]. Appendix 11 gives an evaluation of the fuel swelling vs. the local burn-up. It was found that the swelling rate for uranium dioxide, at low temperatures, is ~ 0.7% for ~ $10^{20}$ fissions/$cm^3$ [16]. In the central zones of fuel, where temperatures are high (> 1970 K) the swelling rate is lower because the fuel material retains negligible gas quantities (there is a high release in this zone). In the intermediate temperature range 1270 – 1970 K, where the fuel plasticity and gas atom mobility are considerable and gas release is lower, the fuel swelling has maximal values (Fig. 9.6 [12]).

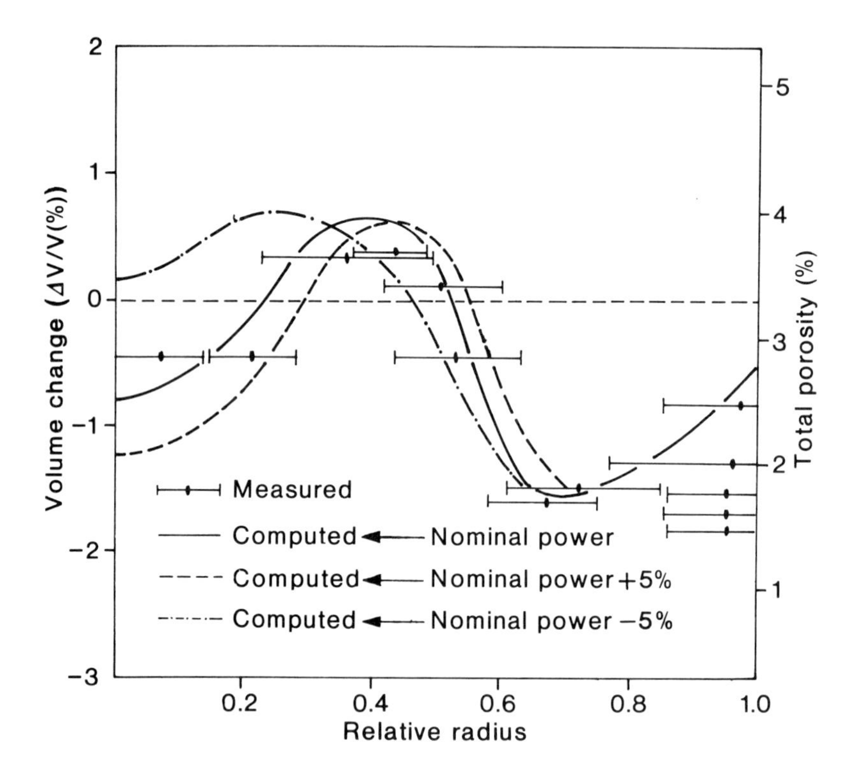


Fig. 9.6. $UO_2$ volume and porosity changes with the fuel element radius.

The data for the relative volume changes as functions of the radial coordinate in the $UO_2$ pellet have been experimentally measured for a PHWR (Pickering) fuel element irradiated up to a burn-up of ~ 200 MWh/kgU (~ 8300 Mwd/tU). The computed values, as per the Notley's model [12], fit well the experimental results. In the same figure the computed curves for a ±5% deviation from the nominal power are plotted.

On Fig. 9.6 one can see that at low temperatures (relatively large radius), as well as at high temperatures (relatively short radius), the specific fuel volume is lower than before irradiation, due to fuel densification, i.e. to a diminishing in the initial porosity. The volume changes are reflective of the variations in the material porosity; one can treat them on an equal footing, since the cell parameter rests unchanged.

### 9.1.7. Mechanical Behaviour of Fuel and Clad during Irradiation

Models of fuel behaviour take into consideration the evolution of the mechanical properties as a result of direct interactions with fast neutrons, the embrittlement due to hydrogen absorption in zirconium alloys, voids formation in stainless steels, dimensional and shape changes in the cladding during irradiation, etc.

In the preceding paragraphs the behaviour of the fuel element components and the main types of mechanical stresses induced upon them during in-pile operation have been discussed; in what follows, their response to stresses — whether applied or due to direct interaction between fuel and cladding — will be tackled. In practice,such an analysis is performed by means of computer programs that model fuel behaviour during irradiation. For computational resons, the rod-type fuel element is divided in several axial zones, each zone being, in turn, divided in several concentric rings (Fig.9.7 [17]).

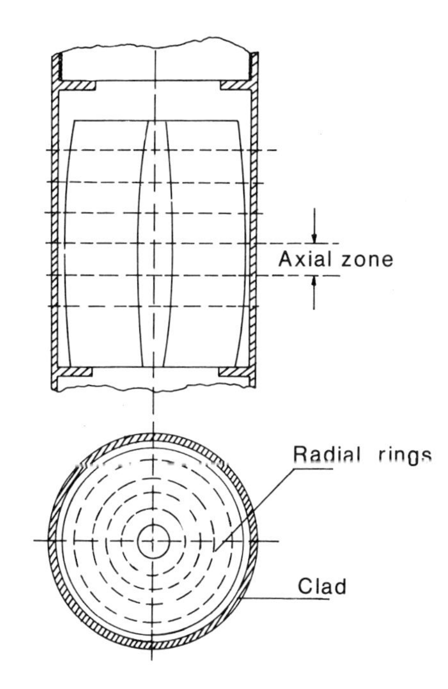


Fig. 9.7. Fuel element subdivision in view of behaviour analysis.

Computer programs [3,17] render the mechanical behaviour by using the following equations reflective of the cylindrical symmetry:

$$\varepsilon_r = E^{-1}\,[\sigma_r - \nu(\sigma_\theta + \sigma_z)] + \alpha T + \varepsilon^s + \varepsilon_r^c\,,$$

$$\varepsilon_\theta = E^{-1}\,[\sigma_\theta - \nu(\sigma_r + \sigma_z)] + \alpha T + \varepsilon^s + \varepsilon_\theta^c\,, \tag{9.16}$$

$$\varepsilon_z = E^{-1}\,[\sigma_z - \nu(\sigma_r + \sigma_\theta)] + \alpha T + \varepsilon^s + \varepsilon_z^c\,,$$

where $\varepsilon_r$, $\varepsilon_\theta$, $\varepsilon_z$ are the components of the total strain, $\sigma_r$, $\sigma_\theta$, $\sigma_z$ are the stress components, $\varepsilon^s$ is the swelling-related strain, $\varepsilon_r^c$, $\varepsilon_\theta^c$, $\varepsilon_z^c$, are the permanent strain components, $E$ is the elasticity modulus, $\nu$ is the Poisson ratio, $\alpha$ is the thermal expansion coefficient, and $T$ is the temperature.

Equations (9.16) apply to both fuel and cladding, providing that appropriate material data ($E$, $\nu$, $\alpha$) and strains that are specific to each material are used.

Permanent strain follows from creep. This can be obtained with the aid of the Prandtl — Reuss equation, written in finite differences:

$$\Delta\varepsilon_r^c = (\Delta\varepsilon^*/\sigma^*)\,[\sigma_r - (\sigma_\theta + \sigma_z)/2],$$

$$\Delta\varepsilon_\theta^c = (\Delta\varepsilon^*/\sigma^*)\,[\sigma_\theta - (\sigma_z + \sigma_r)/2], \tag{9.17}$$

$$\Delta\varepsilon_z^c = (\Delta\varepsilon^*/\sigma^*)\,[\sigma_z - (\sigma_r + \sigma_\theta)/2],$$

where $\sigma^*$ and $\Delta\varepsilon^*$ are the stress, respectively the strain for a triaxial stress:

$$\sigma^* = 2^{-1/2}\,[(\sigma_r - \sigma_\theta)^2 + (\sigma_\theta - \sigma_z)^2 + (\sigma_z - \sigma_r)^2]^{1/2}, \Delta\varepsilon^* = f(\sigma^*), \tag{9.18}$$

$f$ being the specific creep-law of the material.

For a zircaloy cladding, the differential expression on this law [17] is:

$$\frac{d\varepsilon^*}{dt} = 2.83 \times 10^{-17}.\ \exp\left(-\frac{Q}{RT}\right)\Phi^{0.85}\,sh(0.0166\,\sigma^*) \tag{9.19}$$

where $Q$ is the activation energy (58.6 kJ/mol), $\Phi$ is the fast neutron flux for energies higher than 1 MeV (n/cm$^2$ . s) and the units for $d\varepsilon^*/dt$ and $\sigma^*$ are s$^{-1}$ and MPa, respectively.

For $UO_2$ and $(U,Pu)O_2$ one has [17]:

$$\frac{d\varepsilon^*}{dt} = \frac{1.47 \times 10^6}{(D-88)G^2}\,\sigma^*\,\exp\left(-\frac{Q}{RT}\right) + 4 \times 10^{-13}\,P\sigma^*, \tag{9.20}$$

where $Q$ = 376.7 kJ/mol., $P$ is the specific power (MW/tU), $D$ is the density (% TD), and $G$ is the grain size ($\mu$m). Numerical coefficients are given, for $\sigma^*$ — in MPa and for $d\varepsilon^*/dt$ — in $s^{-1}$. The first term in this equation describes the thermal creep, whereas the second describes the irradiation-induced creep. Recent data evolve equations that indicate a more complex nature of the processes involved (see Appendix 11).

The equations describing strain evolution in fuel and cladding can be integrated [17] over space and time so that, finally, one obtains the total cladding strain magnitude, enabling an assessment on whether the cladding, as designed, would preserve its integrity.

Due to the many intricacies that may overwhelm the theoretical approach, the use of semiempirical, experimentally derived formulae describing the evolution of material properties is, in many cases, preferred. Irradiation experiments using trial fuel rods provide useful data for computer programs calibration, which is indispensable in real fuel element performance analysis.

In Romania, the preliminary preparations for testing the behaviour under irradiation of CANDU-PHWR fuel elements have been successfully completed [18] (testing programmes and specifications, design and experimental fuel rod manufacturing, etc.).

## 9.2. CLASSIFICATION OF FUEL ELEMENTS

To date there is a large variety of nuclear fuel elements [19]. These may be classified following such diverse criteria as fuel material composition; fuel element shape; type of fuel-cladding contact; reactor type, etc.

By fuel composition and cladding material, the following fuels may be distinguished:

1. metallic fuel/metallic clad elements (for example, metallic uranium in a magnesium alloy cladding);

2. ceramic fuel/metallic clad elements (e.g. $UO_2$ $(U,Pu)O_2$ or $(U,Th)O_2$ canned in a zirconium alloy or in stainless steel; UC or (U,Pu)C canned in stainless steel);

3. all-ceramic fuel elements (for example, $UO_2$ $(U,Pu)O_2$, $(U,Th)O_2$ grains coated with SiC and pyrocarbon layers and embedded in a graphite matrix.

A classification in frequent use takes into consideration only the fuel nature:

1. metallic fuel elements, where the fuel is a metal or an alloy;

2. ceramic fuel elements, where the fuel is a non-diluted ceramic;

3. dispersed fuel elements, where the fuel is a highly diluted alloy or an all-ceramic material with a low density of fissile material.

By shape, the most common types of fuel elements are the rod, the plate, the tube, etc.

By type of fuel-cladding contact the following may be distinguished:

1. mechanic-contact fuel elements (the contact is achieved by either pressing or rolling);

2. metallurgical-bonding fuel elements (the materials in contact are bond by diffusion);

3. intermediate-layer fuel elements (using an intermediate liquid, solid or gaseous agent for improving fuel-cladding heat transfer).

### 9.2.1. Metallic Fuel Elements

Metallic fuel elements contain the fuel material either as natural uranium or, in some experimental power reactors, as enriched uranium. This type of fuel element is presented as rods, tubes or plates, either single or assembled in bundles. The fuel is canned in clads made of materials that are compatible with both the cooling agent and the fuel itself. The cladding shape and material choice are determined by the coolant's nature. If coolant is a gas, the cladding is provided with flaps, in order to increase the heat transfer surface. In the following, several types of metallic fuel elements are briefly introduced.

The fuel elements for the *EDF* (France) *power reactors* (Fig. 9.8) contain the fissile material (natural uranium) in tubes made of the U + 0.5 — 1% Mo alloy [19]. These "fuel tubes" are canned either only on the outside, or on both sides. The uranium tubes are 540 mm in length, have a wall thickness of 11 mm, an outer diameter of 35 mm and an inner diameter of 14 mm.

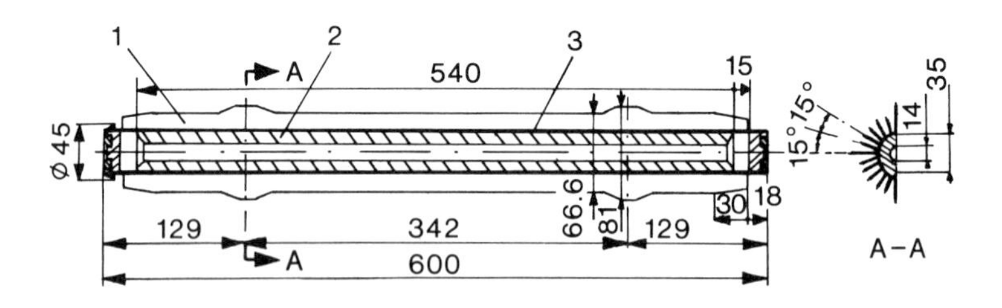


Fig. 9.8. EDF-1 fuel element: 1 - longitudinal flap;
2 - uranium tube; 3 - magnesium alloy cladding.

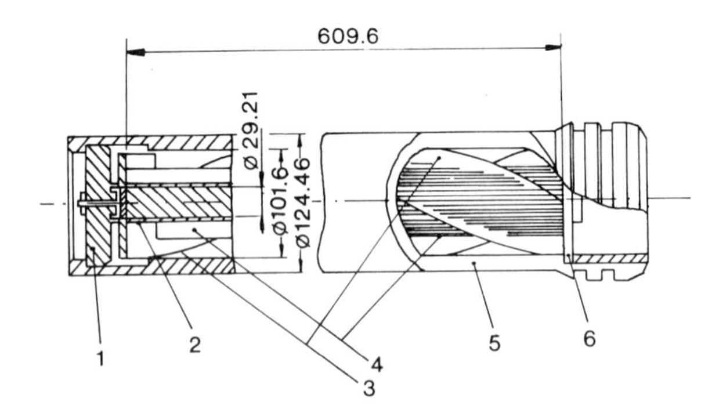


Fig. 9.9. Magnox fuel element: 1 - sintered alumina support;
2 - Magnox alloy cladding; 3 - helical flap;
4 - longitudinal flap; 5 - graphite tube support;
6 - Magnox alloy flange.

The fuel elements for *Magnox reactors* (UK) are shaped as rods of metallic natural uranium containing 600 ppm C, 600 ppm Al and 300 ppm Fe (Fig.9.9) [19].

The fuel is clad in magnesium alloy (magnox), the claddings having on the surface in contact with the cooling agent ($CO_2$) helicoidal wings to enhance the heat transfer coefficient.

### 9.2.2. Ceramic Fuel Elements

As noted in Chapter 3, the ceramic fuels present certain advantages in comparison with the metallic fuels, such as: a higher melting temperature, absence of phase transformations in the working temperature range; an isotropic crystalline structure; a better chemical compatibility with other materials; stability under irradiation, etc., as well as some disadvantages, such as: a lower density, lower thermal conductibility and lower resistance to thermal and mechanical shocks. Despite these disadvantages, the ceramic fuels are, at this moment, the most employed in both thermal and fast power reactors. In the following, several types of ceramic fuels are presented.

The CANDU-type fuel element (Fig. 9.10) consists of 37 rod-shaped elements placed on three concentric circles. The ceramic fuel used is the natural uranium dioxide delivered in sintered pellets of ~ 12.15 mm diameter,~16 mm length and a density of 10.5 — 10.7 $Mg/m^3$. Pellet stacks are canned in zircaloy-4 tubes 495 mm long and 13.08 mm in outer diameter. The sheaths have spacers mounted on their external surfaces to prevent the contact between fuel rods as well as the contact of the fuel rods with the reactor's fuel channel [20].

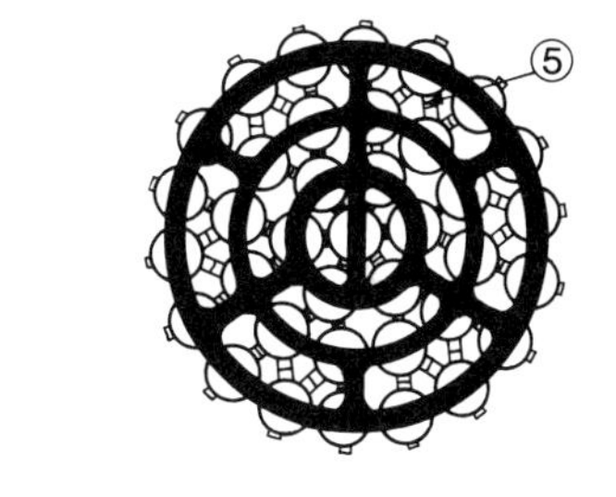


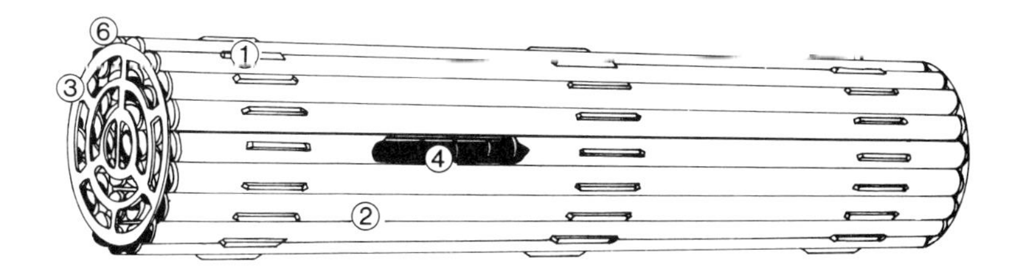


Fig. 9.10. CANDU fuel bundle: 1 - zircaloy pads; 2 - zircaloy clad; 3 - zircaloy grid; 4 - $UO_2$ pellets; 5 - zircaloy spacers; 6 - zircaloy end-plugs.

The bundles are fuelled in the reactor horizontally through the reactor's pressure tubes made of zirconium alloy.

The maximum clad temperature is 600 K and is related mainly to the cooling agent pressure ( ~ 10 MPa). Approximately 50% of the fuel channels have the maximum thermal power of about 6.5 MW, that corresponds to a maximum nominal power for a fuel bundle of 800 kW and to a linear power (for the outer row) of maximum 51 kW/m. The maximum centreline temperature of fuel rods, corresponding to this linear power, is 2170 K. The average burn-up is about 168 MWh/kg U, for an irradiation time of about 270 days at nominal power. The spent fuel still contains 0.23% $^{235}U$ and 0.27% ($^{239}Pu$ + $^{241}Pu$).

In consideration of the fact that uranium resources are limited, current research is conducted to develop new types of fuel for the CANDU power stations, such as a variety based on enriched uranium and thorium, another one combining plutonium as recovered from the conventional fuel with natural uranium, etc. [21].

Present PWRs employ fuel elements having active lengths of 3.8 — 4.3 m, outer cladding diameter of 0.85 — 1.075 cm and clad wall thicknesses of 0.57 0.73 mm [22]. The fuel material consists of enriched uranium dioxide (3%), which provides for higher burn-ups, in the range of 1000 MWh/kg U. Limitations in burn-up are related only to the limited resistance of the cladding materials to prolonged irradiation. The fuel rods are assembled in bundles of 16 x 16 or 17 x 17 rods, of which 20 — 25 positions are left available for control rods. The fuel pellets are made of $UO_2$ and the cladding is a zirconium alloy. The first generations of PWR reactors used to employ fuel claddings made of stainless steel.

After irradiation, the spent fuel still contains 1.5 — 2% fissile material ($^{235}U$ + $^{239}Pu$ + $^{241}Pu$) which must be recovered by fuel reprocessing (see Chapter 10).

Figure 9.11 shows a 15 x 15 rod PWR-type fuel bundle [23].

In Table 9.1 a comparison between CANDU- and PWR-type fuel elements is given.

TABLE 9.1 A Comparison of CANDU and PWR Fuel Element Characteristics

| | *CANDU* | *PWR* | $R^1 = \frac{PWR}{CANDU}$ |
|---|---|---|---|
| Fissile material | Natural U<br>0.7% $^{235}U$ | Enriched U<br>1.5 – 3.5% $^{235}U$ | 3 |
| Fuel cost | Low | High | 3 – 4 |
| Bundle length | Short | Long | 8 |
| Fuel element diameter | Large | Small | 0.7 |
| Sheath thickness | Thin | Thick | 1.45 |
| $UO_2$ density | High | Medium | 0.98 |

[1] R represents the ratio of characteristic variables, of the PWR, to the CANDUs.

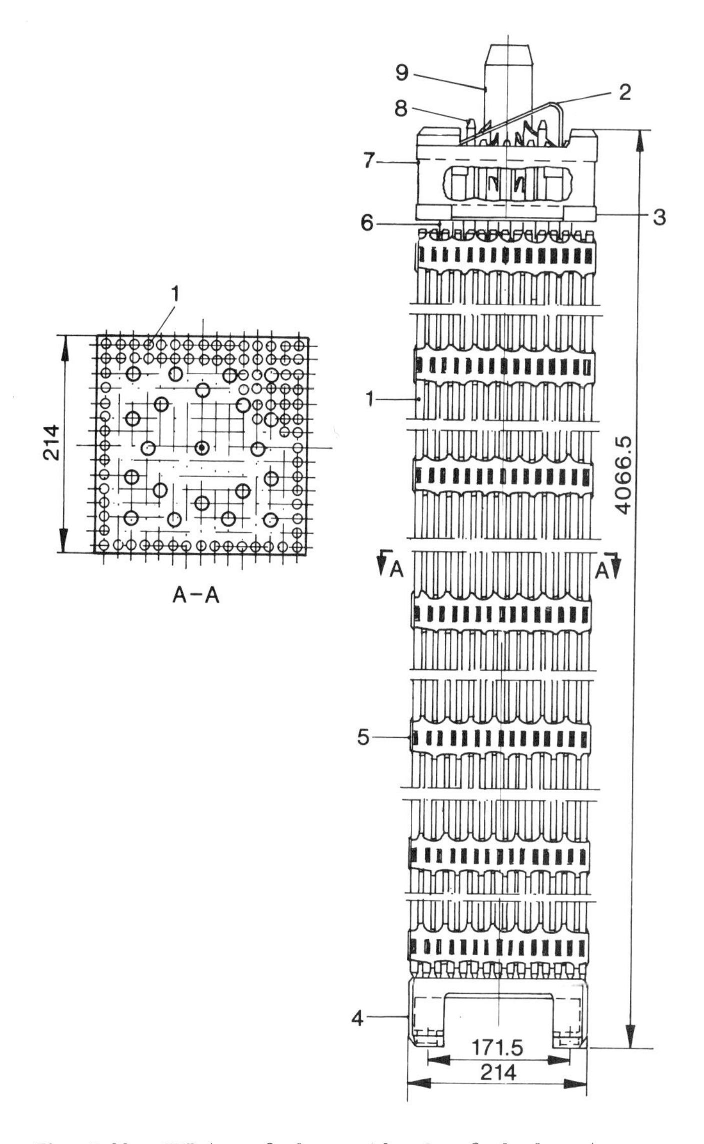


Fig. 9.11. PWR-type fuel assembly: 1 - fuel element; 2 - spring; 3 - adjustment plate; 4 - lower nozzle; 5 - assembly grid; 6 - guiding shaft; 7 - upper nozzle; 8 - control rod; 9 - control rod support.

The fuel elements for the present BWR reactors are similar to those for the PWRs. Yet, dimensions are different: outer diameter (1.23 — 1.25 mm) and wall thickness (0.80 — 0.86 mm) are larger. The BWR fuel bundles are assembled in 8 x 8 rod configurations [22].

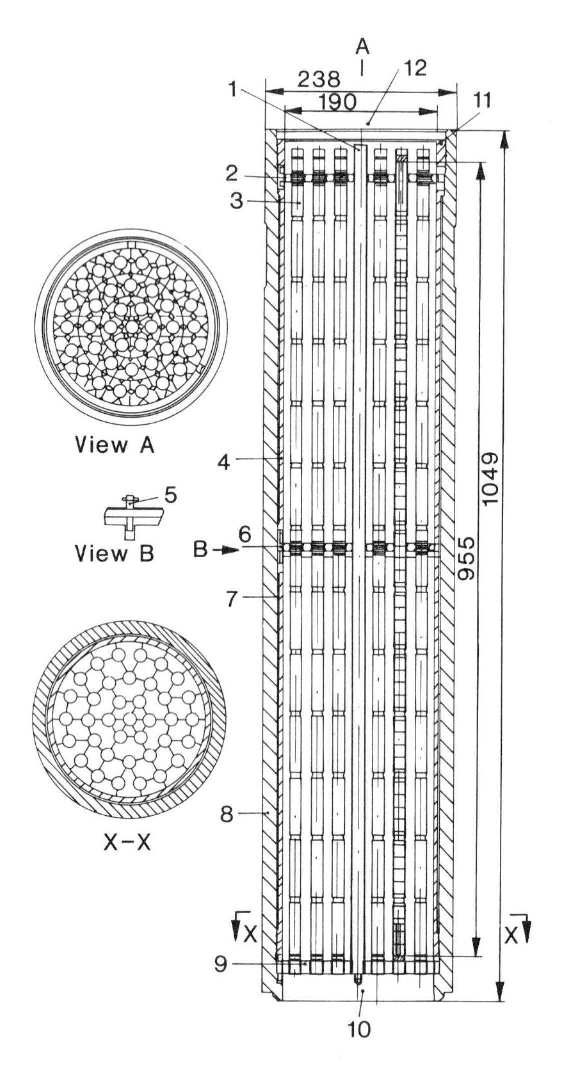


Fig. 9.12. AGR fuel bundle: 1 - guiding shaft; 2 - upper grid; 3 - fuel element; 4 - upper graphite tube; 5 - fixture part; 6 - control grid; 7 - lower graphite tube; 8 - external graphite tube; 9 - lower grid; 10 - lower end; 11 - fixing ring; 12 - upper end.

The fuel elements for AGR reactors (graphite moderated and $CO_2$ cooled, for which the outlet core temperature is 870 K) consist of $UO_2$ hollow pellets with an enrichment at equilibrium of 2.6%. The outer diameter is 14.48 mm and the inner diameter 5.08 mm. The fuel is canned in thin-walled stainless steel tubes (0.38 mm thickness). The bundle, which is about 1 m in length, consists of 36 fuel elements assembled with stainless steel grids and fixed on an supporting tube made of graphite layers with the outer diameter of 238 mm and the inner diameter of 190 mm — Fig. 9.12 [19].

### 9.2.3. Ceramic Fuel Elements for Fast Reactors

The fuel elements for *fast breeder reactors* (FBR) differ from one another as regards the construction pattern and the place alotted to them in the core. The breeding aims at the conversion of fertile into fissile materials, so that the fuel rods placed near the core centre will contain both fissile and fertile materials, while the peripheral ones will contain only fertile material (Fig. 9.13). A fuel bundle, that usually has a hexagonal cut in order to provide for a close-packed arrangement in the core, contains a considerable number of thin rods (~ 6 mm in diameter). In order to limit the inner pressure due to fission gases, at one end of the fuel rod a space is left available (Fig. 9.14 [19]). The length of the fuel rod is 3 – 3.5 m.

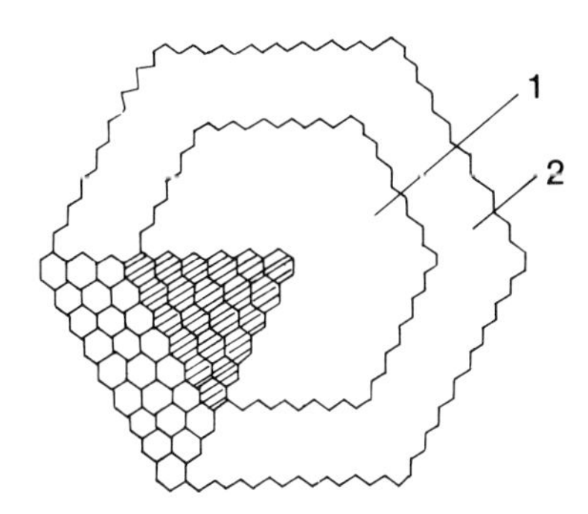


Fig. 9.13. Pattern of fuel elements distribution in a FBR: 1 - zone of fissile bearing fuel bundles; 2 - zone of fertile bearing fuel bundles.

Mixtures of $UO_2$ and $PuO_2$, or UC and PuC with a Pu/(U+Pu) ratio in the range 10 — 20% are used as fuels. Structural components (clad, end-caps, grids, etc.) are made of austenitic stainless steel. Such fuel elements reach high burn-ups — as high as 2400 MWh/kg (U+Pu), the burn-up limitation being imposed by various alterations in the merits of the material during prolonged irradiation. The spent fuel still contains 5 – 10% Pu.

### 9.2.4. Dispersed Fuel Elements

Efforts to create fuel elements able to cope with higher-temperature cooling agents and higher power densities resulted eventually in the manufacture of dispersed fuel elements. These consist of different fissile, fertile, moderating and structural materials, so combined that the heterogeneous assembly thus obtained has better properties than their homogeneous mixture. The fissile material, and sometimes also the fertile one, are dispersed in a non-fissile matrix as agglomerates, mainly spherical, having dimensions of 10 — 1000 μm. The spherical particle may be a simple agglomeration of fissile material or may be covered with several protecting coatings, thus making, at a "microscopic" scale, a fuel element as such. Because the fissile phase is the one that cradles the fission reactions, thus being subject to the most intense irradiation and the highest temperatures, the chemical nature of this phase is so selected that its melting temperature is as high as possible and the resistance to mechanical and thermal stresses maximal. Moreover, the fissile phase must exhibit a small porosity and the highest possible concentration in fissionable nuclei. For this reasons, in all applications only enriched fuel is used.

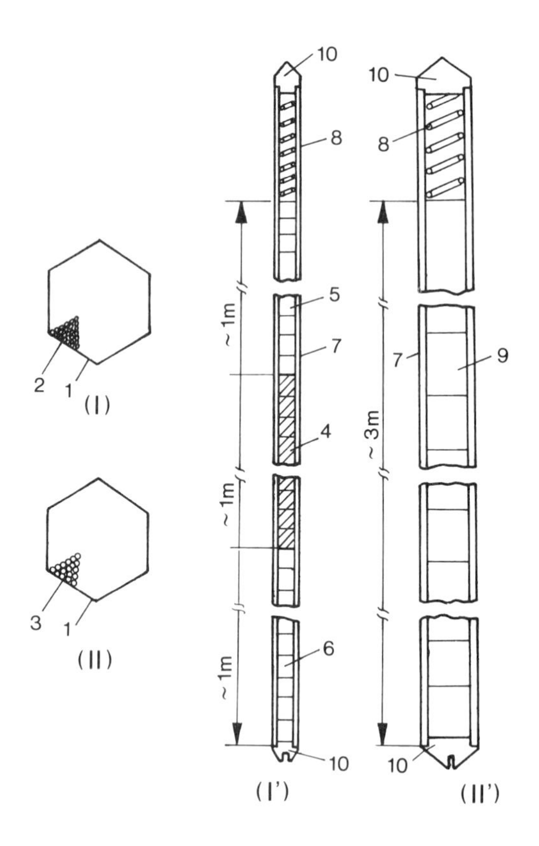


Fig. 9.14. Fuel bundles containing fissile material (I, I') and fertile material (II, II'), respectively: 1 - case bundle; 2 - fissile fuel element; 3 - radial fertile fuel element; 4 - fissile fuel pellets; 5 - fertile fuel pellets forming the upper axial reflector; 6 - fertile fuel pellets forming the lower axial reflector; 7 - stainless steel clad; 8 - spring; 9 - fertile pellets forming the radial reflector; 10 - end plug.

Fissile particle sizes are chosen so that the aforementioned requirements are satisfied. In order to minimize the effect of irradiation on the mechanical properties of the matrix, it is desirable that fuel particle dimensions be as large as possible. In practice, this allows coatings of thicknesses much higher than the critical ones, that would be damaged by fission fragments ($l$ = 15 μm in Fig. 9.15).

On the other hand, a too large increase in diameter would cause an increase in the temperature gradient between the particle's centre and its periphery, which in turn would cause large stresses in both the particle and matrix. The neutron economy also requires the minimization of matrix — fissile material ratio, which worsens the mechanical properties of the matrix. Finally, a compromise dimension of fissile particle is set at 100 — 200 μm, whereas

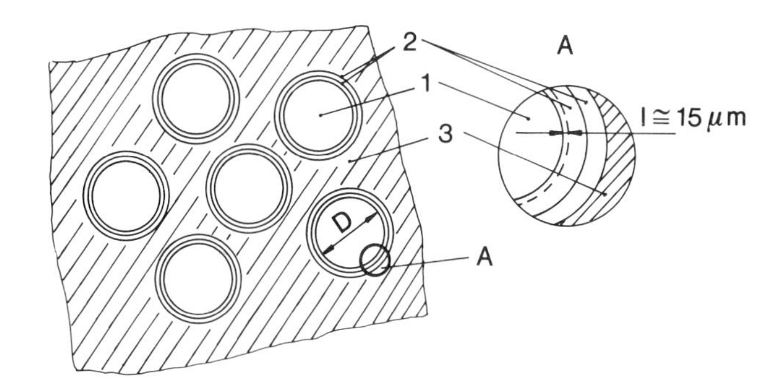


Fig. 9.15. A section through a dispersed fuel: 1 - spherical fuel grains; 2 - graphite coating (two duplex layers); 3 - matrix. Medallion A shows an enlarged view of the coating layers and the layer of $l$ = 15 μm that is affected by fission products.

the matrix/coated particle ratio would depend on the specific, desired structure, and may reach 0.8.

The matrix must be chemically compatible with the fuel, the structural and the cooling materials in contact. Also, the phases in contact must not be mutually soluble, because otherwise a phase homogenization occurs, jeopardizing all advantages obtained by their separation.

According to the nature of the phases, the following dispersions can be distinguished:

1. *metal* dispersions, both the dispersed phase and the matrix being metals;

2. *metal — ceramic* dispersions, the dispersed phase being ceramic and the matrix metallic;

3. *carbo — ceramic* dispersions, the dispersed phase being ceramic and the matrix being the graphite.

In Table 9.2 several important types of dispersed fuels are presented [24, 26].

The metallic dispersion $UAl_4$ — Al is the fuel material for most MTR reactors, both in the USA and Europe. Thin plates of metallic dispersion, with thicknesses of ~ 0.5 mm, are canned in aluminium plates of 0.4 mm. In Fig. 9.16 the MTR and Siloe [27] type fuel elements are presented; they are manufactured of such plates, and have borders sealed with aluminium plates. The enrichment in $^{235}U$ of the dispersion is in the range 20 — 93%. The burn-up reaches values of 5000 — 10000 MWh/kg U without occurrence of dimensional changes. The aluminium matrix is the medium where the fission fragments are stopped. Their free path considerably exceeds the $UAl_4$ grain size.

The $UZr_2$ — Zr dispersion is the fuel material for manufacturing fuel elements for thermal reactors, LWR type, employed in atomic submarine propulsion. The working temperature of these fuel elements must not exceed 770 K, this limitation being caused by excessive swelling at burn-ups higher than 500 MWh/kg U [28]. The fissile phase is, in this case, a saturated solid solution of U — Zr.

In *metallic — ceramic dispersion* (known sometimes as a *cermet*) the fissile phase is an oxide compound or a carbide. These have the advantage of a better irradiation resistance, allowing higher burn-ups, up to 2400 MWh/kgU.

TABLE 9.2 Dispersed Fuel Systems

| *Dispersion type* | *Fissile phase* | *Matrix* |
|---|---|---|
| Metallic dispersion | U | Mg, Th, Zr, zircaloy, ZrH |
| | $UAl_4$ | Aluminium |
| | $UZr_2$ | Zr, zircaloy |
| | $UBe_{13}$ | Be |
| Metal — ceramic dispersion | $UO_2$ | Stainless steel, Fe, nichrome, Mo, Al, Zr, zircaloy |
| | $U_3O_8$ | Aluminium |
| | UC | Stainless steel, Zr, zircaloy |
| | UN | Stainless steel, Zr |
| | $U_3Si$ | Zr |
| Carbon — ceramic | $UO_2$ | Graphite, BeO, $Al_2O_3$, $SiO_2$ |
| | $UO_2$ - $ThO_2$ | Graphite |
| | UC | Graphite |
| | $UC_2$ - $ThC_2$ | Graphite |
| | $UC_2$ - $PuC_2$ | Graphite |

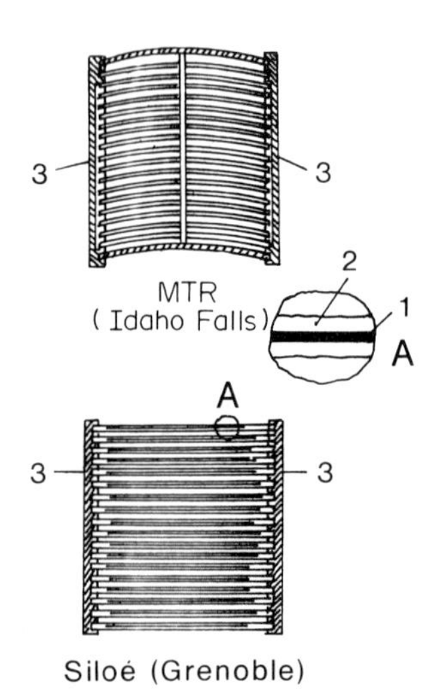


Fig. 9.16. Plate-like fuel assemblies for materials testing reactors MTR and Siloe. **A** is an enlarged view of the plate-fuel structure: 1 - dispersed fuel; 2 - aluminium cladding; 3 - aluminium frame-plate.

Several varieties of metallic — ceramic dispersions have been tested: $UO_2$ — Al, $U_3O_8$ — Al, $UO_2$ — stainless steel, $UO_2$ — Zr, U — $ZrH_{1.6}$,

Fig. 9.17. A section through a coated fuel particle: 1 - porous pyrocarbon layer for fission fragments stopping; 2,3 - dense pyrocarbon layer for fission gas retention; 4 - silicon carbide layer; 5 - porous pyrocarbon layer.

$UO_2$ — refractory metals (Mo, Nb, W) etc. The $U_3O_8$ — Al dispersion canned in Al is used in high flux research reactors (HFIR — the High Flux Intensity Reactor in USA) in the form of plate-shaped fuel, and has, over the U — Al alloy, the advantages of a higher U content at only 30% of the same volume. The $UO_2$ — Zr dispersion was developed in France (the "caramel" fuel) and allows the incorporation of a larger quantity of fuel in a plate. Finally, the dispersion U — $ZrH_{1.6}$ is used in TRIGA type research reactors and SNAP reactors for space applications. The fuel is shaped in rods, sheathed in incalloy 800. Owing to the hydrogen bound in the fuel, this exhibits a strongly negative temperature coefficient of reactivity, thus conferring to the reactor an intrinsic stability. For research reactors, the burn-up is high ($6 - 10 \times 10^3$ MWh/kg U).

The $UO_2$ — stainless steel dispersion is stable up to relatively high temperatures (~ 1300 K), allowing the manufacture of fuel elements that withstand the high temperatures in the fast reactors. The $UO_2$ fuel used for dispersion preparation is a powder with evenly sized, stoichiometric grains (O/U = 2). This type of dispersion contains 30 — 50% $UO_2$ and may withstand irradiation at 870 K up to burn-ups of ~ 2400 MWh/kg (U+Pu) without any sizeable alteration [29].

The *carbo-ceramic dispersion* (no metallic component) makes feasible the manufacture of fuel elements for high-temperature reactors (coolant temperatures in excess of 1020 K). These materials use as dispersed phase a ceramic fuel shaped as microspheres embedded in a graphite matrix — see Fig. 9.17 [29]. To avoid matrix damage by fission fragments, the fuel particles are coated with a porous pyrocarbon layer (1) that plays the stopping medium for fission fragment gases. To retain better the fission products two other dense pyrocarbon coatings (2) and (3) are added over this layer, as tight barriers for fission gases, followed by a silicon layer (4) that confers to (2) and (3) mechanical strength also limiting diffusion of barium and strontium, and sometimes an outer porous pyrocarbon layer (5) — see also paragraph 9.3.3.2.

Carbo-ceramic dispersions are used in the manufacture of fuel elements for high-temperature, gas-cooled reactors.

Prismatic blocks with a hexagonal cross-section opening of 360 mm and a height of 790 mm are manufactured, with cooling channels and holes to be filled with pellets containing a carbo-ceramic dispersion embedded in a graphite matrix, and are finally tightly sealed at the ends. Such elements may reach burn-ups of ~ 2400 MWh/kg (U+Pu) and are discharged from duty after about 6 years of use. Prismatic hexagonal blocks are also used in the reactors for spaceship propulsion (NERVA). In this case, the UC microspheres containing enriched uranium are coated with pyrocarbon and dispersed in graphite, used as both moderator and structural material. The peculiarity of this system consists in its cooling with hydrogen which is, at the same time, the reactive propellant. To protect the hot graphite against corrosion and erosion by the hydrogen jet, the graphite components are coated with niobium carbide layers.

The spherical fuel elements consist of graphite spheres of ~ 60 mm diameter containing a kernel of small spherical particles (~ 1 mm diameter) made of fissile material (U and Th carbide sealed in pyrocarbon) embedded in a graphite matrix. The kernel is covered by a thick graphite coating. The fissile material contains enriched uranium and thorium at a ratio of 1 : 10. Each sphere contains about 1 g $^{235}U$, 10 g Th and 190 g graphite.

The spheres (about 675,000 for the Ventrop-Schmehausen reactor, FRG [30]) are placed in a cylindric vessel 5.6 m in diameter and 6 m in height, containing the reactor core. The balls are gradually unloaded through the bottom;then they are checked for integrity and burn-up and are either fed back into the core or sent to refabrication. The cooling is with helium that flows free through the ball bed at a pressure of 4 MPa and a working temperature of ~ 1020 K.

Other types of dispersions are also tested — among them the ceramic and vitro-ceramic ones.

*Ceramic dispersions* use oxides for both the matrix and the fissile material. As matrix $Al_2O_3$, MgO and BeO are used [31]. These have a good irradiation stability, but a lower thermal conductivity.

*Vitro-ceramic dispersions* are made by melting together a mixture of $SiO_2$ and $UO_2$. Subsequent cooling gives a crystalline $UO_2$ and a vitreous matrix with a high $SiO_2$ content. This dispersion has the advantage of a good irradiation stability up to a burn-up of ~ 400 MWh/kg U [32] and can be conveniently manufactured. It also has a good neutron economy because silicon has a $\sigma_a$ of only ~ 0.16 b and its content is only of 30% (molar).

## 9.3. MANUFACTURE OF FUEL ELEMENTS

The manufacture of fuel and fuel elements requires a highly articulate complex of technological procedures thus conceived as to ensure a safe and efficient utilization of the fuel materials in the reactor. Fuel element manufacturing is a distinct field in nuclear technology, which uses conventional as well as special techniques, and also data and knowledge from such related fields as metallurgy, mechanics, materials physics, etc.

The selection of a manufacturing process for a certain type of fuel element results from an analysis based on technical and economic considerations. The reactor core designer defines the fissile atoms concentration and their ratio to other materials (fertile material, structural materials, etc.). The need to ensure the proper evacuation of the heat generated inside the fuel element determines its geometry and shape, and sometimes the nature of the bonding material between fuel and cladding (mechanical, metallurgical, etc.). Physics of materials defines the choice, actual options, and properties of fuel and clad materials according to requirements imposed upon by the reactor physics, thermo-hydraulics, the irradiated fuel reprocessing layout, and economic considerations. The fuel elements industry has the task of producing, physically, the nuclear fuel in compliance with all the aforementioned requirements, and meeting the diverse demands of currently active power reactors.

Among the procedures ordinarily employed in nuclear fuel manufacture are: melting, casting, lamination, pressing, welding, machine working, sintering, etc. Techniques are very sensitive to the actual properties of the material to be processed — its physical and mechanical properties, chemical compatibility, etc.

Due to the diversity of reactor types and fuel elements, there are many manufacturing procedures; it is as though each type of fuel demands a specific procedure. The following sections review some manufacturing processes, for the fuel elements that have been discussed in the previous section.

### 9.3.1. Manufacture of Metallic Fuel Elements

9.3.1.1. Metallic uranium tubes manufacturing. This process is meant to manufacture uranium tubes of precise shape and dimensions, starting from raw ingots. In selecting the technology one must consider the physical and chemical properties of metallic U. Taking into account the effect of nitrogen and oxygen on its mechanical properties at high temperature (see Chapter 3), the hot processing of metallic uranium must be done under vacuum or in neutral atmosphere (e.g. argon). The presence of three crystalline phases $\alpha$, $\beta$ and $\gamma$, and the fact that $\beta$-phase cannot be processed (due to its hardness and brittleness), complicate the manufacturing process.

Three manufacturing processes are known, for metal uranium rods: (a) *melting and casting* , (b) *plastic deformation*, and (c) *powder metallurgy procedures*. Sometimes the first and second of them are used in succession, melting and casting producing ingots of a convenient shape for the plastic deformation steps that follow.

*Melting* of metallic uranium ingots has the advantage of allowing removal of impurities that were retained in the material during previous technological steps ($H_2$, Mg, etc.). The ingot is placed in a graphite crucible coated with a refractory material (MgO, $Al_2O_3$, $ThO_2$, etc.) with the aim of preventing uranium carbide formation during melting. The metal is heated by induction in an installation that is shown schematically in Fig. 9.18 [27]. During melting the internal pressure is maintained between $10^{-3}$ and $10^{-4}$ mm Hg.

In order to homogenize the melt, the working temperature is set much above the melting point of the metal or alloy, although in some cases mechanical stirring is in order, to speed the homogenization process. When uranium alloys are requested, the "cascade" melting process is employed, where the molten metal is transferred into another crucible that contains the alloying elements. The crucible containing the molten metal has at the bottom a hole for metal evacuation to the moulding forms. This method is used for manufacturing tubes, plates, etc.

Among the *plastic deformation* processes are *forging*, *extrusion* , and *rolling*. These are used either in hot-processing, near the separation of $\alpha$ phase

(~ 870 K), when important deformations are further envisaged, or in cold-processing (~ 570 K) when the accuracy of sizing is important.

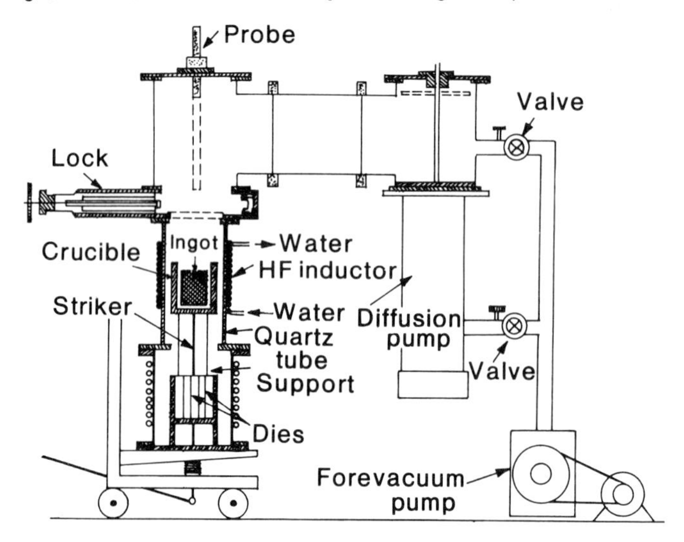


Fig. 9.18. Induction heat equipment for uranium melting.

As is known, *extrusion* is a process by which a mechanical part of given sizes is obtained starting from a metal that "flows" when forced through a drawing die. Uranium can be extruded in α-phase, between 850 and 890 K, "dressed" in a protective coating (for example copper), which avoids uranium oxidation and also die erosion. The larger the envisaged reduction in cross-size, the lower the extrusion speed (it may vary from 0.25 to 1.25 m/min) [27]. At lower temperatures the extrusion needs higher pressures, but allows a better dimensional accuracy that sometimes makes subsequent mechanical processing of the product unnecessary.

*Rolling* allows the manufacture of sheets and rods of different shapes, their cross-section being determined by the rolling cylinder's profile. Uranium rolling is performed generally in α-phase, either at relatively low temperatures (cold-rolling ~ 570 K) or at temperatures near the upper limit of α-phase (hot-rolling, ~ 770 — 870 K). Cold-rolling is usually in order when a higher dimensional accuracy or a certain texture are sought, since this method does not provide for any substantial deformation.

The *forging* process allows one to change the shape and structure of the grains of metallic uranium rods. Yet the process is rather unfrequently used in metallic uranium manufacturing.

*Powder metallurgy procedures* allow the manufacturing of mettalic parts starting from materials with physical and chemical properties which make difficult, or even impossible, the technologies mentioned above. Powder compacting is achieved by grain bonding under the simultaneous or successive action of pressure and temperature. This procedure is recommended when parts of complex shapes are to be manufactured, or when uranium alloys are difficult to obtain by melting and casting. Thus, the alloy is obtained by mixing uranium and alloying powders in definite proportion, followed by pressing and sintering. The method is less used in metal uranium manufacturing, but as will be discussed later, is largely used in manufacturing ceramic and dispersed fuels.

9.3.1.2. Metallic fuels processing before sheathing. The technological procedures for manufacturing uranium tubes induce in the material a crystalline structure that does not meet the dimensional and irradiation stability requirements. Thermal treatments, applied to eliminate this shortcoming, give rise to a fine-grained material structure, with isotropic untextured crystallites. At room temperature, the rolled or extruded uranium is in α-phase, having a large-grain, heavily textured structure.

The β *temper* treatment consists in heating the material up to a temperature corresponding to the β-phase, followed by an abrupt cooling at room temperature. The tempered material is, obviously, in α-phase at room temperature, but preserves the fine-grained structure specific for the β-phase. β tempering is achieved by induction heating of the uranium rod up to 990 — 1000 K, the steep cooling being done by immersion in water or oil. A faster cooling gives smaller grain size. This technological step is followed by *annealing* at temperatures of about 870 K, in order to diminish internal stresses induced through tempering.

To obtain the required grain size, uranium products are machined by conventional techniques, yet with some caution, due to the higher hardness, toxicity, the high pyrophoric nature of the material, etc. After machining, the metallic uranium tubes are cleaned by washing or blasting. Sometimes, electrochemical polishing is in order, with a view to passivating, by oxidation, the material's surface. After cleaning and final drying, the rods are vacuum-degased at a temperature of about 840 K and a pressure of $\sim 10^{-4}$ mm Hg.

9.3.1.3. Cladding procedures. Cladding procedures are numerous and complex, and vary according to the fuel type, and fuel/cladding contact nature (mechanical, metallurgical or with intermediate elements). Mechanical contact between fuel and cladding must be as close and tight as possible. The universally accepted method for assuring mechanical contact consists in applying a high external pressure which, in hot or cold conditions, gives rise to a plastic deformation of the cladding on the uranium tubes (this is the case with the EDF reactors).

The cladding procedure consists in the following steps: after filling the rod with metallic uranium fuel,the end plugs are TIG-welded in an argon atmosphere. The mechanical contact between fuel and clad is achieved in two steps. The preliminary step consists in suppressing the fuel — clad gap, by immersion in oil at room temperature and pressing at ~ 100 — 200 MPa [30]. The final stage is done in a hot gaseous environment ( ~ 670 K and ~ 20 MPa), to achieve a cladding stress relief, thus avoiding cladding — fuel detachment during in-pile operation.

To obtain a metallurgical bonding between fuel and cladding the coextrusion, consisting in a simultaneous extrusion of both materials in contact, is used.

As an example, in Fig. 9.19 a simplified flow-sheet for EDF fuel element manufacture is given [27].

### 9.3.2. Manufacture of Ceramic Fuel Elements

Most of the nuclear power reactors with thermal neutrons use as fuel $UO_2$, sheathed with zirconium alloys: zircaloy-4 for PWR and PHWR reactors; zirconium 1% Nb for VVER reactors [33-36]; zircaloy-2 for BWR; zirconium — niobium alloy for HCPR (the Soviet type, graphite-moderated and water-cooled reactor [37,38]). Sodium-cooled fast reactors use as fuel material $(U,Pu)O_2$ sheathed in stainless steel [39]. A similar solution is used for gas-cooled fast reactors [40], and for various types of fuel elements using carbides, nitrides, etc., containing plutonium [41,42].

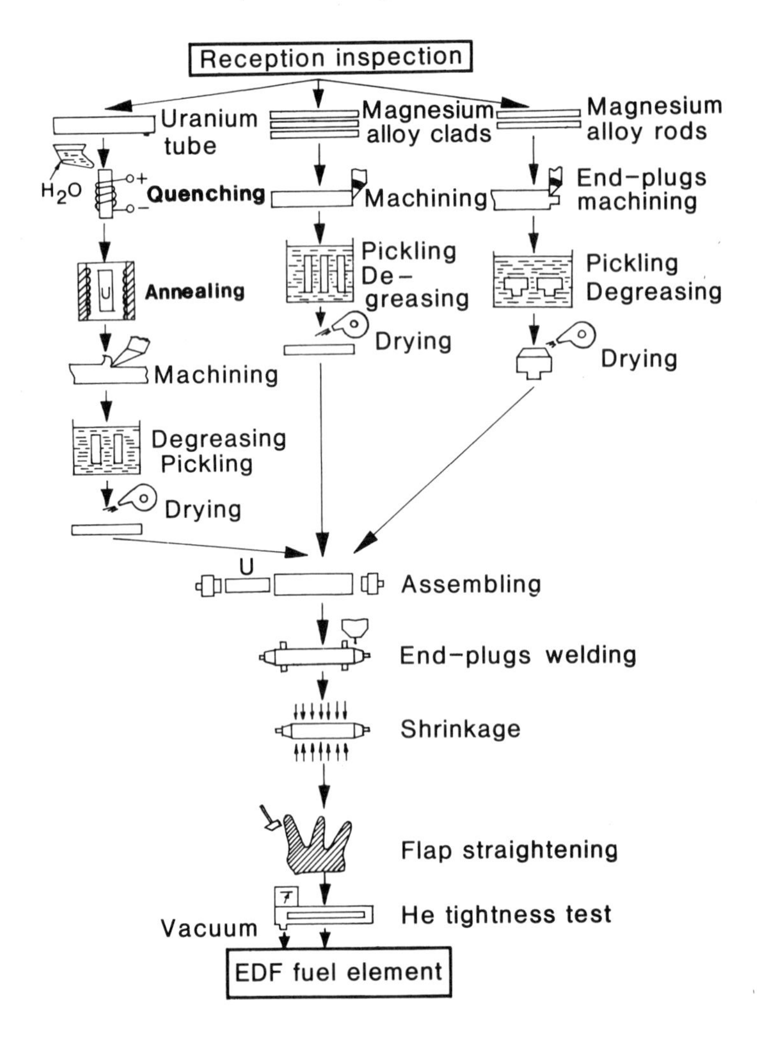


Fig. 9.19. Flow-sheet of EDF-type fuel element manufacture.

9.3.2.1. Ceramic fuels manufacturing procedures. Ceramic fuels are ordinarily obtained by powder compacting procedures, such as: *pressing and sintering*, vibration *compacting*, die *forging*, *melting* and *casting*, etc.

The *melting* — *casting* process is used especially in carbides manufacture; by electron beam melting, a good-quality fuel is obtained [43].

The *pressing* and *sintering* process has the advantage of using lower temperatures than for melting — casting, i.e. approximately 2/3 — 4/5 of the ceramic material's melting point.

*Ceramic powders densification by pressing and sintering* is designed for the manufacture of solid $UO_2$ parts with a density ranging between 8800 and

10800 kg/m$^3$, that is 80 — 98% of the theoretical density, starting from powders with an apparent density of 1500 — 3500 kg/m$^3$. Of consequence for attaining this goal are powder properties (particle sizes and shape, granulometry, surface area, impurity content) and working conditions (pressing procedure and magnitude, sintering temperature and duration, sintering atmosphere, etc. [44-47]).

Before pressing, the powder is mixed with a binder. The binder favours an even distribution of mechanical stresses as induced during pressing, also endowing the compact with a certain mechanical strength.In some cases certain lubricants are used, in order to reduce friction between powder and die. These additions must have such properties that during sintering they would completely vanish from the fuel. Binders ordinarily used are the polyvinyl alcohol, paraffin, kerosene, etc. and as lubricants stearic acid and its salts are employed.

In the following paragraphs the main steps in the manufacture of CANDU-type fuel elements will be introduced. Before starting the pelletizing, each powder batch is tested as to its fulfilment of the specification requirements. After passing the tests, pellet manufacture starts with powder granulation, that means powder compaction at 54 — 84 MPa [44] followed by crushing and grinding the compacts. The material thus obtained is sieved through calibrated sieves. The grained material is mixed with binders and lubricants and then passed at ~ 275 — 550 MPa [44] in special dies, thus obtaining "raw" pellets of specified density and dimensions. At this stage, statistical quality control of "raw" pellets is in order.

The sintering is performed in furnances under a reducing atmosphere ($H_2$), at temperatures ranging between 1920 and 1970 K (cracked ammonia is also used [48]). After sintering, the pellets are checked for dimensions, impurity content, microstructure, etc. The pellet's final diameter is adjusted by grinding down to the design values. After washing (to remove the grinding fluid) the pellets are dried, then dimensionally and visually checked. In Fig. 9.20 a simplified flow-sheet for $UO_2$ pellet manufacturing is given.

Research done in Romania has resulted in the domestic qualification of a manufacturing technology for CANDU-type fuel elements [49-53]. The technology has been tested and is currently operating at a pilot plant scale.

The processes used for manufacturing sintered pellets intended for use in PWR, BWR, FBR fuel elements entail in principle the same sequence of operations as presented before, with some differences in process parameters depending on the finite product's characteristics. In FBR reactors mixed oxides, $(U+Pu)O_2$, are used, the manufacturing process being in this case supplemented with the operation of $UO_2$ (80%) and $PuO_2$ (20%) powders (mechanical) mixing.

The *densification of ceramic powders by vibration* is a process used mainly in the USA, at Hanford and Oak Ridge [54].

The fuel element clads, welded at the bottom, are mounted vertically on a vibrating device that is fed continuously with powder. The powder's granulometric range is quite wide (0.1 μm — 1 mm), to facilitate the compaction process. This method is particularly suitable for plutonium-bearing fuels or for the manufacture of dense and complex ceramic pieces, that cannot be obtained by other methods.

The raw material is the high-density $UO_2$ obtained, for example, by melting. This is ground down to the desired granulometry. The material is fed gradually in tubes, slowly vibrated at the beginning (starting with accelerations of 10 g); then, as the tubes are filled, the acceleration is gradually increased (to reach finally up to 80 — 100 g). The filling and compacting duration for a PRTR (Plutonium Recycle Test Reactor, Hanford, USA) fuel element is about 3 minutes.

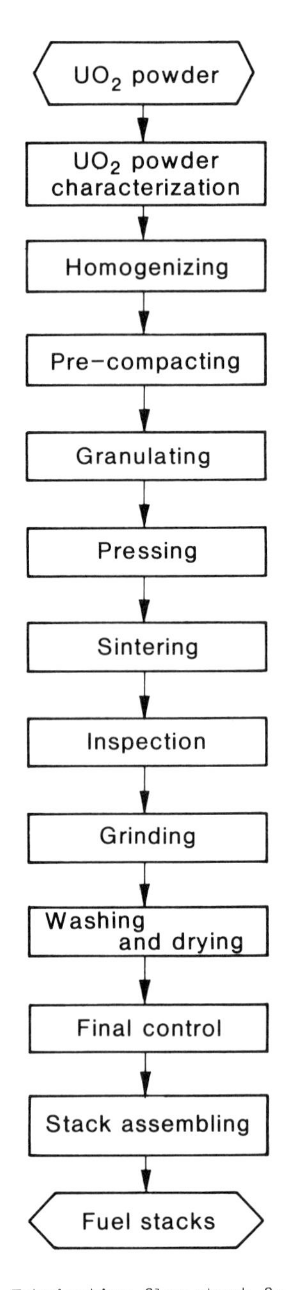


Fig. 9.20. Fabrication flow-sheet for $UO_2$ pellets.

Fabrication of fuel elements with evenly distributed density is possible only if high-density particles in powder (close to TD) are used. The compacting efficiency depends on the particles' granulometry, their shape and orientation. It was established that, when spherical particles in three classes of sizes are used together,a density of 98% of TD [54] may be obtained.

The compacting efficiency depends to a large extent on clad geometry and frequency and acceleration used. As a rule, uniform compacting may be achieved by varying the frequency between 400 and 500 Hz [54] so that some resonance frequencies of the system be excited.

*Ceramic powder compacting by forging* may be performed simultaneously with fuel canning. An option for this process depends on the desired density, on whether appropriate reproducibility is ensured for the product's parameters, and on its implicit irradiation behaviour. The steering parameters are: $UO_2$ ceramic powder characteristics, chemical composition, outer clad diameter and wall thickness, and forging temperature. Using this process, the $UO_2$ powders may be compacted up to densities of 80 — 90% TD (ambient temperature forging) and ~ 92% TD (high-temperature forging), respectively. The reproducibility of final density of any type of $UO_2$ compact is ±0.5% [51].

Before compacting, the $UO_2$ powder is fed into a metallic tube of zirconium alloy, aluminium or stainless steel, which will constitute the cladding. The tube is sealed with provisional end-plugs (for room-temperature forging), or with welded end-plugs (for high-temperature forging). Powder compacting before forging is recommended, in order to avoid large diameter changes that would be required for a high densification.

Forging is performed in a rotating forge where the cladding undergoes radial shocks reducing its cross-section, the reduction depending on the oxide's bulk density and on cladding material nature. During forging, the metallic tube containing $UO_2$ powder behaves initially as a void cylinder; after diameter reduction and powder compaction it behaves like a rigid body. This behaviour affects the clad thickness reduction during forging.

The final density of the compact depends on the powder's characteristics. Generally, high densities are obtained when using powders with low specific surfaces and large particle size [55].

9.3.2.2. <u>Fuel rod manufacturing</u>. The fuel rods that enter a fuel element bundle consist of a number of fuel pellets inserted in a clad. After filling, the clad is sealed by welding, with end-plugs.

The type of rod depends, of course, on the fuel and reactor type. However, regardless of the type of fuel rod, the manufacturing process includes the following standard sequence: raw materials reception cum inspection; clad machining and makeup (cutting to design length, degreasing, graphite coating, etc.); manufacture of structural components (end-plugs, spacers for CANDU fuels, springs for securing the fission gas expansion room in the case of PWRs and FBRs); spacers' brazing on clad surfaces (for CANDU fuels); pellet stack formation and rod filling; end-plugs welding; finite rod surface treatment, etc. Most of these technological steps are accompanied by in-process inspection, in order to guarantee finite product quality. A detailed description of the technological flow-sheet for all types of fuel elements currently in use would go far beyond the scope of this work. In what follows, one confines only to some particulars of the CANDU fuel element technology [56-58].

For spacer manufacturing (pads and spacers) zircaloy sheet is used, thoroughly inspected before effective punching — for dimensions, surface defects and flatness, etc. Generally, spacers are manufactured by punching. The spacer surface is coated with beryllium layers, to provide for metallurgical bonding with cladding, by brazing.

The zircaloy end-plugs are machined to size, of carefully inspected material (ultrasonic testing, mechanical properties confirmation, etc.). Ultrasonic testing is used to identify any void or defect inside the rods, the standards now in effect requiring the detection of voids with diameters lower than 0.025 mm [44] (see Chapter 11). Machining is performed by conventional techniques, followed by dimensional control, cleaning, degreasing, washing and drying.

When the tubes are cut at the design length, the spacers are spot-welded on them, then brazed in vacuum, through induction. Visual inspection and a metallographical examination of the joints follow.

After filling the pellet stacks into the rods, the end plugs are electrically welded under magnetic pressure. Before welding, the rods are filled with a gaseous mixture of helium and argon [44]. The welding quality is metallographically checked using specimens in order to select those process parameters that would assure the best quality to the products.

Since welding is done through plastic deformation of clad and plug, minor superficial irregularities may occur; the outer ones are to be smoothed out by conventional machining. The fuel rods are checked for tightness using the helium leak test method (see Chapter 11), then washed and dried. The manufacturing flow-sheet is schematically shown in figure 9.21 [58].

9.3.2.3. Fuel bundles assembly. Fuel bundles may come in various patterns, depending on the type of reactor they are meant for. As said before, the fuel bundles for PHWRs have circular cross-sections and the fuel rods are assembled on three circles concentric with a central rod. The PWR bundles have a square cross-section, the fuel rods and the control rods forming lattices of 14 x 14 up to 17 x 17 pieces. The assembling technology depends on the bundle configuration. It comprises manufacture of structural components (end-cap grids, spacers, etc.) and fuel rod assembling with them, so that adequate mechanical strength and an optimal exhaust of the heat generated inside the fuel are ensured.

For CANDU-type fuel elements, these demands are properly met by assembling the rods with only end-cap grids, with no other structural component required. The rod position inside the bundle, and the bundle position inside the pressure tube, is assured by spacers and pads brazed on claddings. The end cap grids are manufactured of zircaloy sheet, using conventional processes (punching). After machining, the grids are visually inspected, degreased, pickled, washed, etc.

In view of assembling, the rods are generally placed on an assembly device, the end-cap/grid joints being stiffened by spot welding [44]. When the end-cap/grid welding is completed, the bundle is dimensionally and visually inspected. Sometimes the autoclaving process is used for surface quality and welding control. Autoclaving consists in placing the fuel bundle in autoclaves under conditions similar to those during actual reactor operation (water temperature of 570 K and pressure of ~10 MPa). The surface and welds qualify for an appropriate quality when the resulted oxide layer is black, adherent and shiny. In Fig. 9.22 the assembly flow-sheet for CANDU-type fuel bundles is given [44,58].

In Romania, the manufacture of ceramic-type fuel bundles for PHWR-CANDU power stations will ultimately rely on a domestic technology, developed at the Institute for Nuclear Power Reactors — INPR [56-59].

PWR, BWR and FBR-type fuel bundles assembly is much more complex, because of the large number of fuel rods and structural components of sophisticated shapes involved.

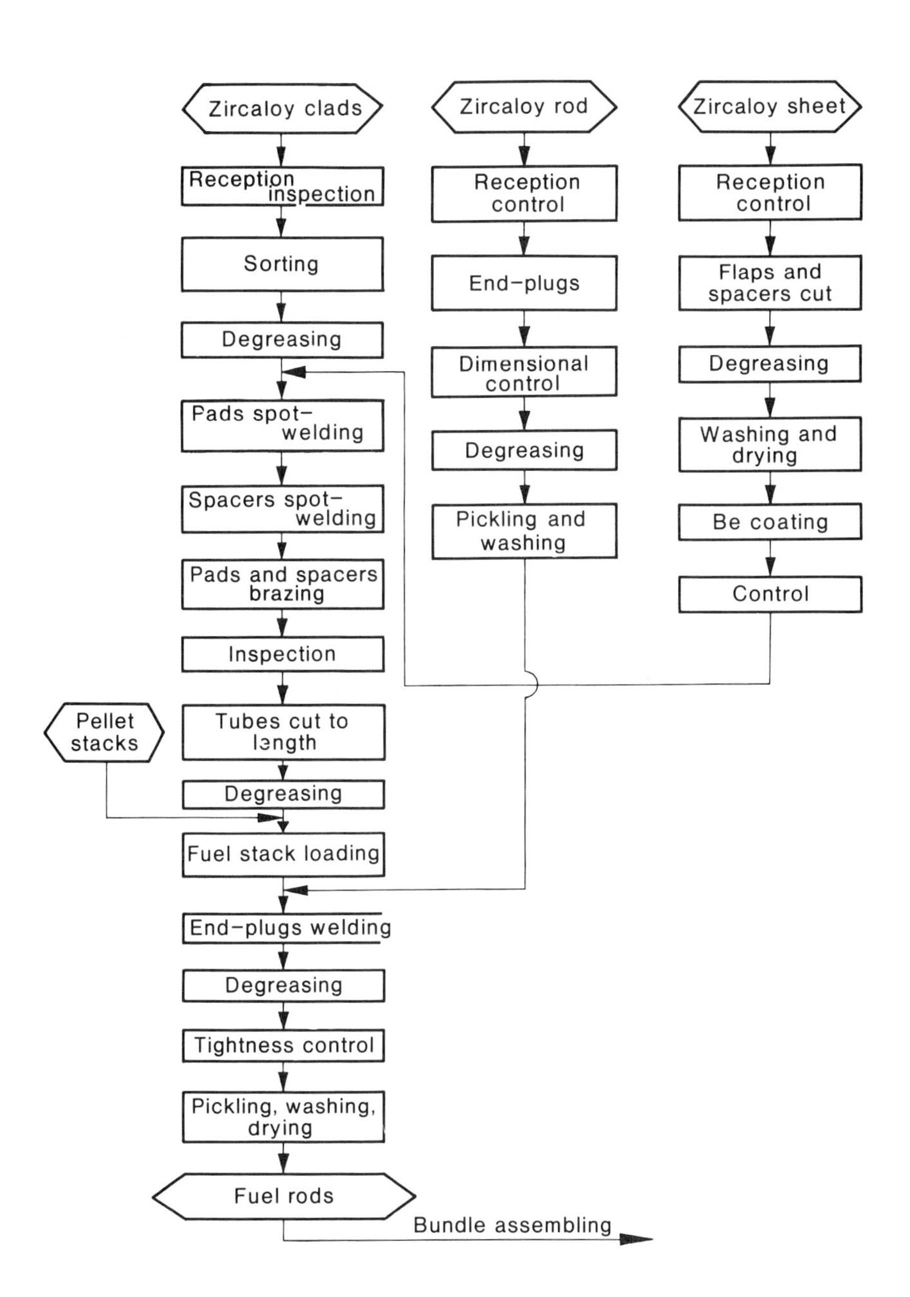


Fig. 9.21. CANDU fuel rods manufacturing flow-sheet.

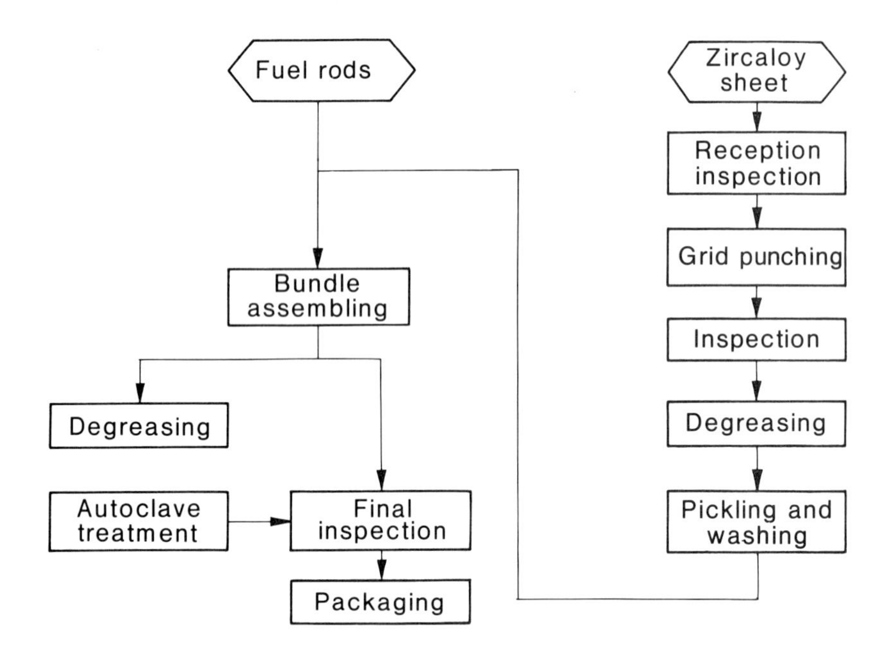


Fig. 9.22. CANDU-type fuel bundle assembling flow-sheet.

### 9.3.3. Manufacture of Dispersed Fuel Elements

9.3.3.1. Metallic and metallic - ceramic dispersions. The fuel elements with dispersed fuel consist of an assembly of thin plates (sometimes tubes) hosting the fuel in between. For fuel manufacture, the fuel *melting* and *powder metallurgy* processes are most frequently used.

The *melting* process generally fits the uranium — aluminium dispersions. Aluminium is melted by induction in a graphite crucible, uranium being added gradually in small quantities and with adequate stirring, in order to obtain a good homogeneity. Uranium may come as $UO_2$ or UN, because the melted aluminium will reduce it to metallic state anyway. Once the specified composition is achieved, the mixture is poured into graphite moulds; after solidification it is rolled down to the desired dimensions and, finally, hot-treated. In this way the intermetallic compound $UAl_4$ is obtained, dispersed in the aluminium matrix [63].

*Powder metallurgy* processes are widely used for manufacturing dispersed fuels. The powder preparation is difficult because particle size, purity and sinterability must be tightly controlled. The grains must come as close as possible to perfect spheres. The $UO_2$ powder is mixed with the metal powder of the matrix, and the powdered mixture thus obtained is submitted to a compacting process (pressing + sintering, rolling, etc.). Powder handling is difficult, because of uranium's toxicity and pyrophoricity. Although no inert atmosphere is required — unlike in the manufacture of the $UO_2$ — stainless steel system — special protection for operators is still in order, which complicates the operation and maintenance of the facility.

*Cladding* of plate-type dispersed fuel is performed by placing the fuel in a hollow plate and applying two other thin plates on the sides (Fig. 9.23) [27].

The assembly thus obtained is rolled at a temperature that depends on the material's nature (720 — 870 K for aluminium, 1070 K for zirconium, 1170 — 1570 K for stainless steel). The required dimensions can be obtained by multiple rollings that alternate with intermediate thermal treatments. Sometimes a final-cold rolling is performed, so that cold hardening enhances the mechanical strength of the assembly. The plates are controlled by X-ray radiography, checking especially the evenness of fuel distribution.

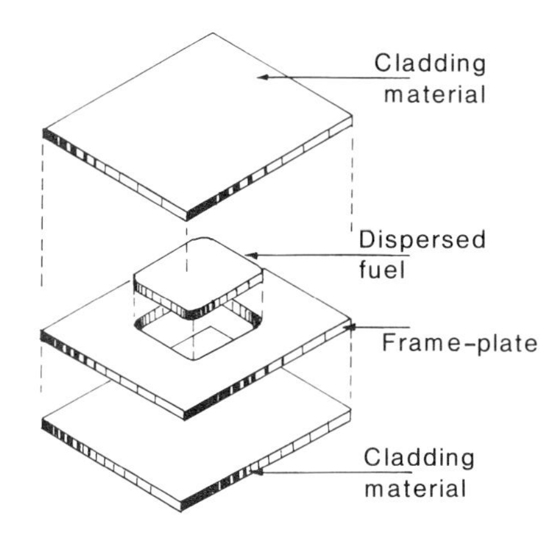


Fig. 9.23. Dispersed plate-type fuel canning.

The fuel element is obtained by assembling the plates with structural parts, by brazing, welding or mechanical joining.

9.3.3.2. Carbon — ceramic dispersions. In section 9.2.3 the fuel element concept using carbon — uranium dispersions for high temperature gas-cooled reactors (HTGR) was discussed.

Spherical fuel particles, of $UO_2$ for instance, are first obtained in raw ("green") state, from $U_3O_8$; then, after sintering, they are effectively reduced to $UO_2$. As indicated, the "canning" of these spheres is achieved by coating with successive layers of pyrocarbon and one layer of silicon carbide.

In Fig. 9.24 a typical furnace for performing this coating process is shown [60].

The fuel particles are dropped, from top to bottom, in a vertical graphite shaft that also serves as heater. From the bottom the "fluidizing" gas (He + $H_2$) is fed in, which generates a fluidized bed of particles and facilitates coating. The pyrocarbon is obtained by thermal cracking of hydrocarbons. A wide range of hydrocarbons (methane, propane, butane, etc.) and concentrations were tested; it was noted that the layers' properties depend upon the gas nature, working temperature and flow conditions across the particle bed [60].

Because the diffusion coefficient of alkali-earth fission products (Sr, Ba) in graphite is much larger than that of the inert gases, it is necessary to introduce additional silicon carbide layers (SiC). Thus, for higher working temperatures, ~ 1520 K, the diffusion coefficient of strontium in pyrocarbon

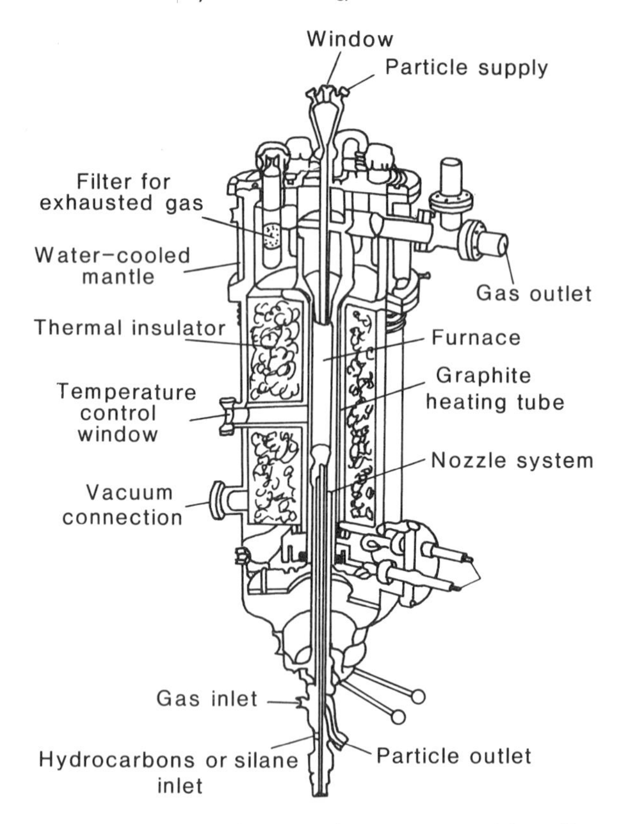


Fig. 9.24. Furnace for $UO_2$ microspheres coating with pyrocarbon and silicon carbide.

is of the order of $10^{-9}$ $cm^2/s$, while that of xenon is ~ $10^{-14}$ $cm^2/s$; therefore, pyrocarbon layers of ~ 100 μm would not significantly retain the fission products such as $^{90}Sr$ [60]. SiC coating is performed by using the same type of furnace as described, but replacing the hydrocarbon by methyltrichlorsilene or silane and the hydrogen by argon as fluidizing agent.

The fuel elements are obtained by assembling the structural parts with the fuel compacts obtained from coated particles mixed with graphite powder and an appropriate binder (the Dragon Project [60]).

### 9.3.4. Manufacture of Fuel Elements for Fast Reactors

As indicated in section 9.2.3 (Fig. 9.14), the core of fast reactors consists of two types of fuel bundles:

1. fuel elements containing fissile material ((U, Pu)$O_2$ mixed oxide) and fertile material ($UO_2$ for the axial blanket);
2. fuel elements containing fertile material ($UO_2$ for the radial blanket).

Both types of bundles have the cladding, spacers, grids and cases made of austenitic stainless steel.

The fuel element dimensions follow from the reactor's characteristics. Thus, the rod diameter is small, to ensure a good cooling at high power densities, and reasonable flow values for the liquid sodium.

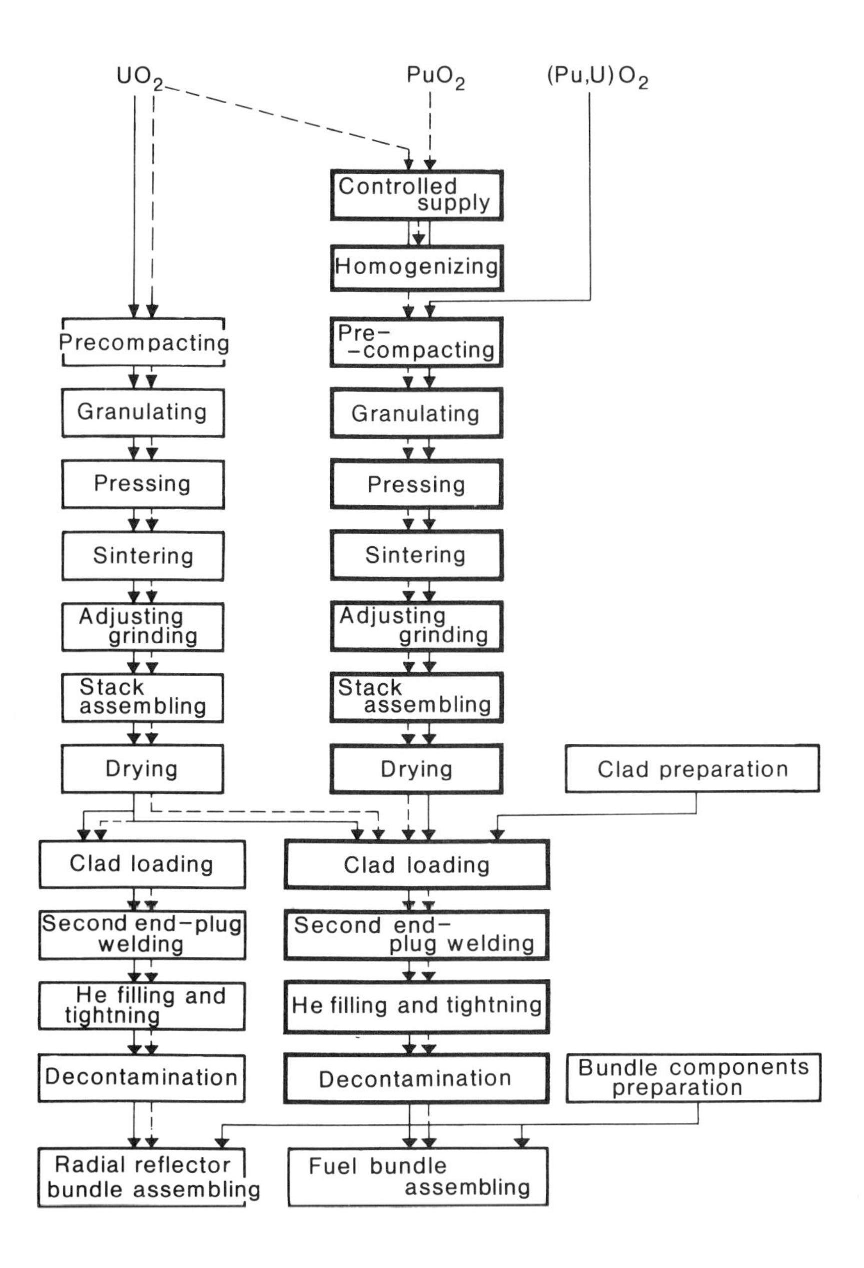


Fig. 9.25. Manufacturing flow-sheet of fissile and fertile fuel bundles for fast reactors.

The pellet density is the result of a compromise between the porosity needed to compensate the fuel swelling and the density needed to ensure a good breeding factor; its magnitude ranges between 80 and 85% TD.

The manufacturing technology of fast reactor fuel elements is similar to that used in the manufacture of $UO_2$ fuel elements for HWR and LWR. Plutonium's extreme radiotoxicity necessitates a tight containment of the manufacturing process in sealed areas (in air-tight boxes) using manipulators and gloves. This constraint adds to the manufacturing cost and, consequently, the technological flow-sheet is so designed as to reduce the number of steps and the amount of materials that are to be handled in glove-boxes. A technological flow-sheet for fissile and fertile fuel elements and bundles is given in Fig. 9.25 [27,61]. For the fuel elements and bundles used in the core two alternatives are available:

1. the raw material is the mixed oxide $(U,Pu)O_2$; the corresponding flow sheet is given as full lines;

2. the raw material is $PuO_2$; then the flow-sheet is given as dotted lines.

TABLE 9.3 French Fast-Reactor Fuel Characteristics

| | *Rapsodie* | | *Phoenix* | | *Nuclear plants* |
|---|---|---|---|---|---|
| | *Initial version* | *Advanced version* | *Initial version* | *Advanced version* | |
| Fuel | $UO_2$-$PuO_2$ | $UO_2$-$PuO_2$ or carbides | $UO_2$-$PuO_2$ | $UO_2$-$PuO_2$ or carbides | $UO_2$-$PuO_2$ or carbides |
| Cladding | AISI 316 X18M(316) | AISI 316 X18M(316) | AISI 316 X18M(316) | AISI 316 X18M(316) | - |
| Pu/U | 1/3 | 1/3 | 1/5 - 1/3 | 1/5 - 1/3 | 1/3 - 1/5 |
| Maximum cladding temperature ($^oC$) | 650 | 700 | 700 | 700 | 700 |
| Maximum linear power ($10^3$ W/m) | 34 - 36 | 40 - 50 | 43 | 40 - 60 | 40 - 60 100 |
| Burn-up (MWd/t) | 30,000 | 30-50,000 | 60-70,000 | 70-100,000 | 70-100,000 |
| Fuel element diameter (mm) | 5.7 | 5.6 | 5.5 | 5 - 6 | 5 - 8 |
| Core height (mm) | 340 | 340 - 600 | 850 | 700 - 1000 | 1000 - 1400 |
| Maximum neutron flux ($n/cm^2 \cdot s$) | $2.5x10^{15}$ | - | $8x10^{15}$ | $8x10^{15}$ | $10^{16}$ |

The steps that require sealed areas, glove-boxes and manipulators are marked as double rectangles.

The mixed oxide powder pre-compacting is performed at a pressure of 50 — 200 MPa. The pellets, of a diameter of about 5 mm, are obtained from granulated powder mixed with a quantity of lubricant, of 0.8% (zinc behenate), by pressing at 100 — 400 MPa. Mixed oxide pellets sintering is performed at 1600 — 1700 °C in an argon — hydrogen atmosphere (90% Ar — 10% $H_2$).

The welding of the first end-plug is done using TIG welding in argon. The welding of the second end-plug ("welding of the 2nd kind", in the figure) is performed using a TIG facility placed inside the glove-box and operated by remote control from outside.

The left side of Fig. 9.25 shows the manufacturing process of $UO_2$ pellets which, together with the $(U,Pu)O_2$ pellets, make up the load of an element in the bundle. The technological line is similar to that used for $UO_2$ pellet manufacturing as described in section 9.3.2.

The manufacturing of fuel elements and bundles for the radial blanket follows closely the flow-sheet for $UO_2$ pellets, and continues with the standard procedure in manufacturing ceramic fuel elements and bundles as discussed in section 9.3.2.

Clad preparation (including end-cap machining and first end-plug welding) as well as structural components manufacturing (spacers, grids, etc.) are similar to that described in relation to the LWR fuel elements.

Table 9.3 displays, as an example, some characteristics of the fuel employed in the French fast reactors [23,62].

## REFERENCES


1. Smith, C. O. (1967). *Nuclear Reactor Materials*, Addison Wesley Publ. Co., Reading (Mass.).
2. Wood, J. C., and others (1980). *J.Nuclear Materials*, 88, 81.
3. Olander, D. R. (1976). *Fundamental Aspects of Nuclear Fuel Elements*, Nat. Techn.Inf.Service, US Dept. of Commerce, Springfield, TID-26711-P1.
4. Asamoto, R. R., and others (1969). *J.Nuclear Materials*, 29, 67.
5. Nichols, F. A. (1966). *J.Appl.Phys.*, 37, 4599.
6. Mac Ewan, J. R., and Moyashi, J. (1967). *Proc.Brit.Ceram.Soc.*, 7, 245.
7. Sens, P. F. (1972). *J.Nuclear Materials*, 43, 293.
8. Gheorghiu, C. (1978). RESTR — A Computer Program for Structural Change Analysis in Sintered $UO_2$ Pellets. In *Proc.Conf.on Nuclear Fuel Element Modelling*, Blackpool.
9. Booth, A. H. (1957). Report AECL-496 (CRDC-721).
10. Matthews, J. R., and Wood, M. H. (1980). A Simple Treatment of Fission Gas for Normal and Accident Conditions. In *Proc.Conf.on Fuel Element Performance Computer Modelling*, Blackpool.
11. Ursu, I., Gheorghiu, C., and Gheorghiu, E. (1978). Efectul modificarilor structurale asupra comportarii gazelor de fisiune in bioxidul de uraniu, In *Proc.Symp.on Physics and Energy*, Pitesti 1978.
12. Hastings, I. J., and Notley, M. J. (1978). A Structure Dependent Model of Fission Gas Release and Swelling in $UO_2$ Fuel. In *Proc.Conf.on Nuclear Fuel Element Modelling*, Blackpool.
13. Notley, M. J. F. (1967). *Nucl.Appl.*, 3, 334.
14. Notley, M. J. F. (1966). Report AECL-2662.
15. Griesmeyer, J. M., and Ghoniem, N. M. (1979). *J.Nuclear Materials*, 80, 88.
16. Notley, M. J. F. (1967). Report AECL-2945.

17. Katsuragawa, M., and Kawamura, H. (1975). *Fuel Performance Code ACTIVE II*. PNCD 841-75, 17; Greenough, G. B., Nairn, J.S., and Sayers, J. B. (1971). Proc.Geneva Conf.A/CONF 49, P/501; Jankus, V. Z. (1972). *Nuclear Engineering and Design*, 18, 83; Verbeek, P., and Hoppe, N. *COMETHE III J — A Computer Code for predicting Mechanical and Thermal Behaviour of a Fuel Pin*. Belgonucléaire Report BN-7609-01, Bruxelles; Hoppe, N. (1980). Improvement to COMETHE III J Fuel Rod Modelling Code. *Nuclear Engineering and Design*, 56, 123.
18. Pascu, A., Gheorghiu, C., Mehedinteanu, S., and Alecu, I. (1979). Report ICEFIZ; Pascu, A., Gheorghiu, C., and Mehedinteanu, S. (1974). Report ICEFIZ.
19. IAEA (1962). *Directory of Nuclear Reactors*, Vol. IV, *Power Reactors*; Vol.V, *Research, Test and Experimental Reactors*, IAEA Vienna.
20. Ingolfsrud, L. G. (1981). Proiectarea centralelor CANDU, In *Rapoartele seminarului tehnic, Bucharest, 15-23 Oct.1981.*
21. Slater, J. B. (1981). Cicluri de combustibil avansat ale viitorului pentru reactoarele CANDU, In *Rapoartele seminarului tehnic, Bucharest, 15-23 Oct.1981.*
22. IAEA (1981). International Nuclear Fuel Cycle Evaluation, *Working Group VIII Summary Report*.
23. IAEA (1976). *Directory of Nuclear Reactors*, Vol. IX, *Power Reactors*; Vol.X, *Power and Test Reactors*, IAEA, Vienna.
24. Bethoux, O. (1963). *Le combustible nucléaire dispersé à phase fissile non métallique*. Report CEA-41.
25. White, D. W., Beard, A. P., and Willis, A. H. (1958). In *Proc.Conf.on Fuel Element Technology* (TID-7546), USAEC, Washington.
26. Weber, C. E. (1959). In *Progress in Nuclear Energy*, series 5, Vol. 2, Pergamon Press, New York.
27. Sauteron, J. (1965). *Les combustibles nucléaires*. Ed. Hermann, Paris.
28. Stubs, F.J., and Webster, C.B. (1959). Report AERE, C/M-372, UKAEA,Harwell.
29. Frost, B. R. T., and others (1964). In *Proc.Third Geneva Conf.*, Vol. 10, p.170; Huddle, R., and others (1969). Influence of Fabrication Parameters on the Incidence of Spearhead Attack on Encoated Particle Fuel. In *Proc.Symp.on Radiation Damage in Reactor Materials*, Vol. II, IAEA Vienna.
30. Gerwin, R. (1971). *Nuclear Power Today and Tomorrow*. Deutsche Verlags-Anstalt, Stuttgart.
31. Williams, J. (1963). In *New Nuclear Materials*, Vol. 2, IAEA, Vienna.
32. Lungu, S. (1963). Romanian Patent No 47 687-29-03.
33. Reshetnikov, F. G., Golovnin, I. S., Bibilashvilli, Yu. K., and Solyanyi, V. I. (1978). *Proc.Conf.on Fabrication of Water Reactor Fuel Elements*, Praga, IAEA-SM-233/44.
34. Denisov, V. P., and others, *Proc.Fourth Geneva Conf.*, A/CONF.49/P/693.
35. Reshetnikov, N. G., Kudryashov, Yu. K., and Panteleev, L. D. (1978). *Proc.Conf.on Fabrication of Water Reactor Fuel Elements*, Praga, 1978, IAEA-SM-233/49.
36. Belevantsev, V. (1978). *Proc.Conf.on Fabrication of Water Reactor Fuel Elements*, Praga, 1978, IAEA-SM-233/47.
37. Petros'yants, A. M., and others. *Proc.Geneva Conf.*, A/CONF.49/P/715.
38. Alshehenkov, P. I., and others. *Proc.Geneva Conf.*, A/CONF.49/P/716.
39. Lejpunskyi, A. I., and others. *Proc.Geneva Conf.*, A/CONF.49/P/750.
40. Krasin, A. K., and others. *Proc.Geneva Conf.*, A/CONF.49/P/750.
41. Zaymorskyi, A. S., and others (1977). *Proc.Conf.on Nuclear Power and its Fuel Cycle, Salzburg 1977*, IAEA-CN-36/353.
42. Reshetnikov, F. G., and others (1977). *Proc.Conf.on Nuclear Power and its Fuel Cycle, Salzburg 1977*, IAEA-CN-36/352.
43. Accary, A., and Trouve, J. (1963). In *New Nuclear Materials*, Vol. 2, IAEA, Vienna.
44. Durant, W. C. (1972). Manufacturing CANDU Fuel. In *IAEA Study Group on Facilities and Technology needed for Nuclear Fuel Manufacturing*, IAEA, Vienna.
45. Timmermans, W., and others (1978). *J.Nuclear Materials*, 71, 256.

46. Mohan, A., and others (1979). *J.Nuclear Materials*, 79, 312.
47. Reshetnikov, F. G., and Kuskakovskyi, V. I. (1980). *Proc.Conf.on Fabrication of Water Reactor Fuel Elements, Prague 1978*, IAEA-SM-233/43.
48. Vlasov, V. G., and Jukovski, V. M. (1963). *Kinetika i Kataliz*, 4, 63.
49. Georgeoni, P., Deju, R., and Dascalu, G. (1974). Report ICEFIZ, 1.
50. Georgeoni, P., Deju, R., and Dascalu, G. (1974). Report ICEFIZ, 2.
51. Mirion, I., and Spinzi, M. (1974). Report ICEFIZ.
52. Mirion, I., Dascalu, G., Pirvu, C., and Serban, C. (1976). Report ICEFIZ.
53. Mirion, I., Spinzi, M., Galatski, G., and Nicolaescu, V. (1976). Report ICEFIZ.
54. Hauth, J. J. (1962). *Nucleonics*, 20, 50.
55. Stenquist, R., and Anicetti, R. J. (1958). Fabrication of Ceramic Fuel Elements by Swaging. In *Proc.Meeting of Metallurgical Soc., Cleveland 1958*.
56. Andreescu, N., and Galatski, G. (1977). UNDP Technical Assistance for the Development of Nuclear Technology in Romania. In *Proc.Conf.on Transfer on Nuclear Technology, Iran 1977*.
57. Andreescu, N., and others (1977). Some Aspects of the Implementation of a Nuclear Technology Project. In *Proc.Conf.on Nuclear Power and its Fuel Cycle, Salzburg 1977*, p. 581.
58. Gheata, V., Galeriu, C., Dobos, I., and Glodeanu, F. (1978). Experience Gained in the Fabrication of Experimental Fuel Rods with Natural Uranium and Zircaloy-4 Cladding for Irradiation Experiments. In *Proc.Conf.on Fabrication of Water Reactor Fuel Elements, Prague 1978*, p. 283.
59. Alecu, M., and others (1975). Romanian Experimental Fuel Rods for Irradiation in MZFR Reactor (Germany), Technical Report, Karlsruhe.
60. Shepherd, L. R. (1970). Development of Coated Particle Fuels for High Temperature Reactor. In *Proc.Joint Meeting of BNES and the Inst.of Fuel, ICE, London 1970*.
61. Jeffs, A. T. (1967). Fabrication of $UO_2$ - $PuO_2$ and $ThO_2$ - $PuO_2$ Experimental Fuels. In *Proc.Symp.on Plutonium as a Reactor Fuel*, Bruxelles, p. 179.
62. Simnad, M. T. (1971). *Fuel Element Experience in Nuclear Power Reactors*. Gordon and Breach Science Publishers, New York, London, Paris.
63. Frost, B. R. T. (1982). *Nuclear Fuel Elements*. Pergamon Press, Oxford, New York, Toronto, Sydney, Paris, Frankfurt.

CHAPTER 10

# Nuclear Material Recovery from Irradiated Fuel and Recycling

## 10.1. GENERAL

In nuclear power reactors, burning of part of the fissile material generates new fissile material, through radioactive transmutations induced by neutrons in the fertile component of the fuel. From the angle of nuclear fuel economy, the most important nuclear transitions are those producing plutonium in uranium-fuelled reactors, as well as those which result in the generation of $^{233}U$ in thorium-fuelled reactors. An overview on major nuclear fuel cycles is given in Appendix 14.

Plutonium is an artificial transuranian element with $Z = 94$, discovered in 1942 by Seaborg and his team. The most abundant Pu isotope in irradiated fuel is the one with $A = 239$, produced from $^{238}U$ by neutron capture followed by two successive β-decays (see Chapter 1):

$$^{238}_{92}U \xrightarrow{n,\gamma} {}^{239}_{92}U \xrightarrow[T_{1/2} = 23.5\ \text{min}]{\beta^-} {}^{239}_{93}Np \xrightarrow[T_{1/2} = 2.37\ \text{d}]{\beta^-} {}^{239}_{94}Pu$$

To date, 15 isotopes of Pu are known, with atomic masses from 232 up to 246. Figure 10.1 presents several reactions resulting in transuranian elements, which are important for nuclear fuel recovery and recycling.

Uranium 233 results from $^{232}Th$, by a neutron capture followed by two successive β-decays:

$$^{232}_{90}Th \xrightarrow{n,\gamma} {}^{233}_{90}Th \xrightarrow[T_{1/2} = 22.2\ \text{min}]{\beta^-} {}^{233}_{91}Pa \xrightarrow[T_{1/2} = 27.4\ \text{d}]{\beta^-} {}^{233}_{92}U$$

Some of the nuclear reactions accompanying generation of $^{233}U$ from $^{232}Th$ are indicated in Fig. 10.2. Production of $^{233}U$ in reactors is particularly interesting because its nuclear properties recommend it as a good substitute for $^{235}U$.

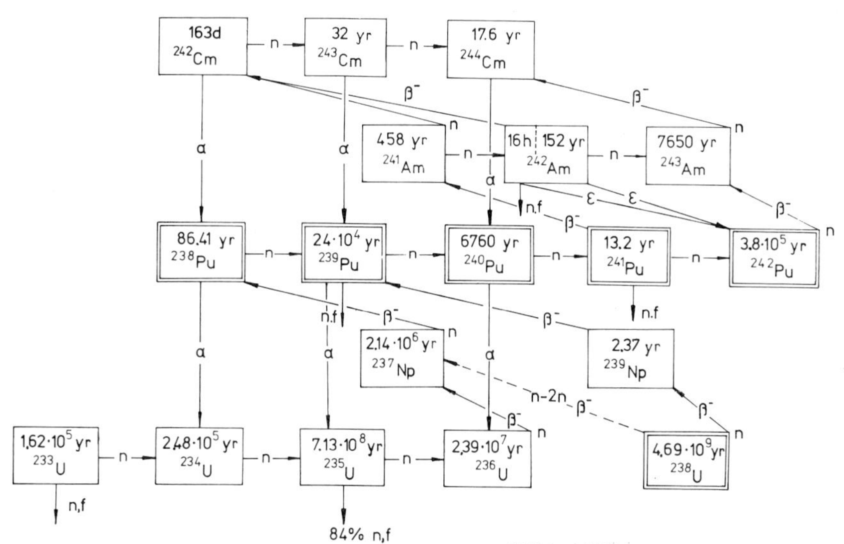


Fig. 10.1. Nuclear reactions in the $^{238}U$ - $^{239}Pu$ cycle and nuclides radioactivity [1]. α - α-decay; ε - electron capture; n - neutron absorption; β - β-decay; f - fission.

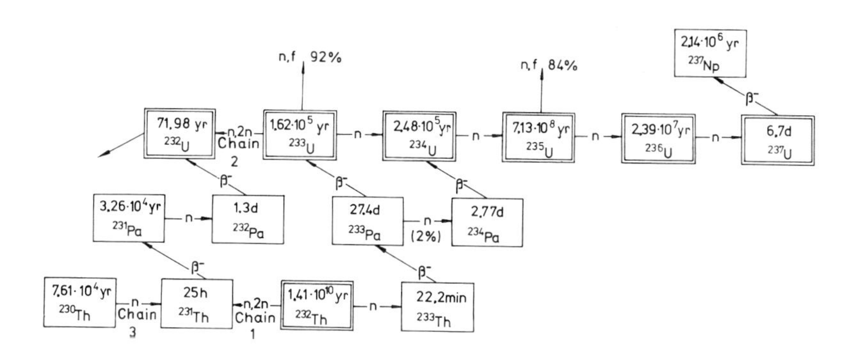


Fig. 10.2. Nuclear reactions in the $^{232}Th$ — $^{233}U$ cycle and nuclides radioactivity. n - neutron absorption; β - β-decay; f - fission.

As nuclear materials, $^{239}Pu$ and $^{233}U$ present fission cross-sections for thermal neutrons of 748.2 b and 524 b, respectively, close to the 586 b cross-section of $^{235}U$.

Table 10.1 shows the concentration of actinides in the irradiated fuel from actual PWRs, HWRs and FBRs. Such data highlight the importance of recycling the spent fuel unloaded from thermal reactors (of PWR and HWR type), and particularly the need to recycle the fuel unloaded from FBR reactors, because in the latter case the concentration of fissile material in the spent fuel is even higher than the concentration in the original, fresh fuel.

TABLE 10.1 Concentration and Composition of Recoverable Actinides in Irradiated Fuel, for Various Types of Commercial reactors

| *Actinides* | *HWR* | *LWR* | *FBR* |
|---|---|---|---|
| | | *(irradiated fuel g/t)* | |
| Uranium - 234 | | | $2.76 \times 10^1$ |
| Uranium - 235 | $2.3 \times 10^3$ | $7.95 \times 10^3$ | $7.51 \times 10^0$ |
| Uranium - 236 | | $4.98 \times 10^3$ | $1.09 \times 10^1$ |
| Uranium - 238 | $9.85 \times 10^5$ | $9.44 \times 10^5$ | $7.19 \times 10^5$ |
| Neptunium - 237 | | $7.62 \times 10^2$ | $1.80 \times 10^2$ |
| Plutonium - 238 | $3.8 \times 10^0$ | $1.66 \times 10^2$ | $1.84 \times 10^3$ |
| Plutonium - 239 | $2.6 \times 10^3$ | $5.38 \times 10^3$ | $1.17 \times 10^5$ |
| Plutonium - 240 | $9.73 \times 10^2$ | $2.17 \times 10^3$ | $5.24 \times 10^4$ |
| Plutonium - 241 | $1.74 \times 10^2$ | $1.01 \times 10^3$ | $1.24 \times 10^4$ |
| Plutonium - 242 | $5.23 \times 10^1$ | $3.49 \times 10^2$ | $9.02 \times 10^3$ |
| Americium - 241 | | $6.18 \times 10^1$ | $1.59 \times 10^3$ |
| Americium - 242 | | $4.14 \times 10^1$ | $2.46 \times 10^1$ |
| Americium - 243 | | $9.04 \times 10^1$ | $7.11 \times 10^2$ |
| Curium - 242 | | $4.52 \times 10^0$ | $3.28 \times 10^1$ |
| Curium - 243 | | $8.72 \times 10^{-2}$ | $2.30 \times 10^0$ |
| Curium - 244 | | $3.07 \times 10^1$ | $4.17 \times 10^1$ |

HWR: $UO_2$ fuel — natural (0.71% — $^{235}U$),
burn-up: 7200 MWd/t U, de-activation time: 180 days

LWR: $UO_2$ fuel — enriched (3.3% — $^{235}U$),
burn-up: 34,000 MWd/t U, de-activation time: 150 days

FBR: $(U, Pu)O_2$ fuel (0.7% — $^{235}U$, 20% Pu),
burn-up: 80,000 MWd/t (U, Pu), de-activation time: 180 days.

The following lines will bring more evidence as to the correlation of recycling and rational utilization of uranium resources. The trends in the development of fission power reactors will then naturally appear.

The large quantities of irradiated fuel unloaded yearly from the reactors of nuclear power plants in operation make increasingly compelling in the near future the large-scale implementation of commercially developed fuel recycling [1-6].

## 10.2. ROLE AND TASKS OF NUCLEAR MATERIAL RECOVERY AND RECYCLING

The turn towards more abundant energy sources and energy systems that would prove more viable than the current ones, based as they are on fossil fuels, as announced by the energy crisis of the 1980s — a time when nuclear power emerged again as a major factor of renewal and hope, resurrected national and international concerns as to the security of nuclear fuel supply.

Reactors of nuclear power plants currently operating worldwide will not burn, over their entire 30-year lifetime, more than 10% of the uranium reserves now known of in the world, and whose mining is considered economically sound.

From world forecasts on the implementation of nuclear power plants up to the year 2000, which foresee growth rates higher than these of the 1970s, it appears that the uranium extraction and processing industry will have to double its capacity every 7 years — which looks rather an ambitious goal. Consequently, the procurement of fissile materials for power reactors, as well as the most efficient utilization of these materials proved to be a problem of critical magnitude.

In this light, one major question is whether it is possible to deliver the uranium in time, in sufficient quantity and at convenient prices. The answer is essentially positive; however, it will, at any rate, require proper identification, technical and economic validation and actual implementation of solutions to the question of a much more efficient utilization of uranium.

To date, the use of uranium to generate power is based essentially on a single performance in the reactor, until the fuel reaches a limited, designed burn-up; what follows is plain storage of the spent fuel, and waiting for innovative decisions concerning its eventual turning to better account.

The prospects of improving the present direct fuel cycles (that do not provide for recycling) are shown in Table 10.2 [5]. One can notice the availability, in principle, of certain solutions which can be applied to the current models of PWRs and PHWRs with the purpose of increasing uranium utilization efficiency.

TABLE 10.2 Uranium Consumption over a 30-year Period, at an Annual Plant Capacity Factor of 70%, for the Present Types of Commercial Reactors (recycling of fissile materials not included)

| *Manufacturer* | *PWR* | | | *PHWR* | |
|---|---|---|---|---|---|
| | *France* | *FR of Germany* | *USA* | *Canada* | *FR of Germany* |
| *Current fuel cycles* | | | | | |
| Average burn-up (MWd/tU) | 31,800 | 33,000 | 30,390 | 7300 | 7400 |
| Required equivalent of natural U (t/GWe): | | | | | |
| Initial core inventory | 324 | 367 | 303 | 130.8 | 112 |
| Loading per year | 140 | 139 | 139 | 121.1 | 118.4 |
| Cumulative, over 30 years | 4380 | 4363 | 4347 | 3716 | 3608 |
| *Prospective, improved, cycles* | | | | | |
| Average burn-up (MWd/tU) | 43,500 | 44,000 | 50,650 | 20,900 | 20,000 |
| Required equivalent of natural U (t/GWe): | | | | | |
| Initial core inventory | 324 | 367 | 314 | 256 | 150 |
| Loading per year | 127 | 128 | 123 | 82.3 | 86.1 |
| Cumulative, over 30 years | 3970 | 4051 | 3823 | 2651 | 2679 |

For PWRs, such improvements would consist mainly in raising the average burn-up, from the current 30,000 — 33,000 MWd/t to 43,500 — 50,650 MWd/t, thus cutting on fresh uranium consumption by approximately (7-12)%. For PHWRs, the cut in fresh uranium consumption resulting from the use of lightly enriched uranium (1.2% $^{235}U$) would reach about 25 — 29%. The potential for plutonium conversion of such a reactor would diminish. Plutonium production would decrease from the current 2.8 kg/t $U_{nat}$ to 1.7 kg/t $U_{nat}$, still remaining superior to that of the PWR, which would decrease from 1.1 kg/t $U_{nat}$ to 0.9 kg/t $U_{nat}$.

It is thought feasible to increase the efficiency of uranium utilization in PWRs by 15 — 30%. This would require, however, besides enhanced burn-up, an increase in the efficiency of extraction of uranium from ore, of the enrichment in $^{235}U$ of the natural U, etc. To a great extent, such improvements would have a similar effect upon the efficiency increment in the utilization of uranium in PHWRs, namely from 25 — 29% to 35 — 38%.

Another avenue which may substantially diminish uranium consumption consists in recycling the fissile and fertile materials contained in the irradiated fuel, and in making good use of thorium in thermal reactors. Table 10.3 [5] presents the effects that may be obtained, if the fissile materials in the current power reactors (PWR and CANDU-PHWR) were recycled; cases (1) and (3) represent the current modes of fuel utilization (without recycling). The rise in uranium utilization efficiency through plutonium recycling is estimated at 32% in the case of PWRs and at 55% for the CANDU-PHWRs, as indicated by (2) and (4) in the table. Case (5) shows the efficiency of bringing thorium into the CANDU-PHWR standard reactor. PWR thorium reactors can also contribute an improvement in uranium utilization efficiency.

TABLE 10.3 Potential for Improvement, through Recycling of Fissile Material, of the Efficiency of Uranium Utilization in Current Thermal Reactors (expressed as annual natural uranium consumption equivalent, in *t/GWe*, at a plant capacity factor of 75%)

| | |
|---|---|
| PWR: | |
| 1. Without recycling | 141 |
| 2. With U + Pu recycling | 87 — 91 |
| CANDU: | |
| 3. Without recycling | 130 |
| 4. With Pu recycling | 59 |
| 5. With Th recycling | 27 |

As a result of the production of enriched uranium, a growing stock of depleted uranium is currently piling up. Turning this material to account would require actual implementation of nuclear power plants equipped with fast breeder reactors. Theoretically, the fast breeder reactors can burn the entire quantity of $^{238}U$ contained in natural or depleted uranium. Thus, the potential utilization of uranium is very high and, consequently, the cost of the energy generated in such reactors would remain practically independent of the uranium price that may eventually increase.

Thermal reactors cannot offer solutions for the utilization of depleted uranium ($^{238}U$) either from the irradiated fuel or resulting from enrichment plants. It is,however, feasible to fuel thermal reactors (especially the HWR type) with thorium. In this case the cost of the energy produced is only slightly affected by fluctuations in the price of the raw material (thorium). Production of $^{233}U$ in a thermal neutron flux is particularly effective, and thus the system may, theoretically, burn the whole stock of thorium while producing $^{233}U$. So, the use and recycling of thorium in CANDU reactors provides, at a constant power rate, for the same virtual independence on resources, as the fast breeder reactors which burn the uranium completely. However, unlike fast breeder reactors, CANDU reactors cannot produce new fissile material to enhance the overall power installed in the system. Both FBR and PHWR thorium reactors imply the existence of an initial stock of fissile material. For FBRs this stock is of ~ 5 t Pu, while for PHWRs it is of ~ 4 t $^{235}U$ (~ 800 t of natural uranium). The source of plutonium resides in the thermal reactors. As discussed before, PHWRs produce 2.8 kg Pu/t $U_{nat}$, whereas PWRs produce 1.1 kg Pu/t $U_{nat}$ [5,6]. Thus, the fuel initially required for the operation of FBRs and PHWRs (Pu-initiated Th cycle) can be provided by thermal reactors. This implies the use of a symbiotic start-up system, in which the fissile material produced by thermal reactors is used as an inventory material for the advanced cycle in the FBRs or PHWRs.

Eventually, FBR — LWR or FBR — HWR symbiotic systems could turn up, in which the fissionable material produced in FBRs will be used to fuel the thermal reactors (LWR or HWR) [7]. The characteristics of two such symbiotic systems in equilibrium are presented in Fig. 10.3. In both systems the fuel is an oxide, namely $(U,Pu)O_2$ for FBR, and $(U,Pu)O_2$ or $(U,^{233}U)O_2$ for LWR.

VARIANT 1: SYMBIOTIC FBR/LWR SYSTEM IN EQUILIBRIUM

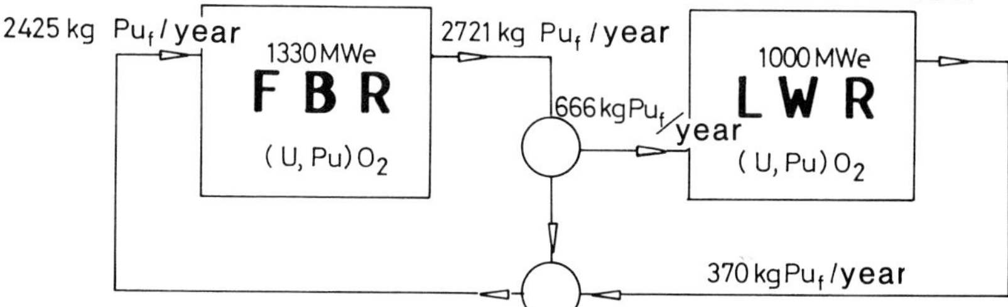


VARIANT 2: SYMBIOTIC $^{233}U$/LWR/$(U,Pu)O_2$ $ThO_2$ FBR SYSTEM IN EQUILIBRIUM

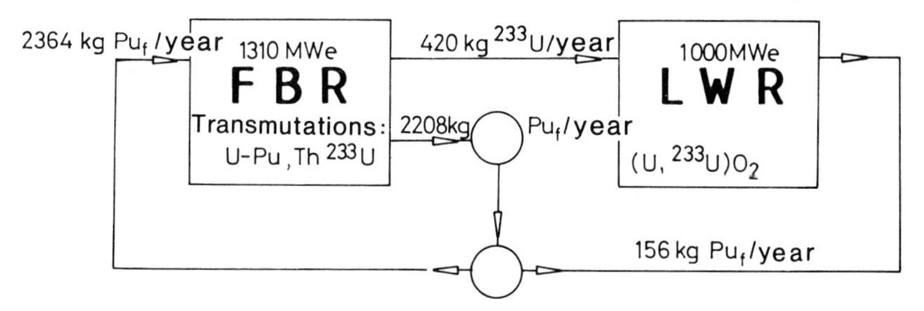


Fig. 10.3. Symbiotic FBR — LWR systems involving the Th cycle [7].

The present CANDU-PHWR model was designed to make efficient use of natural uranium. Under these terms, the system works currently in an open circuit, without recovering and recycling the fissile materials, and the fuel path comes to an end with its storage. An economic alternative aiming at solving the problem of uranium conservation by plutonium recycling and the utilization of thorium would require the adaptation of the current standard of reactor to the new terms of reference.

Tables 10.4 and 10.5 show the uranium consumption and the production of recyclable actinides in the current model of reactor [8,9]. Table 10.6 presents the fuel cycles that can be initiated with CANDU-PHWR reactors which use these recyclable materials.

TABLE 10.4 Typical Requirements of Natural U for 600 MWe CANDU-PHWRs, at an 80% Plant Loading Factor, and at 7000 MWd/t burn-up

| *Parameter* | *Characteristic value* (*t* $U_{nat}$/*MWe*) |
|---|---|
| Initial inventory | 0.134 |
| Annual supply at equilibrium | 0.1425 |
| Total requirement for a 30-year operation of the reactor | 4.36 |

TABLE 10.5 Production of Recyclable Actinides in a 600 MWe CANDU-PHWR

| *Parameter* | | *Characteristic values* | | | |
|---|---|---|---|---|---|
| Fluency | ($10^{21}$ n/cm$^2$) | 1.6 | 1.8 | 2.0 | 2.2 |
| Burn-up | (MWd/t U) | 6715 | 7538 | 8349 | 9150 |
| $^{235}$U | (g/kg) | 2.564 | 2.261 | 1.066 | 0.762 |
| $^{236}$U | (g/kg) | 0.6095 | 0.7337 | 0.7711 | 0.8032 |
| $^{237}$Np | (g/kg) | 0.02332 | 0.02737 | 0.03142 | 0.03544 |
| $^{238}$Pu | (g/kg) | 0.002467 | 0.003236 | 0.004100 | 0.005050 |
| $^{239}$Pu | (g/kg) | 2.513 | 2.582 | 2.632 | 2.668 |
| $^{240}$Pu | (g/kg) | 0.9168 | 1.057 | 1.191 | 1.319 |
| $^{241}$Pu | (g/kg) | 0.1674 | 0.1997 | 0.2315 | 0.2623 |
| $^{242}$Pu | (g/kg) | 0.04379 | 0.06035 | 0.07961 | 0.1014 |
| $^{242}$Cm | (mg/kg) | 0.2397 | 0.3607 | 0.5113 | 0.6911 |
| $^{241}$Am | (mg/kg) | 1.064 | 1.403 | 1.768 | 2.152 |
| Total Pu | (g/kg) | 3.544 | 3.902 | 4.138 | 4.356 |
| out of which fissionale | (g/kg) | 2.680 | 2.782 | 2.864 | 2.930 |

Another variant [10] of a symbiotic system, under investigation in view of applications to CANDU reactors, is shown in Fig. 10.4. A proton accelerator produces, through spalation, neutrons which turn the fertile material into a fissile one. The fissile material is recovered and used in the fabrication of the nuclear fuel that will be loaded in a CANDU-PHWR-type plant. Part of the electric power produced by the turbo-generator is used to feed the proton accelerator.

All the advanced cycles described imply reprocessing of the irradiated fuel, in order to recover the fissile material.

TABLE 10.6 Fuel Cycle Variants for CANDU-PHWRs

| *Cycle* | *Burn-up* (MWd/t) | *Natural uranium requirement for a 30-year operation* (t $U_{nat}$/GWe) |
|---|---|---|
| Natural uranium | 7300 | 4230 |
| Lightly-enriched uranium | 20,800 | 3005 |
| Thorium, with recycling, triggered with highly-enriched uranium | 30,000 | 1640 |
| Thorium, with recycling, triggered with medium-enriched uranium | 30,000 | 1840 |
| Thorium, with recycling, self-sustaining | 10,000 – 15,000 | 870 |

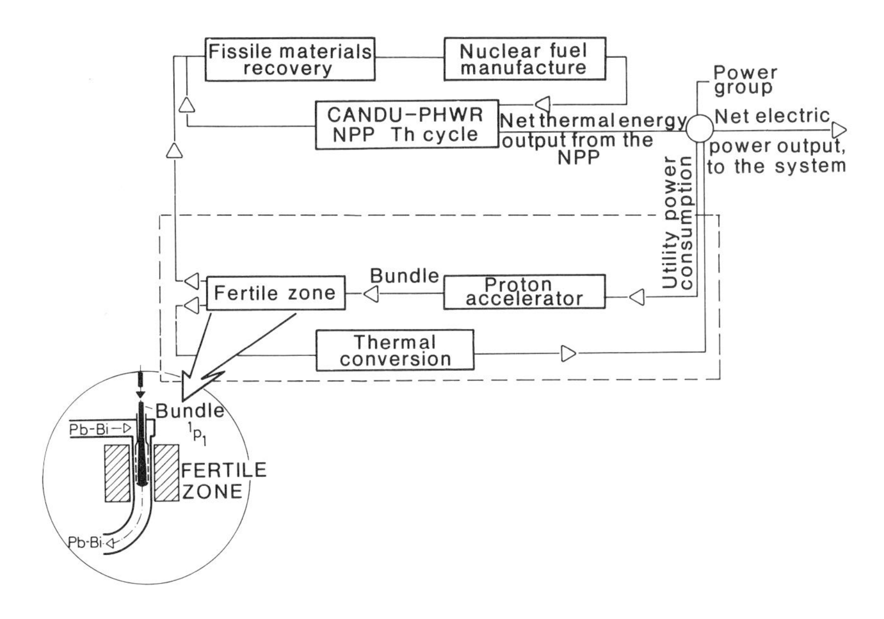


Fig. 10.4. A symbiotic tandem CANDU-PHWR/proton accelerator.

To achieve this, the following requisites are in order:

(a) establishment of technologies that should ensure the efficient recovery of the fissionable material, at commercial scale;

(b) long-term storage of radioactive wastes (especially fission products) resulting from reprocessing. To this purpose, it is necessary that wastes be embedded in matrices that should be stable from the point of view of mechanical, thermal and radiation resistance, then clad, in order to avoid waste dispersion and ensure an appropriate protection of the population and environment.

Studies conducted in the field of nuclear fuel cycle have as an objective the elaboration of pertinent, commercial-scale, methodologies, taking into consideration economics and nuclear security factors, and the provisions of the nuclear weapons non-proliferation treaties now in effect.

The fuel cycle comprises as distinct fields of activity:

1. ore extraction;
2. ore concentration and purification;
3. uranium enrichement (for certain types of reactors);
4. fabrication of fresh and recycled nuclear fuel;
5. fuel-burning, in reactors;
6. storage of radioactive materials;
7. radioactive material transport;
8. recovery of fissile materials by reprocessing of irradiated fuel;
9. treatment and disposal of radioactive wastes.

Figure 10.5 presents schematically these activities for CANDU-PHWRs. In this case, the fuel cycle includes all phases up to the storage of irradiated fuel. It is expected that this cycle be completed in the future.

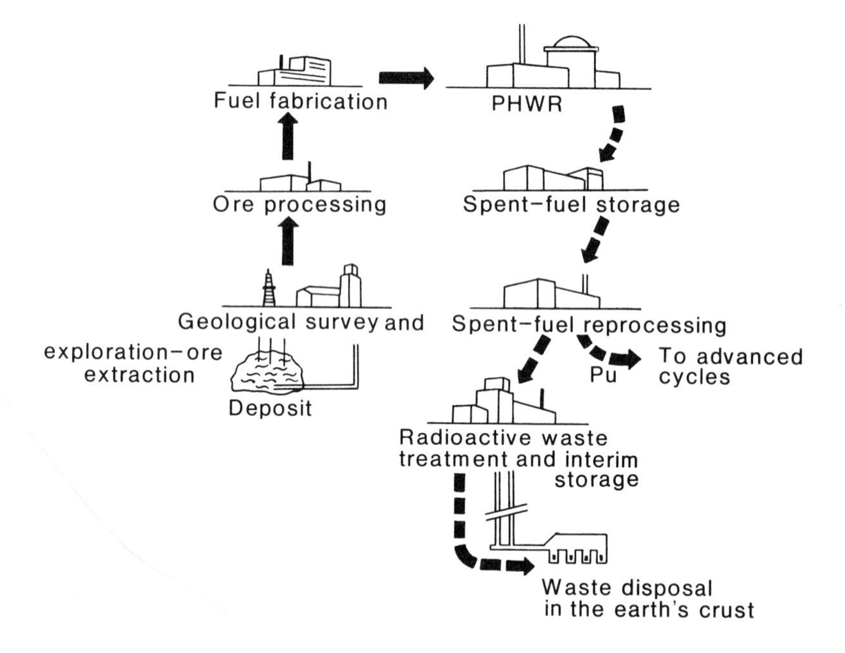


Fig. 10.5. CANDU-PHWR fuel cycle.

The recovery of fissile and fertile materials at industrial scale, by reprocessing, raises special problems because of the risks implied in the handling of large quantities of fission products and actinides. Besides a high $\beta - \gamma$ radioactivity ($10^4 - 10^6$ Ci per irradiated fuel assembly), the irradiated fuel releases important amounts of heat, generated by radioactive self-heating.

The variation in time of radioactivity and thermal energy dissipated per unit-time are indicated in Figs 10.6 and 10.7, for a type of fuel assembly used to date in LWRs. Both the activity and the thermal energy decrease in time according to an $At^{-1.2}$ law, where $t$ is the time counted from the moment of the fission inception.

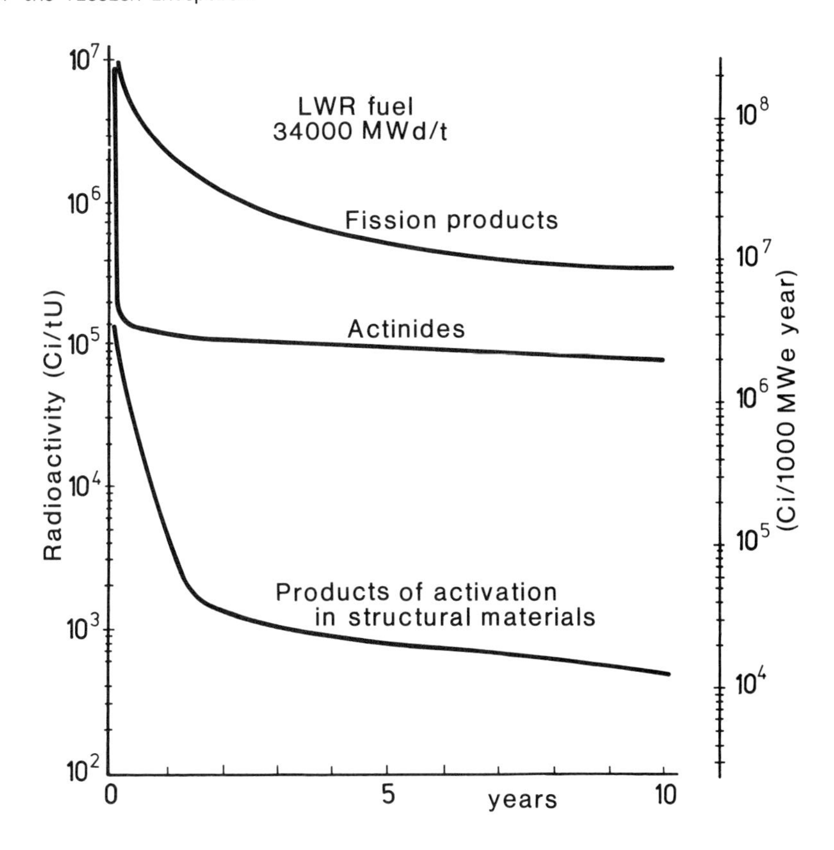


Fig. 10.6. Radioactivity of irradiated LWR fuel after unloading from the reactor [11].

Industrial scale handling of large amounts of radioisotopes and of the associated heat implies the establishment and observance of a detailed check-list that would provide for avoiding:

1. irradiation in excess of the admitted levels, of professional personnel and population;
2. radioactive effluents in the environment, above the admitted limits;
3. overheating of materials, as a result of decays, above the level ensuring the integrity of the geometric shape and the phase ;
4. inadvertent reaching of the critical mass;
5. pollution of the earth's crust (and of the sea) by improper disposal of radioactive wastes;
6. use of recovered fissionable materials with other purposes than those of recycling.

Reprocessing comprises the following operations:

1. bringing the irradiated fuel in a phase that should permit recovery;

2. separation of fissile and fertile materials (heavy elements) from fission products;
3. chemical separation of heavy elements from each other;
4. final purification, in compliance with the recycling requirements;
5. separation, retention and concentration of fission products in view of final disposal.

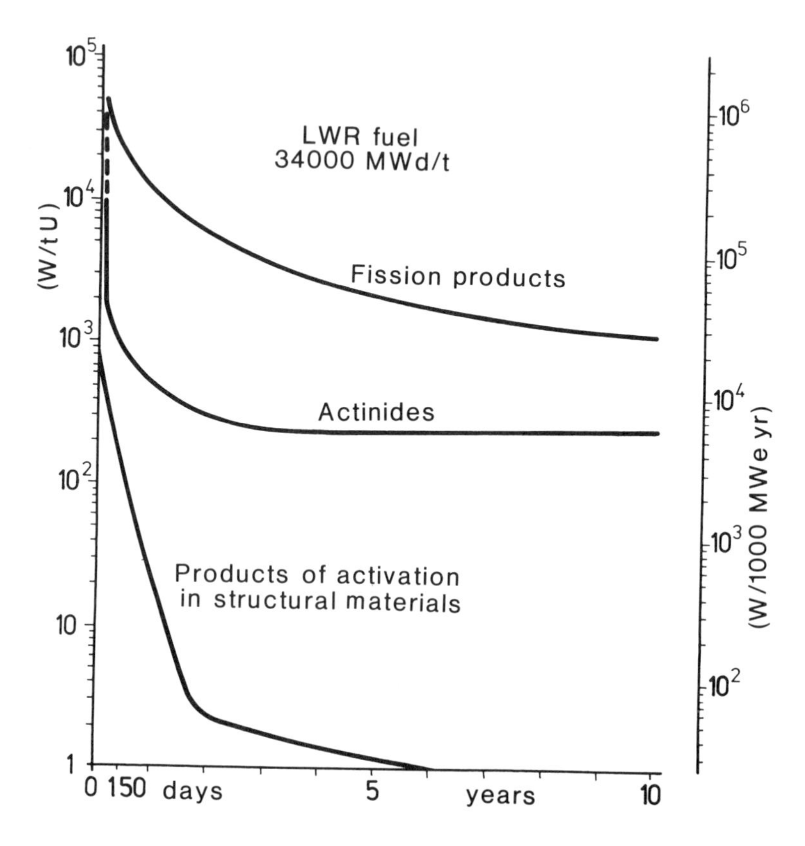


Fig. 10.7. Thermal energy generated by radiative self-heating, in the LWR-irradiated fuel, after unloading from the reactor [11].

The requirements on the radiochemical purification of recovered materials relate to the necessary decontamination level and to the predetermined degree of purity. The first requirement above aims at an easy handling of fuel at fabrication stages, while the second one seeks to reduce to a minimal level the neutron absorption during re-employment of recovered materials in the reactor.

Table 10.7 gives the minimal separation and decontamination ratios needed for the recovered fuel to satisfy the above requirements [2]. The same table also presents the minimal recovery level. Although the present extraction efficiency is only 99 — 99.5%, the requirements pertaining to the reduction of actinides in wastes consisting of fission products imply a future growth in yield up to 99.9% — which is precisely the value given in the table.

TABLE 10.7 Minimal Required Performances, for Fissile Material Recovery Methods

| *Minimal ratios required for:* | |
|---|---|
| Separation of fertile material from the fissile material | $10^7$ |
| Separation of fissile material from the fertile material | $10^6$ |
| Decontamination of fissionable material from fission products | $10^8$ |
| Decontamination of fertile material from fission products | $10^7$ |
| *Recovery level*, of fissile and fertile materials (%) | *99.9* |

Over long storage periods, the radioactivity of wastes is determined by the actinides (especially Np, Am, Cm) as it can be noted from Fig. 10.8.

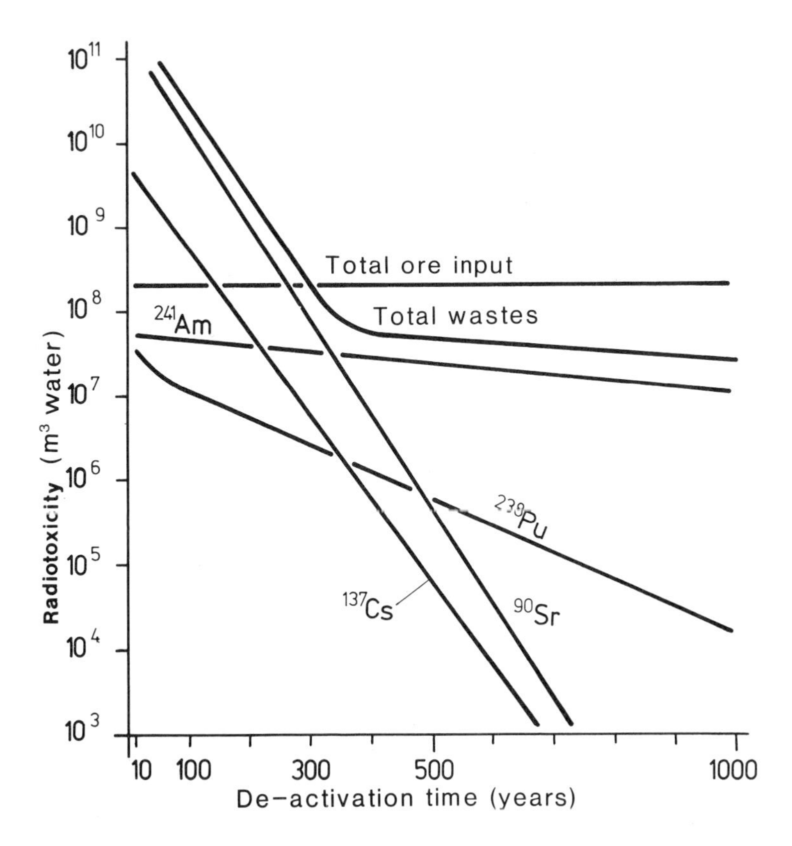


Fig. 10.8. Radiotoxicity of certain radioisotopes contained in the radioactive wastes resulting from retreatment of 1 t of LWR-irradiated fuel [11].

The recovery through reprocessing of fissionable and fertile materials from irradiated fuel is done by chemical methods; for this reason, this activity is also called *chemical processing of irradiated fuel.*

From an inventory management angle, considering the type and level of radioactivity, the half-life and the radiotoxicity, radioactive wastes resulting from reprocessing may be classified as follows:

Group I: Certain isotopes of Ge, As, Br, Rb, Mo, Rh, In, Ba, La, Pr, Nd, Gd, Dy, which, after more than 1 year, turn into stable isotopes, or into isotopes with such a long half-life that they may be found even in nature. Practically, these isotopes could be released into the environment after a 1 year decay period.

Group II: Certain isotopes of Y, Ag, Te, Ce, as well as Mo, Nd and Gd resulting from the decay of Zr, Nb, Ce and Eu; after a period of over 10 years these may also be released into the natural environment.

Group III: Certain isotopes of Sr, Y($^{90}$Sr, $^{90}$Y), Nb, Ru, Sb, Cs, Pm, Sm, Eu and Kr, which need a much longer retention, of more than $10^3$ years.

Group IV: the radioisotopes $^{79}$Se ($6 \times 10^4$ yr), $^{107}$Pd ($7 \times 10^6$ yr), $^{126}$Sn ($2 \times 10^5$ yr), $^{129}$I ($1.6 \times 10^7$ yr), $^{99}$Tc ($2.12 \times 10^5$ yr), and transuranian elements like Am, Cm, Np and Pu, that need permanent retention.

Groups III and IV require cooling, even after a retention of 10 years.

Data shown in Fig. 10.8 emphasize the major contribution of the transuranian elements and other fission products like $^{137}$Cs and $^{90}$Sr to global radiotoxicity. The recovery of $^{237}$Np, Am, Cm, $^{137}$Cs and $^{90}$Sr, as well as their utilization in industry, would simplify the problem of radioactive waste management. Futher on, several possible utilizations of these isotopes will be suggested, that would support the notion of their recovery.

$^{237}$Np — with a half-life of $2.14 \times 10^6$ yr and a specific activity of $2.6 \times 10^{10}$ Bq . kg$^{-1}$, is $\beta$-active and has a spontaneous fission rate of 1.39 fissions . s$^{-1}$ . kg$^{-1}$.

The importance of the recovery of this isotope when reprocessing irradiated fuel has increased with the development of space programmes. $^{237}$Np is the best intermediate product in obtaining pure $^{238}$Pu which, to date, seems to be the best energy source for space applications.

After recovery and purification $^{237}$Np is turned into $NpO_2$, which, subsequently irradiated in the reactor, leads to the generation of $^{238}$Pu.

$^{238}$Pu — with a half-life of 86.41 yr and a specific activity of $6.45 \times 10^{14}$ Bq/kg. The powerful specific activity of $^{238}$Pu recommends it as an intense source of $\alpha$-radiations. It is used as $PuO_2$. Among the most interesting applications of $^{238}$Pu are:

1. ($\alpha$,n) neutron sources, with a conversion efficiency of 70 neutrons/ $10^6$ disintegrations;

2. compact heat sources of high specific power (420 W), used for heating cosmic diver suits;

3. electric power sources based on thermoelectric or thermodynamic conversion of energy (the specific energy released is 0.56 W . g$^{-1}$), to be used in health care and space missions that demand power in the range from a few $\mu$We to several kWe.

$^{241}Am$, $^{243}Am$, $^{242}Cm$, $^{244}Cm$ — are found in very small amounts in the irradiated fuel (see Chapter 10.1). Among potential applications of these are:

1. autonomous heat sources, using the α radiations of $^{242}Cm$ and $^{244}Cm$;
2. (α,n) neutron sources, on $^{241}Am$;
3. fire detectors, relying on the α radiations of $^{241}Am$;
4. production of $^{238}Pu$ (by neutron irradiation of $^{241}Am$, which generates $^{242}Cm$, which turns into $^{238}Pu$ through α-decay).

## 10.3. CLASSIFICATION OF IRRADIATED FUEL REPROCESSING METHODS

As already mentioned in the preceding section, reprocessing of irradiated fuel is done by chemical methods. The principles of separating chemical elements through reprocessing are similar to those employed in standard analytical chemistry.

Any separation stage in analytical chemistry presupposes the existence of two physically distinct phases. The separation stages are established in such a way that the concentration of the material of interest should increase permanently in one of the phases, on the basis of its withdrawal from the other phase. Therefore, a first classification of reprocessing methods could be based on the nature of these phases, i.e.:

(a) solid — liquid (precipitation, ionic exchange, electrolytic refining);
(b) solid — gas (sublimation);
(c) liquid — liquid (solvent extraction);
(d) liquid — gas (fractional distillation, volatilization).

From the point of view of industrial-scale chemical tehnologies one may distinguish:

1. Aqueous procedures, in which the materials are turned into salt solutions, then the substance of interest is extracted through one of the following methods:

   (a) precipitation;
   (b) solvent extraction;
   (c) ion exchange.

2. Pyrometallurgical procedures, in which the materials are first turned to their liquid state by melting, then the materials of interest are extracted through one of the following methods:

   (a) electrolysis;
   (b) evaporation under vacuum;
   (c) extraction with molten metals.

3. Pyrochemical procedures, in which the materials are brought into contact with a chemical agent producing the selective evaporation of the substances of interest.

4. Composite procedures, implying combinations of the above.

A review of these methods, together with their performance, their respective merits and drawbacks, is presented in detail in handbooks on plutonium chemistry [1,2].

Originally, at the inception of the recovery of plutonium from irradiated fuel, that is toward the end of World War II, the interest in plutonium was exclusively of a military nature. The Pu — U separation for the production of the nuclear bombs of 1945 was based on an aqueous precipitation method with bismuth phosphate. The small burn-up of the irradiated uranium ensured the production of sufficiently pure $^{239}Pu$ and permitted a simple technical procedure.

Besides these procedures, aqueous procedures based on solvent extraction were evolved with the development of uranium-fuelled power reactors, the knowledge gained in the reprocessing of reactor fuels permitting the industrial and commercial scaling-up of solvent extraction methods. These methods, which perform the separation and purification of uranium and plutonium from fission products, are much better than precipitation with bismuth phosphate. The thrust towards increasing efficiency in the recovery of fissionable materials stimulated the search for simpler technologies than the aqueous ones, such as pyrochemical and pyrometallurgical processes, especially in view of paving the way for the potential processing of the irradiated fuel coming from future nuclear power plants equipped with fast breeder reactors. Although laboratory, or small-scale pilot-test, facilities have given satisfactory results, economic considerations have stopped research and development. In aqueous procedures the fuel is turned into liquid state by chemical solution in a mineral acid, ordinarily $HNO_3$, used as a salting agent. In any of the solvent extraction methods the reprocessing begins with the preparation of a Pu, U and fission products solution, with $HNO_3$.

Starting from the scope of this chapter, a discussion of the physical and chemical bases of the solvent extraction processes, and of their implementation at industrial scale, is in order.

## 10.4. SOLVENT EXTRACTION

### 10.4.1. Basic Principles of Metal Separation by Extraction

By stirring together the aqueous solution of a metal salt and an organic solvent immiscible with water, the metal salt or one of its constitutive ions is redistributed between the two liquid phases. When equilibrium is reached, the potentials of the solute in the two phases become equal, and their concentrations, $C_{aq}$ and $C_{org}$, no longer change.

If the concentration of the solute in aqueous solution is sufficiently low after equilibrium is reached, the relation between its final concentrations in the two phases is one of proportionality:

$$\frac{C_{org}}{C_{aq}} = D \qquad (10.1)$$

Here $D$ is the *distribution, or partition, coefficient*.

The relation between the different concentrations in the aqueous phase and their correspondents in the organic phase is rendered graphic by the equilibrium isotherm — Fig. 10.9. The slope of the curve represents the ideal value of $D$. In practice the magnitude of $D$ depends on pressure, temperature, nature of phases in contact and, of course, on the nature of the solute.

The proportion in which a certain constituent $A$, in aqueous phase, can be extracted by a single equilibrium step with an organic phase depends, besides the distribution coefficient $D_A$, also on the volume ratio of the two phases, and, respectively, on their volume flows on the assumption of a continuous flow extraction. By increasing the proportion of organic solvent, the extraction efficiency of the $A$ constituent increases, to the detriment of its final concentration in the organic phase.

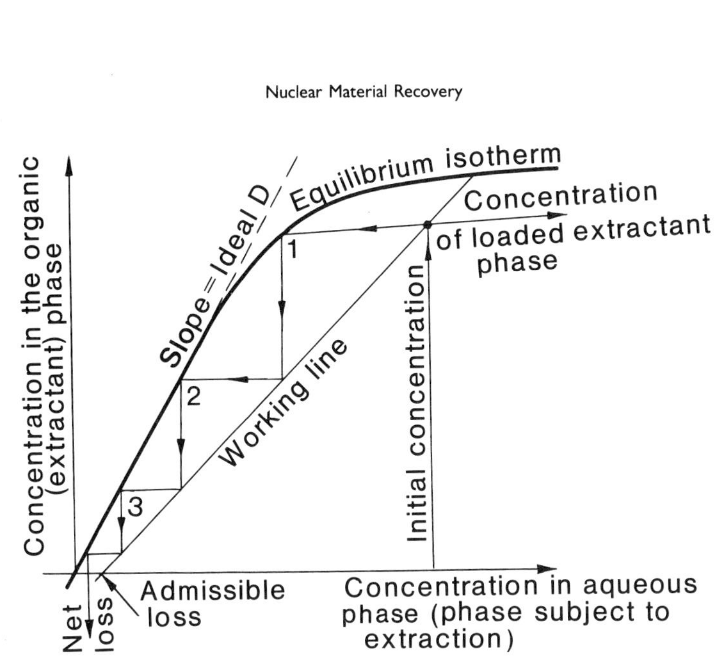


Fig. 10.9. Equilibrium isotherm [13].

Two constituents, $A$ and $B$, to be found simultaneously in aqueous solution are extracted in the organic phase with different efficiencies if their distribution coefficients, $D_A$ and $D_B$ are different. During extraction, the organic phase should be enriched in the solute whose distribution coefficient has a larger value. The efficiency of the separation through one equilibrium step is given by the value of the separation factor $\alpha = D_A/D_B$. A complete separation supposes the exclusive and exhaustive extraction of one of the consituents, be it $A$ or $B$.

Achievement of such a separation through a single-phase equilibration is a rare phenomenon; generally, the desired separation degree can be obtained by repeating the phases equilibration operation, and it is conducted in multistage extractors, in which the two phases move countercurrently. The determination of the number of stages necessary for an extraction battery can be done through one of the following methods [12-14]:

(a) The graphic method (McCabe-Thiele). In Fig. 10.9 nearing the equilibrium isotherm is the working straight line. Its crossing with the abscissa gives the magnitude of the admitted losses in the raffinate. In this way a working range is outlined, in which a step function can be traced, starting from the value of the initial concentration of the desired solute in the aqueous phase. This gives the required number of stages, as well as the concentration of the solute of interest as extracted from the solvent on each stage. Actually the aqueous solution contains a chemical compound of several elements that have mutual effects upon their thermodynamic equilibria, thus modifying the distribution coefficients. To $p$ chemical compounds dissolved together in water would correspond $p$ distribution coefficients $D_1, D_2, D_3, \ldots, D_p$, of the form:

$$\frac{C_{B,i}}{C_{A,i}} = D_i(T, C_{A,1} C_{A,2}, \ldots, C_{A,i}, \ldots, C_{A,p}) \qquad (10.2)$$

where $T$ is the working temperature.

(b) Numerical methods. Computer programs provide general methods to simulate the operation of extraction devices. For the solvent extraction this implies:

1. description of thermodynamic relations at phase equilibrium;
2. solving the non-linear equations that describe the material balance for a system comprising a given number of stages.

Problem No. 1 can be solved by giving the computer an adequate input concerning the thermodynamic equilibrium of phases, at different concentrations.

With this information displayed in tables, or implied in mathematical functional expressions, the system of equations featuring the material balance reads:

$$\begin{aligned} -a^i_{11}x^i_1 + a^i_{12}x^i_2 &= -A^i_1 \\ a^i_{21}x^i_1 - a^i_{22}x^i_2 + a^i_{23}x^i_3 &= A^i_2 \\ \cdots\cdots\cdots\cdots\cdots\cdots & \\ a^i_{n,n-1}x^i_{n-1} - a^i_{nn}x^i_n + a^i_{n,n+1}x^i_{n+1} &= A^i_n \end{aligned} \qquad (10.3)$$

where $x^i_l$ is the concentration of the $i$th material in one of the phases on the $l$th stage and $a^i_{kl}$ are coefficients that depend on the $x^i_l$ concentrations.

To solve system (10.3), standard programs in any computer library are available.

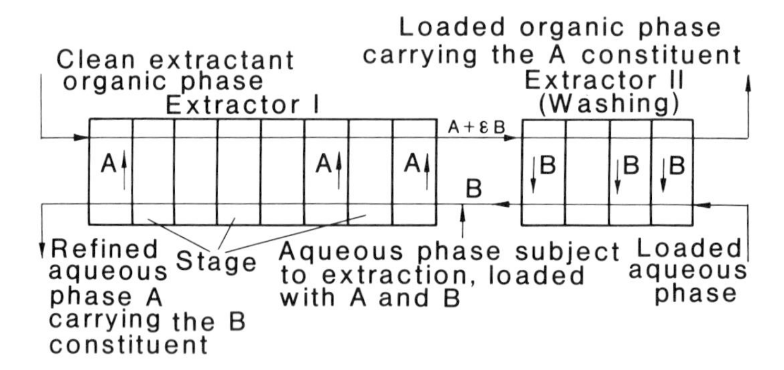


Fig. 10.10. Merger of two extractors to separate two constituents, $A$ and $B$ [13].

If the aqueous phase contains two constituents $A$ and $B$, where $A$ is the substance of interest and $B$ is the material to be evacuated, with extraction efficiencies $D_A >> D_B$, a minute fraction of $B$ will allways be extracted together with $A$. By directing the solvent containing both $A$ and traces of $B$ as evacuated from the first extractor into a second extractor, where an aqueous solution which does not contain the $A$ and $B$ phases circulates countercurrently, the $B$ constituent can be rejected back to the first extractor. By multiplying the number of stages, such a tandem of extractors, coupled as shown in the draft in Fig. 10.10 allows highly efficient separation. The reverse operation, effected according to the same principles, and in the same devices, but with the roles of the liquid phases reversed, is called re-extraction. Figure 10.10 gives a layout of a separation installation with two extraction devices, one being the extractor, the other the re-extractor, with the purpose of eliminating traces of $B$ that accompany the $A$ product to be separated.

### 10.4.2. Nature and Chemical Function of Extraction Solvents

In a two-phase system (aqueous and organic) the preferential distribution of a mineral solute in the organic phase is the result of a strong chemical interaction of the solvent with the solute. By nature of this interaction, the extraction phenomena can be divided into four categories:

1. solvation extraction;
2. cation exchange extraction;
3. anion exchange extraction;
4. chelation extraction.

10.4.2.1. Solvation extraction. Oxygen atoms of certain organic compounds may engage an electronic doublet in a coordinative chemical bond with protons or with certain metal atoms. This solvating potential of the oxygen atoms confers to the organic molecule extractive properties with regard to acids and electrically neutral metal salts.

In a system that comprises the organic extractant $E$, the aqueous phase containing the $m$-charged cation, $M^{m+}$, and the $X^-$ anion, the chemical reaction resulting in the formation of the complex compound through solvation reads:

$$M^{m+}_{aq} + mX^-_{aq} + E_{org} \rightleftharpoons EMX_{m.org} , \tag{10.4}$$

or, in a coordinative bond with a proton:

$$H^+_{aq} + X^-_{aq} + E_{org} \rightleftharpoons EHX_{org} . \tag{10.5}$$

From this expression it appears that, by increasing the concentration of the extractant $E$ or by adding $X^-$ anions to the aqueous phase, the extraction efficiency will increase, because there will be a greater probability that $M^{m+}$ engage in a reaction evolving a complex. The equilibrium constant and the distribution coefficient are, respectively, given by:

$$K_M = \frac{[EMX_m]_{org}}{[M^{m+}]_{aq}[mX^-]_{aq}[E_{org}]} \tag{10.6}$$

and

$$D_M = \frac{M_{org}}{M_{aq}} = \frac{[EMX_m]_{org}}{[M^{m+}]_{aq}} , \tag{10.7}$$

The organic solvating compounds have an important role in nuclear industry, where the oxygen atom is linked to the carbon frame through phosphorus (phosphates, phosphonates, phosphinates, etc.). The most widely used in the reprocessing of irradiated fuel is $n$-tributyl phosphate, known as TBP.

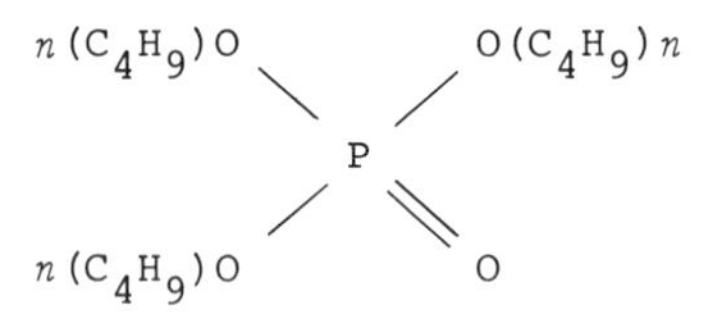


Besides TBP, one can also mention the dibutylcarbitol (dibutyldiethyleneglycol) and the hexones (methylisobutylketone — MIBK) which played an important part in the development of solvent extraction procedures.

10.4.2.2. Cation exchange extraction. In this case the formation of a complex compound results from the cation exchange featuring the reactions between organic acids and an aqueous solution:

$$M^{m+}_{aq} + mRH_{org} \rightleftarrows MR_{m\ org} + mH^{+}_{aq} \tag{10.8}$$

where $M^{m+}$ is an $m$-valency cation and $RH$ is the organic acid.

The equilibrium constant and the distribution coefficient are, respectively:

$$K_M = \frac{[MR_m]_{org}\ [H^+]_{aq}}{[M^{m+}]_{aq}\ [RH^m]_{org}} \tag{10.9}$$

and

$$D_M = \frac{M_{org}}{M_{aq}} = \frac{[MR_m]_{org}}{[M^{m+}]_{aq}} . \tag{10.10}$$

10.4.2.3. Anion exchange extraction. Quaternary amine salts with big molecular masses and whose $R_4N^+$ amine cations show a strong base character are likely to give their anions in exchange for those of the acid, neutral or even alkaline aqueous phases with which they come in contact.

In an acid medium, the nitrogen atom of the amines captures a proton

$$R_3N + H^+ \rightleftarrows R_3NH^+$$

The $R_3NH^+$ ion cannot subsist in an alkaline medium and, therefore, amines can be employed only in the extraction from acid or slightly acid aqueous solutions, as they behave like liquid anion exchangers, according to the following reaction patterns:

$$R_3N_{org} + HX_{aq} \rightarrow R_3NH^+X^-_{org}$$

$$R_3NH^+X^-_{org} + A^-_{aq} \rightleftarrows R_3NHA_{org} + X^-_{aq} .$$

The extraction of a metal can be done only to the extent that its nature or the nature of the chemical substances present in aqueous solution favour its entering a negative charge complex:

$$M^{m+} + nX^{-} \rightleftarrows MX_{n}^{(n-m)}, \tag{10.11}$$

when the complex forming reaction reads:

$$(n-m)R_4N^{+}_{org} + MX^{(n-m)}_{n,aq} \rightleftarrows (R_4N)_{n-m}X_nM_{org}. \tag{10.12}$$

In the resulting organic complex the metal $M$ has the coordination number $(2n-m)$.

Among the heavy amines with good prospects for use in the reprocessing of irradiated fuel is the trilauryl amine:

$$\begin{matrix} CH_3(CH_2)_{11} \\ CH_3(CH_2)_{11} \\ CH_3(CH_2)_{11} \end{matrix} > N$$

10.4.2.4. Chelation extraction. In a system of extraction by chelation the extractant molecule functions at the same time as cation exchanger and as solvent. It must possess, therefore, on the one hand, a slightly acid functional group ( — OH, — SH) and, on the other hand, an electron donor atom (generally oxygen or nitrogen). While the hydrogen atom is replaced by a charge of the metallic cation to be extracted, which leads to the saturation of the electro-valencies, the cation is simultaneously solvated by the donor atom — a phenomenon accompanied by the saturation of the coordination positions.

To secure electric neutrality, the resulting complex must contain a number of chelating molecules equal to the number of electric charges, $m+$, of the $[M^{m+}]_{aq}$ metallic cation.

As in the case of cation exchange, metals are better extracted if their electric charge and hydrolytic capacity are larger.

The property of the chelating substances to accomplish simultaneously the functions of solvent and of cation exchanger acquires a remarkable character when the coordination number of the metal to be extracted is equal to the double of its electric charge. In such circumstances the partners' electro-valencies and coordinations satisfy each other totally. The anhydrous chelate thus formed is particulary stable, and also soluble in organic solvents, thus faciliting extraction.

Considering the weak-acid nature of the chelatants, one finds that markedly-acid aqueous solutions do not favour extraction, but are well suited for re-extraction.

### 10.4.3. Solvent Diluents and their Role

Organic compounds with extractive properties are sometimes in a physical state that is improper to their use in extraction processes. Thus, some are solid, others are liquid, with a high degree of viscosity or with a density close to that of water.

In order to overcome such inconveniences the extractants are dissolved or diluted in a virtually non-inflammable hydrocarbon with a high chemical inertia and which exhibits physical properties favourable to emulsion flow and decanting. Dilution, or dissolution, of the extractant creates the possibility of bringing the concentration to a level corresponding to the optimal extraction efficiency and selectivity during extraction.

The physical qualities which a good diluent should possess are:

1. a departure from the density of the aqueous solution, of 0.12;
2. a boiling temperature of 440 K;
3. solubility in water (0/oo): a few units;
4. interface tension with water: $10^{-2}$ N/m.

### 10.4.4. Solvent Stability

The chemical composition of extractants and diluents, and consequently their physical and chemical properties, are modified under the action of chemical reactants and ionizing radiations. The extent of these phenomena and the effects due to the fragments resulting from breaking of chemical bonds depend on the nature of the extractants and diluents.

Neutral phosphorus compounds and amines present an excellent stability in a radiation field. The hydrolysis of phosphorus compounds is slow; it is sensitive to the amount of ionizing radiations absorbed. Generally, the hydrolysis of these compounds can be slowed down by using aromatic diluents, thus, it also depends on the concentration of extractants in the diluents. Ammonia salts and heavy amines have a lower chemical inertia and a better stability under irradiation than the phosphorus compounds.

Regeneration of worn phosphorus and amine extractants, in order to eliminate all acid product, is done by washing with alkaline agents.

## 10.5. URANIUM AND PLUTONIUM IN AQUEOUS SOLUTION

In solvent extraction procedures, $HNO_3$ is used in order to transform $UO_2$ in aqueous solution [1-3,12]. The dissolution of irradiated $UO_2$ is done, for example, in a $HNO_3$ solution (4.5 M) in a proportion of 3 l/kg $UO_2$. In these conditions the dissolution rate is $5 \times 10^{-3}$ mol $UO_2$/l.min. For a complete dissolution of $UO_2$ it is necessary to introduce in solution 20% excess oxygen. When nitrogen oxides in the form of $HNO_3$ are recovered and then recycled in the process, the dissolution reaction reads:

$$UO_2 + 2HNO_3 + \frac{1}{2} O_2 \xrightarrow{\text{boiling}} UO_2(NO_3)_2 + H_2O$$

At the same time, plutonium and the fissile products are transferred into the aqueous solution.

In aqueous solution, uranium presents ionic species with valencies between +3 and +6: $U^{3+}$, $U^{4+}$, $UO_2^+$ and $UO_2^{2+}$, the last one being the so-called uranyl ion. $U^{3+}$ and $UO_2^+$ species are unstable in aqueous solutions, and $U^{4+}$ and $UO_2^{2+}$ are easy to form complexes with organic compounds.

Plutonium behaves in a similar manner; in the same conditions it presents the same ionic varieties, with valencies between +3 and +6. These are $Pu^{3+}$, $Pu^{4+}$, $PuO_2^+$, $PuO_2^{2+}$; the last one is known as the plutonyl ion. All Pu ions subsist in aqueous state, and $Pu^{4+}$ and $PuO_2^{2+}$ are easy to form complexes with organic compounds, just like uranium.

As a result of dissolution nitrates are formed. The most important are:

$$UO_2^{2+} + 2\ NO_3 \rightleftharpoons UO_2(NO_3)_2$$

$$Pu^{4+} + 4NO_3 \rightleftharpoons Pu(NO_3)_4$$

$$PuO_2^{2+} + 2NO_3 \rightleftharpoons PuO_2(NO_3)_2.$$

In solvent extraction procedures the separation of uranium and plutonium from fissile products starts from the ionic varieties which lead extensively to complexes with the organic compound; as for the separation of uranium from plutonium, it starts from that particular Pu ionic variety that virtually does not react during the formation of the complex, while the uranium is in the ionic state that is the most appropriate to forming a complex with the organic compound.

## 10.6. SOLVENT EXTRACTION PROCESSES

At industrial scale, recovery of fissile and fertile materials from irradiated nuclear fuel is achieved almost exclusively by solvent extraction. This ensures a high recovery efficiency and confers a high degree of purity to the material.

Among the commercially proven extraction processes are those based on organic complex formation through solvation: Redox, Butex, Purex and Thorex. The Purex process is currently employed in industrial reprocessing facilities like those at Windscale (England), La Hague (France), and Barnwell (USA).

In the following sections a general description of the extraction processes is given. For further information, the reader is referred to Refs 1-3.

### 10.6.1. The Redox Process

Redox process employs as solvent non-diluted methylisobutylketone (hexone). The process is based on the co-extraction of uranium and plutonium in the $UO_2^{2+}$ and $PuO_2^{2+}$ ion state, followed by re-extraction of plutonium by selective reduction [4]. Due to hexone's lability in the presence of nitric acid even in moderate concentrations, it is necessary to use a supplementary salting agent in the form of a metal nitrate.

The solution containing uranium, plutonium and fissile products, resulting from dissolution, is treated with an oxidizing agent ($Na_2Cr_2O_7$) in order to bring all plutonium to the form $PuO_2^{2+}$:

$$3Pu^{4+} + Cr_2O_7^{2-} + 2H^+ \rightarrow 3PuO_2^{2+} + 2Cr^{3+} + H_2O.$$

Pu re-extraction is then made possible, by reduction with a reducing agent (ferrous sulphamate) to the $Pu^{3+}$ state. The resulting solutions, separately containing uranium and plutonium, must be concentrated and then purified. The most efficient purification is done by ion exchange.

### 10.6.2. The Butex Process

The Butex process employs two solvents: non-diluted dibutylcarbitol (butex) to separate uranium and plutonium from fissile products (co-decontamination) and then for purifying uranium; and a 20% TBP solution in kerosene for plutonium purification. Butex efficiently extracts $UO_2^{2+}$, $PuO_2^{2+}$ and $Pu^{4+}$ and, to a lesser extent, $Pu^{3+}$. It is stable in the presence of nitric acid, thus qualifying for use as a salting agent.

Reduction of plutonium to the $Pu^{3+}$ state, for the separation from uranium by re-extraction, is also done with ferrous sulphamate as reducing agent.

### 10.6.3. The Purex Process

The Purex process uses in all phases a solution of TBP in a diluent (e.g. kerosene) as extractant, and $HNO_3$ as salting agent.

Plutonium is most efficiently extracted in the $Pu^{4+}$ state. To this effect the Pu, U and fissile products solution is treated with a reducing agent ($NaNO_2$ or gaseous $NO_2$) in order to turn plutonium into a tetravalent state:

$$PuO_2^{2+} + NO_2^- + 2H^+ \rightarrow Pu^{4+} + NO_3^- + H_2O.$$

The first phase of the process consists in separating uranium and plutonium from fissile products; then an immediate plutonium re-extraction from TBP follows, by reduction to the trivalent state with an aqueous solution of ferrous sulphamate:

$$Pu^{4+} + Fe^{2+} + NH_2SO_3^- \rightarrow Pu^{3+} + Fe^{3+} + NH_2SO_3^-.$$

Ferrous sulphamate reduction presents the inconvenience of introducing iron and the sulphamate radical in solution, which may cause fast equipment corrosion.

As an alternative, instead of ferrous sulphamate $U^{4+}$ is used as reducing agent, in a 1.5 — 2 M $HNO_3$ solution stabilized with hydrasine in order to prevent the re-oxidation of $Pu^{3+}$ to $Pu^{4+}$ by the $HNO_3$ present in solution.

After separation of uranium from plutonium their purification cycles follow.

Uranium re-extraction from the organic phase is done by using lightly acid $HNO_3$ solutions.

At present the Purex process is the most efficient among aqueous technologies, and has surpassed in performance the Redox and Butex processes due to the special TBP characteristics. Outstanding merits of Purex versus the Redox process are:

1. the fact that the lower volatility and higher ignition temperature of TBP make it safer to use than the hexones employed by Redox;

2. the fact that $HNO_3$ as a salting agent can be recovered by distillation, thus reducing the amount of liquid wastes and vehiculated materials;

3. the quality of the process of being particularly flexible; indeed, it can be adjusted to meet certain specific requirements.

### 10.6.4. The Thorex Process

The Thorex process is employed in reprocessing thorium fuel, in order to separate thorium and uranium, respectively, from fissile products. It involves two or more extraction and re-extraction cycles, followed by final purification by ion exchange.

A variant of this process aims at the extraction of thorium, uranium and plutonium from fuel elements in which plutonium was used as triggering (and/or additional) fissile material. The process is potentially interesting for CANDU-PHWRs that would employ the thorium fuel cycle. After decontamination, thorium, uranium and plutonium are separated from the other fissile products. The solution containing actinides is concentrated and then, by addition of ferrous sulphamate, the plutonium is brought to the (III) valency state. This allows its separation from thorium and uranium. Finally, uranium is separated from thorium [14].

## 10.7. EQUIPMENT FOR SOLVENT EXTRACTION

The development of extraction equipment for aqueous solution — solvent diphasic distribution has followed step by step the developments with the extraction processes. The solvent extraction equipment must meet certain basic requirements, such as:

(a) assurance of a good mass transfer between the two phases;
(b) use of simple and reliable designs;
(c) equipment should be compact and the phases' contact duration must be short;
(d) equipment should be easily adaptable to stepwise processes (in the form of countercurrent cascades);
(e) equipment should provide for expeditious maintenance;
(f) low capital costs and operating expenses.

The extraction equipment most frequently encountered in the industrial enterprise comprise:

(a) continuous exchange extractors: columns; centrifugal extractors;
(b) separate stage extractors: mixer – settlers.

### 10.7.1. Continuous Exchange Extractors

These extractors appear either as vertical columns of various types, with or without mechanical stirring, or in compact form, as centrifugal extractors.

Generally, for continuous exchange extractors two characteristic parameters are defined [13]:

1. the transfer unit height (TUH);
2. the transfer units number (TUN).

The product of the two determines the column height, $Z$.

In order to establish the values of the TUH and TUN parameters one must start from the diagram in Fig. 10.11.

The charged aqueous phase transfers the solute to the solvent. The specific material flux is given by the equation:

$$\varphi = K_S(Y_1 - Y) = K_A(X - X_1), \qquad (10.13)$$

where $K_A$ and $K_S$ are the transfer conductances for each phase (length/unit-time), and $Y_1$ and $X_1$ are the concentrations in the two phases at equilibrium (mols/unit-volume).

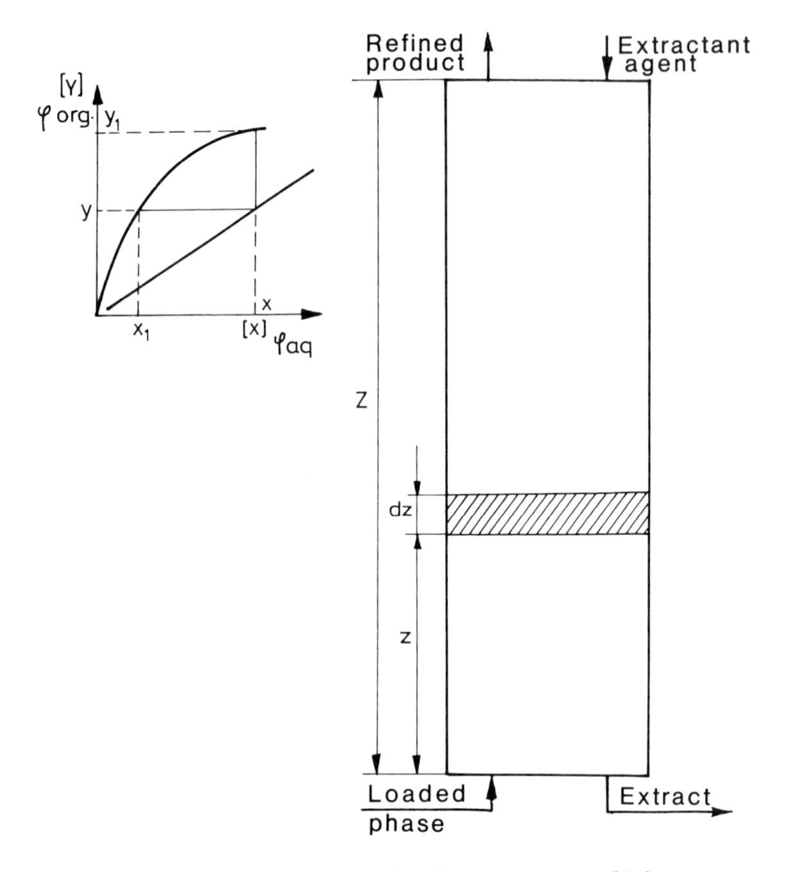


Fig. 10.11. Diagram to determine TUH and TUN [13].

In the element $dz$ the flux of transferred material is:

$$d\Phi = A\ dX = S\ dY \ , \tag{10.14}$$

where $A$ and $S$ are the volume flows of the aqueous phase and of the solvent, respectively.

Alternatively, the expression for the material flux is:

$$d\Phi = \varphi a \Omega\ dz \ , \tag{10.15}$$

where $a$ denotes the specific exchange area ($m^2/m^3$) and $\Omega$ is the column cross-section area:

$$K_S(Y_1 - Y)a\Omega\ dz = S\ dY \ . \tag{10.16}$$

Integrating over the column length one obtains:

$$Z = \frac{S}{K_s a\Omega} \int_{Y_0}^{Y_z} \frac{dY}{Y_1 - Y} . \qquad (10.17)$$

Repeating the procedure for the aqueous phase, it results:

$$Z = \frac{A}{K_A a\Omega} \int_{X_0}^{X_z} \frac{dX}{X_1 - X} . \qquad (10.18)$$

The first factor in (10.17) and (10.18), which has the dimension of a length, defines the TUH, while the second, which defines a non-dimensional factor, stands for the TUN. Their product gives the effective column height $Z$.

The transfer unit height depends on the chemical system in question ($K_A$), and on the operating conditions of the device ($K_A a$). Factors $K_A$ and $a$ are difficult to calculate; there is no general method for determining them; their evaluation is done experimentally.

An important problem in the design of industrial continuous extractors is the determination of their dimensions. The extrapolation of TUH to large extraction columns using the evaluations done on experimental models proves useful only in a limited number of cases. Scaling-up the diameter, that comes with the scaling-up of the overall sizes, leads to the TUH modification, as a consequence of the axial diffusion phenomenon which tends to equalize the material concentrations in each phase along the column axis.

### 10.7.2. Separate Column Plants

These installations comprise several ordinarily identical stages, each of them equipped with a stirring device actuated by a turbine or another kind of engine, and a settler where the separation of phases takes place. As far as decanting, two types of devices can be distinguished; the natural decanting, and the centrifugal decanting type.

The equipment most frequently used in current solvent extraction is of the mixer — settler type. Generally these present a horizontal layout, and are grouped in batteries.Pumping and mixing of phases is ensured by a turbine, and the natural phase decanting is done (in most cases) at a pace of the order of minutes. Figure 10.12 shows a section through a mixer — settler stage. Extraction batteries are obtained by cascading such elementary stages. A drawback with these installations is the large decanting areas required, which entails:

1. a long stagnation time in decanting, which gives way to chemical and radiolytic solvent degradation;

2. the immobilization of large quantities of material and solvent;

3. long response time at various perturbations.

In order to avoid such inconveniences, the centrifugal decanting method was tried. To this purpose, single-stage as well as multi-stage equipment was conceived.

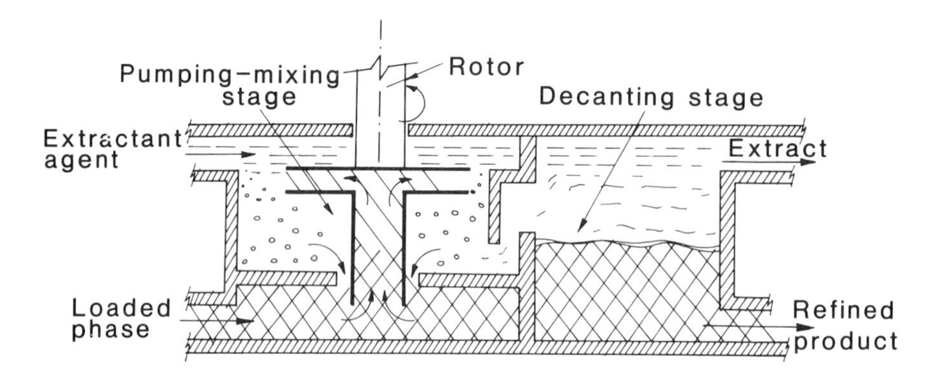


Fig. 10.12. Section through a mixer — settler [13].

### 10.7.3. Selection Criteria for Extraction Equipment

Factors that may influence the choice of the extractor type are [13]:

1. The physical and chemical characteristics of the solvent-solution system. The more important parameters are:

   (a) the density difference between the two phases;
   (b) the viscosity of the two phases;
   (c) the interface tension;
   (d) the separation characteristics (distribution coefficient, solvent saturation);
   (e) the transfer kinetics.

The first three of them determine whether the dispersed phases coalesce, whereas the last two determine the performance with the material transfer.

2. The extraction devices performance, particularly as far as:

   (a) flow admitted in the extractor (expressed, for continuous extractors, by the specific flow);
   (b) device efficiency.

## 10.8. INDUSTRIAL-SCALE IMPLEMENTATION OF THE SOLVENT EXTRACTION METHODS FOR REPROCESSING OF THE IRRADIATED NUCLEAR FUEL RESULTING FROM NUCLEAR POWER PLANTS

### 10.8.1. General

As indicated already, the industrial scale U and Pu extraction from irradiated fuel is generally performed by the Purex process, which employs a solution of n-tributyl phosphate diluted in a hydrocarbon as extracting agent. This reprocessing procedure was developed soon after World War II, and rapidly expanded to replace gradually other reprocessing procedures, that were:

1. plutonium precipitation with bismuth phosphate — a process used at the Hanford Laboratories (USA), in 1945, with the purpose of separating this fissile material for the manufacture of the first atomic bombs;

2. the hexone extraction process (Redox) and the dibutyl carbitol extraction process (Butex) experimented in the USA and Europe in the 1950s.

The beginnings of Purex date back in 1954, when it was introduced as reprocessing procedure at the Savannah River (USA) and Marcoule (France) facilities. At present this process is currently employed in the W.A.K. pilot reprocessing facility (FR of Germany), J.A.E.R.I. (Japan), Eurochemique (Belgium), and in the commercial facilities at Windscale (UK) and La Hague (France). It is expected that the process will be used also in the commercial facilities of over 1000 t . $yr^{-1}$, commissioned for the present decade.

Certain characteristics make Purex outstanding in comparison with other solvent extraction methods; as said already, these regard mainly:

1. the lesser volatility and higher ignition temperature of the organic solvent (TBP);

2. the use as salting agent of the nitric acid, which is recoverable and thus reduces the amount of waste.

The tributyl phosphate (TBP) can be diluted either in odourless kerosene (OK), or in n-dodecene, ordinarily at a 30% concentration in volume; at this particular concentration it presents an optimal behaviour with respect to extraction, phase disengagement, and risk of fission-related criticality.

The LWR system is concerned mainly with the recovery of residual uranium with sizeable amounts of $^{235}U$. The spent fuel from this system comprises the bulk of the entire fuel calling for reprocessing to recover and recycle uranium and plutonium, as well as to concentrate the fission products prior to long-term storage. It is for these reasons that the parameters of industrial-scale Purex reprocessing, as developed for the LWR-type fuel, provide the terms of reference, and the reprocessing of HWR-type and FBR-type irradiated fuel should be assessed in comparison with them.

The chief similarity between these fuels is that they are all oxides, whereas the difference consists in their different burn-ups and, consequently, in their different fission products and actinides content, as well as in the amounts of irradiated fuel unloaded yearly from the nuclear power plants equipped with such reactors (per 1000 MWe).

Thus the reprocessing of the HWR-type irradiated fuel does not require changes in the technology corresponding to the LWR-type fuel. It is, however, necessary to take into consideration the following factors:

(a) The uranium as recovered by reprocessing HWR fuel is depleted uranium (0.22 — 0.23% $^{235}U$), whereas in case of LWR fuel it is a slightly enriched uranium (the $^{235}U$ content is 0.8 — 0.9%). Thus, the final chemical form would also differ: oxide, for the uranium recovered from HWRs, and hexafluoride for the uranium recovered for LWRs.

(b) The amount of plutonium recovered per generated energy unit is larger for the HWRs than for LWRs.

(c) The amount of irradiated fuel unloaded yearly from a HWR is 3 — 4 times higher than the corresponding amount from a LWR, rated at the same electric power (PWR — 1000 MWe, 28 t/yr; BWR — 1000 MWe, 38 t/yr; PHWR — 1000 MWe, 100 t/yr). Since the uranium content in the irradiated fuel is the parameter which ultimately defines the required capacity of a reprocessing facility, the reprocessing demand is 3 — 4 times higher in the case of the PHWR fuel as compared with the case of the LWR fuel, relatively to the same amount of electric energy produced.

Industrial implementation of processes to recover U and Pu, as well as other potentially useful by-products,from irradiated fuels is confronted with several special problems caused by the extremely high global radioactivities, of the order of $10^5 - 10^7$ Ci for the whole fuel assembly; among these are:

1. the degradation of chemical agents and particularly of organic solvents, due to irradiation;

2. the criticality hazards, due to the growth in fissile material concentration along the technological flow-chart;

3. problems with the long-term storage and the appropriate management of radioactive wastes resulting from the process;

4. health hazards for the operating personnel, due to current exposure to radiations;

5. the irradiation risks for the operating personnel related to longer radiation exposure during decontamination operations and the repairing — maintenance of contaminated components;

6. environmental pollution through current radioactive effluents, as well as the potential danger of massive, accidental, environmental pollution.

To mitigate such problems, besides a judicious sizing of the facility's capacity, special attention is in order, with regard to the following aspects:

1. current monitoring of the radioactivity levels of the irradiated material, to choose properly the best moment when recovery and recycling are in order – technology and economics considered;

2. proper options for process equipment and size, as well as proper positioning of equipment within the facility, so that the radiolytic degradation of chemical agents be minimal and the criticality risks minimized;

3. provision of sufficient interim and ultimate storage capacities all over the flow-chart, and for a maximum recovery of chemical agents, with the purpose of their recycling in the process;

4. appropriate sizing of biological protection and choice of radioactive materials handling procedures, so that the irradiation doses and the inhalation contamination levels of the operating personnel be reduced to a minimum

The correct sizing of reprocessing installations and equipment, of the handling devices and biological protection, as well as the definition of the most suited procedures of handling the fissile and radioactive materials at minimal risks require an accurate knowledge of the amount and concentrations of the fissile elements, fission,and activation, products contained in the irradiated fuel and a detailed assessment of the criticality level in various points of the installation. The solution at hand is the calculation of the relevant parameters, followed by their comparison with the results from experimental measurements.

Calculation of the concentrations of fissile materials, fission products and activation products observes, ordinarily, the following pattern:

The concentration in the irradiated fuel of the radionuclide in question is denoted with $N(t)$ atoms . $m^{-3}$. Then its variation in time through neutron capture is described by the equation:

$$\frac{dN(t)}{dt} = -\sigma_c \Phi N(t), \qquad (10.19)$$

and, through radioactive decay, by the equation:

$$\frac{dN(t)}{dt} = -\lambda N(t) \tag{10.20}$$

where $\lambda$ is the decay constant $[s^{-1}]$, $\sigma_c$ is the capture cross-section $[m^{-2}]$, and $\Phi$ is the neutron flux density $[m^{-2} . s^{-1}]$.

The occurrence of any radionuclide $N_1$ as a result of the decay of the nuclide $N_2$ is described by the generation equation:

$$\frac{dN_1(t)}{dt} = \lambda_2 N_2(t). \tag{10.21}$$

Considering the most important nuclear reactions induced by neutrons in the reactor as well as the radioactive decays which occur, one could write a system of hundreds of such coupled differential equations.

All the $\lambda$ values are constants for this system, while the reaction rates $\sigma\Phi$ depend both on $\Phi$, and $\sigma$; consequently, they depend on the power levels monitored according to the reactor's operation diagram and, thus, on the neutron spectrum and radionuclide concentrations.

If one takes $\mathbf{N}(t)$ as the vector of the radionuclide concentrations $N_i(t)$, and $A = \{A_{i,k}\}$, $i, k = 1, 2, ...n$ as the coefficients (transition) matrix, then the system of differential equations reads, in matrix notation:

$$\frac{d\mathbf{N}}{dt} = A\mathbf{N}(t), \text{ where } A = A(\mathbf{N}(t)). \tag{10.22}$$

If the matrix elements are constant — a condition which can be met over time intervals in which the flux and the spectrum do not vary to any sizeable measure — or when $\Phi = 0$, the above system is linear, and its solution has the following form:

$$\mathbf{N}(t) = \mathbf{N}(0) + tA\mathbf{N}(0) + (t^2/2)A[A\mathbf{N}(0)] + ... = \mathbf{N}(0)\exp(At). \tag{10.23}$$

The expression (10.23) allows the numerical solution of the system of differential equations. The time interval chosen for the calculation of the concentration evolution determines the accuracy. If the occurrence of radionuclides is due to forerunners with a much shorter lifetime than the chosen time span, other algorithms must be used.

The problem is to be given a computer treatment and, to this effect, the ORIGEN [15] or FISSPROD [16] codes may be employed.

The curves in Figs 10.6 and 10.7 plot the results of the computation by the ORIGEN code,for a LWR fuel assembly irradiated up to a burn-up of 34,000 MWd/t.

The curve in Fig. 10.13 repeats the one in Fig. 10.6, with the moments at which various operations in the LWR fuel cycle are effected being marked.

In order to evaluate by computation the criticality state of a device or of a part of installation, i.e. the magnitude of $k_{eff}$, the geometric, mass and concentration parameters for $k_{eff} < 1$ should be determined.

A computer program pertinent to this purpose is a variant of those currently employed in nuclear reactors design [17]; the object-configurations are assumed to be composed of simpler geometrical forms, such as spheres, cylinders or straight prisms, randomly arranged. For the configurations which cannot be modelled through such simpler geometrical forms, the programs use arbitrarily oriented sets of planes that intersect each other at random. Besides the description of a configuration by elementary forms the materials themselves are expressed as tables of constituents, concentrations, and microscopic cross-sections, on the basis of which the corresponding macroscopic cross-sections are calculated [13].

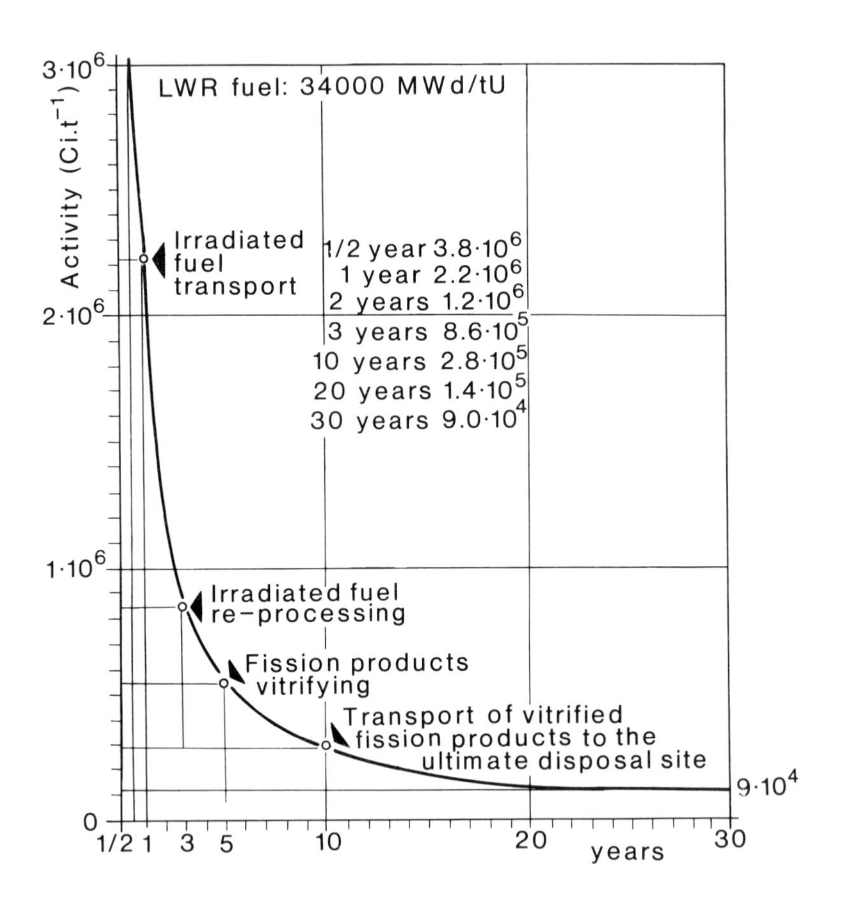


Fig. 10.13. Optimum moments for execution of certain phases of the fuel cycle [11].

The simpler forms are positioned in space with reference to a unique orthogonal system of coordinates, and their relative positions are described by means of mathematical operators. In principle, the configuration decomposition can be represented as follows:

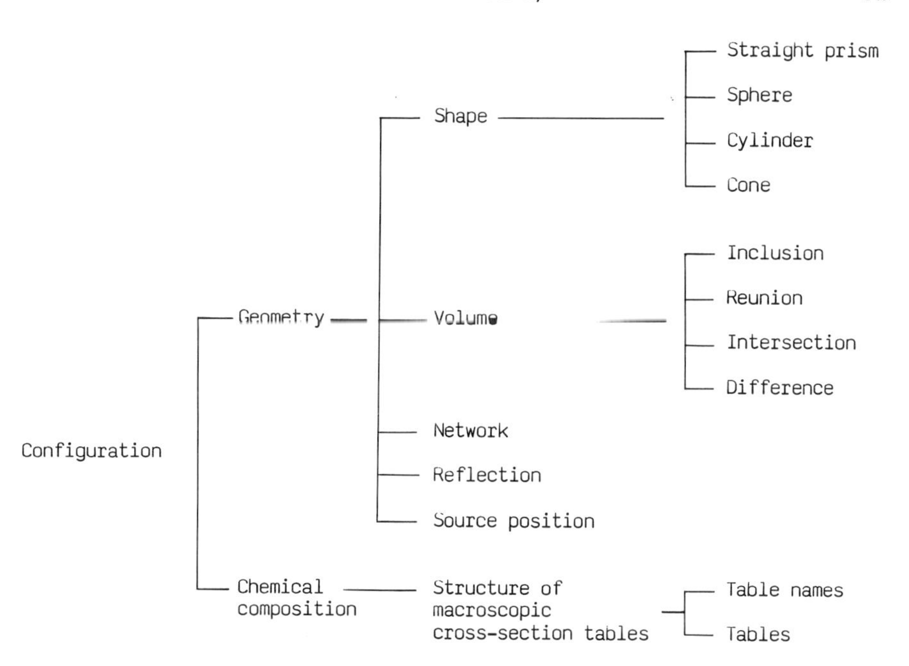


As a general method for solving the problem, criticality programs make use of a multigroup treatment in the diffusion approximation, taking into account several generations of a constant number of neutrons. The life of each neutron consists of a succession of events, separated in space, like: birth, collision, crossing of a separation surface between two media, reflection, escape. Factor $k_{eff}$ is determined by the number of probable collisions in each stage, and the corresponding expression is:

$$k_{eff} = \frac{\sum_V \sum_i \nu\Sigma_f(V,i) \cdot P(V,i)}{\text{leaks} + \sum_V \sum_i \Sigma_a(V,i) \cdot P(V,i)}$$

where $\Sigma_f(V,i)$ and $\Sigma_a(V,i)$ are the macroscopic effective fission and absorption cross-sections of the volume $V$ relative to the neutrons from the $i$th energy group, $P(V,i)$ is the total distance covered by all the neutrons of the $i$th group during their stage inside the volume ; $\Sigma_a(V,i) \cdot P(V,i)$ represents the probable number of these neutrons which are absorbed, and the product $\Sigma_f(V,i) \cdot P(V,i)$ is the probable number of neutrons produced by fission. Such an analysis of the problem is characteristic for the Monte Carlo method. To effectively solve the problem, computer codes like KENO-IV [17,18] may be employed.

Figures 10.14 and 10.15 present the material balance characteristic for PWR-1000 MWe and CANDU-PHWR-600 MWe irradiated fuel, while Table 10.8 gives the composition of fission products contained in a PWR-1000 MWe fuel assembly irradiated to approximately 34,000 MWd/t.U [16].

A recent IAEA study concerning the technological and economic aspects of the feasibility of multinational nuclear fuel recovery and recycling centres [19] points out that only the ones rated at 300 — 3000 t . $year^{-1}$ in the 300, 750, 1500 and 3000 range would work on an economically sound basis. Figure 10.16 shows the material balance for a reprocessing unit for LWR irradiated fuel (1500 t . $yr^{-1}$).

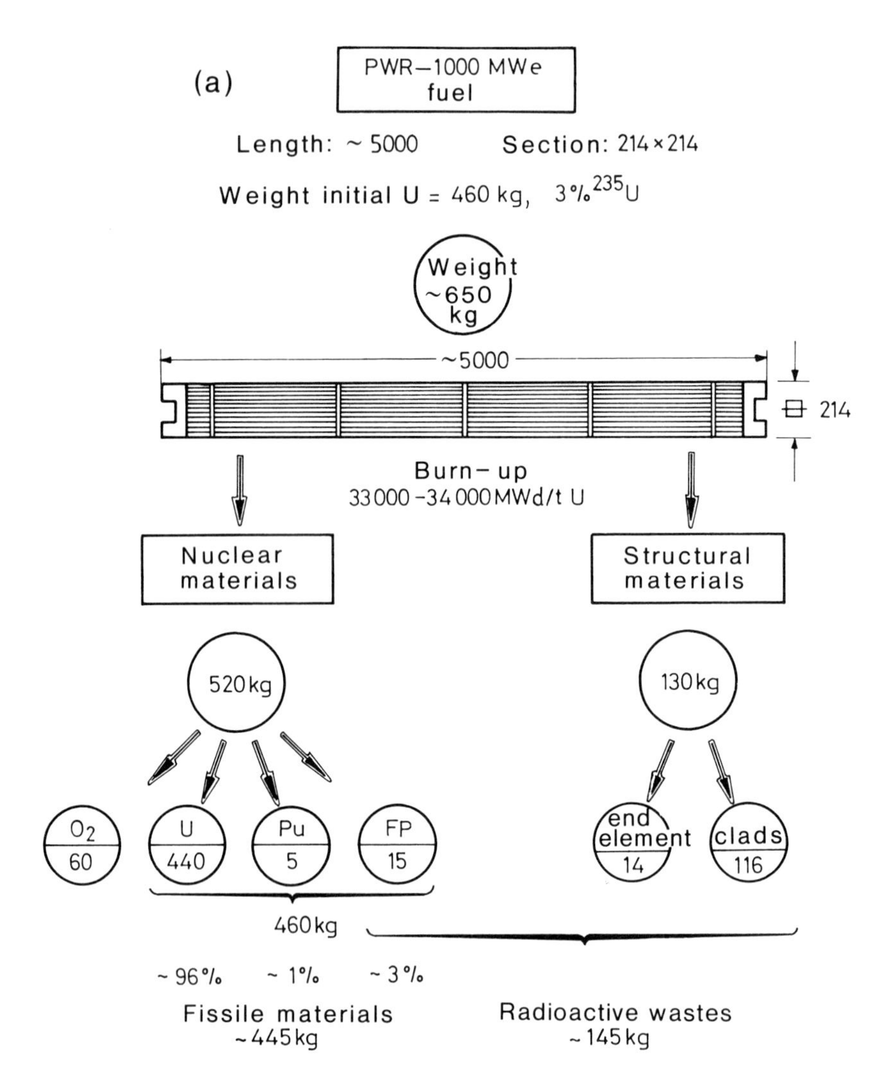


Fig. 10.14. Gross material balance corresponding to a re-treatment unit of 1500 t . $yr^{-1}$, for PWR-1000 MWe fuel elements.

In what follows the sequence of technological operations characteristic for an industrial unit based on the Purex method is briefly introduced.

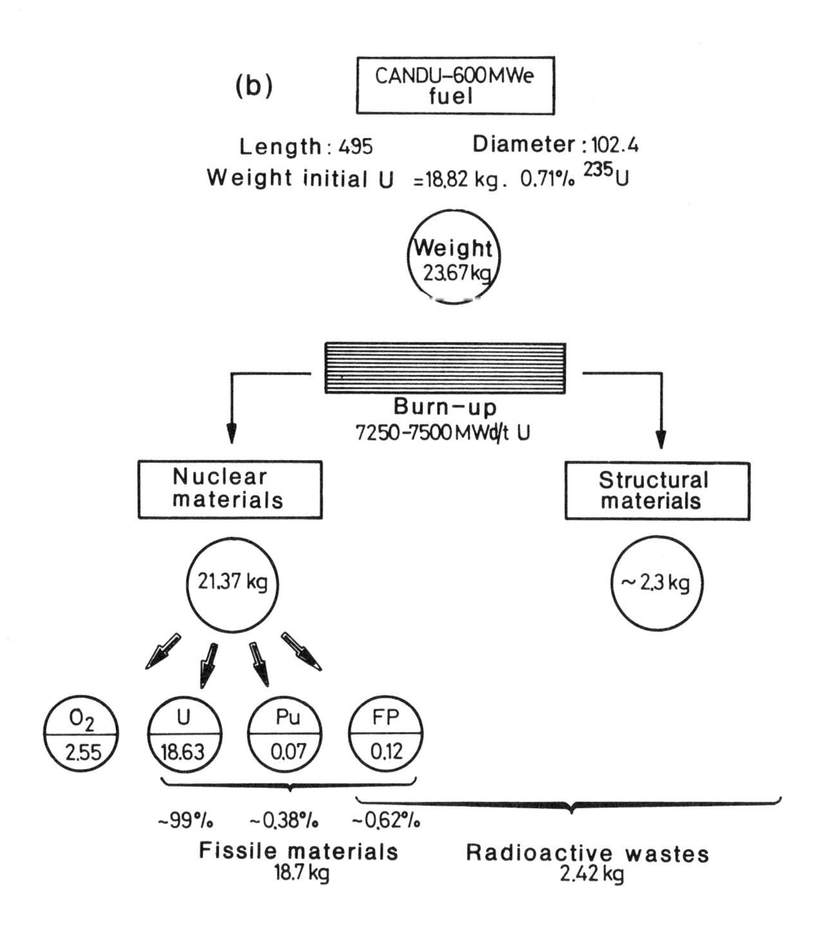


Fig. 10.15. Gross material balance corresponding to a re-treatment unit of 1500 t . $yr^{-1}$, for CANDU-600 MWe fuel elements.

## 10.8.2. General Description of the Process

The flow-chart presented in Fig. 10.17 characterizes the industrial reprocessing based on the Purex method [19]. The main stages of the process are:

(a) cutting the fuel assembly;
(b) solution of fuel and adjustment of pH;
(c) co-decontamination of uranium and plutonium from fission products and their separation;
(d) two consecutive purification cycles for uranium, to separate it from plutonium and fission products;
(e) two consecutive purification cycles for plutonium, to separate it from uranium and fission products.

TABLE 10.8 Fission Products and Tritium Concentrations in PWR-1000 MWe Irradiated Fuel
(burn-up: approx. 34,000 MWd/t U;
de-activation time: 3 years)

| *Isotope* | *Half-life* | *Ci/t* | *Isotope* | *Half-life* | *Ci/t* |
|---|---|---|---|---|---|
| $^{85}$Kr | 10.6 years | $1.1 \times 10^{4}$ | $^{90}$Sr | 28 years | $1.22 \times 10^{5}$ |
| | | | $^{95}$Zr | 65 days | $1.2 \times 10^{1}$ |
| $^{131}$Xe | Stable | - | $^{95}$Nb | 35 days | $2.3 \times 10^{1}$ |
| | | | $^{106}$Ru | 1 year | $1 \times 10^{5}$ |
| $^{129}$I | $1.6 \times 10^{7}$ years | $4 \times 10^{-2}$ | $^{106}$Rh | 30 s | $1 \times 10^{5}$ |
| | | | $^{137}$Cs | 30 years | $1.74 \times 10^{5}$ |
| $^{3}$H | 12.3 years | $3.7 \times 10^{2}$ | $^{144}$Ce | 284 days | $8.25 \times 10^{4}$ |
| | | | Various | | Others |
| Total | | $1.17 \times 10^{4}$ | Total | | $8.6 \times 10^{5}$ |
| Total grams/t: | | | | about | 30,000 |

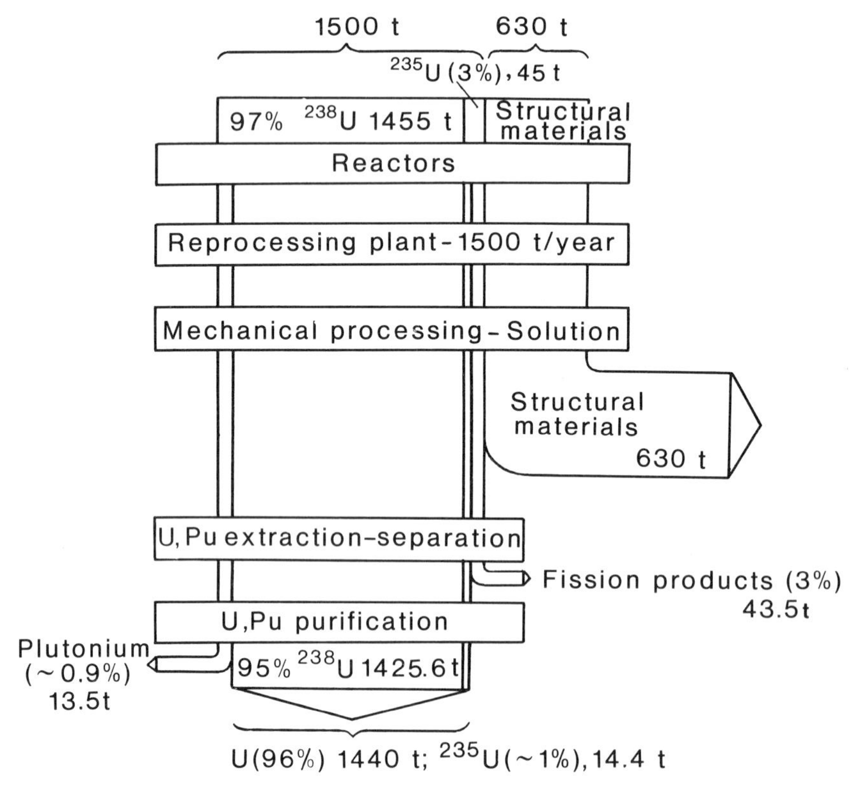


Fig. 10.16. Material balance over a Purex flow-chart, with reflux purification of Pu [19].

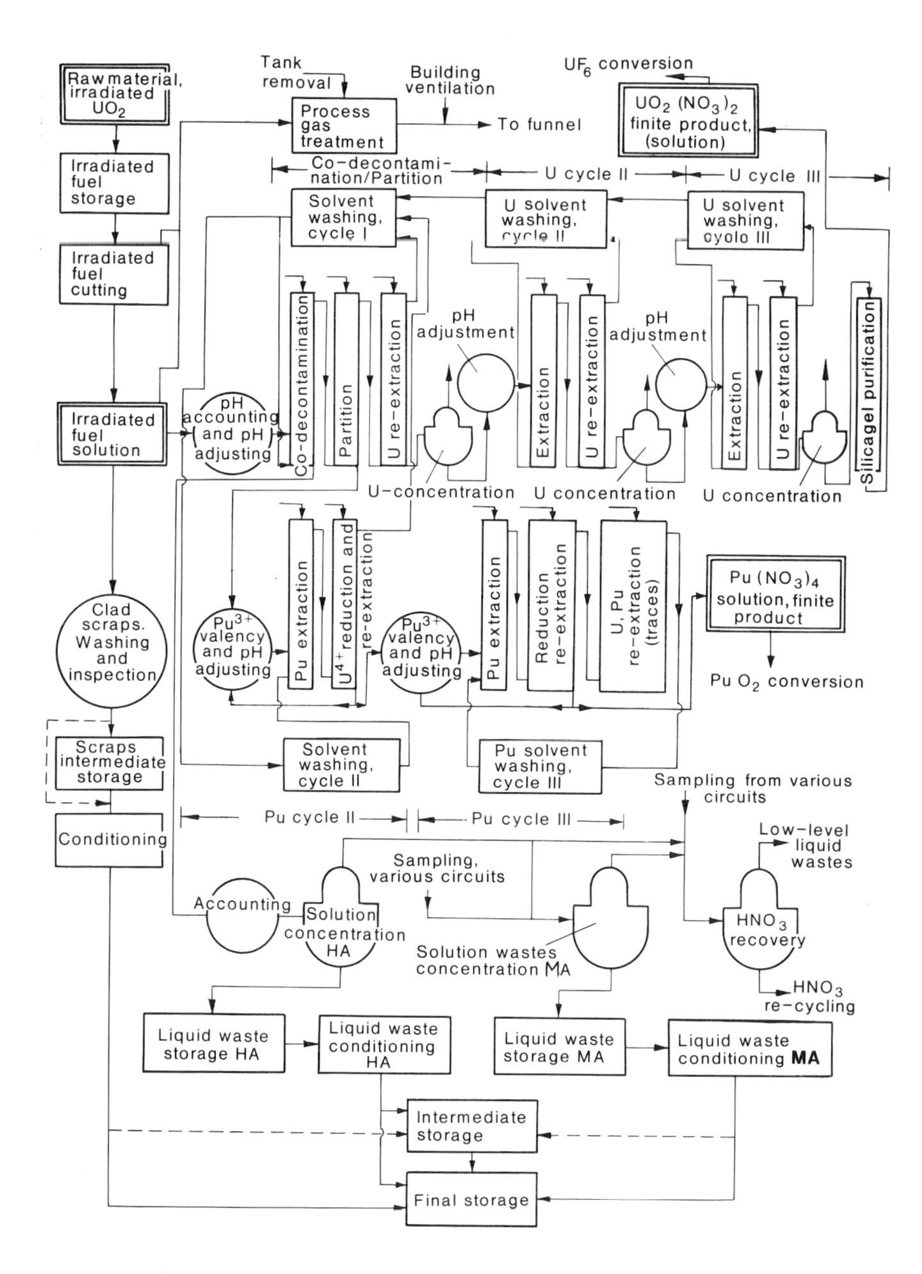


Fig. 10.17. Purex process flux-chart [19]: HA = high activity; MA = medium activity.

At the co-decontamination from fission products of uranium and plutonium contained in the aqueous phase, more than 99.5% of U and Pu are extracted in the organic phase, while approximately 99% of fission products remain in the aqueous phase. Table 10.9 lists the elements, or their inorganic compounds which, owing to their chemical properties, are effective in obtaining the necessary product purity by the solvent extraction process [13]. One notes that Zr and Ce are elements that may raise problems in uranium and plutonium decontamination and purification; the magnitude of the problems is closely related to the concentrations of these elements in the irradiated fuel, and thus to fuel burn-up.

TABLE 10.9 Behaviour of some Chemical Elements and Compounds in Solvent Extraction Processes

| *Category* | *Chemical elements or compound* | *Ability to form nitrates* | *Distribution coefficients in TBP* |
|---|---|---|---|
| A | $UO_2^{2+}$, $PuO_2^{2+}$, $Pu^{4+}$, $U^{4+}$, | Relatively good | High |
| | $Zr^{4+}$, $Ce^{4+}$ | | Low |
| B | $Pu^{3+}$, $Y^{3+}$, $Ce^{3+}$, $La^{3+}$, $Pr^{3+}$, $Nb^{3+}$, rare earths | Moderate | Low to negligible |
| C | Cs, Sr, Ba, Mo, Te, $Ru^{4+}$, Rh | Very low | Virtually nil |

During the 2nd and 3rd plutonium purification cycles — Fig. 10.17 — part of the extracted plutonium is directed back to the beginning of the cycle, thus obtaining its concentration in the extraction phases. Because of this product-recycling, the procedure is also called *with Pu reflux circulation*. This provides for an operation of the plutonium purification cycles II and III at a constant level of concentration, independently of the concentration in the irradiated fuel. In this way, a plutonium concentration factor of the order of 10 is obtained between the inlet and outlet points. Thus, the concentration by evaporation of plutonium in solution is avoided, which gives the process even more efficiency and safety.

Solvent regeneration, that is in order in every cycle, as well as nitric acid recovery, ensure important savings on reactives and a considerable reduction in the volume of radioactive liquid wastes.

Other details concerning various technical, economic and nuclear security aspects of the industrial scale implementation of the Purex process are given by Cleveland [2] and in the review work [19].

### 10.8.3. Specifications for Final Products

The final products resulting from the process, as indicated also in Fig. 10.17, are $Pu(NO_3)_4$ and $UO_2(NO_3)_2$; Tables 10.10 and 10.11 give the chemical compositions and purities of these [19]. The values as presented have only an illustrative character, and differ from country to country.

TABLE 10.10 Specifications for Uranyl Nitrate

| | |
|---|---|
| U concentration | 200 – 400 kg . $m^{-3}$ |
| Free $HNO_3$ | 1 M |
| Impurities (ppm) | |
| Cr | 100 |
| Fe | 300 |
| Ni | 100 |
| Halogens, total | 300 |
| Boron equivalent[1] | |
| The amount measured for each of the indicated elements multiplied by the boron equivalent factor as indicated shall not exceed 8 ppm | |
| B | 1.000 |
| Cd | 0.3904 |
| Co | 0.0090 |
| Dy | 0.0815 |
| Eu | 0.4124 |
| Gd | 4.4384 |
| Fe | 0.0007 |
| Li | 0.1457 |
| Sm | 0.5313 |
| Fission products | |
| Total measured activity of following isotopes shall not exceed 0.5 µCi/g U | |
| $^{95}Zr$, $^{95}Nb$, $^{103}Ru$, $^{106}Ru$, $^{137}Cs$, $^{144}Ce$ | |
| Alpha activity | |
| Total measured activity of following elements shall not exceed 15,000 dis . $min^{-1}$ (g U)$^{-1}$ | |
| Am, Cm, Np, Pu, | |

[1]The natural boron concentration ensuring the same absorption cross-section for thermal neutrons as the concentration of a specified impurity element.

TABLE 10.11 Plutonium Nitrate Specifications

| | |
|---|---|
| Pu concentration (kg . $m^{-3}$) | 180 – 250 |
| Free $HNO_3$ (M) | 2 – 10 |
| Metallic impurities (excluding Pu and U) (ppm) | 5000 |
| U as impurity (ppm) | 5000 |
| Total impurity content expressed in boron equivalent (ppm) | 10 |
| Total halogen content (ppm) | 150 |
| Sulphate content (ppm) | 1000 |
| Gamma activity due to fission products with half-lives < 30 days (µCi (g Pu)$^{-1}$) | 40 |
| $^{95}Zr$, and $^{95}Nb$, gamma activity (µCi (g Pu)$^{-1}$) | 5 |
| $^{241}Am$ content, 9 months after delivery to conversion installation (ppm) | 5000 |

### 10.8.4. Losses in the Process – Admitted Values

The losses occurring in the technological process are evaluated by comparing the uranium and plutonium amounts obtained with the amounts in the irradiated fuel — the latter as estimated by calculations with an adequate code (e.g. ORIGEN). The input is taken from the operation diagrams of nuclear power plants (power levels, number of days at various power levels, number of reactor shut-down days, coolant and moderator operating temperatures, etc.). Computed results are validated by physical and chemical analyses (gamma-, alpha- and mass-spectrometry, X-ray fluorescence) performed on samples taken from the solvent solution. The permitted losses are, at maximum, 0.5% for uranium and 1% for plutonium.

Much of the losses can be recovered by adequate treatment of liquid radioactive wastes resulting from the process: 50% uranium and 75% plutonium.

## 10.9. RADIOACTIVE WASTES

### 10.9.1. Classification

Irradiated nuclear fuel reprocessing generates a large variety of wastes which may be classified into:

1. gaseous wastes;
2. high-level liquid wastes;
3. medium-level liquid wastes;
4. low-level liquid wastes;
5. high-level solid wastes;
6. medium- and low-level solid wastes;
7. organic wastes.

Figure 10.18 presents the origins of the various types of wastes. As it appears from the diagram, the flow-sheet of fuel reprocessing includes radioactive waste treatment operations, so that these can be considered as part of the fuel reprocessing [19].

### 10.9.2. Gaseous Wastes

Such wastes come from:

(a) gaseous emissions during fuel cutting and digestion operations;
(b) air aspirated from the installation tanks;
(c) air aspirated from inside the biologically protected chambers (hot cells) which contain process equipment and from the rooms allotted to process control and to surveillance, auxiliary, equipment.

The radioactive constituents of gaseous emissions — $^{85}Kr$, traces of $^{129}I$ and tritiated water vapours (see Table 10.8) — come from the fuel cutting – dissolving operations.

In current practice these volatile and gaseous radionuclides are evacuated towards the ventilation funnel of the installation. Prior to funnel evacuation, the gaseous flow is washed by countercurrent circulation of a natrium hydroxide solution and is then passed through filters with an aerosol retention power in excess of 99.99%(absolute filters). With the aim of bringing noxious emissions from nuclear installations to a reasonably attainable minimal level, treatment techniques for effluents are now studied; the decontamination factors anticipated for the raffination of gaseous effluents are given in Table 10.12.

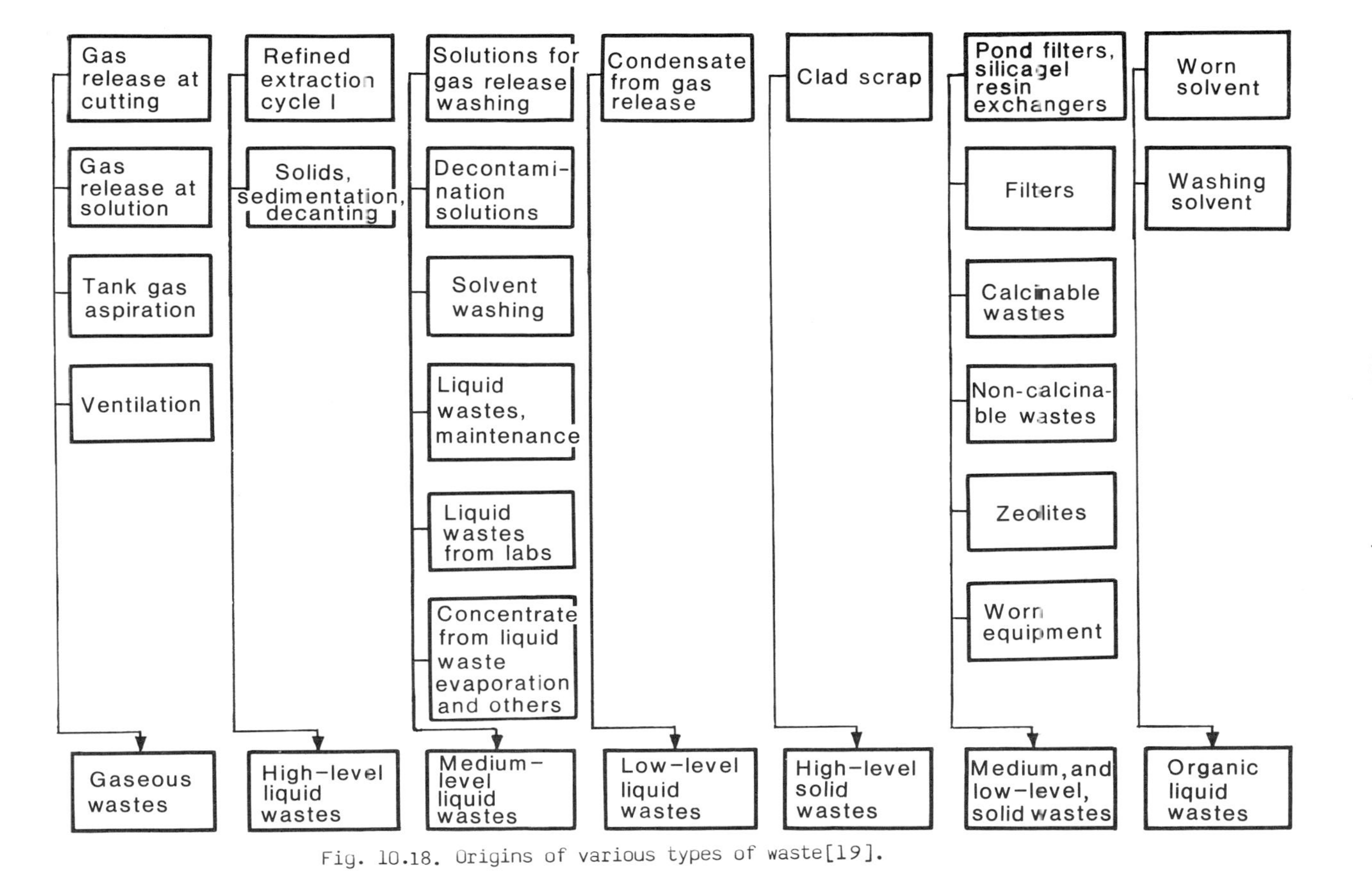


Fig. 10.18. Origins of various types of waste[19].

TABLE 10.12 Decontamination Factors, DF[1] anticipated for the Rafination of Gaseous Effluents

| *Noxious isotope* | *DF* |
|---|---|
| $^{131}I$ | 500 |
| $^{129}I$ | 500 |
| $^{85}Kr$ | 50 |
| $^{14}C$ | 50 |

[1] DF represents the ratio of contamination values measured in imp . $s^{-1}$ with a single-channel spectrometer, upstream and downstream with respect to the radioactive gaseous products retention installation.

As general methods of retention of the iodine, iodine compounds ($I_2$, HI, HIO, etc.), noble gases and tritium, one may mention:

1. absorption on active carbon beds and the reaction with silver or silver zeolites;
2. cryogenic distillation, absorption in liquids, separation through selectively-permeable membranes. Among these techniques, cryogenic distillation seems to be closest to industrial application. The method starts from the industrial experience gained in the separation of noble gases. For krypton and xenon retention in reprocessing, new problems occur because of the Kr/Xe molar ratio which differs from the one existing in nature, namely:

Kr : Xe = 0.104, in reprocessing;
Kr : Xe = 13.1, in nature.

The thermal desorption of tritium from the fuel is achieved by the latter's heating before dissolution, thus avoiding contact between tritium and the aqueous solution, and, consequently, the formation of HTO [19]. Such a procedure allows the collection of tritium and its adequate conditioning.

The methods used at present require tritium confinement during the first reprocessing stages, by the following methods:

1. limitation of the fresh water addition to the process, using the extensive recycling of existing material;
2. preventing the tritiated water transfer during the process, at the level of the first extraction cycle, by washing the organic phase with non-tritiated water;
3. evacuation of small fractions of tritiated water,either for conditioning and final storage, or directly in the environment (sea or ocean), after diluting the tritium down a permitted concentration level of ($3 \times 10^{-3}$ Ci . $m^{-3}$).

### 10.9.3. Liquid Wastes

(a) High-level liquid wastes. These are generated especially in the process of co-decontamination/partition, and contain at least 99.5% of the fission products that are originally in the irradiated fuel. The refined product resulted from these processes is concentrated by evaporation and transferred in storing tanks. The volume of concentrated, high level,liquid waste,vary between

0.7-1.0 $m^3$ per t U for fuels de-activated for 2 years and 0.25 $m^3$/t U for those de-activated for more than 5 years. The volume of high level waste is directly proportional to the reprocessed fuel quantity.

(b) Medium level liquid wastes. These result from the evaporation of various liquids produced by chemical processes, such as: washing of gaseous emissions, products concentration, and as liquids from the decontamination and maintenance of the installation. Most of the wastes appear as $NaNO_3$, their activity being of a few tenth Ci per litre. The amounts of liquids produced during maintenance are about 3 — 5 $m^3$/t U and may contain traces of uranium and plutonium up to 0.2% of the total reprocessed quantities.

(c) Low level liquid wastes. These consist especially of water vapours condensed in various evaporators and contain radionuclides such as: $^{90}Sr$, $^{137}Cs$, $^{106}Ru$, tritium as HTO, and — in conditions of light accidental contamination — $^{60}Co$, U and Pu as well as other fission products. The accumulated condensed vapours may amount to 2 — 3 $m^3$/t U. For this class of wastes, certain technological provisions are contemplated, to the effect that their total quantity would not exceed 100 $m^3$/t U, and of creating the potential of their being recycled, to reduce the conditioning effort that is in order prior to their release in the environment.

### 10.9.4. Solid Wastes

(a) High-level solid wastes. These contain residues of structural materials from the irradiated fuel, and sediments of solid undissolved materials consisting of fine particles of structural material and hardly soluble fission products, such as: Ru, Rh, Mo, Te, Pd.

Structural material residues amount to a volume of ~ 0.4 $m^3/t_{fuel}$ and contain less than 0.1% of the processed quantity of uranium and plutonium.

(b) Medium and low level solid wastes. These consist of contaminated materials resulting from decontamination operations, as well as from worn, contaminated, parts replaced during repair — maintenance operations. This category comprises the equipment and contaminated protection clothing of the personnel performing decontamination and maintenance. It is generally admitted that the quantity of such wastes is about 5 — 6 $m^3$/t U.

The wastes in this category are subdivided into wastes of calcinable and non-calcinable nature, and are treated accordingly. The wastes that come from the fuel are 60 — 70% of the total waste volume; they contain 0.5 — 1% and 1.5% of the total quantity of uranium and plutonium, as vehiculated. Such a content in fissionable materials necessitates a calcination processing and chemical recovery of uranium and plutonium from ashes.

### 10.9.5. Organic Liquid Wastes

Besides the solvent worn in the process (TBP-kerosene), the wastes in this category also comprise the kerosene used to remove the solvent held in the aqueous solutions of the final extraction cycles. The quantity of organic liquid waste may reach a maximum of 0.8 $m^3$/t of spent fuel.

## 10.10. IRRADIATED FUEL TRANSPORT

Reprocessing makes necessary the transportation of considerable amounts of irradiated fuel from the nuclear power plants to the reprocessing facilities. Yearly transported quantities are determined by the reprocessing capacity.

As a rule, radioactive materials are transported in compliance with strict regulations, in containers that are designed, manufactured and tested according to well-enforced specifications [20]; details of concept and design follow from the particulars of the radioactive material to be transported, namely:

1. intensity and type of radioactivity;
2. the amount of heat generated by radiative self-heating;
3. the aggregation state and chemical composition.

For the irradiated fuel, the transport containers provide for appropriate gamma and neutron protection as well as for an adequate capacity of fuel-cooling. Transportation of the fuel from the nuclear power plant to the reprocessing facility may be performed only after a minimum period of de-activation, of 150 — 200 days, depending on the fuel type and burn-up. The on-site de-activation, at the nuclear power plant provides for the de-activation of certain volatile and radiotoxic fission products, such as $^{131}I$ ( $T_{1/2}$ = 8.05 d), down to negligible levels of radioactivity.

For the containers in current use, the ratio of the weight of biological protection (lead, steel) to the weight of the transported uranium is ~ 30 : 1.

By coolant employed during transportation, containers are divided into two types:

1. water-cooled ("wet") transports;
2. air- or noble gases-cooled ("dry") transports.

Containers belonging to the first category are meant to convey large quantities of irradiated fuel; they do not need additional neutron protection, but create problems as far as nuclear safety during transportation, owing to the risks entailed by criticality, incremental internal pressure, and the loss-of-coolant prospects.

Containers in the second category do not raise special safety problems, but require neutron protection in addition to the gamma protection and imply higher temperatures of the fuel during transportation.

None of these types is better, by any technical and economic standard. Table 10.13 and Fig. 10.19 display some characteristics of the containers for the transportation of irradiated fuel, as employed in Europe [19]. Figure 10.20 shows a transverse cut through a transport container designed for the present PWR fuel. Containers having a weight of less than 40 t are ordinarily meant for road transportation, while those having a superior weight call for railway transport. Both road trailers and railway transports have a special construction; they provide for a tight fastening of the container during transportation, as well as for control over operating parameters such as temperature and pressure.

The concept and design of a container call for consideration of the following factors:

(a) content of fission products, actinides and activation products in the irradiated fuel;
(b) heat generated through radiation self-heating;

as well as for the solving of the following problems:

(a) a geometry and a disposition of the materials that would ensure the necessary gamma and neutron protection, while also maintaining the configuration in a subcritical condition ( $k_{eff} < 1$);
(b) heat transfer modes that would comply with the established configuration and the required heat-release rates.

TABLE 10.13 Irradiated Fuel Containers, as in service in Europe

| *Container type* | *NTL 2* | *NTL 3* | *NTL 4* | *NTL 5* | *NTL 8* | *NTL 9* | *NTL 10* | *NTL 11* | *NTL 12* | *NTL 14* |
|---|---|---|---|---|---|---|---|---|---|---|
| Capacity: | | | | | | | | | | |
| PWR (ass./sect) | 4/200 | 7/200 | 7/200 | 7/200 | 3/215 | - | 12/230 | 7/215 | 12/215 | 5/230 |
| BWR (ass./sect) | 9/114 | - | 19/114 | 12/140 | - | 7/140 | - | 17/140 | 30/140 | - |
| Heat capacity(kW) | 15 | 30 | 35 | 35 | 35 | 25 | 100 | 42 | 100 | 50 |
| Total weight (t) | 32 | 52 | 65 | 69 | 36 | 34 | 104 | 75 | 95 | 82 |
| Payload (t U) | 1.1 | 2.0 | 2.3 | 2.3 | 1.4 | 1.4 | 6.2 | 3.3 | 5.7 | 2.7 |
| Shaft length (mm) | 3875 | 3380 | 4370 | 4675 | 4280 | 4520 | 5050 | 4630 | 4580 | 5160 |
| Shaft diameter (mm) | 440 | 864 | 864 | 864 | 3x230 | 474 | 1220 | 914 | 1220 | 914 |
| Heat carrier | Air | Water | Water | Water | Air | Air | Air (water) | Water | Air (water) | Water |
| Transport mode | HW | RW | RW | RW | HW | HW | RW | RW | RW | RW |

HW = highway; RW = railway.

Determination of the proper design parameters must duly apply to computer codes, of the kind mentioned before.

## 10.11. IRRADIATED FUEL STORAGE

The spent fuel that is periodically or continually unloaded from the reactor is stored for de-activation in storage ponds that are part of the infrastructure of any nuclear power plant.

The storage ponds provide for continuous evacuation of the heat generated in the irradiated fuel by radiation self-heating, as well as for the biological protection of the personnel during fuel handling and storing. Water is used, as a biological protection material and heat-carrier.

Storage aims mainly at ensuring the necessary time for the de-activation of the irradiated fuel to get down to such reactivity levels as to allow its further handling for transportation and reprocessing.

Spent-fuel ponds at the site of a nuclear power plant ensure proper protection to the fuel, preventing mechanical and chemical damage, thus assuring its safe preservation in time — for the preservation of the fuel element cladding integrity (the first physical barrier against uncontrolled release of fission products in the environment) is, by all measures, an important problem.

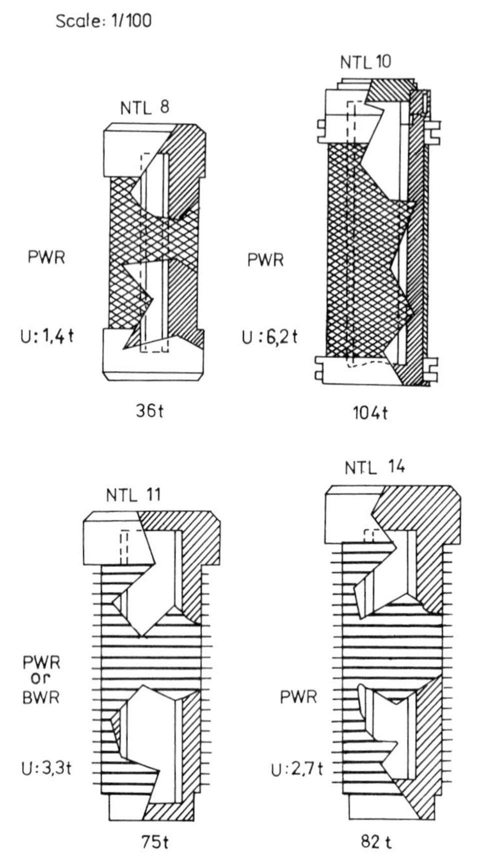


Fig. 10.19. Containers for irradiated fuel transportation [19].

Storing capacity of the ponds at the nuclear power plant site is so designed as to provide for a fuel de-activation of 5 — 15 years. The capacity of ponds at the reprocessing facilities is mainly determined by facilitiy's yearly production. As mentioned in section 10.10, the minimum storage stage at the nuclear power plant site is of 150 days.

For storage of irradiated fuel at high-capacity reprocessing facilities (1500 – 3000 t/year) two competing solutions are contemplated:

(a) the classic, storage pond solution, with water as heat-carrying agent;
(b) the storage bin solution, using air as heat-carrying agent.

The first solution has all advantages of a proven technology, in use since more than 35 years, while the second looks more economic, the global costs (capital, operation) being approximately 22% lower [20,21].

The problems which may arise with the concept and design of irradiated fuel stores, as well as their solution, are to a large extent identical to those discussed in section 10.10.

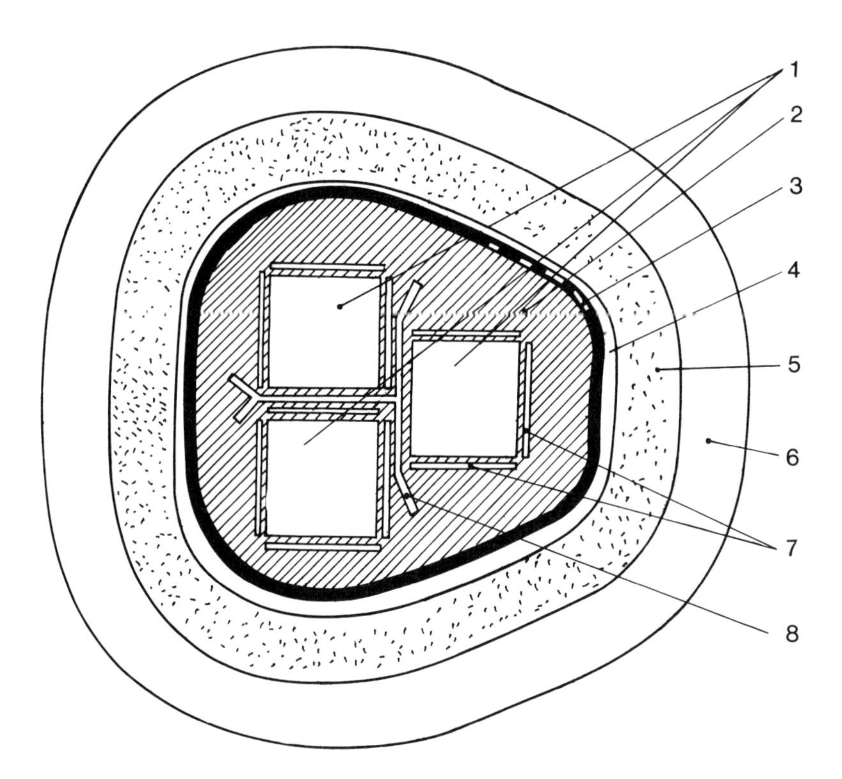


Fig. 10.20. Section through the NTL 8 transport, in service for PWR fuels: 1 - shaft (one for each LWR fuel assembly); 2 - lead; 3 - insulator (about 10 mm thick); 4 - stainless steel ring (about 20 — 30 mm thick); 5 - homogeneous mixture, cooling flaps of boron resins — copper (about 15 — 20 cm thick); 6 - reflector (water); 7 - neutron absorbent shield; 8 - copper plates.

## 10.12. RECYCLING OF THE RECOVERED FISSILE MATERIALS

### 10.12.1. Plutonium Recycling

As was emphasized earlier, plutonium has a very high radiotoxicity ($10^4$ times higher than uranium's), which requires that its handling be done in special conditions, in sealed chambers (glove boxes), in order to avoid personnel's direct contact with this material. Plutonium handling for mixed fuel fabrication entails high costs.

The alternatives with the use of Pu in the production of fuel elements assemblies are [22]:

(a) use Pu in *all* fuel elements over the core (in PHWRs, PWRs and FBRs);
(b) use Pu *only* in the central fuel elements (in BWRs).

The adoption of either of the above solutions is determined by the actual modes to control reactivity in the respective reactors.

On the assumption of an even distribution, Pu concentrations used (per unit-mass of uranium) are 0.3 — 1.5% for CANDU-PHWRs (case (4) in Table 10.3), 3% for LWR reactors, and 15 — 20% for FBR reactors.

Besides radiotoxicity, there are also α and β radioactive decays, neutron emission caused by spontaneous nuclei fission, and γ and X photon emissions accompanying radioactive decay phenomena that raise radioprotection problems in plutonium handling. Beta and α-specific activity, as well as neutron and photon emissions, depend on plutonium's isotopic composition and on the content in the actinides that are generated from plutonium isotopes.

### 10.12.2. Plutonium Isotope Composition and related Radioprotection Problems

The isotopic composition of plutonium found in irradiated fuels unloaded from thermal reactors depends mainly on the following parameters:

(a) reactor type;
(b) burn-up;
(c) duration of recovered plutonium storage;
(d) number of previous recycling campaigns.

Table 10.1 lists information that clarifies the dependence on parameters (a) and (b). The modification in time of plutonium's isotopic composition is due particularly to β-decay of the isotope 241, which leads to generation of $^{241}Am$. This circumstance demands, in the case of high $^{241}Pu$ content fuel (LWR and FBR), the shortest possible time span for the fuel unloading — reprocessing — mixed fuel fabrication cycle. Table 10.14 presents the X and γ radioactivity levels for plutonium obtained from LWR irradiated fuel, immediately after spent fuel unloading and minimally required storage, and after a 3-year intermission before reprocessing, respectively; the growth in activity, especially in the 20 — 60 keV energy range, is due to the X emission of $^{241}Am$ [22].

TABLE 10.14 Total X and γ Emission from 1 gram of Recovered Plutonium

| | *Radiation energy* (keV) | | | | |
|---|---|---|---|---|---|
| | *20-30* | *40-60* | *100-200* | *300-400* | *500* |
| LWR — freshly recovered plutonium ($10^{10}$ photons) | 4.19 | 0.04 | 0.01 | 0.006 | 0.003 |
| LWR — Pu, after 3 years storage ($10^{10}$ photons) | 6.95 | 5.52 | 0.01 | 0.006 | 0.003 |

Neutron emission caused by spontaneous fission is highest for $^{238}Pu$, followed in order by $^{242}Pu$ and $^{240}Pu$. The latter has the greatest contribution to plutonium's neutron emission, because of its higher abundance as compared to the other two isotopes. In the case of oxide fuels, neutron emission is amplified by the (α,n) reactions with light elements — a fact evidenced by $PuO_2$ neutron emission, which is 30% higher than that of metallic Pu.

From these data, it follows that the radioprotection measures required in mixed fuel fabrication depend on plutonium's isotopic composition and concentration. The magnitude of needs and costs of solutions increase in direct proportion with plutonium's concentration.

Handling of important quantities of plutonium in the industrial fabrication of mixed fuels calls for intensive process mechanization and automation, in order to avoid personnel contamination and direct exposure to radiations.

### 10.12.3. Uranium Recycling

As yet, the fabrication of $(Th,^{233}U)O_2$-type fuel elements is a problem of the future in nuclear technology [23]. $^{233}U$ handling raises important radio-protection problems, due to the high α and γ radioactivity of certain offspring of successive α-decays of $^{232}U$, which accompanies $^{233}U$.

If one envisages the use of $^{233}U$ in CANDU-PHWR reactors in a self-sufficient thorium cycle at equilibrium, the whole uranium concentration is 2.6%, as referred to thorium, and its isotopic composition is:

Uranium - 232, traces
Uranium - 233, 61%
Uranium - 234, 23%
Uranium - 235, 6%
Uranium - 236, 9%
Uranium - 238, traces.

The high γ radioactivity necessitates adequate biological protection as well as advanced mechanization and automation (robotization) of the fabrication process.

A recent study concerning $(Th,^{233}U)O_2$ fuel fabrication for CANDU-PHWR reactors [22] presents the structure of a pilot installation of a 100 t year capacity, also outlining the problems raised by the handling of $^{233}U$ in the manufacture of thorium fuel elements.

### 10.12.4. Choice of Matrix Material for Manufacture of Dispersed Fuels

As base-materials for mixed fuel fabrication, the following can be considered:

natural uranium: 0.71% in weight $^{235}U$,
depleted uranium: 0.2 — 0.3% in weight $^{235}U$,
recovered uranium: 0.75 — 0.95% in weight $^{235}U$,
natural thorium
recovered thorium.

In most plutonium programmes natural uranium is the matrix material. Accordingly, plutonium fuel elements design for CANDU-PHWRs might well make use of the experience gained with the use of natural uranium in these reactors.

The design of plutonium fuels for LWRs must consider utilization, as matrix material, of the lightly enriched uranium obtained from these reactors' spent fuel reprocessing.

REFERENCES

1. Pascal, P. (Ed.) (1970). *Nouveau Traité de Chimie Minérale* (tome XV, Cinquième Fasc.). Masson et Cie, Paris.
2. Cleveland, J. M. (1970). *The Chemistry of Plutonium*. Gordon and Breach Science Publishers, New York.

3. Galateanu, I. (1981). *Procese si metode radiochimice in energetica nucleara*, Edit. Academiei, Bucharest; Bunus, F. (1981). *Actinidele si aplicatiile lor*, Edit. Stiintifica si Enciclopedica, Bucharest.
4. CMEA (1975). Isledovanya v oblasty pererabotky oblutchenovo toplyva. In *Proc. 3rd Symp. Marianske Lazne, 24-26 March 1974*, Vol. III. A CMEA Publication; Zilberman, Ya. I. (1961). *Osnovy himicheskoy technologyi iskustvennyh radioaktivnyh elementov*. Gos.Izd.Literatury Atom.Nauky i Techniky, Moskow.
5. IAEA (1980). *INFCE - International Nuclear Fuel Cycle Evaluation*, Final Report IAEA.
6. Mooradian, A. J. (1978). *New Technology and Fuel Cycles*, AECL-6535.
7. Praboulos, J. J., and Noyes, R. C. (1979). *Economic Implications of Symbiotic $^{235}U$ Fast-Thermal Fuel Cycles*. Trans. ANS, 31, 297.
8. Banerjee, S. and others (1976). *Prospects for Self-Suficient Equilibrium Thorium Cycles in CANDU Reactors*, AECL-5501.
9. Slater, J. B. (1981). *Cicluri de combustibil avansat ale viitorului pentru reactoarele CANDU*, a Technical Seminar, Bucharest.
10. Fraser, J. S. and others (1981). *A Review of Prospects for an Accelerator Breeder*, AECL-7260.
11. IAEA (1976). Interregional Training Course, Nuclear Power Planning and Implementation, *Spent Fuel Storage*, Karlsruhe.
12. Mac.Cready, W. L, and Wethington, Jr., J. A. (1981). *Nuclear Technology*, 53, 280.
13. CEA (1973). Report BIST No 184, Sept. 1973.
14. Grant, C. R. (1980). *Heavy Element Separation for Thorium-Uranium-Plutonium Fuels*, AECL-6712.
15. Bell, M. J. (1973). *ORIGEN — The ORNL Isotope Generation and Depletion Code*, ORNL- 4268.
16. Walker, W. H. (1975). *FISSPROD — An improved fission product accumulation program*, AECL-5105.
17. CEA (1975). Report BIST No 208, Nov.1975.
18. Whitesides, G. E., and Cross, N. F. (1969). *KENO-IV-A Multigroup Monte Carlo Criticality Program*, CTC-5.
19. IAEA (1977). *Regional Nuclear Fuel Cycle Centers*, Vol. I and II.
20. IAEA (1976). *Safe Transport of Radioactive Materials*, Safety Series No 6.
21. Mooradian, A. J. (1976). *CANDU Fuel Cycles - Present and Future*, AECL-5516.
22. van den Benden, E. (1981). *Nuclear Technology*, 53, 186.
23. Feraday, M. A. (1981). *Nuclear Technology*, 53, 176.

CHAPTER 11

# Quality Control of Nuclear Materials

## 11.1. GENERAL

As in many other fields, *quality control*, that extends conspicuously over the entire research, development and production areas of nuclear materials, is meant to certify the full compliance of each material with the designed requirements, thus warranting confidence in its safe performance under anticipated conditions. When in service, nuclear materials undergo much more severe thermal, mechanical and other compound stresses than the common ones. Moreover, as a result of irradiation, their properties are subject to alterations; also, neutron economy at core level requires the use of only high (nuclear) purity materials. Reactors' safe and sound operation, as well as their remarkable record of minimal net impact on the environment, owe much to the outstandingly special nature of the technology devised to obtain, process and manufacture them — one that gives the closest to 100%-proof confidence in their predicted in-service behaviour — and, on the other hand, to the careful definition and extensive use of a coherent and adequately redundant quality control system.

The obvious intrinsic importance of the quality control when it comes to nuclear facilities — considerations of nuclear safety and sound economic performance included — together with the awareness that a special, very high, level of technical skill and a demanding discipline are required to conduct such a control properly, all these coined the expression "Nuclear Quality", in relation to which such a high authority as the International Atomic Energy Agency identifies two prominent, intertwinned though clearly distinguishable preocupations: "quality assurance", and "quality control", currently recognized by the acronym "QA/QC".

In Romania, a Law on Quality Assurance of Nuclear Facilities was enacted in 1982. The law regulates licensing of all enterprises that participate in the production of any nuclear facility; establishes QA provisions for design, manufacturing, civil engineering codes, building/assembling, operation, handling, transport and storage of all related materials, as well as the obligations and liabilities of all departments — government and others; provides the regulatory terms of reference to the specialist QC entities; and enacts appropriate penalties for non-compliance with the legal regulations.

The law details the nuclear facility-related products and services that it covers. These are:

1. Products manufactured in accordance with the contractor's specification for use in nuclear facilities: materials, components, subassemblies, equipment, subsystems, systems and buildings.

2. Services on contractor/owner's order, concerning these products: design, supply, marketing, import/export, shipping, delivery, storage, etc.

3. Services on contractor/owner's order, concerning the erection and operation of nuclear facilities: besides the above — consulting, technical assistance, personnel training, building/assembling, commissioning, operation, maintenance, routine repairs and overhaul, and decommissioning.

All these products and services, as well as the operation of the nuclear facilities, are subject to licensing and authorized inspection, as provided by law. Erection, operation, product manufacturing and services as related to nuclear facilities develop in accordance with quality assurance programmes.

*The quality assurance programme* establishes the necesary technical and organizational framework in each enterprise in order to assure product and service quality, to prevent and detect any departure from technical specifications and regulations, to correct all non-conformities, providing for nuclear safety, operational reliability and sound economic performance of the nuclear enterprise. QA programmes comprise: (1) the QA manual; (2) the QA procedures; (3) the QC inspection, verification and testing check-list.

1. *The quality assurance manual* is the document introducing the quality assurance concept as adapted to the profile of each unit that supplies products and services to the nuclear facilities. It defines, *inter alia*: (a) the managerial requisites that are in order, to cope with the quality assurance requirements; (b) the procedure of periodic updating of the quality assurance manual; (c) the procedure of elaboration of the QC inspection, verification and testing check-list; (d) the procedure for the quality assurance technical examinations; (e) the list of other quality assurance procedures, also defining their role within the QA programme as well as manners and modes in which they are to be implemented.

2. *The quality assurance documented procedures* are the documents in accordance with which all activities — functional, quality control, verification and test qualification – are performed.

3. *The quality control verification and testing check-list* is the document which lists and describes all acts of quality control, verification and testing required for each execution phase for a product or service.

*The technical examination and inspection plan* is the document by which the QA/QC activity is planned from the inside of an enterprise in relation with suppliers' duties. Effective control consists of:

1. *Quality assurance technical examination.* This is targeted mainly towards assessing QA effectiveness and efficiency. The technical examination is performed by the unit itself, on its production processes, as well as by its contractors and by the State Inspection for Nuclear Activity Control and Nuclear Quality Assurance, to all nuclear facilities.

2. *Quality assurance inspections.* These are performed either extemporaneously, or in pre-established physical stages by the executing unit itself, by the contractor and by the State Inspection for Nuclear Activity Control and Nuclear Quality Assurance, to all nuclear facilities.

As per the law, quality control of nuclear materials and components covers research, design and operation stages on an equal footing.

QA/QC objectives on nuclear materials are attained through a compound of destructive and non-destructive, conventional and non-conventional tests that are adapted to each material. These tests are complemented with material behaviour analysis in conditions relevant to their future performance in nuclear reactors. To this effect, simulation tests of in-pile behaviour and irradiation tests are in order. Irradiation tests, as performed in material testing reactors (MTRs) and other special devices, provide for assessment of materials' merits, of trial- and commercial reactor components (fuel elements, control rods, etc.).

In the following sections some of the current non-destructive methods employed in the control and characterization of nuclear materials are presented. Then, as an illustration, the modes of the quality control in the particular case of the ceramic fuel elements clad in metallic materials will be reviewed.

## 11.2. NON-DESTRUCTIVE CONTROL METHODS

Non-destructive inspection methods allow determination of a material's physical characteristics without altering its integrity and properties, thus preserving its potential for effective utilization. The main targets of these methods are to qualify a material's homogeneity, as well as to identify the nature, location and size of the defects.

Defects that can be identified by non-destructive methods fall into three classes:

(a) primary defects, induced during the preparation of the material;
(b) processing defects, induced during processing;
(c) operational defects, induced in the material during its service.

Among the defects and structural inhomogeneities in these three categories are: surface and internal cracks, porosity, discontinuities, inclusions, variations of the crystallite sizes, modifications induced by thermal treatment, changes in chemical composition, etc.

There are many non-destructive methods employed in the characterization of materials; among them are visual methods, thermal methods, radiographic methods, ultrasonic methods, and electric methods [1].

### 11.2.1. Visual Methods

These methods generally imply sample exposure to a light beam and its examination, either visual or by means of optical instruments (magnifying glasses, periscopes) and electronic devices (photo-electric cells etc.). Surfaces of the samples must be carefully cleaned, in order to alleviate any chance of disturbance on the observation or the missing of some surface defects. The minimal size of the defects that may be easily seen depends on the quality of the examined surface, on the intensity of light, on the contrast between sample and environment. The eye sensivity differs with the wavelenght $\lambda$, showing a maximum at $\lambda \sim 550$ nm. Once the surface is prepared, one may examine surface defects (cracks or inclusions). Various areas which require further destructive examination (metallographic preparation and examination) can also be identified and zeroed-in.

The potential for insight of such investigations can be expanded by use of some optical instruments (magnifying glasses, microscopes). The photographic technique is ordinarily used in order to put on the record the anomalies on the material surface.

In order to qualify the flatness of the surface, one may use an examination method based on light interference. The light coming from a monochromatic source crosses a perfectly flat glass plate, placed on the surface of the part under investigation. Any departure from flatness generates interference fringes. According to the number of fringes on the unit-length, one may obtain the local pitch angle of the surface.

Surface roughness is determined by means of optical and electric methods. In the optical method, the roughness of the sample is compared to roughness standards. Such an optical device is introduced in Fig. 11.1. An optical system illuminates the surfaces of a standard sample and the surfaces of a mechanically processed sample. The finishing degree is established by comparing the aspect of the two surfaces, zeroing-in the standard exhibiting the roughness closest to that of the examined part.

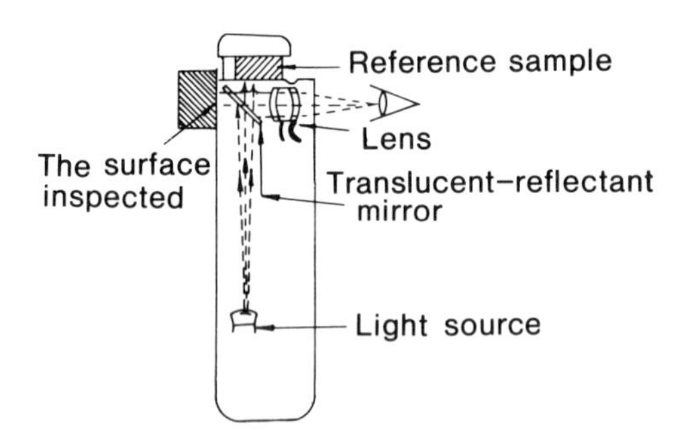


Fig. 11.1. Optical device for roughness determination by comparison with a reference sample.

Another method in the quality inspection of materials uses piezoelectric transducers. A transducer with a pin-point end explores the surface, its movement following the sample microprofile. Its movements are converted by the piezoelectric transducer into electric signals, and then amplified and recorded. The microprofile investigation allows the determination of the mean value (arithmetic or proportional) of the absolute departure from perfect, microscopic, flatness. The transducer's pin is cone-shaped, and ends with a tiny spherical surface, 12.7 μm in radius [2]. The force providing the contact between transducer and sample is ~ $10^{-8}$ N.

Surface roughness is of considerable consequence in maintaining an adequate fuel — clad contact, which is required to obtain a good heat transfer. A good-quality surface is needed for assembling the mechanically processed parts (for instance, at the tube flanging assembly of pressure tube/end fittings), for limiting the fluid pressure drop, as well as to achieve a good tightness by "O"-sealing rings.

### 11.2.2. Thermal Methods

The principle of these methods consists in heating the sample and analysing the distribution of temperatures generated in the material. The defects are "obstacles" for the heat propagation and thus they affect temperature distribution. Heating may be done by direct thermal contact with a heat source, by Joule effect, by induction or by infrared irradiation.

Temperature distribution may be detected by using substances cast on the material surface, which modify their properties at well-determined temperatures (melting, phase transformations, etc.) as well as temperature-sensitive phosphorescent substances; infrared-sensitive films; thermocouples; photo-conductive materials.

In the following, two control procedures are presented, based on the thermal method.

In the thermographic method,the sample is covered with a film of phosphorescent material, ~ 0.025 mm thick, and then submitted to sudden temperature changes (heating and cooling). The sample is then beamed with ultraviolet radiation ($\lambda$ = 365 nm) from a source located at a distance of 60 cm. The light beam makes an angle of 30° with the optical axis of a camera facing the sample. Sample brightness variations at temperature changes, generated by alternating cold ( ~ 0 °C) and hot ( ~ 55 °C) water currents are recorded. Brightness variations are a consequence of the material's thermal conductivity variations, which in turn can be related to the brightness modification of a material with standard defects, placed in the same experimental set-up. Thus, one may obtain data concerning the nature and localization of the defects in the material.

The inspection of the material by thermal methods can also be performed by other procedures involving coloured, temperature-sensitive substances (which modify their colour at certain temperatures); infrared photography (for the fast determination of temperature distribution in heated bodies [1]), etc.

### 11.2.3. Liquid Penetrant Methods

Liquid penetrant methods detect and localize cracks and pores existing on the surface of welds, cast or forged parts, cracks resulting from mechanical stresses, etc. [3]. There are two kinds of penetrant liquids: colouring and fluorescent.

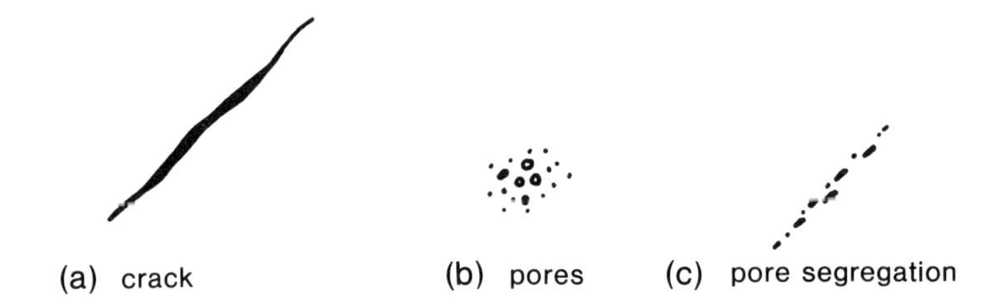


Fig. 11.2. Characteristic pattern of defects, as evidenced by the liquid penetrant method.

The surface to be examined is cleaned by degreasing, then the penetrant liquid is applied by sample immersion in the liquid, painting, or spraying. The excess liquid is removed by means of certain solvents; special care must be exercised in performing this particular operation, so that the excess liquid removal is limited to the sample surface, avoiding removal of the liquid from pores, cracks or from the other defects. A developer is then spread on the surface; it removes the penetrant liquid from the defects and, at the same time, provides a strong contrast with the penetrant liquid. Then the surface examination follows, which is performed visually. Figure 11.2 displays some characteristic images produced by the liquid penetrant methods, that are typical for cracks, pores or pore segregation.

### 11.2.4. Magnetic Particle Methods

These methods are based on the effect upon fine magnetic particles of a non-uniform magnetic field at the level of a magnetized sample surface. The principle is used in identification of cracks or other linear discontinuities. Sensivity is maximal for surface defects, and diminishes as the distance of the defect to the surface increases.

The following methods are used to magnetize the sample [4]:

(a) Passing an electric current between two electrodes, pressed on the sample, at distances of 75 — 200 mm from each other. Depending on the sample thickness, the current intensity $I$ is 3.5 — 5 A/mm; a local magnetization of the sample appears.

(b) Passing a direct or pulsed current between the ends of the part. In this case, the lines of the magnetic field are circular, in planes perpendicular to the sample axis (circular magnetization). For round parts with $D < 125$ mm, the ratio $I/D$ is 28 — 35 A/mm, and for $D > 250$ mm this ratio is 12 – 20 A/mm.

(c) Passing a direct current through a coil winding the part (longitudinal magnetization). The current values are from 45,000 $D/L$ for parts with $L/D$= 2 – 4 to 35,000/($L/D$+2) for parts with $L/D \geq 4$.

(d) Magnetization with permanent magnets or electromagnets.

After cleaning, the part is magnetized; then fine ferromagnetic particles are spread on it. There are three types of particles:

1. "dry" particles, usually coloured, in order to provide a contrast with the sample surface; the method may be used up to $T$ = 590 K;
2. "wet" particles, in suspension in a fluid, applicable up to $T$ = 330 K;
3. particles coated with phosphorescent paint, ultraviolet light-sensitive, at $\lambda \simeq 365$ µm.

Discontinuities on the surface determine local agglomerations of particles. Examination is usually performed by magnetizing the sample in two normal directions, in order to identify the defects with different orientation.

The method is used in the inspection of cast products and forgings, and in the identification of the cracks produced by fatigue.

### 11.2.5. Defectoscopic Methods with Penetrant Radiations

Penetrant radiation defectoscopy is one of the most frequent non-destructive methods now in use to obtain information on the material's inner structure. The radiographic methods employ electromagnetic radiation and particle beams (X-rays, $\gamma$-rays,electrons,neutrons). Knowing the incident beam characteristics and analysing the emergent beam, one may decide on the material quality. Radiography involves three items: the radiation source, the tested material, and the detector.

As radiation source, one may use X-ray generators (Roentgen tubes, tuned transformer generators, linear accelerators, electrostatic generators, betatrons); $\gamma$-ray sources, natural or artificial; a nuclear reactor or other neutron sources for neutron radiography. The emergent beam is recorded on radiographic films or is displayed on fluorescent screens, electron-optical amplifying tubes, or closed-circuit television systems. When passing through the material, the incident radiation intensity is modified by the presence of defects. The differential absorption is illustrated in Fig. 11.3.

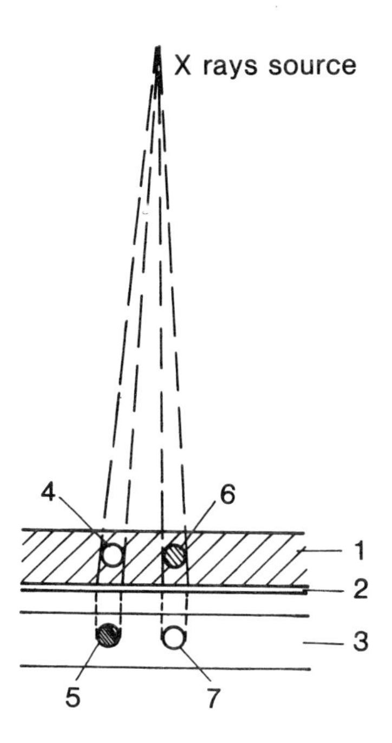


Fig. 11.3. Imaging by X ray radiography: 1 - part to be inspected; 2 - film; 3 - image on film; 4 - pore (low-density inclusion); 5 - pore image; 6 - high-density inclusion; 7 - inclusion image.

In the case of film radiography, the material's internal structure results from the analysis of the blackening density $D$, defined as the decimal logarithm of the opacity (the opacity being given by the ratio of the incident to the transmitted light). Radiographic imaging quality depends on many factors, related to exposure technique (geometrical factors, source dimensions and shape, source — object distance, object — radiofilm distance, object dimensions and shape, film orientation); radiation quality; film quality, etc.

Evaluation of the defect dimensions and verification of contained structures is performed by the examination and interpretation of the radiographs. In analysing radiographs, one must keep in mind that they are only the projections on the film plane of the true defects, their magnitude depending on the geometry of the experimental set-up.

The film radiography method is particularly useful in the quality inspection of weldings. Among the internal defects existing in these one notes gaseous inclusions, non-penetrated welds, non-melted zones, cracks, inclusions, etc. Gaseous inclusions appear on the radiographs as black spots with a regular contour and a high contrast (Fig.11.4).

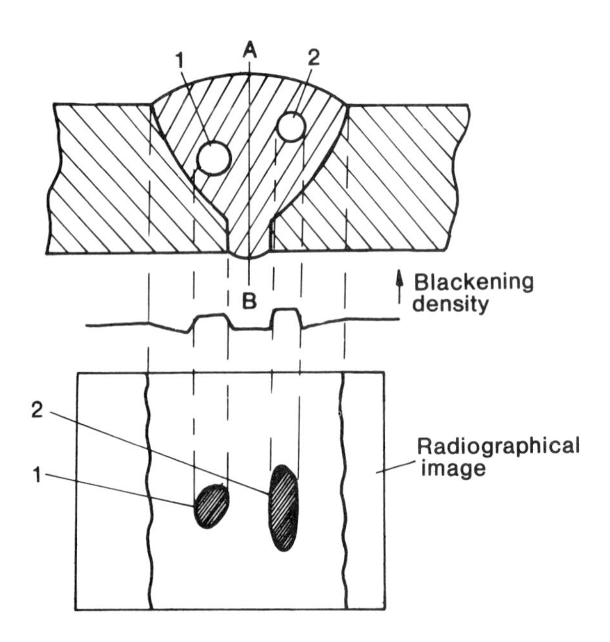


Fig. 11.4. Radiographic image of a welded joint with gas inclusions: 1 - spherical pore; 2 - oblong pore.

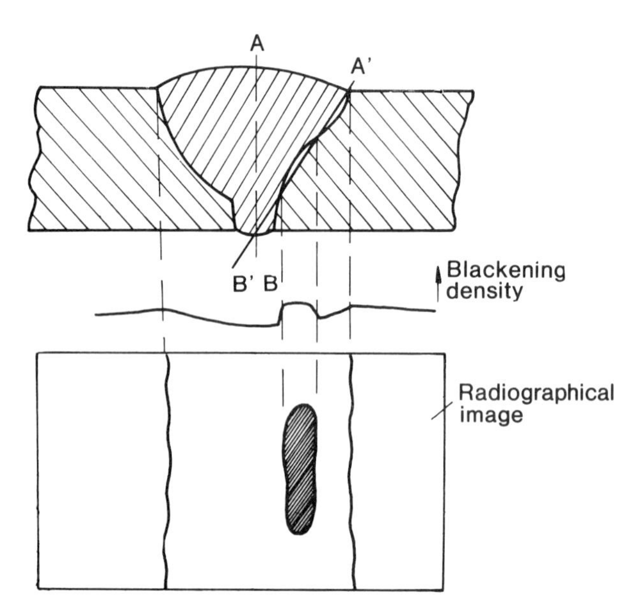


Fig. 11.5. Radiographic image of a welded joint with unmolten zones: AB — joint axis; A'B' — defect plane.

Surfaces with non-penetrated welds or incomplete welded zones are also easily identifiable. In this case, the radiographs show black zones that are parallel with the welding seam (Fig. 11.5 [5-7]).

Among particle-beam methods, neutron radiography has many applications in nuclear material inspection. The method widens the possibilities of radiographic investigation, facilitating the examination of very thick parts (the neutrons having a great power of penetration) and of high-level irradiated materials, which are otherwise difficult to examine by electromagnetic radiations.

Ordinarily the neutron source is a nuclear reactor. The configuration of a set-up for neutron radiography [8] is given in Fig. 11.6.

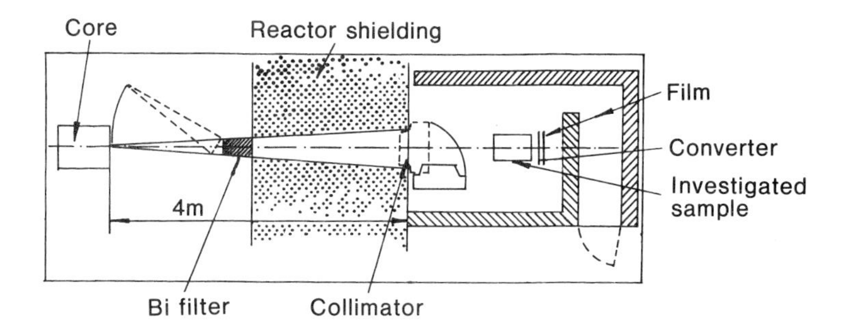


Fig. 11.6. Radiography facility, with neutrons from a reactor.

Radiography is performed by placing the object in a thermal neutron beam, and analysing the emergent neutrons by means of a plate made of a material with large neutron capture cross-section; $\gamma$-radiation as generated in the plate eventually impresses a film in contact with the plate. The method offers a very high resolution and allows an appraisal on the developments with the physical properties during the irradiation of reactor components (fuel elements, control rods) in irradiation experimental devices. Figure 11.7 displays the neutron radiography of a fuel rod ($UO_2$ — zircaloy), before and after irradiation.

Another example is the control of the fuel material's correct positioning in the sheath.

An alternative method is penetrant radiation autoradiography. This allows the fuel material homogeneity control, etc.

### 11.2.6. Ultrasonic Methods

The high efficiency of the ultrasonic method in defectoscopic inspection of materials is due to the fact that ultrasound wavelength is short enough to be compared with the dimension of the defects in materials. Many methods are available for the generation and receiving of ultrasounds; among them are piezoelectric, magnetostrictive, mechanical and thermal methods. Piezoelectric generators are frequently used in defectoscopy. In such generators the transducer is obtained by sintering together strontium and barium titanate. The piezoelectric transducer generates ultrasounds and also receives their echo. A set-up for ultrasonic defectoscopy is illustrated in Fig. 11.8.

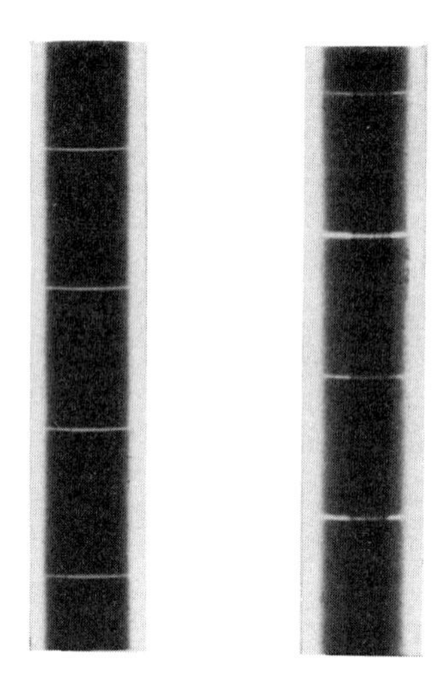


Fig. 11.7. Neutronographic image of a fuel rod [45].

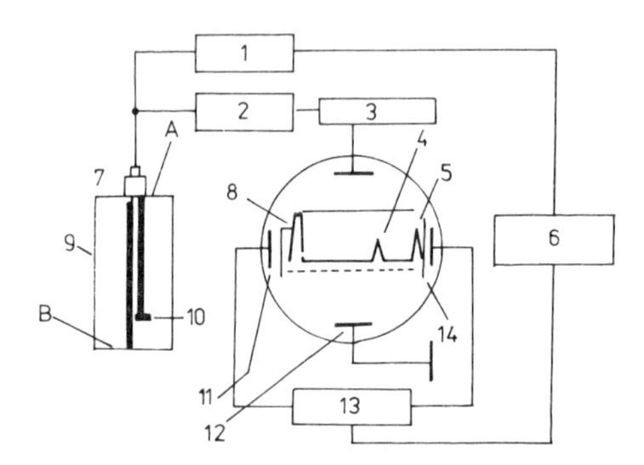


Fig. 11.8. Set-up for ultrasonic control by echoing method: 1 - emitter; 2 - receiver; 3 - amplifier; 4 - signal reflected on defect; 5 - signal reflected on surface (B); 6 - impulse generator; 7 - transducer; 8 - signal reflected on surface (A); 9 - inspected part; 10 - inhomogeneity; 11 - horizontal deflection plates; 12 - vertical deflection plates; 13 - time base; 14 - display.

Ultrasonic propagation in materials comprises three types of waves: longitudinal, or compression (vibrations parallel to the propagation direction); transversal, or shear (vibrations perpendicular to the propagation direction); and Rayleigh, surface waves (a combination of the two above, which propagate across the surface of the sample, regardless of its shape). The choice of one of these modes is made by adjusting the ultrasonic beam incidence angle. Propagation in a single mode is always preferred. In the case of a stainless steel tube, if the incidence angle is between $0^{o}$ and $15^{o}$, one may encounter both transversal and longitudinal waves; between $15^{o}$ and $28^{o}$ only transversal waves exist (the $15^{o}$ angle represents the limit angle for the total reflection of the longitudinal waves). According to the type of waves, one may select the testing frequencies. Detectors with nominal frequencies between 0.5 and 10 MHz are used on longitudinal waves, while those with frequencies between 2 and 2.5 MHz are used on transversal waves.

The ultrasounds propagate very well in liquids and in metals; at the frequencies used in defectoscopy air behaves as a screen, obstructing propagation. Thus, a coupling medium is necessary to assure propagation continuity from the detector to the material. The best medium is mercury; but water or oil are more frequently used in practice, for economic reasons.

Several ultrasonic methods are in use today: the visual method, the shadowing method, the ultrasonic resonance method, the pulse methods [9], the transparency method, and the echoing method. The transparency method uses two transducers: the transmitter and the receiver, the part being placed in between. The echoing method uses only one transducer. The emitted ultrasonic wave train will be received after its reflection on the opposite surface of the investigated part, or on material defects.

Ultrasonic defectoscopy has many applications in the study and quality inspection of materials; it provides for identification of defects (location, size, shape, orientation), quality inspection of the welded joints, dimensional control (determination of the material thickness), quality inspection of thermal treatments and many others.

The procedure used in the identification of defects by the echoing method is illustrated in Fig. 11.9. This method, at frequencies of 2 — 10 MHz, is used in the inspection of stainless steel or Zr alloy clads [6] with diameters of 5 – 15 mm.

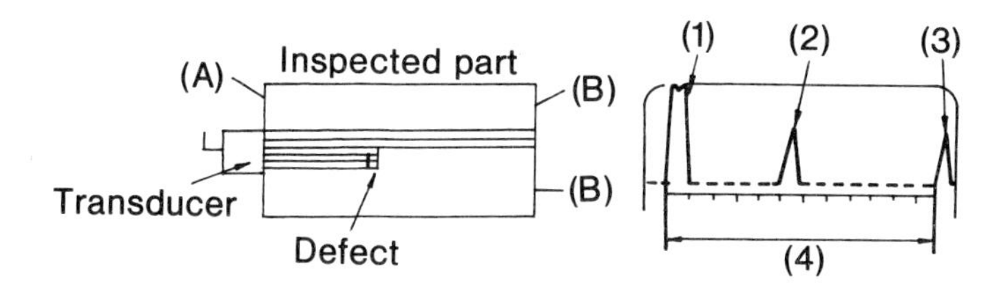


Fig. 11.9. Detection of defects in a material by the echoing method: 1 - initial impulse (signal reflected on surface (A)); 2 - signal reflected on defect; 3 - signal reflected on surface (B); 4-distance equivalent to the piece dimension.

Ultrasonic examination also facilitates the evaluation of material thickness. Accuracy depends on the part's thickness. For parts 0.2 — 0.5 mm thick, the accuracy is 1% [10].

Ultrasonic inspection of end-cap/clad welded joints employs the echoing technique, in a transversal wave mode. Inspection is performed with a focused detector at a frequency of 10 MHz [11] in contact with a tank containing water in which the fuel element in rotary motion is immersed. Echoes are recorded on a tape that is synchronized with the rotation of the fuel element. Fig. 11.10 presents the recorded signal pattern in the case of a non-penetrated joint.

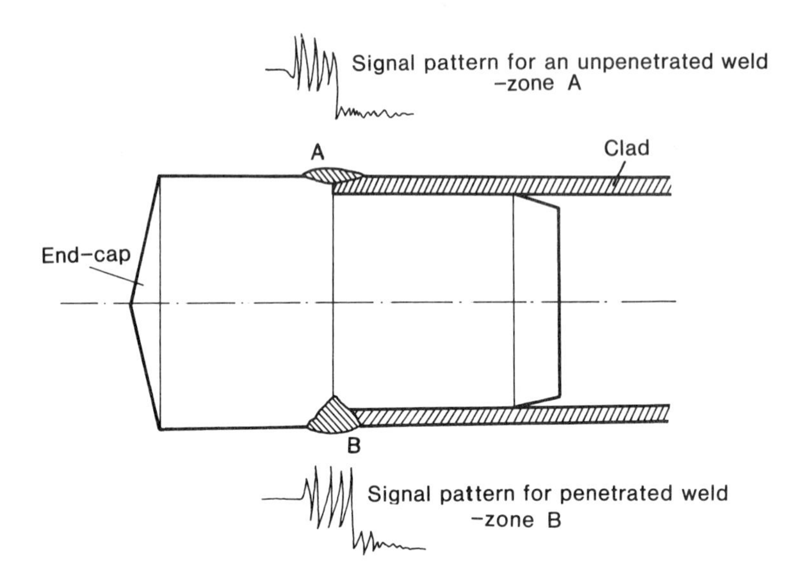


Fig. 11.10. Patterns of signal recorded in ultrasonic control of an end-cap/clad weld, unpenetrated and penetrated, respectively.

The use of ultrasonics in structural characterization of thermally treated materials is based on the comparison of the ultrasonic energy transmitted through the material with that of a standard having the same geometric shape and known grain dimensions. The method was used in the inspection of structural homogeneity of the metallic uranium contained in the EDF fuel elements, after the thermal treatment of the finite element was performed [12]. Inspection is usually carried out by total imersion of the sample in water, at frequencies of 3 — 5 MHz (depending on the grain size).

Ultrasonic methods in material characterization are reviewed e.g. in Ref. [13].

### 11.2.7. Eddy Current Methods

The eddy current method is used in the inspection of conductive materials to identify defects, structural aberrations and compositional changes. Among the applications of this method are: cracks and inclusions identification; measuring of plate and tube thicknesses; evaluation of the thickness of conductive, as well as non-conductive, layers which are applied on a material with conductive properties, etc. The method is especially useful in the study of defects existing near (or on) the material surface.

Its principle is as follows: if an alternative current flows in a coil, the generated variable magnetic field will induce eddy currents in a material near the coil; their magnitude depends on the alternative current frequency, as well as on the electric conductivity, magnetic permeability, and shape of the piece, on the relative coil — piece position, on the material's discontinuities etc. Eddy currents induce a magnetic field in the metallic piece of oposite sign compared to the initial field; thus, the "excitation coil" impedance and that of any probing coil are affected by this magnetic field. The path of eddy currents is disturbed by the presence of defects or inhomogeneities in the material and, consequently, the apparent impedance is modified. Such impedance changes can be measured and used to obtain information concerning the defects and other physical, chemical and metallurgical irregularities in the material [14]. For a plane conductor the induced eddy currents are concentrated near the material surface; their intensities decrease exponentially with the distance from the surface.

There are three types of excitation coils: concentric, probing, and internal (Fig. 11.11) [2].

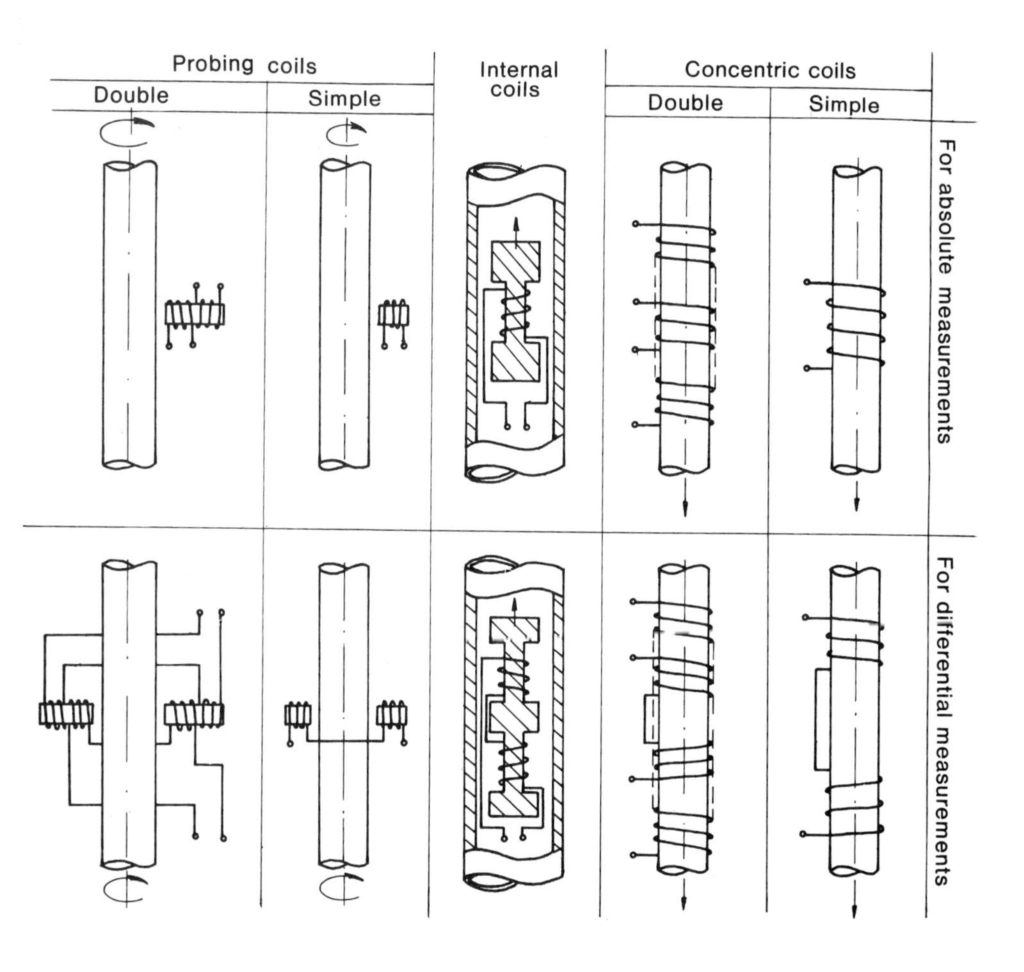


Fig. 11.11. Typical coils for control by eddy currents.

Eddy current inspection is performed by several measuring methods [2,15]. In the multi-frequency method [16], eddy currents of two different frequencies are induced in the sample. The high-frequency current allows compensation of variations in the distance between the coil and the sample, while the low-frequency current is used to investigate the sample.

The modulation of a high-frequency, sinusoidal oscillation, with a signal that is proportional to the resistivity of the controlled part, is used in the oscillograph method [2]. The measuring head is a coil. This coil and the entrance capacities in the measuring device form an oscillatory circuit. For a surface defect, the local resistivity increases and thus the eddy current damping effect on the oscillatory circuit decreases. This method is used in the inspection of metallic uranium rods or steel tubes.

An important application of this method is the determination of the thickness of metallic layers, which are electrolytically applied on other metals. Thus, nickel layers of a thickness between 2.4 and 50 $\mu$ m can be measured with an accuracy of ±5%.

The method can also be used in the determination of the thickness of tubes with thin walls, as used in fuel element cladding [17,18].

Eddy current methods have been used in electric conductivity measurements at high temperatures, in the determination of inter- and intra-granular corrosion, and in the identification of gaseous inclusions in welded joints of some types of fuel elements, etc.

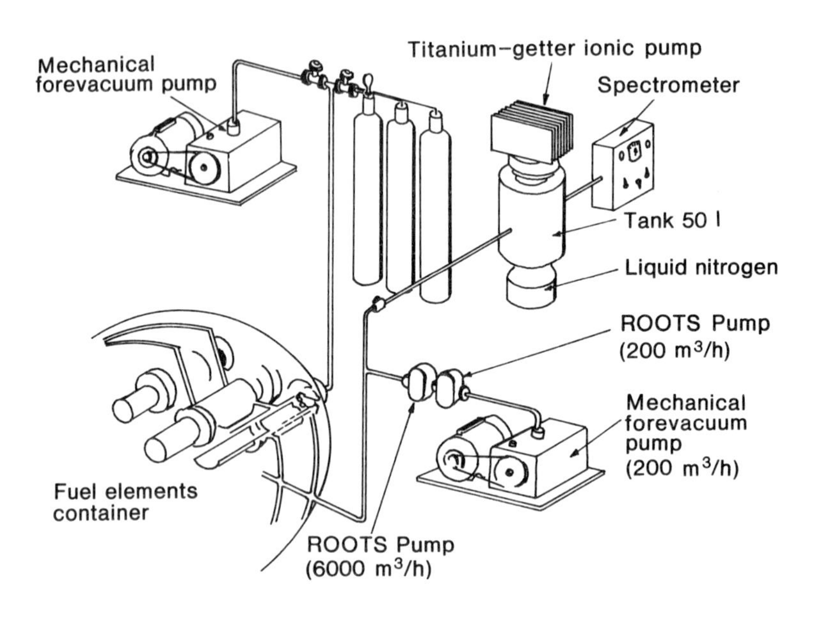


Fig. 11.12. Facility for fuel element tightness control.

## 11.2.8. Tightness Inspection

The high tightness standards in nuclear technology require adequate methods for the identification and pinpointing of leaks. Such methods are frequently used as final inspection operations in some nuclear reactor component manufacture (in the case of fuel rods, for example, the leakages through the end-cap/clad welding imply its rebutting).

In fuel element manufacture, the method for identification of tightness defects consists in the detection of helium leakages [19].

A typical installation for the inspection of the fuel element tightness for the EL4 reactor is presented in Fig. 11.12 [12]. The fuel element is introduced in a container which is evacuated, then pressurized with helium at 3.5 MPa. After depressurization and helium removal from the container, the presence of the helium in the remaining gas is checked by means of a high-vacuum device. If the fuel element is not tight, the helium diffuses through its material and may be detected. When the filling gas of the fuel element contains helium the pressurization is no longer necessary.

Tightness inspection by helium detection does not localize the defect or defects which cause the leak; it only indicates their presence. Yet this is sufficient, because in fuel element manufacture repairs are not admitted, so that the defective element is simply rebutted. There are cases when it is necessary to find the defect location (e.g. for irradiated fuel elements which are submitted to a similar test after irradiation). In such cases simple localization methods are used: fluorescent and colour penetrants, immersion of the irradiated rod in liquids, etc.

## 11.3. DESTRUCTIVE CONTROL METHODS

Destructive methods are in order when one seeks to determine nuclear material properties in the elaboration phase (composition, mechanical and physical properties, structure, metallurgical state, corrosion behaviour, etc.), to verify the material characteristics and to simulate the material's performance in nuclear reactor conditions.

There is a large variety of destructive testing methods. Some examples of fuel element material testings are presented below:

1. composition homogeneity control (Mo in UMo alloys with 0.5 — 1%, used in the fuel element fabrication);

2. detection of impurities that spoil core structural materials (e.g. hafnium in zirconium);

3. control of plutonium accumulation in the fuel material, as a result of reactor operation;

4. determination of fissile isotope concentration in U – Pu fuels;

5. determination of the enrichment in $^{235}U$ of $UO_2$ powders with enriched uranium;

6. determination of impurities in $UO_2$ powders or pellets.

Since the methods are generally conventional, we will confine ourselves to only a brief description of a few of them.

### 11.3.1. Composition

The nuclear purity required by the low neutron absorption cross-section criterion implies, besides knowledge of the composition (alloying elements), accurate information on the impurity content. Hence, the methods used in composition analysis should allow both the determination of the major components (that may account for several per cent) and of the impurity levels (weighed in ppm).

The techniques used are diverse and are based on physicochemical and chemical methods. Among these, are: quantitative chemical analyses, mass spectrometry, analysis by activation, emission spectroscopy, atomic absorption spectroscopy, etc. Analysis methods are generally standardized. For instance, chemical composition determination of the $UO_2$ powders is performed in accordance with the ASTM C-753 [20], etc.

Composition determination is particularly relevant in the evaluation of the quality of the moderator materials for thermal neutron reactors; among these materials is the heavy water (see Appendix 8).

In what follows, some methods used for determining material composition are briefly described.

*Mass spectrometry* [21] is a very accurate and sensitive isotopic analysis method. It may be used both for gaseous and solid samples.

Uranium isotopic analyses may be performed on solid $U_3O_8$ samples (in ion source spectrometers with thermal ionization) or on gaseous $UF_6$ samples. The accuracy is around 0.2% for solid samples, and 0.02% for gaseous samples. The high accuracy gain with the $UF_6$ gaseous samples is counterbalanced by handling difficulties, since uranium hexafluoride is extremely corrosive.

Mass spectrometry may also be used for the radioactive determination of the age of rocks. To this purpose the relative isotopic abundances are determined for $^{206}Pb/^{238}U$; $^{207}Pb/^{235}U$; $^{207}Pb/^{206}Pb$; $^{210}Pb/^{206}Pb$, etc.

Another possible application of mass spectrometry consists in the evaluation of some representative fission product concentrations that may supply data about nuclear fuel burn-up. It was established that the stable isotope $^{148}Nd$ is most suitable for this effect. This isotope has the same yield from $^{235}U$ fission as from $^{239}Pu$ fission. Its fission yield does not depend on neutron energy [22].

There are several other features that make neodymium suitable for the characterization of the fission process. Out of the seven stable isotopes of Nd, only six are produced by fission (see Fig. 11.13), so that $^{142}Nd$, which is not produced by fission, may be used as reference for the calculation of the relative abundance of the other isotopes. This method allowed the identification of some zones in the uranium mines in Gabon, where natural nuclear reactors had operated in the distant past.

Mass spectrometry is also employed in $D/H$ isotopic analysis, that is relevant in the quality control of heavy water. To this purpose the water is decomposed and the resulting hydrogen is analysed.

*Chemical methods* make use of analytic chemistry procedures, either for determination of the presence of some chemical elements (qualitative analysis) or for determination of an element concentration in a certain material (quantitative analysis).

Chemical methods are accurate; they also have a good selectivity, but generally are destructive methods, also difficult to automate.

*Radiochemical methods* consist of dissolving the sample, followed by extraction, by a chemical process, of the component to be examined and then in the measurement of the natural radioactivity of the isotope [23]. These methods are extremely diverse, depending on the considered elements. They were

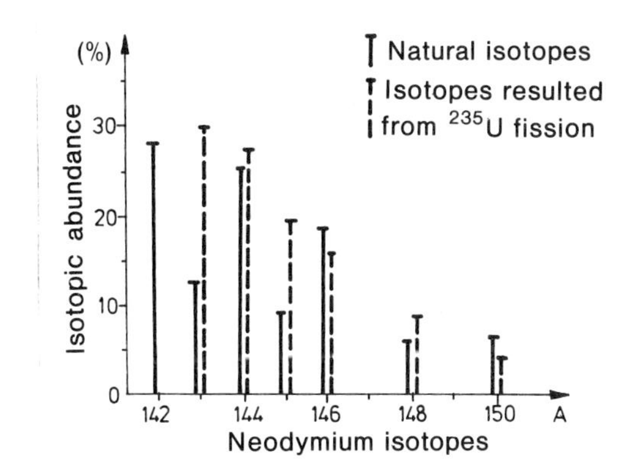


Fig. 11.13. Abundance of natural neodymium, in comparison with the abundance of the Nd isotopes resulted from $^{235}$U fission.

selected from analytic chemistry and were applied in the radioelements' chemistry by Marie Curie. As an example, radium is co-precipitated with barium sulphate; the separation efficiency is established by means of $^{133}$Ba; the amount of $^{266}$Ra may then be determined by measuring the activity and taking into consideration the radioactive chain products.

Thorium is co-precipitated with calcium and magnesium phosphate, from which is then separated by ion-exchange techniques.

*Optical spectroscopy* methods are based on the isotopic shift of spectral lines. In Table 11.1 the wavelenghts and the isotopic shifts for some elements in the visible and ultraviolet range are given [24].

Isotopic shifts are very small. One of the main merits of the spectroscopic method resides in its high sensitivity, the resolution limit in optimal conditions reaching $10^{-12}$ g.

In the study of uranium isotopic abundance, the problems with the reproducibility of the excitation conditions and of the compound's volatilization are not of consequence, since both isotopic constituents of U belong to the same chemical species. Difficulties occur, however, as a result of the complexity of uranium's optical spectrum. For the purposes of uranium isotopic analysis by atomic emission spectroscopy, the 424.437 nm line is employed; it presents an important enough isotopic shift. A high resolution is usually obtained with Fabry — Perot standards; for the hydrogen isotopes, where resolution may be lower, diffraction lattices with a 0.5 — 1 nm/mm dispersion may be used.

For water and HDO or $D_2O$ molecular species analysis, absorption or emission molecular spectroscopy can be employed. Some data concerning the analysis by molecular spectroscopy of the isotopes H and D are given in Table 11.2.

The use of tuned dye lasers as radiation sources will in the future improve spectroscopic sensivity and accuracy [24]. The spectral domains as discussed here may be covered by two types of lasers:

1. dye lasers, in the range 340 – 1200 nm;
2. solid-state lasers, in the range 630 – 3400 nm.

TABLE 11.1 Isotopic Displacements of Spectral Lines used in Isotopic Analysis

| *Element* | *Wavelength* (nm) | *Isotopes* | *Isotopic displacement* (nm) |
|---|---|---|---|
| H | 656.279 | H/D | 0.1784 |
| | 486.133 | | 0.1320 |
| | 121.568 | | 0.0329 |
| He | 667.815 | $^{3}He/^{4}He$ | 0.0501 |
| | 587.562 | | 0.0042 |
| | 501.568 | | 0.0213 |
| | 388.865 | | 0.0212 |
| C | 247.857 | $^{12}C/^{13}C$ | -0.0009 |
| | 283.671 | | -0.0049 |
| Ar | 451.073 | $^{36}Ar/^{40}Ar$ | 0.0010 |
| | 457.938 | | 0.0021 |
| Th | 377.138 | $^{232}Th/^{230}Th$ | 0.0068 |
| | 439.111 | $^{232}Th/^{229}Th$ | 0.0075 |
| U | 460.986 | $^{238}U/^{235}U$ | 0.0077 |
| | 424.437 | | 0.0248 |
| | 472.273 | | -0.0058 |

TABLE 11.2 Hydrogen Isotopes Analysis by Molecular Spectroscopy

| *Isotope (atomic concentration)* | *Analysed band* (nm) | | *Accuracy (absolute values)* | *Molecular spectroscopy* |
|---|---|---|---|---|
| 0.03% D | 3980 | $D_2O$ | 0.003% | |
| 0.8% D | 3980 | $D_2O$ | 0.01% | |
| 0 — 5% D | 3980 | $D_2O$ | 0.05% | Absorption |
| 99.5 — 100% D | 2940 | HOD | 0.003% | |
| 0 — 5% $H_2O$ in $D_2O$ | 1470 | $H_2O$ | 0.5% | |
| 0.015% D | 306.4 | OH | 0.0006% | |
| 4% D | 306.4 | OH | 0.3% | |
| 20% D | 306.4 | OH | 0.4% | Emission |
| 60% D | 306.4 | OH | 0.3% | |

The gamma spectrometry method is based on the comparison of the γ or X spectral line intensities of the investigated isotopes with the same lines as emitted by standard samples. For this, one may use spectrometers fitted with multichannel analysers and scintillation detectors (NaI (Tl) crystals) or semiconductor detectors (Ge or CdTe). Devices are easy to handle and transport, and do not require calibration prior to each measurement. With this type of facility, one may study:

(a) the radiations emitted by the isotopes in the material, either directly, or by measuring the gaseous radioactive effluents' activity;
(b) the X-ray fluorescence, excited by beaming the material with charged particles (electrons, $\alpha$-radiations, $\beta$-radiations, emitted by natural sources, or protons and heavier ions accelerated in particle accelerators);
(c) the radiation emitted by the radioactive isotopes generated in the material by particle beaming (radioactivation). Radioactivation may be achieved by using thermal neutrons, fast neutrons, protons, deuterons, $\alpha$ particles or $\gamma$-photons.

Fissile U and Pu concentrations in fuels can be determined by $\gamma$-spectrometry. The enrichment in $^{235}U$ of the uranium dioxide as low enriched powder or pellets may be determined in the same way. To this purpose, the 180 keV line of $^{235}U$ is monitored. The method may be directly used in uranium dioxide pellet production (the process can be thus automated and computer-controlled).

X-ray fluorescence may be employed in the control of Pu enrichment of (U,Pu) fuel materials. To this purpose, the Pu/U ratio of the intensities is analysed. The plutonium's and uranium's $K_\alpha$ radiation (99.5 keV and 94.7 keV, respectively) is excited by the $^{57}Co$ $\gamma$-radiation of 122 keV. Uranium's and plutonium's $K_\alpha$ fluorescence radiation penetrates the clad, thus allowing determination of Pu/U concentration. The accuracy obtained is 1% for measurements that do not last more than 1 — 5 minutes. The equipment has to have a high resolution and, for this, semiconductor detectors are used.

The activation method may be used for the detection of fluorine traces in zircaloy clads. To this purpose, the nuclear reaction $^{19}F(p,\alpha)^{16}O$ [25] is of assistance.

*Magnetic resonance methods* are based on quantum absorption of electromagnetic radiation by atoms or nuclei. The energy levels between which the transition phenomena take place are Zeeman levels of the nuclear magnetic moment or of the magnetic moment of the atom placed in a magnetic field.

Nuclear magnetic resonance (NMR) may be directly performed on all the nuclei that have a non-zero magnetic moment. Investigations have considered the nuclei $^{1}H$, $^{2}D$, $^{11}B$, $^{14}N$, $^{13}C$, $^{15}N$, $^{19}F$, $^{27}Al$, $^{35}Cl$, $^{51}V$, $^{103}Rb$, $^{127}I$, $^{119}Sn$ [26]. Direct NMR on $^{233}U$ and $^{235}U$ nuclei can be performed only with difficulty because they exhibit very low nuclear magnetic moments.

Electron paramagnetic resonance (EPR) can be performed on all systems exhibiting unpaired spins.

These methods prove very useful in determining isotopic or atomic concentrations, as well as the mutual positioning of atoms in molecules or solids.

Nuclear magnetic resonance may be considered as a passive investigation method, of an extreme selectivity, good sensitivity and accuracy. One great advantage of the method is that it can be effectively implemented on-line, at any separation plant.

It is believed that the study by NMR of nuclei with which the uranium gives compounds could offer good prospects for uranium isotopic composition analysis. Thus, the nuclear magnetic resonance on $^{19}F$ in $UF_6$ reveals interesting isotopic effects [27]. The enrichment effect is emphasized by the analysis of the relaxation time, and is determined by modifications in the $^{19}F$ — $^{235}U$ interaction. The effect has been evidenced on the liquid and gaseous phases of uranium hexafluoride (Fig. 11.14 [28]).

Under special conditions electron paramagnetic resonance can prove a very accurate method of analysis for the uranium ions isotopic composition (valences +3, +5), when the ions are hosted in solid matrices. Their

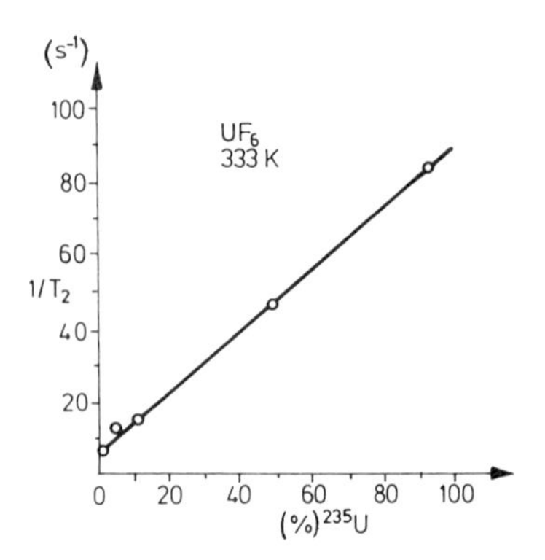


Fig. 11.14. Isotopic enrichment effect of $^{235}U$ in $UF_6$, as evidenced by alterations in the relaxation time.

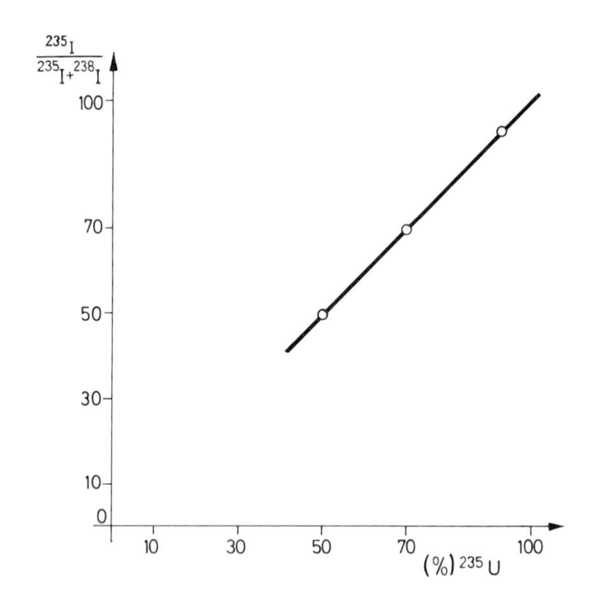


Fig. 11.15. Dependence upon the content in $^{235}U$, of the relative line intensity of $^{235}U$ in sodium fluoride.

concentration is determined from the ratio of the $^{238}U$ central signal intensities to the intensities of the hyperfine components from $^{235}U$ [29,30]. To this effect, the pentavalent uranium — sodium fluoride system was found to be the most adequate, when measurements were performed at liquid nitrogen temperature. A linear dependence on the relative concentration of the uranium isotope, of the ratio of line intensities of a given isotope to the intensity of all the lines was clearly evidenced (Fig. 11.15). For abundances higher than 75% in $^{235}U$, the absolute accuracy has reached 0.1%. One merit of the EPR method consists in the fact that it allows the isotopic content analysis for almost every uranium compound, except for those which evaporate at temperatures lower than the melting point of sodium fluoride.

On these results, Ursu *et al.* [31] have been the first to propose the utilization of magnetic resonance methods for the $^{235}U$ analysis and enrichment control in various phases of uranium's nuclear fuel cycle.

Structural and magnetic properties of several uranium compounds are presented in Appendix 10.

### 11.3.2. Mechanical Properties

The methods used in the determination of mechanical properties of materials give the elasticity modulus [32], the fracture strength, the yield strength, and the elongation [33], in conditions almost similar to those of the reactor operation. These methods are employed both in the selection of materials for fabrication processes and in the material reception control, care being taken to ensure properties compatible with a good in-pile service. They are also in order for the study of irradiation effects on materials, when the analyses are usually performed in special facilities known as "hot cells", that ensure radiation protection to the personnel.

Several methods, either special or conventional, are available, among which are tensile strength tests (to determine fracture resistance, elasticity modulus, elongation, etc.) resilience tests (impact strength determination), fracture resistance under biaxial stress (explosion under internal pressure) [33], hardness determination, etc. Performing these tests requires adequate sample preparation in accordance with national or international standards, availability of some specific devices operating in different ranges of mechanical characteristics of the sample, provision of monitoring capability to record the conditions in which the tests take place (temperature, pressure) as well as the material's response to the stresses it is subject to.

As an illustration, in the following lines the breaking test under biaxial stress (explosion) for the zircaloy-2 tubes [33,35] is briefly described. The test aims at determining the effect on the material's mechanical properties of the circumferential precipitation of zirconium hydride plaquettes. Samples are first hydrided up to ~500 ppm $H_2$, then sealed by end-cap welding through which a capillary tube passes in order to enable introduction of the fluid under pressure. Testing is performed in an oven, and the temperature is measured by means of a cromel — alumel thermocouple. Samples are rapidly heated at the testing temperature, and kept for 15 minutes at this temperature; after that, the internal pressure is applied, by pumping in argon. The explosion resistance ($\sigma_e$) is determined with the expression:

$$\sigma_e = (PD)/(2t), \tag{11.1}$$

where $P$ is the internal pressure at the intensity that produces explosion, $D$ is the average internal diameter of the tube and $t$ is the tube wall thickness. The variations of the explosion resistance and of the total circumferential yield are presented in Figs 11.16 and 11.17 [33]. The material's resistance decreases by ~ 50% when the temperature increases from the ambient temperature to ~ 400 °C.

The resistance of the hydrided tubes ( ~ 460 ppm $H_2$) is greater than that of the non-hydrided material, this being determined by the material's hardening, as a result of the dispersion of the zirconium hydride precipitates.

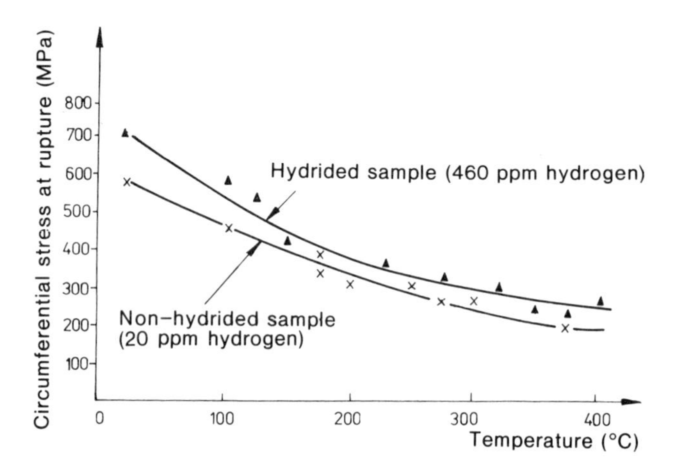


Fig. 11.16. Strength variation with temperature, for a zircaloy-2 tube.

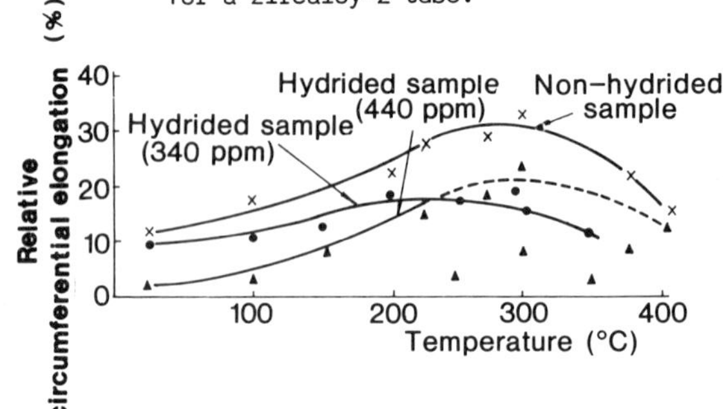


Fig. 11.17. Variation with temperature of the relative circumferential elongation per unit length, for a zircaloy-2 tube.

### 11.3.3. Metallographic Methods

Ordinarily, the methods used in the metallographic analysis are conventional [34]. Their purpose is to determine the material's microstructural characteristics on the knowledge that their macroscopical properties are conditioned by these characteristics. The material's structural characterization is performed in order to determine the crystallites (grains) dimensions, the presence and extension of various phases (precipitates) [36], the induced modifications, the plastic deformation (to distinguish the crystals and the slip bands twinning), as well as the effects produced near the welded zones or the welding quality [37].

These objectives are achieved by standard metallographic techniques, which include selection and incorporating operations, mechanical grinding, mechanical and electrolytic polishing, chemical and ionic etching and microscope examination. The operations preceding the examination are sample selection and preparation. Sample incorporation in an easily fusible alloy, in order to obtain perfectly planar surfaces, is a common operation in metallography, and so are also the other preparative operations. The metallographic microscope provides for the study of materials under optimal conditions: illumination in normal or polarized light, variable incidence angles of the incident beam, 1000x magnifications, photographic facilities, sample heating during examination for monitoring structural changes, etc. Some microscopes are even fitted with devices for material microhardness determination.

In Fig. 11.18 the structure of a zircaloy clad sample is presented, that was etched prior to examination in order to exhibit the grain limits. The image is used to determine the average grain size.

Fig. 11.18. Metallographic structure of a sample from a zircaloy clad [37].

Use of metallographic methods in the analysis of the quality of joins and thermally-affected zones is illustrated in Figs 11.19 and 11.20. In Fig. 11.19 a section through the end of an experimental fuel element manufactured at INPR-Pitesti (Romania) and irradiated in the MZFR (FR of Germany) facilities is presented. The sequence (a), (b), (c) of images presents details at different magnifications. The cut was done in order to analyse the irradiation behaviour of the end-cap/clad weldings. The presence in the material (Zircaloy-4) of some α-Zr large grains in the zone of the welding seam is notable, as well as the display of the initial granulation of the material. Figure 11.20 presents the joint quality as obtained by brazing a spacer on a fuel element clad.

The possibilities of investigating the material structure are further expanded when one employs high-resolution electron microscopes, which facilitate the identification of defects in the crystalline lattice (vacancies, dislocations, defects, agglomerations, etc.).

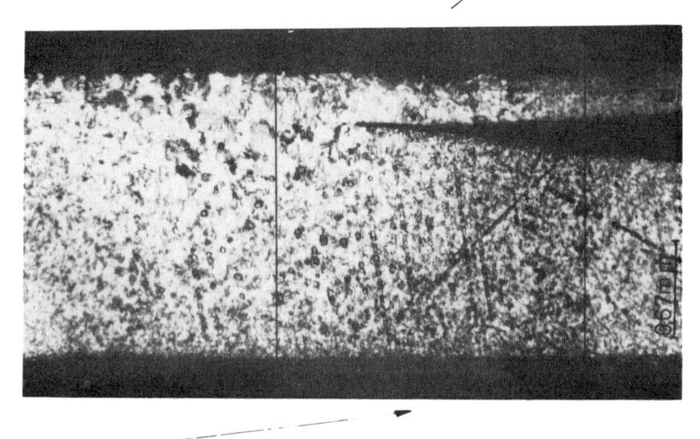

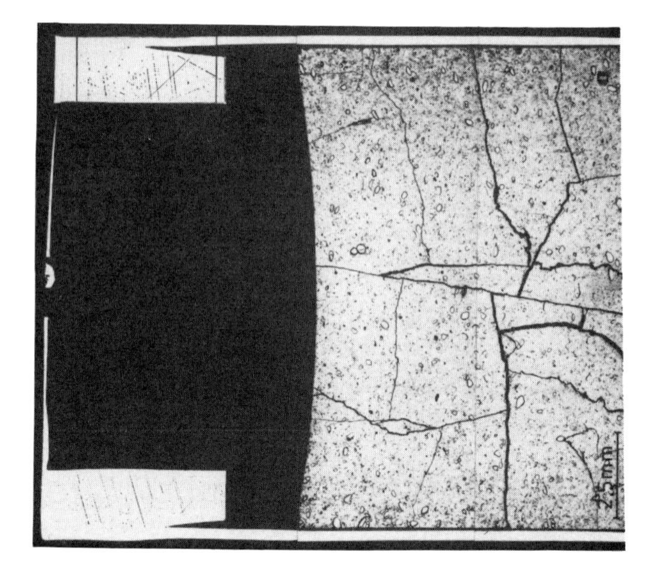


Fig. 11.19. Metallographic structure of an end-cap/clad weld [45].

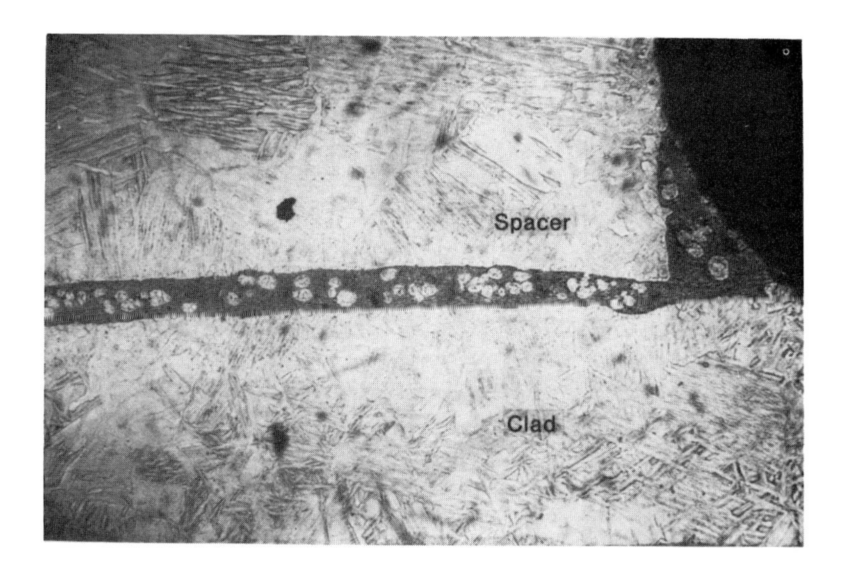


Fig. 11.20. Metallographic structure of a brazed joint [37].

11.3.4. Monitoring of Corrosion Effects

Nuclear material corrosion is studied in facilities which offer conditions similar to those in the reactor, in terms of corrosive environment and operating parameters (pressure, temperature test, duration). Corrosion studies in aqueous media for nuclear metallic materials are usually performed in autoclaves [38].

The testing is preceded by sampling and sample preparation. In the case of zircaloy-2 fuel clads, the sample is first etched in a solution of 67% $H_2O$ + 30% $HNO_3$ + 3% HF [39,40]. Samples are then introduced in an autoclave which contains de-ionized and de-oxygenated water. The exposure temperature is correlated to the phenomenon to be analysed (450 °C for the dissolved hydrogen) while the pressure depends on the type of the reactor considered. The exposure time varies according to the test objective, and may last up to 1000 hours. During the operation the parameters that characterize the testing conditions are controlled: water pressure, temperature, electric resistivity, etc. [40]. Corrosion behaviour is assessed by the weight increment of the tested sample or by metallographic studies which determine the thickness of the resulted oxide layer, the configuration of zirconium hydride plaquettes (Fig. 11.21), etc.

## 11.4. QUALITY CONTROL IN THE MANUFACTURE OF NUCLEAR FUEL ELEMENTS

The manufacture quality inspection in pilot plants or in the commercial production of fuel elements is a continuous testing and verification activity which keeps under control all the processes implied and gives, successively, green lights throughout the flow-chart only to those materials and components that meet specified requirements. Because any fuel element failure in the reactor would ultimately increase the cost of the nuclear power delivered [16], the need for thorough and adequate inspection procedures cannot be exaggerated.

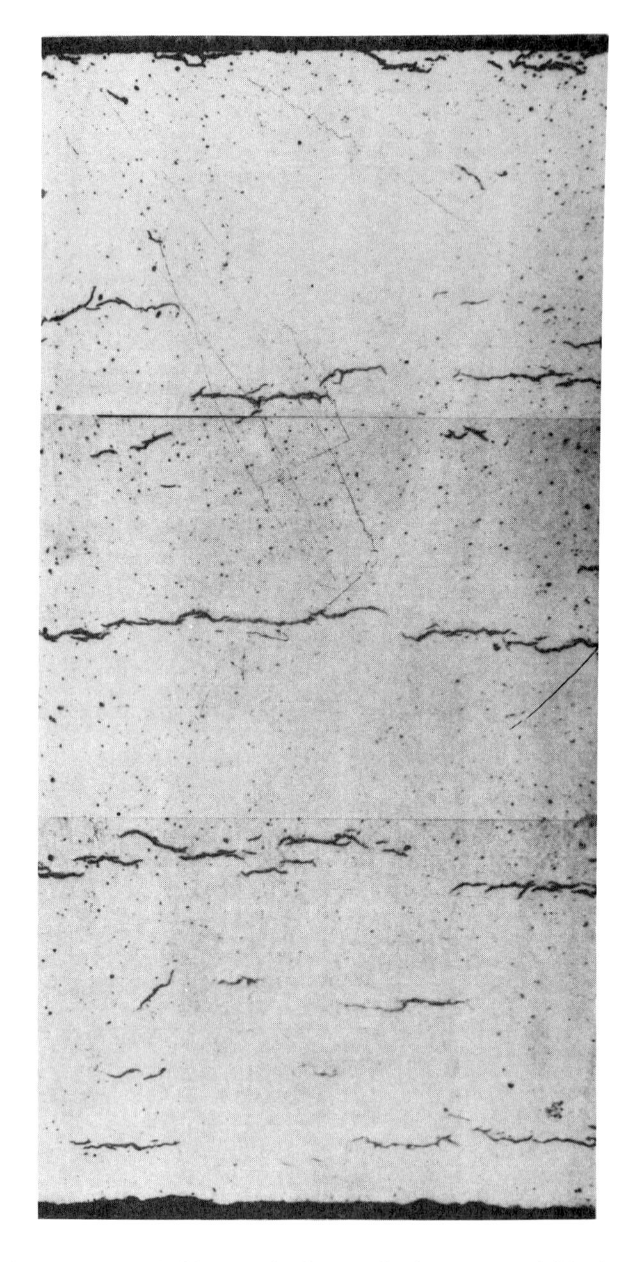

Fig. 11.21. Metallographic image of zirconium hydride [3,44]. plaquettes disposition.

The quality assurance concept is defined as the body of systematic actions required to warrant satisfactory confidence that a product behaves as designed, in the service conditions designed [20]. The quality assurance system implies the continuous evaluation of the effectiveness of the adopted control programme and the implementation of corrective measures, when required. These goals are achieved by a combination of destructive and non-destructive tests, which are systematically applied after performing the manufacturing operations according to the technological flow-chart [18,19,41,42,43]. As an illustration, the sequence of control operations performed throughout the manufacturing flow-sheet of the fuel bundles of ceramic fuels clad in metallic rods [20] is displayed in Table 11.3.

TABLE 11.3 Sequence of Quality Control Stages, by Operation and Overall, in Ceramic Fuel Elements Manufacturing

| *Operations* | *Quality control procedures* |
|---|---|
| I. *Powder preparation* | |
| A. Characterization | A.1. Enrichment control<br>A.2. Chemical analyses<br>A.3. Physical properties |
| II. *Pellet manufacture* | |
| A. Pressing | |
| B. Sintering | B.1. Sintering tests<br>B.2. Microstructure analysis |
| C. Calibrated grinding | C.1. Hardness determination<br>C.2. Density measurements<br>C.3. Dimensional determinations<br>C.4. Chemical analyses and moisture determination |
| III. *Fuel rod components reception* | |
| A. Tubes for clads | A.1. Chemical analysis<br>A.2. Surface quality control<br>A.3. Mechanical tests<br>A.4. Corrosion resistance determination<br>A.5. Defects analysis (non-destructive tests)<br>A.6. Dimensional control<br>A.7. Metallographic control |
| B. End-caps and other inner components | B.1. Material quality control<br>B.2. Superficial defects analysis<br>B.3. Dimensional control |
| IV. *Fuel rod assembly* | |
| A. Superficial treatment (degreasing, passivation, autoclaving) | A.1. Surface quality control<br>A.2. Dimensional measurements<br>A.3. Length measurement |
| B. Welding of the first end-cap | B.1. Non-destructive testing<br>B.2. Metallographic control* |
| C. Pellet stacks forming | C.1. $H_2O$ (moisture) content determination |

Table 11.3 (continued)

| *Operations* | *Quality control procedures* |
|---|---|
| D. Pellet stacks loading in clads | D.1. Loading control |
| E. Second end-cap welding (including injection of filling gas) | E.1. Non-destructive testing<br>E.2. Metallographic control* |
| F. Final preparations | F.1. External clad surface contamination control<br>F.2. Final dimensional control<br>F.3. Weighing<br>F.4. Fuel enrichment control<br>F.5. Tightness control<br>F.6. Visual inspection |
| V. *Fuel element bundles assembly* | |
| A. Components reception (grids, spacers, etc.) | A.1. Visual inspection<br>A.2. Dimensional control<br>A.3. Specific tests required by specifications (non-destructive and destructive, corrosion behaviour, etc.) |
| B. Bundle assembly | B.1. Dimensional control<br>B.2. Final tightness control<br>B.3. Final visual inspection |
| *Finite fuel bundle* | |

*Metallographic control is performed on samples preceding the end-cap/clad welding.

A detailed description of all these control procedures is beyond the scope of this book; some of them are standardized.

The ASTM standards for the inspections that are in order with the manufacture of ceramic fuel powders, fuel pellets and structural components are listed in Table 11.4.

TABLE 11.4 Several ASTM[1] Standards in Fuel Element Manufacturing Control [20]

| *Standard* | *Method denomination* |
|---|---|
| A. *Ceramic powders quality control* | |
| ASTM C.696-80 | Standard Methods for *Chemical, mass spectrometric, and spectrochemical analysis on nuclear-grade uranium dioxide powders and pellets* |

[1]ASTM: American Society for Testing and Materials

Table 11.4 (continued)

| Standard | Method denomination |
| --- | --- |
| ASTM C.753-81 | Standard Specification for *Nuclear-grade, sinterable uranium dioxide powder* |
| ASTM B.213-77 | Standard Test Method for *Flow rate of metal powders* |
| B. *Pellets quality control* | |
| ASTM E.3-80 | Standard Methods of *Metallographic specimens* |
| ASTM E.112-81 | Standard Methods for *Estimating the average grain size of metals* |
| ASTM E.45-81 | Standard Practice for *Determining the inclusion of steel* |
| ASTM B.329-76 | Standard Test Method for *Apparent density of powders of refractory metals and compounds by the Scott volumeter* |
| ASTM E.453-79 | Standard Practice for *Examination of fuel element cladding including the determination of the mechanical properties* |
| C. *Fuel element components quality control* | |
| ASTM E.8-81 | Standard Methods for *Tension testing of metallic materials* |
| ASTM E.21-79 | Standard Methods for *Elevated temperature tension tests of metallic materials* |
| ASTM E.139-79 | Standard Recommended Practice for *Conducting creep, creep — rupture and stress — rupture tests of metallic materials* |
| ASTM E.453-79 | Standard Practice for *Examination of fuel element cladding including the determination of the mechanical properties* |
| ASTM E.146-68 (rev.1979) | Standard Methods for *Chemical analysis of zirconium and zirconium alloys* |
| ASTM G.2-80 | Standard Practice for *Aqueous corrosion testing of samples of zirconium and zirconium alloys* |
| ASTM E.29-67 (rev.1980) | Standard Recommended Practice for *Indicating which places of figures are to be considered significant in specified limiting values* |
| ASTM E.3-80 | Standard Methods of *Metallographic specimens* |
| ASTM E.112-81 | Standard Methods for *Estimating the average grain size of metals* |
| ASTM E.94-77 | Standard Recommended Practice for *Radiographic testing* |

## 11.5. IRRADIATION EFFECTS INSPECTION OF NUCLEAR MATERIALS

The control of nuclear material irradiation behaviour is done by irradiation in experimental devices that are part of the materials testing reactors, offering conditions almost similar to those in a power reactor.

Characterization of the irradiation effects in these materials is performed by the so-called post-irradiation examination, generally after a certain "cooling" period (for activity diminution). The activity of the irradiated materials — sometimes very high — requires that the post-irradiation tests are performed in special facilities (hot cells),which ensures personnel protection. Hot cells allow the remote handling of the instrumentation placed inside.

Destructive and non-destructive methods, similar to those described above, are used for irradiation behaviour characterization. Some methods in the post-irradiation examination of the metal-clad ceramic fuel elements are presented in the following paragraphs.

Post-irradiation examination of fuel elements implies a sequence of analytic techniques adapted to the type of information that is to be obtained from each irradiation experiment. The test types, their timing, their character and complexity, the adopted procedures are to a large extent influenced by the way in which irradiation evolved and by the abnormalities noticed through a permanent monitoring of the irradiation parameters [44,45].

The sequence of post-irradiation examination operations is presented in Fig. 11.22. Non-destructive control operations may begin directly with the fuel element control (dotted line), in the following cases: (a) if the irradiations in the material testing reactor were performed only on fuel elements; (b) if a preliminary post-irradiation examination of the fuel bundles was performed in the storage tank, in order to select some interesting cases. Destructive control begins with sample delivery in accordance with a sampling plan, drawn on the basis of the findings from the non-destructive control operations.

### 11.5.1. Non-destructive Testing

11.5.1.1. Visual inspection. This consists of a thorough scrutiny of the clad's external aspect, aiming at the identification of peculiar occurrences, like the clad and joint corrosion, abnormal deposits on the clad, micro-cracks, etc. The operation is performed by a binocular periscope whith magnifications up to 10x.

11.5.1.2. Dimensional control. This identifies the dimensional modifications induced by irradiation in the fuel elements. The operation verifies the clad profile, searching for changes in element length and curvature. Profile measurements are performed on special benches, by moving two differential detectors on the clad surface.

11.5.1.3. Tightness control. This has the same purpose as in manufacture control, and is performed with the same procedures.

11.5.1.4. "$\gamma$ Scanning". This has as a purpose the determination of the fission products' $\gamma$-activity distribution, and is performed by the scanning of the irradiated fuel element with a $\gamma$-ray detector of NaI(Tl) crystals or with semiconductor detectors — Ge(Li). In this way one may obtain information on the actual irradiation conditions (power distribution throughout the rod, burn-up distribution, etc.).

11.5.1.5. Neutronography. This is performed in order to identify internal geometry and structural modifications of the irradiated fuel element

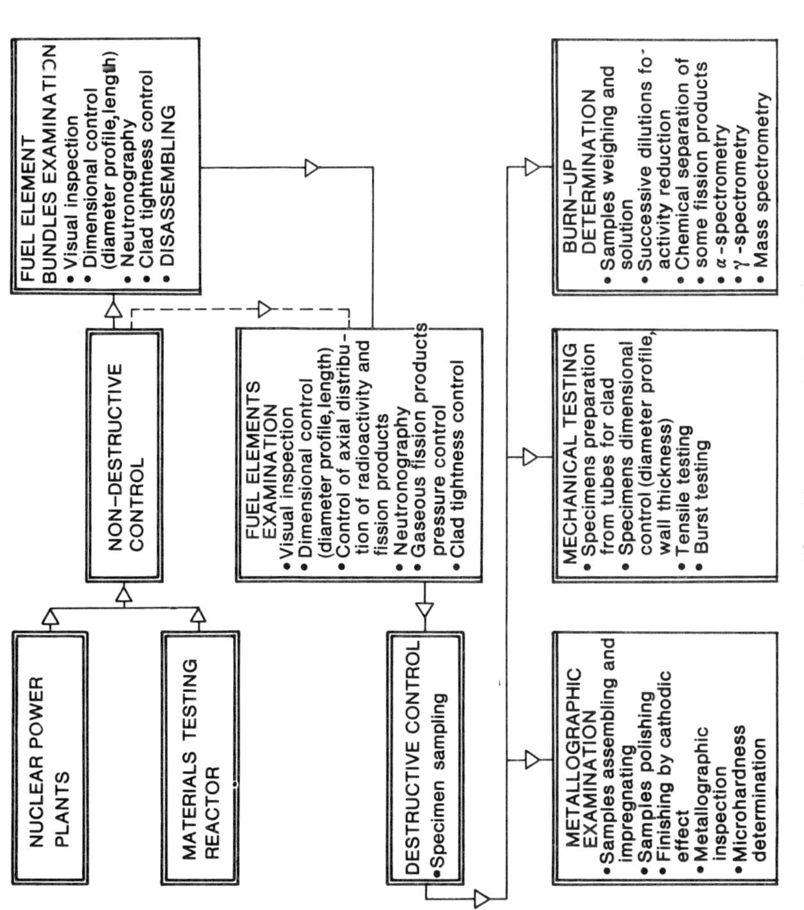


Fig. 11.22. Post-irradiation examination.

components. Thus, one may zero-in the fuel cracks, structural modifications, changes in the fuel element's internal geometry, the zones with local agglomeration of hydrides in the clad, etc.

The results of non-destructive testing are useful in the identification of the zones of interest, wherefrom samples will be selected for further, destructive, investigation.

### 11.5.2. Destructive Testing

11.5.2.1. Fission gas analysis. This is performed in order to determine the volume, pressure and composition of the fission gas released from fuel during irradiation. The operation is performed by drilling a hole in the clad and recovering the gas in a vacuum ampoule.

11.5.2.2. Fuel metallographic analysis. This is performed in order to determine the modifications induced by irradiation in the fuel structure: crystallite growth, porosity redistribution, fission gas bubble distribution (by electron microscopy), and the distribution of the solid fission products, precipitates, etc. The technique used is the conventional one and it is preceded by a consolidation of the fragments of material cracked during irradiation.

11.5.2.3. Clad metallographic examination. This supplies data regarding:

1. oxidation of the clad internal and external surfaces;
2. determination of the dissolved hydrogen and of the concentration and orientation of hydride-plaquettes (for zirconium alloy clads);
3. the mechanical and chemical fuel-cladding interaction;
4. the identification of abnormalities: cracks, "hot" points, local hydriding, etc.

11.5.2.4. Self-radiography. Self-radiography of the longitudinal and transversal sections in the fuel rods has as a purpose the determination of the fission products' axial and radial distribution (gammagraphy) and the plutonium redistribution (alphagraphy). The operation is performed by putting the sample surface in contact with a film which is impressed by the $\alpha$ or $\gamma$ rays emitted by plutonium, and by the fission products, respectively.

11.5.2.5. Determination of the clad mechanical properties. This has as a purpose to obtain information on the global irradiation effects on clads. Results are correlated with the specific conditions under which the irradiations were performed (temperature, integrated neutron flux, etc.).

## 11.6. SAFEGUARDS OF NUCLEAR MATERIALS

Article 12 of the Statute of the International Atomic Energy Agency (IAEA), indicates that one major task of the Agency resides in providing, *inter alia* through appropriate control, for ensuring peaceful utilization of nuclear materials, especially of the fissionable materials. This article defines the terms under which the IAEA should accomplish this task; accordingly, the Agency has to adopt and implement control measures, in order to ensure that any assistance given to member states (materials, services, equipment), either directly or through the IAEA or under its guidance or supervision, will not be used for military purposes.

In pursuance of its tasks, the Agency has designed and adopted a Safegurads and control system and has created within its Secretariat a Safeguards Inspectors Group [46]. The IAEA Safeguards system has been established on the understanding that its applicability should not impede the technological and economic development of the interested member states and should be compatible with the state's evidence and control domestic regulations, so that it should not disturb the effective and sound operation of nuclear facilities. A document pertaining to the inspectors' regime was added to the Safeguards system providing clarification on the terms under which Safeguards inspections may be carried out, and terms of reference for inspectors' activity (reporting procedures, observance of proprietary, professional secrets, etc.).

Considering the prospects with the power installed in nuclear power plants (Table 11.5 [47]), and the plutonium production which may result from the nuclear plants operation, the Agency is constantly preoccupied to improve on control activities and their technical and economic performance.

TABLE 11.5 Dynamics of Electric Power of Nuclear Origin

| *Year* | *1975* | *1980* | *1990* | *2000* |
|---|---|---|---|---|
| Nuclear-electric power (GW) | 110 | 145 | 375 | 830 |

On the grounds of the IAEA Statute, an Agreement between the Agency and the State Required in Connection with the Treaty on the Non-Proliferation of Nuclear Weapons is now in effect, providing for Safeguards and control of nuclear materials and related nuclear installations. The Agreement contains two parts: the first deals with general, as well as with political and legal aspects; the second deals with technical aspects of Safeguards control. Romania is a member of the IAEA and a signatory to the Non-Proliferation Treaty.

### 11.6.1. Safeguards of Nuclear Fuel Cycle

IAEA Safeguards rest upon three concepts:

1. inventory accounting for nuclear fissionable materials;
2. confinement;
3. surveillance.

*Inventory accounting* means keeping a strict quantitative control over pertinent outputs and inputs. *Confinement* refers to measures concerning the structural integrity of the ducts, vessels, buildings, and other structural items etc., meant for nuclear material storage and transport, in view of limiting material, equipment and personnel movements in and out of these structures; confinment also includes methods and devices employed to detect loss of this integrity. *Surveillance* may be performed directly by inspectors, or by special devices. It may be temporary, periodical or permanent.

The control performed by the Agency to make sure that nuclear materials are exclusively used for peaceful purposes, is done in accordance with the information given by the member states and relies upon the verification of the information's authenticity, especially with those concerning nuclear material quantities and emplacements. The Agency's information comes from four sources: the installations design; the installations inventory accounting systems; the reports sent to the Agency; the Agency's in-field inspections.

*Design analysis* is performed in order to establish the characteristics of the installations which use fissionable materials important from the control's standpoint; the checkpoints of material balance; the strategic points for sampling and measurements; nuclear materials flows and the adequate confinement and surveillance measures.

*Inventory accounting system* comprises both nuclear material recording and operational parameters recording, for the most important installations under control. These data are also used in the elaboration of reports as required by the Agency.

*Reports* contain information sent to the Agency by the state that agreed upon being controlled; their format is standardized; reports are presented as a system of forms that are filled in separately — for the initial report, for the inventory accounting reports, and for the special reports.

*Inspections* are performed by Agency inspectors in order to check the accuracy of the information sent by states through reports as well as the accuracy of the measurement methods.

The importance granted to the technical aspects of the control of fissionable materials and nuclear installations follows naturally from the exceptional importance attached to these materials, from the awareness of the highly technical content of the problems with their utilization and, on the other hand, from the fact that only a sound technical approach may provide for minimizing errors.

In short, the technical objective of nuclear materials control is to establish and verify the following characteristics of these materials from the control's standpoint: composition (identity); quantity (effective); location (utilization). According to these data and to the stage in the nuclear fuel cycle where the nuclear materials are used, one assesses:

1. the strategic value, inferred from the number of transformation stages that the material would have to undergo in order to be used for explosive purposes;

2. the critical time, i.e. the time required for a diverted nuclear material be used for explosive purposes; strategic value is inversely proportional to critical time.

In observance of these characteristics, one establishes the control and recording methods, as well as the inspections and reports frequency. They also allow the definition of a control system efficiency, by the credibility degree $x$ (in %) that the nuclear material inventory accounting is performed with the accuracy $y$ (in %), in a given time. The closer to 100% $x$ and $y$, the more efficient the system is believed.

### 11.6.2. Inspection Methods in the Nuclear Fuel Cycle

As indicated in Chapters 3 and 10, fissionable materials undergo a series of transformations, from ore extraction to service in nuclear reactors and then to reprocessing in order to recover uranium and accumulated plutonium. All these make up the nuclear fuel cycle. A whole-cycle control has the merit that it is more accurate than the isolated control of some nuclear installations, because control results may be appropriately correlated, thus improving statistics and overall assessment.

Although there are certain differences in defining the stages (phases) of the nuclear fuel cycle, one may settle for the following flow-chart:

1. extraction and processing of nuclear-sensitive ores;
2. concentration installations;
3. conversion plants (chemical transformation);

4. isotopic enrichment plants;
5. nuclear fuel manufacture plants;
6. nuclear reactors;
7. reprocessing plants (chemical separation);
8. research and development facilities;
9. transfers;
10. wastes (recoverable and non-recoverable, disposable).

An upgrading, by raising the content in the useful component up to 90%, is achieved in the ore processing and concentration plants. When isotopic enrichment is required, the concentrate is transferred to the conversion plants where one may obtain uranium hexafluoride ($UF_6$) from $U_3O_8$ or $UO_3$. In this form the product may be enriched, then transformed into $UO_2$ or UC. When isotopic enrichment is not required, the concentrate is transformed directly in $UO_2$. These uranium compounds are transferred to the fuel element manufacturing plants. The fuel elements go into the power reactor from where, after utilization, they are unloaded, only to be immediately stored in special ponds and then directed to the reprocessing plants in order to recover the unburned uranium and the plutonium generated during operation.

The dynamics of the control activities performed by IAEA between 1969 and 1971 is illustrated in Table 11.6 [48].

TABLE 11.6 Nuclear Installations under Safeguards Inspection

| *Installation* | *1969* | *1970* | *1971* |
|---|---|---|---|
| Nuclear power plants | 6 | 9 | 13 |
| Chemical treatment plants | 3 | 3 | 4 |
| Fuel elements plants | 1 | 5 | 9 |
| Research reactors | 50 | 54 | 65 |
| Critical assemblies | 9 | 9 | 11 |
| Subcritical assemblies | 2 | 2 | 4 |
| Other research facilities | 59 | 74 | 78 |
| Total | 130 | 156 | 184 |

In order to perform the control, national systems are provided with nuclear material analysis laboratories for each phase in the cycle. IAEA inspectors may verify the methods and devices employed, or may use their own devices, according to the installation type and the phase which is controlled.

Control measurement methods have to meet the following requirements:

1. to provide for high accuracies;
2. to allow appropriate discrimination between U, Pu and other active or stable elements;
3. to be fallacy-proof, and free of systematic errors;
4. to be simple, fast, and economic, allowing the use of automated and/or portable instrumentation;
5. to guarantee non-interference with the built-in monitoring and other devices in the controlled installation; also their implementation should not require highly-trained personnel;
6. their application should not disturb production.

Among the uranium and plutonium analysis methods that comply with the above, notable are:

*A. Destructive methods*

I. Determination of U and Pu concentration:

1. electrochemical methods (potentiometric, voltametric, ampermetric,etc.);
2. physicochemical methods (spectrophotometry, X ray spectroscopy, etc.);
3. radiochemical methods.

II. Determination of U and Pu isotopic content

1. mass spectrometry;
2. gamma and alpha spectrometry.

*B. Non-destructive methods* (analyses of isotopic concentration and composition)

I. Passive interrogation methods:

1. calorimetry;
2. gamma spectrometry;
3. neutron transmission;
4. neutron coincidences (spontaneous fissions);
5. magnetic resonance.

II. Active interrogation methods:

1. reactivity measurements;
2. slow-down time methods;
3. neutron interrogation;
4. photon interrogation.

In order to draw a nuclear materials balance account, the large installations, and especially the production installations, are conventionally divided into smaller areas called material balance areas (MBA).

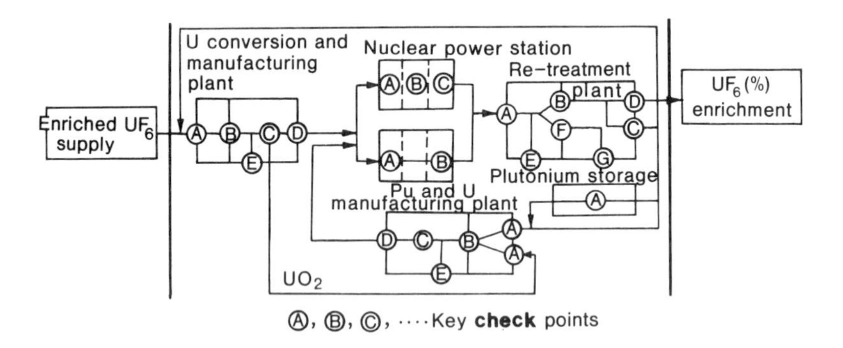


Fig. 11.23. Key check-points in nuclear materials control.

Within each MBA several key points are designated (Fig. 11.23), where measurements are performed on the nuclear materials, because all the materials in the technological flow-sheet are bound to pass through these points [48]. In the key points periodic measurements are performed, both for technological and for Safeguards purposes. Considering these measurements one may conclude on certain significant losses of material, that may be due to natural causes in the technological process, or to unusually large incidental leaks, or to illegal diversions. The chief criterion used in the identification of such losses is based on the comparison of one important parameter (MUF — material unaccounted for) with the limit of the errors. This parameter is defined as being the difference between the recorded and the physical stock for a certain period, which reads:

MUF = (initial stock + outputs - inputs) - (physical final stock).

When establishing the MUF magnitude one uses the data from measurements performed at the key points as well as inventory-accounting information to be found in the documents regarding imports, exports, transfers and manufacturing.

If there are no significant loss sources and the measurments are accurate enough, then, within the limits of the chosen credibility (e.g. 95%), the MUF value has to be as close to zero as possible. An important departure from this value is a warning that calls for additional efforts aiming at satisfactory clarifications. The MUF magnitude is affected by all sorts of ordinary errors, either accidental or systematic. Under such circumstances a careful statistical analysis of the data is essential.

If one admits that major errors on MUF come from three sources: measurement errors (ME); losses of material that cannot be measured but are normal during operation (normal operation losses — NOL); possible materials diversion (D), then one may write:

$$MUF_{TOTAL} = MUF_{EM} + MUF_{NOL} + MUF_{D}.$$

While steps may be taken to diminish measurement errors, the other two components are still of concern. By a careful analysis of the operating conditions, the NOL value may be determined or estimated, thus isolating $MUF_D$; this fact should allow the identification — with a high probability — of possible diversions. The problem with the errors is longstanding. Errors may always be reduced, but at the cost of serious efforts and expense, that may sometimes exceed technical and economic common sense.

Some examples follow, illustrating the importance and implications of the errors in determining certain parameters that are specific for Safeguards control [48].

If a state owns 1000 kg of plutonium of a certain quality and, at an inspection, the quantity is determined with an unacceptably high error, e.g. 1013 ± 42 (which means that the true value can be anywhere between 971 and 1053 kg), it would follow that an important quantity of Pu could be illegally diverted without being detected. The personnel at the nuclear facilities usually employ more accurate measurement methods and have knowledge of Pu stock with a lower error — e.g. 1000 ± 15 kg (which places the total between 985 and 1015 kg). Even so, a quantity of 10 kg Pu may still be taken out illegally, without notice; note that 10 kg Pu is more than enough for manufacturing an atomic bomb. As one can see, in such a case a wrong measurement may have serious consequences.

Another example pointing out the importance of measurement errors may be taken from the field of chemical reprocessing plants for irradiated nuclear fuel, where the $^{235}U$ and $^{239}Pu$ of different concentrations are temporarily stored, in tanks of various capacities. If the measurement errors are too big — e.g. instead of 20 l estimated value the real quantity is 25 l — that amount could exceed the critical mass, which would generate a serious nuclear incident. Sometimes such situations are prevented by limiting the tanks' capacities to values that would not permit, in any circumstance, that the critical mass be exceeded.

REFERENCES


1. Andreescu, N. (1968). Incercarile nedistructive ale materialelor si pieselor pentru centrale nuclearoelectrice. In *Proc.CNIT Colloque 1968*, Bucharest; Dubrovskiy, V. B. (Ed.) (1979). *Stroitelstvo atomnyh elektrostantsyi*. Energya, Moskow.
2. Mc Gonnagle, W. J. (1967). *Essais non destructifs*. Edit. Eyrolles, Paris.
3. Hyatt, B. Z. (1982). *Metallographic Standard for Estimating Hydrogen Content of Zircaloy 4 Tubing*. Report WAPD-TM-1431, Bettis Atomic Power Lab., Pittsburgh.
4. ASME (1977). *Nondestructive Examination*. ASME, New York.
5. Vladescu, M., and Doniga, N. (1968). *Tehnica defectoscopiei cu radiatii penetrante*. Edit. Tehnica, Bucharest.
6. Brinzan, C., and Radu, R. (1975). *Controlul nedistructiv al materialelor prin metode radioactive*. Edit. Tehnica, Bucharest.
7. Popa, V. (1978). *Probleme practice ale radiografiei industriale cu radiatii X si* $\gamma$. Edit. Tehnica, Bucharest.
8. IAEA (1976). *Proc.IAEA Seminar on Nuclear Fuel Quality Assurance, Oslo 1976*.
9. Negoita, C. (1965). *Masurari si control cu ultrasunete*. Edit. Tehnica, Bucharest.
10. Prot, A. C. (1971). In *Non-destructive Testing for Reactor Core Components and Pressure Vessels*. Proc.IAEA Conf., Vienna 1971.
11. Pillet, C. (1971). In *Non-destructive Testing for Reactor Core Components and Pressure Vessels*. Proc.IAEA Conf., Vienna 1971.
12. Alain, C. (1966). In *Proc.Symp.on Non-destructive Testing in Nuclear Technology, Bucharest 1965*. IAEA, Vienna.
13. Miloserdin, Yu. V., and Baranov, V. M. (1978). *Vysokotemperaturnye ispytanya reaktornyh materialov*. Atomizdat, Moskow; Cahn, R. W. (1980). Characterization of Solid Materials: Summary and Bibliography. In *Physics of Modern Materials*, Vol. I, IAEA, Vienna.
14. Andreescu, N., and Andreescu, O. (1974). Consideratii asupra variatiilor de impedanta produse de defecte in controlul nedistructiv prin curenti turbionari, In *Comunicarile primei conferinte de control nedistructiv, Bucharest 1974*; Andreescu, N., and Andreescu, O. (1967). L'effet de la forme et l'effet de skin dans le controle des echantillons cylindriques par courants de Foucault. In *Proc.8th World Conf.on Non-destructive Testing, Cannes 1976*.
15. Cecco, V. S., and Van Drunnen, G. (1981). *Eddy Currents Testing*, Vol. 1, Manual of Eddy Currents Methods. Report AECL 7523.
16. Renken, C. I., and Myers, R. G. (1960). Report ANL-5861, Argonne National Laboratory.
17. Andreescu, N., Baltac, I., and Andreescu, O. (1971). *Rev.Roum.Phys.*,16,915.
18. Andreescu, N., Andreescu, O., Baltac, I., Roiban, F., and Cirstoiu, B. (1971). Report ITN.
19. Dobos, I. (1974). Report ICEFIZ.
20. IAEA (1976). *Quality Assurance and Quality Control in the Manufacture of Metal-Clad* $UO_2$ *Reactor Fuels*, Technical Report Series no 173. IAEA, Vienna.
21. Hilpert, K. (1979). *J.Nuclear Materials*, 80, 126.
22. Frejaville, G., and Lucas, M. (1972). Report CEA-R-4265.
23. Nascutiu, T. (1973). *Metode radiochimice de analiza*. Edit. Academiei, Bucharest.
24. Vasaru, Gh. (1968). *Izotopi stabili*. Edit. Tehnica, Bucharest; Popescu, I. M., and others (1979). *Aplicatii ale laserilor*. Edit. Tehnica, Bucharest.
25. AECL (1978). Atomic Energy of Canada Ltd., Progress Report, Oct.-Dec.1977. Report AECL-6090.
26. Doniach, S. (1972). In *Magnetism and Magnetic Materials*, AIP Conf.Proc. no 5 (1971), New York, p. 549.
27. Ursu, I., Demco, D. E., Simplaceanu, V., and Vilcu, N. (1974). *Rev.Roum.Phys.*, 19, 605.

28. Ursu, I., Demco, D. E., Simplaceanu, V., and Vilcu, N. (1975). Romanian patent no 60813.
29. Ursu, I., and Lupei, V. (1981). In *Proc.EPS Int.Conf.on At.and Mol.Phys., Bucharest 1981*.
30. Ursu, I., and Lupei, V. (1976). Report at the ICEFIZ Symposium, Bucharest 1976.
31. Ursu, I. (1979). *Rezonanta magnetica in compusi cu uraniu*. Edit. Academiei, Bucharest.
32. Northwood, D. O., and Rosinger, H. E. (1980). *J.Nuclear Materials*, 89, 147.
33. Gheata, V., and others (1974). Report ICEFIZ; Gheata, V., and Galeriu, G. (1979). Report ICEFIZ.
34. Geller, Yu. A., and Rakhshtadt, A. G. (1975). *Materialovedenye*. Metalurgya, Moskow.
35. Slattery, G. F. (1969). The Mechanical Properties of Zircaloy-2 Tubing Containing Circumferentially Aligned Hydride. In *Application-Related Phenomena in Zirconium and its Alloys*. American Soc. for Testing and Materials.
36. Atrens, A., and Cann, C. D. (1980). *J.Nuclear Materials*, 88, 42.
37. Gheata, V., Sabau, C., and Dobos, I. (1973). Report ICEFIZ; Gheata, V., Sabau, C., and Dobos, I. (1979). Report ICEFIZ no 546/I/1979.
38. Dickson, I. K., and others (1979). *J.Nuclear Materials*, 80, 223.
39. Johnson, A. B. (1969). Corrosion and Failure Characteristics of Zirconium Alloys in High-Pressure Steam in the Temperature Range 400 to 500 $^{o}C$. In *Application-Related Phenomena in Zirconium and its Alloys*. American Soc. for Testing and Materials.
40. Tuta, G., and others (1974). Report ICEFIZ.
41. Tsykanov, V. A., and Davydov, E. F. (1977). *Radiatsionnaya stoykosti teplovydelyayuschyh elementov iadernyh reaktorov*. Atomizdat, Moskow.
42. Galatchi, C., and others (1973). Determination of Mean Diameter and Density of Powder Aggregates by Gas Permeability. In *Proc.Symp. Structure and Properties of Materials, Prague 1973*.
43. Gheata, V., Galeriu, C., Dobos, I., and Glodeanu, F. (1978). Experience Gained in the Fabrication of Experimental Fuel Rods with Natural Uranium and Zircaloy-4 Cladding for Irradiation Experiments. In *Proc.Conf.on Fabrication of Water Reactor Fuel Elements, Prague 1978*.
44. Ursu, I., Tuturici, I., and Gheorghiu, C. (1979). *Examinarea post-iradiere a elementelor combustibile iradiate; tehnici experimentale, analize si interpretarea datelor* — a Workshop Report, INPR, Pitesti 1979.
45. Tuturici, I., and Gheorghiu, C. (1977). *Examen post-iradiere al barelor combustibile 9-07 si 8-03 iradiate in reactorul BR-2 (Belgia)* — Report at the Workshop on Nuclear Reactors Physics, Design and Engineering, Pitesti 1977; Tuturici, I. (1981). *Post-irradiation Examination of Romanian Fuel Elements Irradiated in MZFR Reactor*. Report ICEFIZ 849/1981.
46. IAEA (1970). The Structure and Content of the Agreement between the Agency and the State Required in Connection with the Treaty on the Non-Proliferation of Nuclear Weapons. INFCIRC/153, IAEA, Vienna.
47. IAEA (1978). Int.Nuclear Fuel Cycle Evaluation WG 5, Final Report.
48. Ursu, I. (1973). *Energia atomica*, Edit. Stiintifica, Bucharest.

# CHAPTER 12

# Materials for Fusion Reactors

The preceding chapters dealt with the fission nuclear reactors, as a central element in the current nuclear power infrastructure. Appendix 12 lists the nuclear reactors now in service and under construction, thus highlighting the privileged place of this solution in contemporary energy systems.

Nevertheless, as it is currently believed, a full turning to good account of nuclear power would still have to wait for commercial, safe and sound *fusion* nuclear reactors. A discussion of the technological problems entailed by fusion reactors may seem premature, since too many questions on the technical and commercial feasibility of the controlled fusion are not yet satisfactorily answered. However, such an exercise is worth doing, for several good reasons:

1. first, it is a fact that recent years have brought about several important advances in the field of controlled thermonuclear fusion;

2. second, it is almost certain that, if feasible, the fusion reactor will represent the outstanding energy megasource of tomorrow, regardless of the highly probable diversity of future energy mixes;

3. third, even if we are still ignorant of the ways and means of solving certain perplexing problems with controlled fusion, a general pattern of the field is already emerging, including good many clear-cut requirements on the materials to be used.

Moreover, in an intermediate variant that cradles the notion of hybrid fusion – fission reactors – great producers of fissionable material ($^{239}Pu$, $^{233}U$) out of fertile materials ($^{238}U$, $^{232}Th$) – controlled nuclear fusion does not appear too remote an objective.

After a brief discussion of the fundamental processes related to the fusion reaction, physical processes in the potential fusion reactors will be described, as well as current fusion reactor concepts; then the characteristics of the materials in use, or candidates for use, with this kind of reactor will be tackled.

## 12.1. THERMONUCLEAR FUSION REACTION

Nuclear fusion is a reaction of two light nuclei, that results in the formation of a heavier nucleus. Emission of neutrons, protons, tritons, helions, etc., can take place during this process, and also a certain amount of energy may be released and then taken over by the newly formed nucleus and the other particles, as kinetic energy. As per unit-mass, the released energy exceeds the one obtained with the fission reaction. Due to the immense stocks of light fusionable elements that exist in nature — especially heavy hydrogen — the fusion reaction may be used as one major source of energy. The example at hand of such a process of energy "production" is the energy generation in stars.

Fusion reactions of special practical interest are the following:

$$ {}^{2}_{1}D + {}^{2}_{1}D \rightarrow {}^{3}_{2}He + n + 3.2\ MeV \qquad (12.1)$$

$$ {}^{2}_{1}D + {}^{2}_{1}D \rightarrow {}^{3}_{1}T + p + 4.0\ MeV \qquad (12.2)$$

$$ {}^{2}_{1}D + {}^{3}_{1}T \rightarrow {}^{4}_{2}He + n + 17.6\ MeV \qquad (12.3)$$

$$ {}^{2}_{1}D + {}^{3}_{2}He \rightarrow {}^{4}_{2}He + p + 18.3\ MeV \qquad (12.4)$$

$$ {}^{6}_{3}Li + n \rightarrow {}^{3}_{1}T + {}^{4}_{2}He + 4.8\ MeV \qquad (12.5)$$

$$ {}^{7}_{3}Li + n \rightarrow {}^{3}_{1}T + {}^{4}_{2}He + n + 2.5\ MeV \qquad (12.6)$$

It can be noticed that, as far as released energy, the most interesting reactions are (12.3) and (12.4). Actually, a D — T blend may also engage in some of the other reactions, such as (12.1) and (12.2). Reactions (12.5) and (12.6) are also worth consideration, for their production of tritium that is required to feed the leading reaction D — T. The factual frequency of the reactions above are determined by their cross-sections, given in Fig. 12.1 [1].

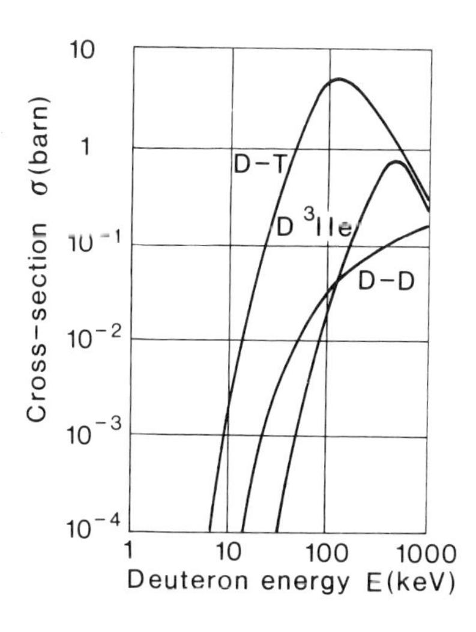


Fig. 12.1. Cross-sections for fusion reactions D — D, D — T, and D — $^{3}$He.

The main difficulties in obtaining the controlled fusion have to do with the following requirements:

(a) to provide a high kinetic energy to the participating nuclei, so that the fusion reaction cross-section allows a sufficiently high reaction rate;
(b) to provide a sufficiently high atom density, so that a considerable number of fusion acts be obtained;
(c) to limit the energy losses of the reactive medium.

The first requirement relates to *plasma heating*. The second and third imply securing of a high ion density for a sufficiently long time, so that the reaction can be self-sustaining. This can be achieved by holding the plasma in a given volume — an act known as "*plasma confinement*".

An energy in excess of 10 keV per deuteron (Fig. 12.1), that is necessary to obtain a significant cross-section of the D — T fusion reaction, can be reached by raising the temperature of the D — T mixture to ~ $10^8$ K (1 eV corresponds to $1.1605 \times 10^4$ K). At such a temperature the atoms are completely ionized and the resulting ion and electron mix makes a high-temperature ("hot") plasma. A minimally required temperature, $T_{trigg}$, can be established for the thermonuclear reaction commencement, corresponding to the temperature for which the energy generated by fusion exceeds the energy losses by bremstrahlung. The triggering temperatures for D — D and D — T plasmas are, respectively:

$$T_{trigg} \simeq 4.1 \cdot 10^8 \text{ K} \quad (D - D)$$

$$T_{trigg} \simeq 4.6 \cdot 10^7 \text{ K} \quad (D - T)$$

In order to be self-sustaining, thermonuclear plasma should satisfy one more requirement, which combines the requirements (b) and (c) above — known as the "Lawson criterion" [2]. This considers the lowest admissible value of fusionable nuclei density ($n$) and plasma life-time ($\tau_c$) at a temperature superior to $T_{trigg}$, and reads:

$$n \cdot \tau_c \geqq (n \cdot \tau_c)_{min} \tag{12.7}$$

For D — D and D — T plasmas the Lawson criterion leads to the following characteristic values, respectively:

$$(n \cdot \tau_c)_{min} = 2 \cdot 10^{15} \text{ s/cm}^3 \quad (D - D);$$

$$(n \cdot \tau_c)_{min} = 6 \cdot 10^{13} \text{ s/cm}^3 \quad (D - T).$$

The most convenient thermonuclear plasma is, therefore, the deuterium — tritium mixture.

To assess the grounds for optimism concerning the development and utilization of the thermonuclear reactions, it is instructive to compare, in Table 12.1, the 1980 actual performances in relation to the parameters anticipated for a full-scale reactor. Data in Table 12.1 refer to results obtained with TOKAMAK equipment.

Several ways are, in principle, available at present to obtain controlled fusion. They can be classified into two main categories:

(a) magnetic confinement systems:
- closed systems (toroidal) – as, for instance, TOKAMAK;
- open (linear) systems – as, for instance, magnetic mirror systems;

b) inertial confinement systems:
- charged particle beam systems;
- laser beam systems.

TABLE 12.1 Minimum Requirements for the Parameters of a Fusion Power Reactor

| *Parameters* | *Confinement time* (s) | *Ionic temperature* (K) | *Lawson law* ($s/cm^3$) | *Pulse duration* (s) |
|---|---|---|---|---|
| Full-scale reactor requirements | ~ 1 | ~ $10^8$ | ~ $10^{14}$ | ~ 10 |
| 1980 results | $10^{-1}$ | $8 \times 10^7$ | $3 \times 10^{13}$ | 10 |

For each of the above connections there are problems with the specific processes, both principial and of a technological nature. To solve them theories have been evolved concerning hot plasma, and processes that take place in various situations have been simulated by means of computers, in the search for optimal results; also, appropriate experiments have been performed (Fig. 12.2).

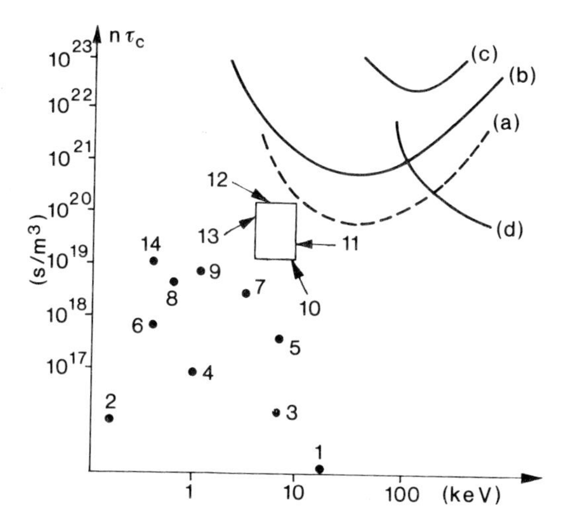


Fig. 12.2. Parameters of leading fusion facilities: 1 - "Phoenix" magnetic mirror; 2 - Stellarator; 3 - "2X" magnetic mirror; 4 - Helical pinch; 5 - θ-pinch; 6 - TOKAMAK "ST"; 7 - TOKAMAK "PLT"; 8 - TOKAMAK "T10"; 9 - TOKAMAK "ALCATOR"; 10 - TOKAMAK "TFTR"; 11- TOKAMAK "JT-60"; 12 - TOKAMAK "JET"; 13 - TOKAMAK "T-15"; 14 - Laser giant impulse. Positive power balance condition for: (a) D – T fusion; (b) D – T reactor; (c) D – D reactor; (d) direct conversion D – T reactor.

The following section will briefly review the physical questions raised by the two confinement systems, as well as some fusion reactor projects. First, the processes taking place in the plasma confined by magnetic fields will be discussed, then those in the inertially confined plasma.

## 12.2. PHYSICAL PROCESSES IN FUSION REACTORS

### 12.2.1. Basic Processes

To initiate a fusion reaction, plasma must be heated in a thermonuclear reactor. Some of the energy used to that purpose is wasted by radiation. Radiation losses are caused mainly by two processes:

1. bremstrahlung — the electromagnetic radiation emitted by electrons when slowed down in electric fields;
2. cyclotron radiation, emitted as a result of the electrons' movement on helical paths in the magnetic fields ensuring plasma confinement.

For a self-sustained fusion reaction, it is at least necessary that the power produced by fusion be equal to the power lost by radiation. The power generated by the fusion reactions (D — D) is:

$$P_{\mathrm{reaction}} = (1/2)n_{\mathrm{D}}^2\langle\sigma v\rangle W_f , \qquad (12.8)$$

where $n_{\mathrm{D}}$ is the deuterium nuclei density, $\langle\sigma v\rangle$ is the mean value of the product of the cross-section and speed, and $W_f$ is the energy produced per one reaction act.

Power lost by radiation [W/cm$^3$] for deuterium plasmas is given by [2]:

$$P_{\mathrm{rad}} = 0.54 \cdot 10^{-30}\, n_{\mathrm{D}}^2 T_e^{1/2} \qquad (12.9)$$

where $n_{\mathrm{D}}$ is the deuterium nuclei density (equalling electrons' density $n_e$) given in particles/cm$^3$, and $T_e$ is the electronic temperature of the plasma, given in keV.

The reaction is self-sustaining when

$$0.54 \cdot 10^{-30}\, T_e^{1/2} = (1/2)\langle\sigma v\rangle W_f . \qquad (12.10)$$

The necessary ionic temperature $T_i$ is 35 keV for the reaction D — D and only 4 keV for the reaction D — T (it has been assumed that $T_i = T_e$).

In order to reach these temperatures, plasma must be confined in a closed space for a time stipulated by the Lawson criterion, and thus it should not come in contact with the walls which could otherwise melt, thus cooling down the plasma. To that purpose, the seemingly only available solution is to use a magnetic field to thermally insulate (confine) the plasma.

In effect, this issue is central to all research carried out on fusion performed on magnetic confinement systems (magnetic fusion): to design and obtain an adequate configuration of magnetic fields capable of thermally insulating the thermonuclear plasma. In such a confinement system the principal force that, in stationary conditions, prevents the plasma touching the walls is the Lorentz force:

$$\vec{F} = \vec{j} \times \vec{B}, \tag{12.11}$$

where $\vec{B}$ is the magnetic induction, and $\vec{j}$ is the plasma current density.

As a consequence of its propensity to expand, the plasma would fill up the whole volume and, therefore, would rapidly destroy the walls. To maintain the plasma in a given volume, an induction magnetic field is used to counterbalance, by its active power, the expansion tendency of the plasma. Magnetic pressure, $p_m = B^2/2\mu_o$, operates on a direction perpendicular to the field lines. The pressure describing the plasma propensity for expanding is $p_p = (n_e + n_i)k_BT$, where $n_e$ and $n_i$ are the number of electrons and ions, per unit-volume, respectively. To exemplify, for a neutral plasma ($n_e = n_i = n$) at $T = 10^8$ K and $n \simeq 10^{20}$ m$^{-3}$, it results $p \simeq 2.8\times10^5$ Pa.

To obtain plasma confinement in an ideal case and on a plane geometry, from the condition $p_p = p_m$ an induction field $B = 0.9$ T results. With the actual plasmas the magnetic field is inhomogeneous. Consequently, in the expression used for the estimation of $B$ a factor $\beta = p_p/p_m$, given by the ratio of the plasma pressure to the magnetic pressure, should be inserted, to account for the magnetic field efficiency in the confinement process. Indicating by $B_e$ the induction field around the plasma, one gets:

$$\beta = \frac{2\mu_o p_p}{B_e^2} . \tag{12.12}$$

The magnetic field-generating system is one of the most expensive components of a fusion reactor; an increase in $\beta$ leads, at given plasma parameters, to a decrease in magnetic field, in the stored electromagnetic power and in the forces generated in the reactor's structure.

In order that a fusion power system become economically competitive, it is necessary that $\beta \geqq 0.1$. Yet, the $\beta$ factor is important for a fusion reactor not only from an economic standpoint; the physical properties of plasma depend on $\beta$. The higher $\beta$, the more difficult — both theoretically and experimentally — the problem of obtaining a stable confinement configuration (since there is more energy to cause plasma instability).

Magnetic confinement may be approached in two distinct ways:

1. considering the plasma as a good electric conductor subject to the effects of electromagnetic fields;
2. considering the plasma as a system of electrically-charged particles, and analysing the path of the ions and electrons moving around the magnetic field lines.

The choice between these models depends on the parametres of the confined plasma and on the characteristics of the available kinds of magnetic confinement systems — the open and the closed one.

In open systems the magnetic field lines ultimately leave the confined plasma. The best-known instances of such configurations are: magnetic mirror systems [3], plasma-focus facilities [4-6] and "theta-pinch" linear systems [7]. In a closed system the magnetic field lines stay in the confinement area. Such configurations are produced with the TOKAMAK [8] and STELLARATOR [9] systems.

Priority in the world nowadays goes with the closed systems with magnetic confinement (especially the TOKAMAK systems); plasma confinement in open systems is limited by the losses at the ends, losses that exceed those occurring in closed systems. On the other hand, from the angle of a power reactor, the open systems are much simpler technologically than the closed systems which, as it will be shown, are bound to have a toroidal configuration.

To describe the equilibrium configuration of plasma in a closed confinement system, the ideal magneto-hydrodynamic model of a plasma should be used. In such a model plasma is considered a fluid having a macroscopic velocity $\vec{v}$, pressure $p$ and density $\rho$. Pressure is supposed to be a scalar, and plasma behaves like an infinite electric conductor. In this case the plasma motion equation is:

$$\rho \frac{\partial \vec{v}}{\partial t} + \rho \vec{v} \cdot \nabla \vec{v} = \vec{j} \times \vec{B} - \nabla p. \tag{12.13}$$

Magnetic induction $\vec{B}$ and current density $\vec{j}$, satisfy Maxwell equations:

$$\nabla \times \vec{B} = \mu_0 \vec{j}, \qquad \nabla \vec{B} = 0. \tag{12.14}$$

If plasma is stationary, equation (12.13) becomes:

$$\vec{j} \times \vec{B} = \nabla p \tag{12.15}$$

Thus, the plasma equilibrium configuration is described by equations (12.14) and (12.15) and they impose very restrictive conditions upon the geometry of the magnetic fields for confinement.

If equation (12.15) is scalarly multiplied by $\vec{j}$ and then by $\vec{B}$, and using equation (12.14) it results:

$$\vec{B} \cdot \nabla p = 0. \qquad \vec{j} \cdot \nabla p = 0. \tag{12.16}$$

Equations (12.16) state that magnetic induction and current density are normal to the pressure gradient. Supposing that pressure is a scalar value, in a closed system $p = C$ (where $C$ is a constant) defines a closed surface, characterized by a constant pressure. Then pressure gradient $\nabla p$ is normal to the $p = C$ surface, and lines of magnetic induction $\vec{B}$ do not intersect this surface. Similarly, surface $p = C$ can be considered as consisting of magnetic induction lines and is called a magnetic surface. In order that these constant pressure surfaces describe an electric fluid conductor confined within, some additional requirements must be observed: surfaces should be confined in a limited volume, they should have no edges, and one must have $\vec{B} \neq 0$ on the surfaces. In such conditions — states a theorem in topology dated as far back as 1935 – these surfaces should be toroidal.

The requirement to obtain such toroidal magnetic surfaces by an adequate distribution of physical conductors around plasma brings about considerable technological disadvantages that can be overlooked only due to the remarkable confinement performances of these systems. Most technological features of the main components (combustion chamber, blanket, magnetic coil system) of the toroidal fusion reactors are a consequence of certain particular conditions imposed by the very toroidal geometry (problems concerning the access, installation and maintenance, the spatial force distribution, etc.).

In the case of inertial systems, the chief physical processes are related to the interaction of a laser beam with a D — T target, usually of a spheric shape.

The energy supplied to the target must exceed a certain limit-value. Consider $E_L$ the energy carried by the laser beam, $\eta$ — the efficiency of generating the laser energy $E_L$, $E_f$ — the energy generated by fusion in the D — T target, and $\eta_f$ — the overall efficiency of conversion of the fusion energy in the systems. Then, for a positive power balance of the fusion system, the following condition is required:

$$(E_L + E_f)\eta_f \geqq E_L/\eta, \tag{12.17}$$

Considering $M$ — the multiplication factor expressed by:

$$M = E_f/E_L, \tag{12.18}$$

with (12.18), the requirement that the system perform will be:

$$M \geqq [1/(\eta\eta_f)] - 1. \tag{12.19}$$

The physical requirement for a reactor to perform is given by $M \geqq 1$ and corresponds to: $\eta \cdot \eta_f \geqq 0.5$.

To obtain a compression factor of the order of $10^4$, an energy of several tens of kJ is necessary. It is important that the time in which energy $E_L$ is absorbed be as short as possible. It has been demonstrated that, to this purpose, the power $P$ transferred by laser to one drop of fusion material should vary in time following the law:

$$P = P_o(1 - t/t_f)^{-2}, \tag{12.20}$$

where $P_o$ is the initial power per unit surface and $t_f$ is the time taken by the shock-wave resulting from the impact of the beam on the drop to reach the target's centre.

For a target radius of 100 $\mu$m and $P_o = 10^3$ W . cm$^{-2}$, $t_f$ is about $10^{-9}$ seconds. In actuality, the power does not become infinite for $t = t_f$ but only very high, somewhat about $10^{16}$ W . cm$^{-2}$.

If the maximal value of the instant power exceeds $10^{15}$ W, at a frequency of pulse repetition of at least one pulse per second, the multiplication factor reaches 100, for a laser energy of $10^3$ kJ.

To exemplify, here are several characteristics of the Los Alamos reactor project:

*$CO_2$ Laser*

| | |
|---|---|
| Total energy on target (J) | $10^6$ |
| Laser beam | 8 |
| Energy per beam (J) | 0.125 . $10^6$ |
| Repetition frequency (Hz) | 30 |
| Impulse duration (s) | $10^{-9}$ |

*D – T Target*

| | |
|---|---|
| Mass (kg) | 0.98 . $10^{-6}$ |
| Radius (m) | $10^{-3}$ |

*Power balance*

| | |
|---|---|
| Thermal power (MWt) | 3744 |
| Conversion efficiency | 0.40 |
| Total electric power (MWe) | 1500 |
| Fraction re-injected in the reactor | 0.33 |

The section to follow covers the technological requirements to be met in view of the two types of fusion reactors, which entail provision of several characteristics of the materials for building such reactors.

### 12.2.2. Fusion Reactor Projects

The magnetic confinement most studied to date is the TOKAMAK system, built for the first time at the "I. V. Kurtcheatov" Atomic Energy Institute in Moskow, under the guidance of L. A. Artsimovich [10]. Figure 12.3 is indicative of the configuration of this toroidal system. Table 12.2 indicates the main characteristics of the TOKAMAK facilities under construction or in service.

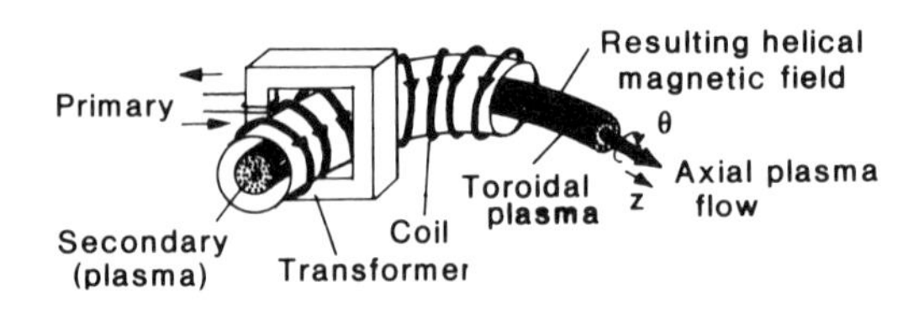


Fig. 12.3. Principle of a TOKAMAK facility.

A TOKAMAK generates a plasma by ionizing the low-pressure gas in the toroidal combustion chamber. Plasma stands for the secondary coil of a transformer whose magnetic flux runs through the toroidal chamber. The winding of the chamber makes still another coil generating a toroidal magnetic field. The magnetic field system bears a stabilizing effect on plasma that inhibits magneto-hydrodynamic instability. The energy necessary for the formation, heating and, to some extent, for the confinement of plasma is transferred in the form of impulses from the electric power supply of the transformer's primary circuit.

Large facilities have actually been built, able to test physically the conditions required for the reactor's operation. The great many problems still under debate with the TOKAMAK and other types of thermonuclear reactors call for considerable design and technological efforts in top industrial fields [11].

Here are some of the technological problems in the construction of a thermonuclear reactor:

1. obtaining materials adequate to the combustion chamber (*inter alia* – first-wall materials) so as to provide for a long and safe reactor service life;

2. obtaining superconducting materials featuring high critical temperatures and high critical magnetic fields;
3. overcoming the technological difficulties raised by superconducting magnetic coils of special dimensions and characteristics;
4. control of plasma impurities;
5. tritium handling and storage;
6. technologies for obtaining and machining the diverse materials meant for use in the thermonuclear reactors, which are not currently produced in sufficient quantity.

TABLE 12.2 Characteristics of TOKAMAK Facilities

| *Facility* | *Plasma current* (kA) | *Torus outer radius* (m) | *Plasma radius* (m) | *Toroidal magnetic field* (T) | *Ionic temperature* (keV) | *Plasma density* ($10^{20}$ $m^{-3}$) | *Life-time* (ms) | *Remarks** |
|---|---|---|---|---|---|---|---|---|
| T-4 (USSR) | 300 | 0.9 | 0.17 | 4.5 | 0.7 | 0.1-1 | 20 | Dc |
| ORMAK (USA) | 225 | 0.8 | 0.23 | 3 | 1.5 | 0.6 | 17 | Dc |
| ATC (USA) | 60-150 | 0.90-0.38 | 0.17-0.1 | 1.5-3.4 | 0.2-0.6 | 0.2-1 | - | Dc |
| TFR-600 (France) | 600 | 0.98 | 0.22 | 6.0 | 2 | 2 | 35 | Op |
| ALCATOR-C (USA) | 500 | 0.64 | 0.16 | 14.0 | 1.4 | 8 | 30 | Op |
| PLT (USA) | 500 | 1.32 | 0.41 | 3.2 | 7.1 | 1 | 100 | Op |
| T-10 (USSR) | 650 | 1.5 | 0.39 | 5.0 | 1.1 | 0.6 | 60 | Op |
| TFTR (USA) | 2500 | 2.48 | 0.85 | 5.2 | 15 | 0.5 | 300 | Op |
| JT-60 (Japan) | 2700 | 3.03 | 0.95 | 4.5 | 5-10 | 0.2 | 200-1000 | Op |
| JET (Western Europe) | 3000 (5000) | 2.96 | 1.25/2.1 | 2.7 (3.5) | 5-10 | 0.3 | 300-2000 | Op |
| T-15 (USSR) | 1400 | 2.4 | 0.75 | 3.5-4.5 | 5-7 | 1 | 300 | Cn |
| INTOR (International) | 6400 | 5.2 | 1.2/1.9 | 5.5 | 10 | 1.4 | 1400 | Pr |

* Status as from January 1985. Dc = discontinued; Cn = in construction; Op = in operation; Pr = project [22].

Besides the magnetic confinement thermonuclear reactors, among which the TOKAMAK type is outstanding and the most advanced, research also seeks developments with the inertial confinement thermonuclear reactors, among which the laser fusion facility is the most studied type [12,13].

For this type of TNR (thermonuclear reactor) the deuterium — tritium fuel is delivered in the form of solid microspheres (pellets) which are injected periodically into the laser beams that are focused on the centre of the combustion chamber. Here the implosion of the pellets takes place, corresponding to an increment in pressure capable of triggering the thermonuclear reaction. The condition for a positive energy balance — the equivalent of the Lawson criterion for the magnetic confinement of a D - T reaction — is:

$$\rho r \geq 1 \ (\mathrm{g/cm^2}), \tag{12.21}$$

where $\rho$ is the D — T compound density and $r$ is the pellet radius.

The D — T quantity necesary to release 100 MJ is 0.3 mg which, at the liquid D — T density ( $\rho \sim 0.214$ g/cm$^3$), corresponds to a pellet radius of 0.7 mm. Assuming that the pellet is bound to be heated at a temperature of approximately $1.2 \times 10^8$ K ( $\simeq$ 10 keV), in order to generate the reaction the minimal energy necessary is 200 kJ. At such a density, the pressure reaches $\sim 10^{14}$ Pa. The duration of this process is of the order of $10^{-9}$ s. Hence we are dealing with a series of microexplosions, induced by the laser radiation, which lead to shock waves and to remarkably high temperatures. Maintaining the combustion chamber in an operable condition for a sufficiently long time requires special technologies and materials. Technology and materials are, therefore, of essence, throughout the time of bringing fusion energy to the marketplace.

Experiments have demonstrated, in principle, the feasibility of obtaining fusion along these lines. A convincing achievement with this type of TNR depends critically on the forthcoming improvements with the laser systems, that must provide higher laser performances. In the context, it is believed that a fusion-grade laser equipment should exhibit the following parameters:

| | |
|---|---|
| energy efficiency | $\geq 10\%$ |
| beam power | $\geq 10^{15}$ W |
| beam energy | $\geq 10^6$ J |
| repetition frequency | $\geq 1$ pulse per second. |

To give a better idea of the would-be designs of some TNRs, Figs 12.4 and 12.5 display two such projects [14]. Figure 12.6 suggests how a set-up that would integrate such a fusion facility in a TN power station might eventually look. For the sake of the example, the concept in this figure refers to a magnetic confinement fusion equipment. In the case of a laser fusion equipment, the injector would correspond to the fuel pellet pulsed supply device, and instead of the equipment generating the magnetic field there would be a whole compound of lasers and radiation focusing equipment.

A complementary approach to thermonuclear reactors is found in the proposals of fusion — fission hybrid reactors [15,16]. Such projects are based on the use of 14 MeV neutrons generated by D — T thermonuclear reactions that could be used for a multiplication of neutrons and for extracting energy through fission from a material blanket with a large $Z$ surrounding the fusion plasma. In this way, fission energy can be obtained together with fusion energy, or fissionable materials can be generated from fertile materials in the said blanket.

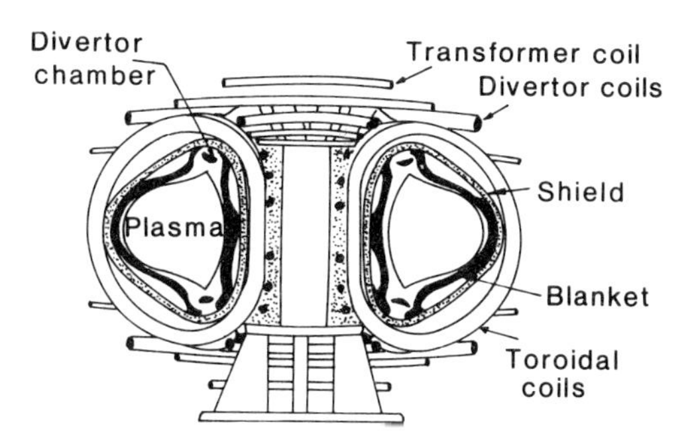


Fig. 12.4. A section through a TOKAMAK TNR.

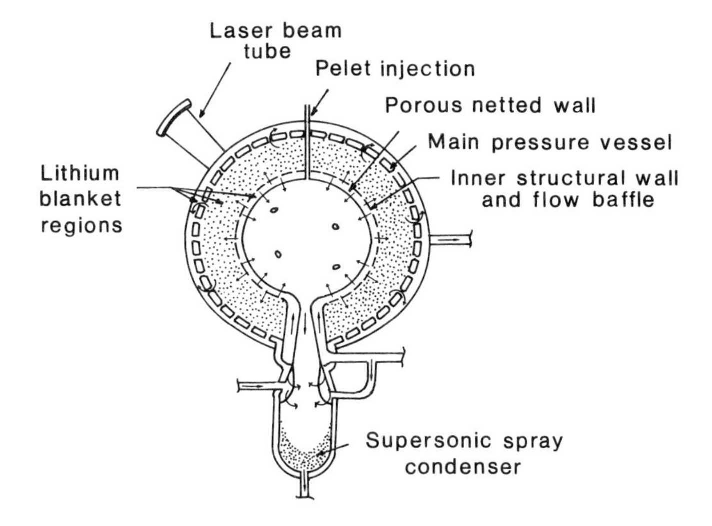


Fig. 12.5. A section through a laser-triggered TNR.

In a hybrid reacator, some layers of fertile material would cover the cooling mantle in order to obtain fissionable material.

The operating parameters of a TOKAMAK reactor [16] using metallic uranium as a fertile material would read:

| | |
|---|---|
| Fusion power | 690 MW (thermal) |
| Total power | 5100 MW (thermal) |
| Maximal neutron flux at the first wall | $4.6\times10^{13}$n/cm$^2$s (100 W/cm$^2$) |
| First wall surface | 410 m$^2$ |
| Tritium breeding ratio | 1.2 |
| Average power in the blanket | 90 W/cm$^2$ |
| Maximal concentration of Pu in the blanket | 1.1% |
| Total duration of Pu cycle | 3 years |
| Annual production, Pu | 4.1 t |
| Specific production, Pu | 0.8 t/GW (thermal) per year. |

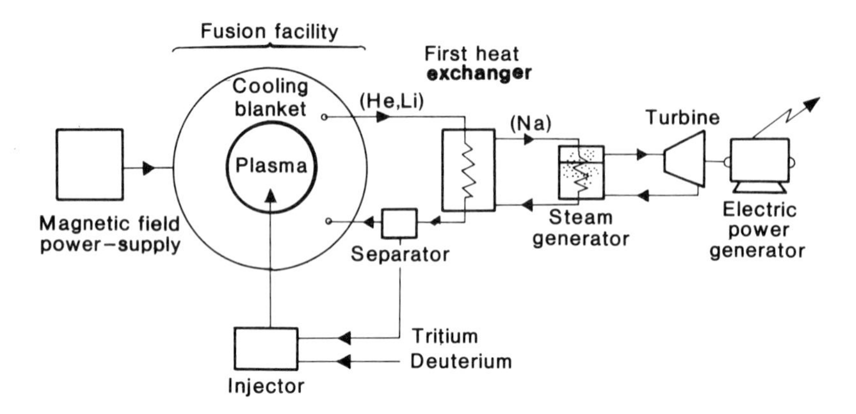


Fig. 12.6. Layout of a power thermonuclear reactor.

This example is reflective of the interest taken in the hybrid reactors as generators of fissionable material; the specific production of plutonium is, in this case, 10 times larger than with the fast reactors. The hybrid reactors are likely to represent an intermediate solution of great interest, until the advent of the TNRs as such.

## 12.3. FUEL MATERIALS

### 12.3.1. Fuel Cycle

Considering the fusion reactions (12.1) through (12.6) one can conceive, in principle, three fuel cycles [14]:

(a) *The D – T – Li cycle*. In this cycle D and Li are primary fuels. Tritium is produced through reactions (12.5) and (12.6) in the reactor blanket, which contains Li. In this cycle some 80% of the fusion energy is delivered as the energy of the neutron:

$$D + T \rightarrow {}^4He\ (3.5\ MeV) + n\ (14.1\ MeV)$$

Due to the fusion's comparatively larger cross-section (Fig. 12.1) this is viewed as the most convenient cycle for power TNRs.

(b) *The D – D – T cycle*. D is the primary fuel in this cycle. Tritium generated by the reaction D — D is recycled in plasma, therefore this cycle does not require regeneration. Since the cycle includes the fusion reactions (12.1), (12.2) and (12.3), the fusion energy is a total sum of the contributions of the three reactions; only approximately 35% of the energy is carried by neutrons, which limits the impact of the neutron radiation onto the combustion chamber wall. This cycle is also called the *catalytic D – D cycle*.

(c) *The D – D – $^3$He cycle*. As with cycle (b), D here is the primary fuel. $^3$He generated by reaction (12.1) is recycled, while the tritium as generated is collected and, after decay into $^3$He, is used in the reactor.

For this reason, this cycle does not require *T* regeneration. The exoenergetic reactions of this cycle are (12.1), (12.2) and (12.4), so that only ~ 15% of the energy goes with the neutrons.

Tables 12.3 and 12.4 list the chief material characteristics to be considered for a TNR. Due to the gaseous condition of tritium and deuterium at working temperatures, their injection in plasma would not imply special difficulties. Since deuterium can be obtained from heavy water, the issues regarding its extraction and purity are those discussed in Chapter 5. Consequently, tritium and lithium only will be of concern in the following.

TABLE 12.3 Nuclear Characteristics of Several Materials of Special Interest for the TNR

| *Isotope* | *Natural isotopic abundance* (%) | *Atomic weight* (u) | *Half-life* | *Radiation emitted* | *Capture cross-section for thermal neutrons* (mb) |
|---|---|---|---|---|---|
| $^1H$ | 99.985 | 1.007825 | - | - | 0.3326 |
| D | 0.015 | 2.0140 | - | - | 0.51 |
| T | - | 3.01605 | 12.26 years | β(16.61 keV) | $6 \times 10^3$ |
| $^3He$ | 0.00013 | 3.01603 | - | - | 0.1 |
| $^4He$ | 99.99987 | 4.00260 | - | - | ≈ 0 |
| $^6Li$ | 7.42 | 6.01512 | - | - | 45[1] |
| $^7Li$ | 92.58 | 7.01600 | - | - | 37 |
| Li (nat.) | | | | | $72 \times 10^3$ |

[1] Owing to the reaction (n,α) the absorption cross-section of $^6Li$ for thermal neutrons is 910 b.

TABLE 12.4 Physical Characteristics of Several Materials, of Special Interest for the TNR

| | *Molecular weight* (kg/kmol) | *Density* (kg/m$^3$) | *Melting temperature* (K) | *Boiling temperature* (K) |
|---|---|---|---|---|
| $H_2$ | 2.0159 | 0.0899[1]<br>70[2] | 14.01 | 20.65 |
| $H_2O$ | 18.0153 | 1000 | 273.15 | 373.15 |
| $H_2O_2$ | 34.01 | 1442.2 | 272.74 | 423.35 |
| $D_2O$ | 20.031 | 1105 | 276.97 | 374.57 |
| He | 4.0026 | 147[3] | 0.95 | 4.55 |
| Li | 6.939 | 534<br>515[4] | 452.15 | 1590.15 |

[1] gas, in normal conditions; [2] liquid; [3] at ~ 3 K; [4] liquid, at the melting point.

### 12.3.2. Tritium

As shown in Table 12.3, tritium is β-radioactive, with a half-life of 12.26 years. Since tritium atoms have small radii, this gas is expected to diffuse easily at reactor temperatures, which, considering its radioactivity, raises environmental protection problems.

Due to isotopic exchange processes, the biological half-life of tritium does not exceed 19 days, which means that the actual period to be considered in the calculation of the radiation dose should be of no more that 19 days. Because of these, and due to the low energy of β-radiations emitted, tritium is considered a low-radiotoxicity isotope. The maximally allowed concentrations are given in Table 12.5. Still, the amount of tritium necessary in the reactor being is excess of 1 kg, its activity reaches ~$10^7$ Ci, which calls for special precautions.

TABLE 12.5 Limits of Exposure to Tritium

| | *Restricted areas* | | *Unrestricted areas* | |
|---|---|---|---|---|
| | *Air* ($Ci/m^3$) | *Water* ($Ci/m^3$) | *Air* ($Ci/m^3$) | *Water* ($Ci/m^3$) |
| HTO or $T_2O$ | $5\times10^{-6}$ | $1\times10^{-1}$ | $2\times10^{-7}$ | $3\times10^{-3}$ |
| HT or $T_2$ | $2\times10^{-3}$ | - | $4\times10^{-5}$ | - |

Thus, tritium comes with its problems; extraction from Li, storage, preventing leaks in the environment, preparedness for limiting consequences of an accidental, massive leak [7].

Some of the most powerful fission reactors nowadays are provided with T-retention equipment. This may be an alternative to the generation of tritium. In TNRs the generation of tritium from lithium would take place in the blanket surrounding the combustion chamber, under the effect of the neutrons, through the reactions (12.5) and (12.6): $^6Li(n,\alpha)T$, $^7Li(n,n'\alpha)T$. To enhance the amount of neutrons in the area where tritium is generated, beryllium may be used as a neutron multiplier [by the reaction (n,2n)].

Table 12.6 lists the lithium compounds serving as tritium sources, as well as the areas where such processes can occur.

The production of tritium in the reactor is indicated by the "breeding ratio". Let $n$ be the amount of tritium in the blanket, apart from the tritium in plasma, and let us indicate by + and - the T generating process and the T consumption process, respectively. In these terms, one gets the kinetic equation [2]:

$$\frac{dn}{dt} = \frac{dn^+}{dt} - \frac{dn^-}{dt}. \qquad (12.22)$$

Breeding ratio $T_r$ can be defined therefore by:

$$T_r = \frac{dn^+}{dt} \Big/ \frac{dn^-}{dt} \qquad (12.23)$$

TABLE 12.6 Generation of Tritium from Lithium

| *Source material* | *Breeding* | *Cooling* | *Moderator* |
|---|---|---|---|
| Liquid Li | * | * | |
| Li molten salts | | | |
| $Li_2BeF_4$ | * | * | |
| LiF | * | * | |
| Ceramic compounds | | | |
| $Li_2O$ | * | | * |
| $Li_2C_2$ | * | | * |
| $Li_4SiO_4$ | * | | |
| $Li_2SiO_3$ | * | | |
| $LiAlO_2$ | * | | |
| Alloys | | | |
| LiAl | * | | |

Denoting with $\eta$ the relative tritium consumption per unit-time:

$$\eta = \frac{1}{n}\frac{d\bar{n}}{dt}, \tag{12.24}$$

the tritium rate equation becomes:

$$\frac{dn}{n} = \eta(T_r - 1)dt, \tag{12.25}$$

which, assuming $\eta$ = constant, leads to a time interval in which the tritium amount is doubled, given by:

$$t_2 = 0.693/\eta(T_r - 1). \tag{12.26}$$

Specific reactor power, $P_S$, can be defined by:

$$P_S = \frac{P_t}{M_c} \tag{12.27}$$

$P_t$ is the thermal reactor power, and $M_c$ is the burned fuel mass. Tritium mass $M_S$ burned to generate 1 MW(t) in a time $t_D$ is:

$$M_S = \frac{M_c}{P_t t_D} \tag{12.28}$$

One has

$$\eta = P_S M_S. \tag{12.29}$$

Using the values given by Kammasch [2], namely $P_S$ = 500 MW(t)/kg and $M_S$ = 6.4x10$^{-6}$ kg MW(t)h, the time $t_2$ required to double the amount of tritium can be obtained:

$$t_2 = 2.47 \cdot 10^{-2}/(T_r - 1) \qquad (12.30)$$

where $t_2$ is expressed in years.

To improve tritium's breeding ratio, some projects suggest the use of lithium, enriched in $^6$Li up to 90 — 95%.

Extraction of tritium from the reactor, both to reuse it and to prevent "pollution", is mainly done in two working areas: the plasma and the breeding area.

For continuous operation, plasma has to be maintained at a certain concentration in useful components and also must be cleared of impurities. When recycling the process gas (fuel) for purification, tritium is also recovered.

Tritium concentration in the recycled process gas depends on the fuel cycle. To exemplify, Table 12.7 shows the relative tritium concentration for three fuel cycles, for a magnetic mirror reactor operated at temperatures corresponding to a 300 — 400 keV ion kinetic energy [14]. The process gas can be recycled and purified either through the reactor vacuum system, or by capturing plasma on a solid surface. Separation of the various gas components (helium, tritium, deuterium) should be achieved by distillation at low temperatures, or by means of stacks of thin palladium films.

TABLE 12.7 Obtaining Tritium, in three Fuel Cycles

| *Cycle* | *Kinetic ionic temperature* (keV) | *Proportion in plasma* | *Relative tritium concentration* | *Remarks* |
|---|---|---|---|---|
| D — T — Li | 300 | 60% D; 40% T | 1.00 | T-breeding material in the blanket |
| D — D — T | 400 | 88% D; 12% T | 0.30 | T generated in the reaction D(D,H)T is recycled |
| D — D — $^3$He | 400 | 80% D; 20% $^3$He | 0.02 | T generated by D(D,n)$^3$He is recycled, while T obtained by D(D,H)T is stored |

Obtaining tritium from plasma is not likely to be a special problem. The question of obtaining tritium from the fertile material (lithium) in the blanket look much more difficult. When Li is in the cooling system, tritium extraction is in order, so as to minimize its diffusion in the steam circuit. Diffusion velocity $R$ of tritium in this circuit depends on a single material parameter, which is the wall permeability $P$ [14]:

$$R = AP(p_r^{1/2} - p_a^{1/2})/d, \qquad (12.31)$$

where $A$ is the area of the cooling agent/steam contact surface in the heat exchanger, $d$ is the exchanger's wall thickness, and $p_r$ and $p_a$ are the partial tritium pressures on one side and the other of the heat exchanger wall, i.e. on the coolant side and on the steam side, respectively.

Because of the quick hydrogen — tritium isotopic exchange,one gets $p_a \simeq 0$, so that the maximum allowed partial pressure of tritium in the cooling agent shall be [14]:

$$p_{rm} = (R_m d/(AP))^{1/2}, \tag{12.32}$$

where $R_m$ is the maximum velocity allowed for tritium diffusion in the steam circuit.

Since, in principle, $d$ and $A$ are determined from hydraulic and thermal consideration, the only parameter upon which one can work is the permeability $P$. The question rests, therefore, with finding the adequate materials, in order to meet the requirements as specified. Permeability $P$ can, generally speaking, be modified in two ways:

1. by decreasing the heat exhanger's operating temperature, and

2. by covering the walls with materials constituting a shield against tritium penetration.

A high value of permeability should be provided in areas specially designed within the cooling circuit in order to isolate the tritium and use it in the fuel compound (for the D — T cycle).

The maximal tritium quantity allowed in the environment limits the T value to approximately 1 mg per day — an amount that is evacuated with the turbine cooling water; a maximum activity of ~ 10 Ci of tritium per day corresponds to this quantity. The following must be taken into consideration in order to reach levels below this maximal one:

1. use of materials characterized by a high hydrogen absorption power;
2. use of oxygen or of an oxidizing agent to obtain $T_2O$, to which metals are not permeable;
3. use of cold "traps", for tritium to condense;
4. covering the pipes through which the tritium compound flows with shielding materials.

Future research will certainly evolve other, better, solutions.

*Materials characterized by a high hydrogen absorption power* are listed in Table 12.8 together with some physical properties of special interest.

*Tantalum* is capable of absorbing hydrogen 55 — 700 times its own volume. Considering the temperature variation of the absorption rate, the absorption optimal temperature is 600 °C. Hydrogen absorption affects tensile properties. If heated in vacuum, tantalum transfers the absorbed hydrogen and regains its original tensile properties. Ta easily absorbs both oxygen and nitrogen, while being extremely reluctant to absorb noble gases. It is easily affected by alkaline hydrate melts or solutions.

*Palladium* is capable of absorbing large quantities of hydrogen, which are released when the metal is heated under high vacuum. It is also extremely permeable for hydrogen, especially in the range 100 — 300 °C and under a pressure of 100 — 700 torr.

TABLE 12.8 Hydrogen-absorbing Materials

| *Material* | *Symbol* | *Molecular weight* | *Density* (kg/m$^3$) | *Melting temperature* ($^{\circ}$C) | *Boiling temperature* ($^{\circ}$C) | *Remarks* |
|---|---|---|---|---|---|---|
| Tantalum | Ta | 180.948 | 16,600 | 2996 | 5425 | |
| Niobium | Nb | 92.906 | 8,570 | 2468 | 4927 | |
| Zirconium | Zr | 91.22 | 6,490 | 1852 | 3578 | |
| Thorium | Th | 232.038 | 11,700 | 1700 | 4000 | Radioactive |
| Titanium | Ti | 47.90 | 4,500 | 1675 | 3260 | |
| Palladium | Pd | 106.40 | 11,400 | 1552 | 2927 | |
| Erbium | Er | 167.28 | 9,164 | 1497 | 2900 | |
| Yttrium | Y | 88.905 | 4,340 | 1495 | 2927 | |

Metallic *yttrium and erbium* are high absorbents of hydrogen, and therefore, are used in some tritium absorption projects.

Figure 12.7 illustrates the capability of various metals to absorb hydrogen, as functions of temperature [18].

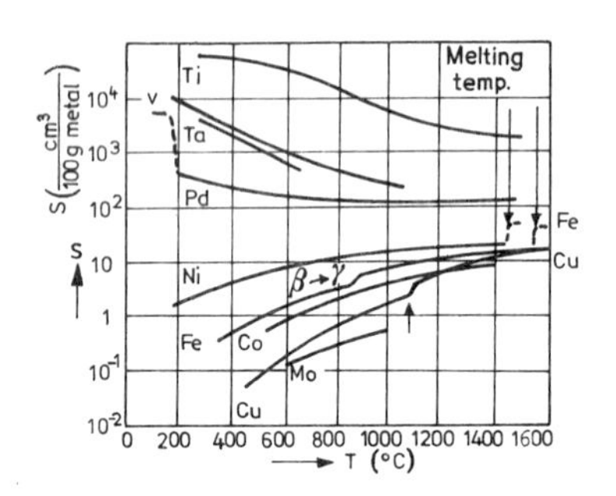


Fig. 12.7 Hydrogen amount absorbed by 100 g of metal, vs. temperature.

Figure 12.8 shows a flow-chart of a tritium recovery system comprising, in general lines, all processes discussed above [19].

*Using oxygen to transform tritium into $T_2O$* seems a promising method for the gas-cooled systems (He, Ar). Partial pressure of oxygen in the cooling gas is about $10^{-2}$ torr and may reduce the partial pressure of tritium under $2 \times 10^{-13}$ torr. $T_2O$ is to be absorbed on molecular sieves to maintain $T_2O$ steam pressure at $10^{-3}$ torr. Calculations [20,21] demonstrate the need of a daily sieve supply of 0.07 m$^3$(640 kg/m$^3$).

The efficiency with the use of cold traps along the cooling circuit depends on their temperature — the lower the temperature, the higher the efficiency — but due to the limitations with lowering the temperature this method is still to be complemented by the use of other processes. The use of a shield to lower

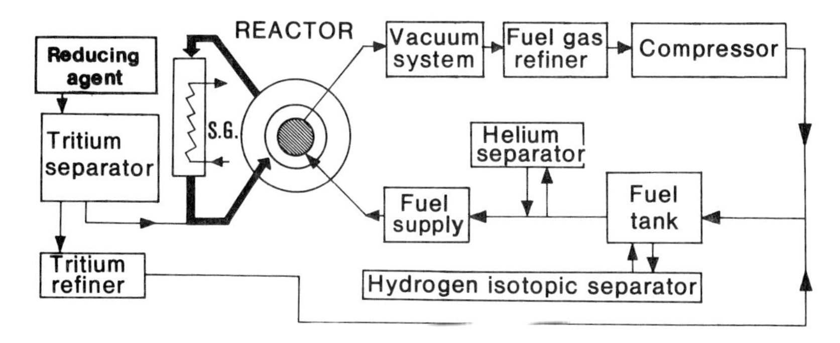


Fig. 12.8. Tritium recovery system (SG — steam generator).

the permeability to tritium can be a solution, in the future. The coating in vacuum of the heat exchanger tubes with a *W* layer 0.3 mm thick is one suggestion. Other projects recommend formation of compact oxide shields on these surfaces. Unfortunately, none of these methods has been tested in actual service and, therefore, cannot be assessed as to its effectiveness.

At the other end is the equally important question of the *materials permeable for tritium and deuterium, and opaque for helium* and for metallic impurities. Such materials, in the form of films, are meant to separate tritium and purify the fuel gas. To illustrate this, Fig. 12.9 gives the permeability for hydrogen of some metals. Among the metals lately under trial, the most efficient ones seem to be the niobium and vanadium alloys [18].

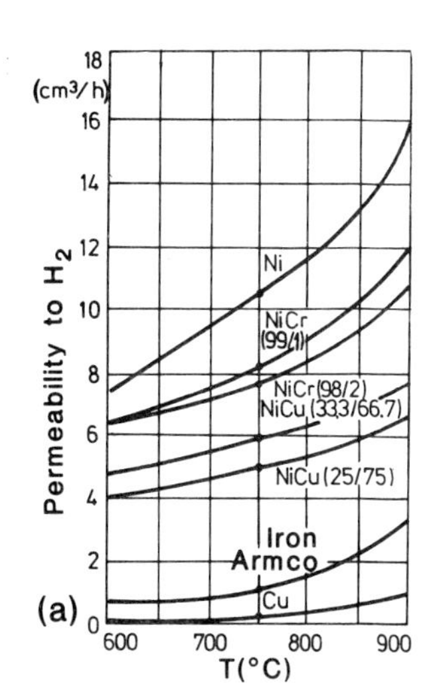


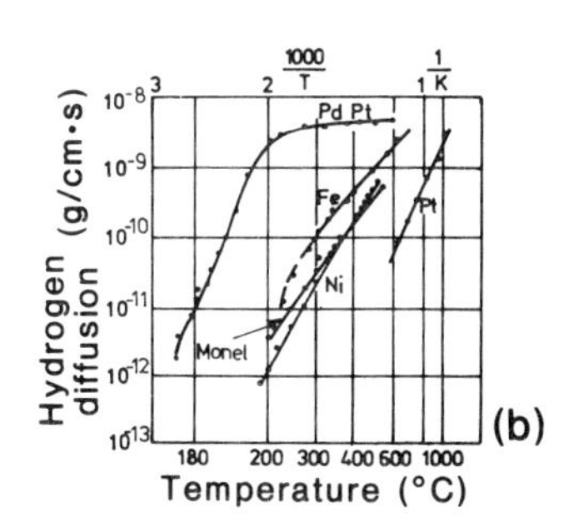


Fig. 12.9. Permeability to hydrogen of several metals and alloys [18]: (a) volumetric outflow through a wall 7.5 mm thick and 48.5 $cm^2$ wide, at normal pressure; (b) mass of the hydrogen that penetrates the wall unit-thickness (1 cm) in the unit-time (1 s) and on the unit-area (1 $cm^2$).

The matter is still more difficult with the solid materials containing lithium for tritium reproduction, as is the case with $LiAlO_2$ or $Li_2O$. The choice for $LiAlO_2$ is based on its high melting temperature (1900 — 2000 °C) and satisfactory behaviour with respect to tritium diffusion. To facilitate diffusion, $LiAlO_2$ is bound to be used in the form of microspheres with a radius of ~20 μm. Anyway, the question of these materials swelling under the effect of the gases is still to be investigated. Operation at high temperature is believed to favour diffusion, since $LiAlO_2$ exhibits a low thermal conductivity, and so one may expect large temperature gradients in the material. When using $Li_2O$ as fertile material, as a result of the nuclear reaction the compound LiOT will be obtained, which, under a sufficiently high operating temperature ( ~ 700 °C) will dissociate yielding tritiated water. Helium (or an equivalent gas) flow providing the cooling will carry the tritium which will then pass through the purification, or separation, stages already discussed.

### 12.3.3. Fuel Supply

Another essential concern with the TNRs is the continuous supply of the reactor with fuel, to maintain the power rate at a given level. Several procedures are foreseen:

1. Injection of fuel under the form of neutral atom beams, focused on the plasma centre. Yet, to penetrate thermonuclear plasma, an energy of 1 MeV is needed (for neutral atoms) and so the equivalent currents would reach considerable values (much higher than those required for plasma heating). Consequently, for the time being, no physical and technological solutions are available to implement this procedure;
2. Injection of solid hydrogen spheres into the plasma. The drawback with this would be a lowering in the plasma temperature, and the implication that the fuel spheres would have to be "shot-in" at a high speed ( ~ $10^4$ m/s), in order for them to survive erosion while reaching the plasma centre;
3. A D — T gaseous blanket at a higher pressure, and surrounding plasma. Concentration gradients would then drive D and T inward and He outward. Problems with plasma stability are likely to occur.

To understand better the solutions proposed concerning inertial confinement TNR fueling, let us review briefly the physical processes likely to occur in the D — T fuel pellets. The positive energy balance condition requires that the product (density) x (radius) be above 1. Considering the mass that can be accommodated in the corresponding volume, this implies a density of ~ $10^6$ kg/m$^3$, i.e. approximately 5000 times the density of the D — T liquid mixture. To obtain this increment in density, the pellet must be compressed under a pressure of ~ $10^{10}$ MPa. To this effect a sudden adiabatic temperature burst outside the pellet is envisaged, up to several keV. Pressure will then go up, until the thermal expansion of the material will tend to limit it. This process would allow a jump in pressure to ~ $10^7$ MPa. The process was made possible with the advent of laser beam pulses.

Pressure increment is a necessary, but not sufficient condition to obtain compression, since its effect on density depends upon the fluid's thermodynamic properties (specific heat, thermal conductivity, etc.). To obtain a further increase in density, two methods may be followed:

1. use of a pulse sequence of accurately controlled intensity, duration and repetition frequency;
2. use of some highly structured fuel pellets, in order to avoid occurrence of non-symmetric implosions, which could compromise the efficiency.

The first method requires a laser technique that looks forbidding, for the time being. The process has been studied mostly by computer simulation; the second method looks closer at the presently available possibilities. The use of microspherical pellets provided with inside cavities instead of compact ones would be less demanding in power — by one order of magnitude. Use of multilayered spheres, formed of several materials, allows an even greater effect on the quality of the microexplosion. Such a multilayered microsphere is shown in Fig. 12.10. The outer layer, of a low-$Z$ material, absorbs the laser radiation and explodes as a consequence of the sudden burst in temperature. The explosion generates a shock wave with a powerful compression effect on the high-$Z$ material inside. Simultaneously, the low-$Z$ material becomes a plasma that allows an even distribution of the implosion impulse, thus alleviating the mischance of asymmetric implosion.

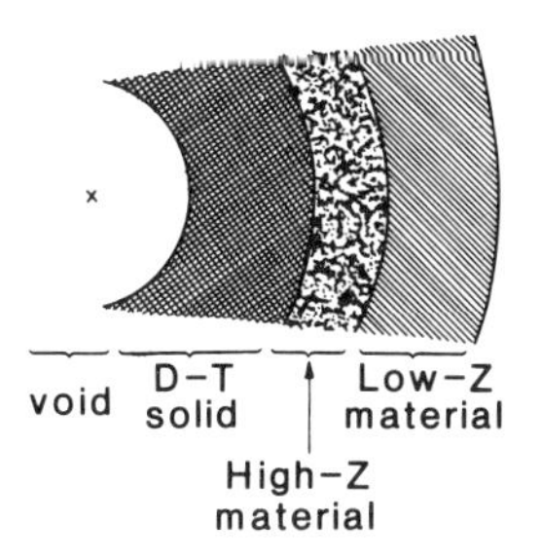


Fig. 12.10. The structure of a multilayer fuel pellet, for laser fusion.

The high-$Z$ material is meant both to confine the area where the fusion takes place, and as a "piston", to amplify the implosion pressure. As the temperature decreases, the D — T fuel compound solidifes. One obtains, therefore, as a starting point, a nuclei density much higher than the one attainable in TOKAMAK facilities.

Materials used to date for the manufacture of fuel pellets are solid D — T, included in a glass microsphere, or a compact microsphere made of solid $CD_2$.

### 12.3.4. Fuel Materials for Hybrid Reactors

The question of the *hybrid fusion – fission reactors* has been less studied. Only a few studies [15] concerning the operating conditions and the energy efficiency are available. Even these would start rather from traditional assumptions regarding the neutron flux and system's geometry, so that the calculations are but illustrative. As with the other cases discussed, two major impediments are the lack of live experience in the field, as well as the lack of detailed knowledge on material behaviour in such unparalleled conditions [23].

Again for illustrative purposes, Table 12.9 lists the fuel materials that could be used in the fission area of hybrid reactors. Note the diversity of trial compositions, which show that no optimal solution has been agreed upon, as yet.

Recently, an interesting project was evolved — to couple a CANDU-PHWR to a fusion reactor. The system would be thorium-fuelled. The fusion reactor, of the D — D or D — T type [24], operating as a powerful neutron source, would convert thorium into fissile material, providing the necessary fuel for the fission CANDU reactor. Some parameters featuring this system, with a designed power of 1000 MWe, are listed in Table 12.10.

TABLE 12.9 Fission-section Materials in Hybrid Reactors (concept)

| *As designed in the year* | *Blanket and coolant composition for the fission section* | *Authors of concept* | *Laboratory of origin* |
|---|---|---|---|
| 1957 | $^{238}U$ — $^{6}Li$ | P. C. Thonemann<br>J. D. Lawson *et al.* | UKAEA<br>Harwell |
| | $UO_2SO_4$ - cooling agent | L. G. Barrett | KAPL<br>(Knolls Atomic Power Lab.) |
| 1965 | $LiF$ — $BeF_2$ — $UF_4$ in blanket<br>$LiF$ — $UF_4$ | L. N. Lontai<br>L. N. Lontai | MIT<br>(Mass.Inst.of Tech.) |
| 1969 | $LiF$ — $BeF_2$ — $ThF_4$ in proportion of: 71% — 2% — 27% | L. M. Lindsky | MIT |
| 1970 | Study, on various blanket compositions: $^{235}U$; $^{233}U$; $^{239}Pu$; $^{6}Li$ | J. D. Lee | LLL<br>(Lawrence Livermore Laboratory) |
| 1972 | BeO — moderator<br>U — natural in blanket<br>He — coolant<br>Li — outside<br>Non-homogeneous structure of Li and UC, in Nb spheres or rods; He-cooling | B. R. Leonard | PNL<br>(Pacific North Lab.) |

TABLE 12.10 Characteristic Parameters of a Fussion — Fission Hybrid Reactor Designed for Symbiotic Operation in Tandem with a CANDU-PHW Reactor (thorium[1] - fuelled)

| | *Options for the fertile zone* | | |
|---|---|---|---|
| | *Th* | *Th — U* | *U* |
| Number of fissile atoms produced per fusion event | 0.68 $^{233}U$<br>- | 0.76 $^{233}U$<br>0.38 $^{239}Pu$ | -<br>1.64 $^{239}Pu$ |
| Number of fissile atoms produced per fusion event (in $^{233}U$ equivalent) | 0.68 | 1.08 | 1.37 |
| Number of $^{3}T$ atoms produced per fusion event | 1.075 | 1.075 | 1.075 |
| Thermal power produced by fusion ($MW_f(t)/GWe$) | 57.5 | 36.2 | 28.5 |
| Global thermal power produced by the hybrid ($MW_H(t)/GWe$) | 126 | 166 | 211 |
| Energy multiplication factor, owing to the fertile zone | 2.5 | 5.5 | 9.0 |

[1] Capacity factor 80%; net thermal efficiency 29.2%; average burn-up 29.3 MWd/kg; fertile - fissile conversion factor 0.88.

One noteworthy remark is that the D — T — Li cycle, of special interest for fusion reactors, is ultimately a hybrid reactor cycle as well, since the reactions (12.5) and (12.6) with Li are actually fission processes.

Some physical properties of the potential fuels for hybrid reactors are displayed in Table 12.11.

TABLE 12.11 Physical Properties of some Prospective Fuel Materials

| *Material* | *Molecular mass* (kg/kmol) | *Density* ($kg/m^3$) | *Melting temperature* (°C) |
|---|---|---|---|
| $UF_4$ | 314.02 | 6700 | 1036 |
| $ThF_4$ | 308.03 | 6320 | 1114 |
| $UC_2$ | 262.05 | 11,280 | 2470 |
| UC | 250.04 | 13,630 | ~ 2370 |

## 12.4. MATERIALS FOR BLANKET AND COOLING SYSTEM

Grouping together of these materials was determined by the fact that, in many TNR designs, the same material is used both for the manufacture of the blanket and for the cooling system. Beyond the scope of this section will fall the *structural* materials for the blanket and cooling system, which will be dealt with in the next section. The main physical characteristics of the materials proposed for the blanket and cooling system are given in Table 12.12.

As expected, lithium and its compounds play an important part in TNR design, and therefore require special consideration.

### 12.4.1. Lithium

Lithium is the lightest metal, its density being approximately half that of water. There is no free lithium in nature. The most important minerals containing lithium are:

Lepidolite: $KLi_2Al[(F,OH)_2/Si_4O_{10}]$ — Theoretic content in useful substance: 1.2 – 5.6% in weight $Li_2O$

Spodumen: $LiAl[Si_2O_6]$ — Theoretic content in useful substance: 3.7% in weight Li

Petalite: $(Li,Na)[AlSi_4O_{10}]$
Ambligonite: $LiAl[F,OH)/PO_4]$

In minute amounts, lithium is also found in most of the eruptive rocks, in the water of mineral springs, or in seawater. Table 12.13 shows the adundance, and the estimated reserves, of lithium [14].

TABLE 12.12 Physical Characteristics of Blanket and Cooling System Materials

| *Material* | *Function* | *Molecular mass* (kg/kmol) | *Density* ($kg/m^3$) | *Melting temperature* (°C) | *Boiling temperature* (°C) |
|---|---|---|---|---|---|
| Li | Breeding<br>Cooling | 6.939 | 554 | 179 | 1317 |
| K | Cooling | 39.102 | 860 | 63.65 | 774 |
| Na | Cooling | 22.9898 | 970 | 97.81 | 892 |
| LiF | Breeding<br>Cooling | 25.94 | 2300 | 842 | 1676 |
| $Li_2O$ | Breeding<br>Moderation | 29.88 | 2013 | 1700 | |
| $Li_2C_2$ | Breeding<br>Moderation | 37.90 | 1650 | > 1000 | |
| $LiAlO_2$ | Breeding | 65.92 | 2550 | 1700 | |
| $BeF_2$ | Neutron multiplication<br>Cooling | 47.01 | 1086 | Sublimates at 800 | |
| Be | Multiplication<br>Moderation | 9.0122 | 1820 | 1283 | 2970 |
| BeO | Multiplication<br>Moderation | 25.01 | 3030 | 2520 | 4260 |
| $Be_2C$ | Multiplication | 30.04 | 2440 | 2400 | |

TABLE 12.13 Lithium Abundance

| *Realm* | *Abundance* |
|---|---|
| Cosmic[1] | 49.5 |
| Terrestrial (crust) | 65 g/ton |
| Total amount in the first 10 m of crust | $2 \times 10^{14}$ kg |
| Seawater | 0.1 g/ton |
| Total amount in seawaters | $2 \times 10^{14}$ kg |
| World reserves as estimated in 1970 | $0.8 \times 10^{9}$ kg |
| Projected demand in the year 2000 (in the USA) | $1.5 \times 10^{7}$ kg |

[1] Cosmic abundance is related to silicon's abundance, at a standard value of $10^6$.

Lithium can be obtained by electrolysis, from its molten chloride. It reacts with water, even if not so energetically as natrium does. Since it has the highest specific heat of all solid metals, it may be efficiently used in heat exchange. Lithium has a high electric conductivity in the molten state, which makes possible its recycling by means of electromagnetic pumps. Still, this feature raises some questions as to the use of metallic Li in magnetic confinement TNRs, due to the electrodynamic interaction with confining magnetic fields. Both Li and its salts in molten state are highly corrosive. Quartz, asbestos and zirconium oxide can stand the corrosive action of Li; yet, at higher temperatures Li affects quartz, and even platinum. When liquid, its density (between 200 °C and 1600 °C) varies according to the equation [43]:

$$d = 0.515 \times 10^3 - 0.101\,(T - 200)\ (\mathrm{kg/m^3})$$

where $T$ is the temperature, in °C.

From the above, it results that an *increment in the capacities for Li production* is of essence for the future. One suggestion is to isolate it from seawater using energy supplied by the thermonuclear power station itself. It is believed that a 5000 MWt TNR will be able to meet Li demands by extracting it from $\sim 10^5$ tons of seawater a day. Such an amount of water is still small compared to the quantity usually required for cooling, so that the procedure looks feasible.

Another problem comes with lithium's *isotopic content*. As discussed in previous sections, in fission nuclear reactors — where the neutron economy is of essence — one can use as cooling agents only those Li compounds that are $^6$Li-free, since this particular isotope is a powerful absorbent of thermal neutrons, and thus the $^7$Li isotope is preferred. With the TNRs, the odds are reversed: $^6$Li is essential for breeding tritium. Table 12.3 displays some important nuclear parameters of Li. Figure 12.11 shows cross-section dependence on neutron energy, for several important materials.

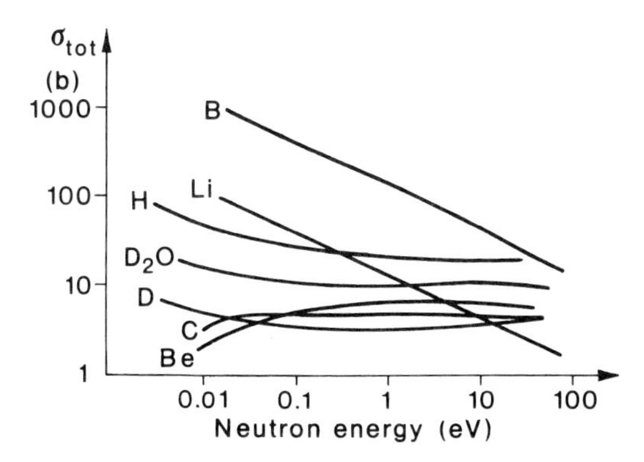


Fig. 12.11. Total cross-section dependence on neutron energy, for certain materials.

*Li enrichment* in one of its isotopes can be performed through various methods. By isotopic exchange between $^7$Li and $^6$Li the enrichment in $^6$Li may reach 16%.

By centrifuging, high enrichments can be obtained through only a small number of steps. A LiCl aqueous solution is used and a magnetic field is applied, antiparallel to the angular speed of the centrifuge. By distillation of metallic Li at 670 — 870 K, the per-stage separation factors are 1.08 at 720 K, and 1.055 at 820 K. The feasibility of isolating Li by thermal diffusion of molten salts has also been studied. The demand for enriched Li will bring about further studies concerning also the economic feasibility of various competing processes.

The main difficulty with using molten Li or molten Li salts in the cooling system (as, in general, with all alkaline metals) resides in their high corrosive potential. The importance of this question makes it worth further consideration in one of the subsequent sections.

## 12.4.2. Lithium Compounds and their Properties

Lithium can be used in reactors as liquid compounds (melts) or solid compounds as well.

The most common molten salt that can be used for cooling and for tritium breeding is the eutectic compound (2LiF) . $BeF_2$, i.e. $Li_2BeF_4$. The main physical properties of the constituents of this compound are given in Table 12.11. $Li_2BeF_4$ offers advantages over the liquid Li, because of its less corrosive character. As for neutron absorption, the use of $Li_2BeF_4$ is, again, preferable leading to a lower demand of cooling agent — as low as 20% in comparison with plain Li. Yet, on the other hand, the breeding ratio is lower, which requires enrichment in $^6Li$. An improvement in breeding ratio without using enriched Li calls for the insertion of a beryllium layer as a neutron multiplier. Another advantage is a more convenient extraction of tritium. As a result of nuclear reactions with neutrons, this compound modifies its chemistry:

$$^6LiF + n \rightarrow {}^4He + TF,$$

$$^7LiF + n \rightarrow {}^4He + TF + n,$$

$$BeF_2 + n \rightarrow 2n + 2.{}^4He + 2F \text{ (or } F_2).$$

Another salt mixture to be considered for cooling and breeding is LiCl . LiF. As far as Li concentration, this compound is more convenient, but raises some questions related to its higher melting temperature and to the production of $^{35}S$ isotope via the reaction: $^{35}Cl(n,p)^{35}S$. Isotope $^{35}S$ is $\beta$-active, with an energy of 0.17 MeV and a half-life of 87.1 days, its activation cross-section for thermal neutrons being 0.13 b.

Among lithium's solid compounds considered for breeding, the most important are $Li_2O$ and $LiAlO_2$. Some of their physical properties are given in Table 12.11. Tritium's breeding ratio is superior in $Li_2O$, as compared to $LiAlO_2$. On the other hand, studies on thermal stability of these materials at high temperatures prove that $LiAlO_2$ is more stable. Instability expresses in the thermal decomposition of these materials, accompanied by a predominant production of gases ($O_2$), besides Li, at temperatures from 700 through 1400 $^oC$. The need to use the tritium thus generated, and tritium's low solubility in these materials, determine their disposition in an inhomogeneous structure (for instance included in graphite or distributed in a porous compound $Li_2O - Al_2O_3$). This calls for a deeper knowledge of the reactions that may occur. The following values have been obtained [25] for the formation heat $\Delta H_r^o$ of the compounds $\beta$-$Li_5AlO_4$, $\gamma$-$LiAlO_2$ and $LiAl_5O_8$ through the reaction $Li_2O + \alpha Al_2O_3$:

| | $\Delta H_r^o$ (298.15 K) |
|---|---|
| $Li_2O + (1/5)\alpha\text{-}Al_2O_3 \rightarrow (2/5)\beta\text{-}Li_5AlO_4$ | -4 kcal/mol |
| $Li_2O + \alpha\text{-}Al_2O_3 \rightarrow 2\gamma\text{-}LiAlO_2$ | -25 kcal/mol |
| $Li_2O + 5\alpha\text{-}Al_2O_3 \rightarrow 2LiAl_5O_8$ | -40 kcal/mol |

The contact with graphite gives the following reaction:

$$\gamma\text{-}LiAlO_2 + (2/5)C \rightarrow (4/5)Li\ (gas) + (2/5)CO + (1/5)LiAl_5O_8.$$

Aluminate decomposition reads:

$$\gamma\text{-}LiAlO_2 \rightarrow (4/5)Li\ (gas) + (1/5)O_2 + (1/5)LiAl_5O_8.$$

The same decomposition process, accompanied by lithium production, has been studied for other compounds as well. The pressure of gaseous lithium corresponding to some of the compounds is given in Fig. 12.12, versus temperature [25]. The diagram shows that, as far as the maintenance of the Li concentration is concerned, the most convenient material is the one that has a minimal Li concentration: $LiAl_5O_8$. Considering also the amount of lithium required, $LiAlO_2$ looks a good choice. As far as neutron activation, the only interesting component is aluminium. On irradiation, nuclei of $^{28}Al$ or $^{24}Na$ are formed, of a short half-life — as shown in Table 12.14. Consequently, this compound raises no problems of lasting radioactive pollution. Also, the activity of the aluminium compounds decreases rapidly, which allows an expeditious intervention in the blanket in case of abnormal occurences. One unfortunate characteristic of $LiAlO_2$ resides in its low thermal conductivity, which may lead to important temperature gradients.

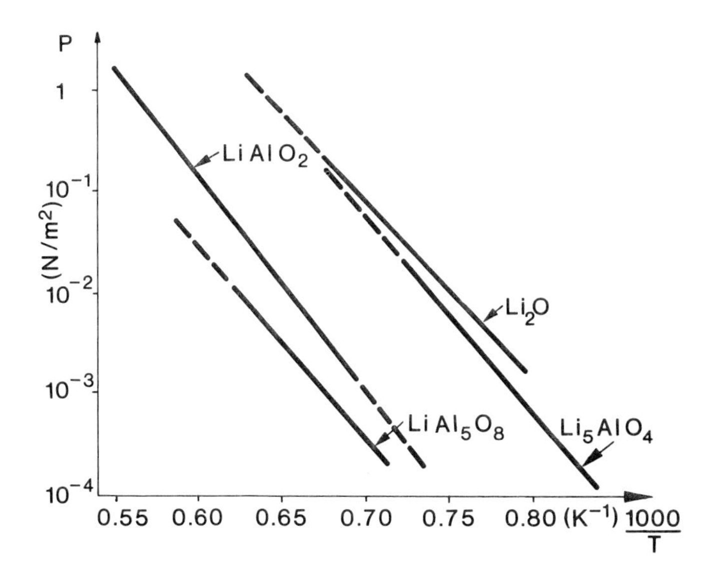


Fig. 12.12. Gaseous lithium pressure as function of temperature at the $LiAlO_2$, $Li_2O$, $Li_5AlO_4$ and $LiAl_5O_8$ decomposition. Dotted lines indicate extrapolation of measured curves [25].

TABLE 12.14 Nuclear Data of Aluminium and Offspring Radionuclides

| *Abundance* | $\sigma_{abs}$(b) | $\sigma_{act}$(b) | *Offspring radionuclide* | *Remarks* |
|---|---|---|---|---|
| $^{27}$Al 100% | 0.240 | 0.21 (n,γ) | $^{28}$Al | For thermal neutrons |
| | | 0.121 (n,α) | $^{24}$Na | For fast neutrons $E_n$ = 14.1 MeV |

Offspring radionuclides

| *Decay mode* | *β-energy* | $T_{1/2}$ | *Associated* $E_\gamma$ | *Resulted nuclide* |
|---|---|---|---|---|
| $^{28}$Al $\beta^-$ | 2.8 MeV | 2.3 min | 1.78 MeV (100%); - by disintegration 7.72 MeV (20%); - by the capture reaction $^{27}$Al(n,γ) | $^{28}$Si — stable |
| $^{24}$Na $\beta^-$ | 1.4 MeV | 15.0 h | 1.37 MeV (100%) - by disintegration 2.75 MeV (~ 100%) | $^{24}$Mg — stable |

### 12.4.3. Beryllium

Current fusion reactor concepts consider the application of beryllium as neutron moderator and neutron multiplier [26]. The most prominent physical properties of beryllium are listed in Table 12.12. Owing to the fact that this material has been dealt with in the context of fission reactor nuclear materials, we shall not dwell much longer upon it (see Chapter 5.4). Table 12.15 gives the nuclear properties of beryllium.

TABLE 12.15 Nuclear Properties of Beryllium ($^9$Be)

| *Abundance* (%) | $\sigma_{abs}$ (mb) | $\sigma_{reaction}$ (mb) | *Resulted nucleus* | $T_{1/2}$ | *Remarks* |
|---|---|---|---|---|---|
| 100 | 10 ± 1 | 9.2 ± 0.5 (n,γ) | $^{10}$Be | 2.7 x $10^6$ yr | For thermal neutrons |
| | | 10 ± 1 (n,α) | $^6$Li | 0.83 s | For fast neutrons $E$ = 14.0 MeV |
| | | (n,2n) | 2 $^4$He | Stable | Multiplication |
| | | (γ,n) | 2 $^4$He | Stable | Photoneutrons production |

As with lithium, the future will probably witness a considerable increase in the production of beryllium, in connection with TNRs. According to the structure suggested for the TNR types mix, it is estimated that, for a capacity of $10^6$ MWe of fusion power, an amount of approx. $4.6 \times 10^7$ kg beryllium will be required. The data regarding beryllium's abundance in nature are listed in Table 12.16 [14]. Beryllium deposits are smaller than those of lithium — a fact also reflected in beryllium's cost, that is about 40 times higher than that of Li, as per the same weight. The most important minerals containing beryllium are:

| | |
|---|---|
| beryl | $3BeO . Al_2O_3 . 6SiO_2$ (contains 6% Be) |
| crysoberyl | $Al_2BeO_4$ |
| fenakite | $Be_2[SiO_4]$. |

TABLE 12.16 Beryllium Abundance

| *Realm* | *Abundance* |
|---|---|
| Cosmic[1] | 0.81 |
| Terrestrial | 6 g/ton |
| Total amount in the first 10 m of crust | $1.5 \times 10^{13}$ kg |
| In seawater | - |
| World reserves as estimated in 1970 | $0.9 \times 10^8$ kg |
| Projected demand in the year 2000 (in the USA) | $2 \times 10^6$ kg |

[1]See Table 12.13.

Beryllium is a light metal. It has an elasticity modulus slightly superior to that of steel. It does not oxidize when in contact with air. Beryllium and its salts are toxic and, therefore, have to be carefully handled. The maximal beryllium concentration allowed in the form of aerosol, dust, for an 8 hours service a day, is 2 $\mu g/m^3$, and the concentration in areas that are not under permanent control should not exceed 0.01 $\mu g/m^3$.

### 12.4.4. Corrosion caused by Liquid Lithium and its Molten Salts

Corrosion induced by liquid metals when coming in contact with structural materials stands out as an important technological problem (see also section 2.3). The use of liquid metals as cooling agents in fission reactors allowed a significant gain in experience.

Thus, it is well known that the usual structural materials (Ni, Cr, Fe, Nb) as well as ordinary non-metallic impurities (C, N, H, O) are generally more soluble in lithium than in liquid natrium, which raises difficult problems of compatibility. Judging by the data available [14], the maximal admissible temperatures for the use of structural materials in liquid lithium are:

| | |
|---|---|
| Ni alloys | 450 °C |
| Ferritic steel | 500 °C |
| Carbon steel | 550 °C |
| Austenitic steel | 670 °C |
| Nb - 1% Zr | 900 °C |

Several studies consider that the maximal temperatures should be much lower. For instance, according to these, austenitic steel should not be used at temperatures over ~500 °C.

Consequently, tests are required in order to find appropriate solutions. Table 12.17 gives some data on the corrosion of structural materials by lithium and its salts. Molten salts compounds seem to be less corrosive and, therefore, more compatible with structural materials.

TABLE 12.17 Corrosion of Structural Materials by Lithium and its Salts

| | *Structural material* | *Austenitic steels* | *V-alloys* | *Mo-alloys* | *Nb-alloys* | *Graphite* |
|---|---|---|---|---|---|---|
| Liquid lithium | Maximal temperature in service | ≤ 500 °C | ≤ 900 °C | ≤ 1100 °C | ≤ 1300°C | Incompatible |
| | Effect of dissolved oxygen (impurity) | | no | no | Extremely annoying | - |
| $Li_2BeF_4$ liquid | Corrosive action | no | no | no | | no |

Yet, other phenomena may also arise: $Li_2BeF_4$ is a powerful deoxidizing agent, thus "cleaning" structural metals of any oxide, and making them extremely sensitive to any corrosive agent impurity. If the compound is transferred through a powerful magnetic field, it may be chemically destabilized as a consequence of the electric effects induced — a fact that is accompanied by an increase in its corrosive potential. Finally, although lithium is inert to graphite, it cannot be used together with other refractory structural materials, since it may carbonize them through carbon transfer.

In conclusion, only more research can decide upon the optimal choice in solving such tough technological and economic problems.

## 12.5. STRUCTURAL MATERIALS

This group includes all material which is used in the construction of the combustion chamber, thermal insulation, reinforcement structure as well as the materials meant to protect, against corrosion or erosion, certain parts in the blanket and cooling system [40,41].

The most difficult service conditions are cast upon the materials destined to the combustion, or reaction, chamber [27,42]. The combustion chamber's first wall material must preserve its tensile properties and integrity, even when subject to complex strains, extremely high temperatures and radiation levels (a great diversity of radiations: neutrons, protons, deuterons, tritons, He ions, Li ions, ions of heavier elements — impurities, X rays), corrosion due to solid materials, liquids or gases. The material may therefore change its crystalline or chemical structure and become brittle and porous; it may be corroded or eroded, it may swell and peel; finally, it may become radioactive. For these reasons the choice of the appropriate combustion chamber material is a key problem, still unsettled, and also one of the top problems in material technology. In the absence of satisfactory experimental data the design will allow a work stress half the value of the stress necessary to generate a plastic deformation of 0.2% over 100,000 hours.

Since temperatures in structural materials exceed 550 °C or may even reach 1800 — 2000 °C for the combustion chamber, some temperature resistant materials have been suggested, such as: (a) alloys based on refractory metals (N, Nb, Mo); (b) Fe alloys; (c) Ni alloys; (d) Al-based materials; (e) C-based materials; (f) others.

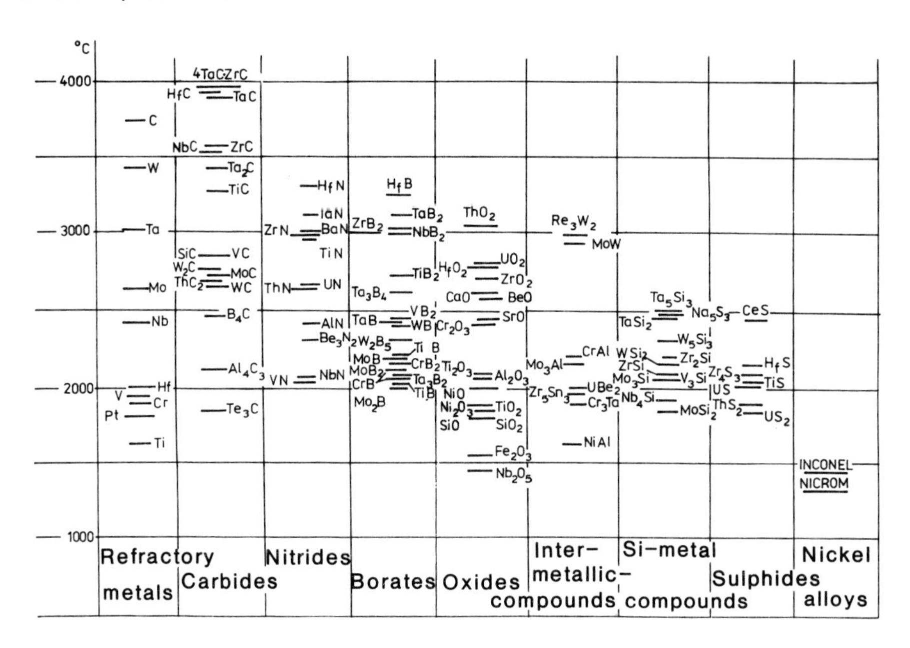


Fig. 12.13. Melting temperatures of various materials.

Figure 12.13 displays the main refractory materials, or temperature-resistant materials, together with their melting temperatures. Taking into account the conditions in the reactor, only a few of these materials are actually applicable. Let us examine this issue in more detail.

## 12.5.1. Refractory Metals and their Alloys

The chief refractory metals in use are: Nb, V, Mo and Ti. Their physical properties are listed in Table 12.18, together with various data regarding available reserves and estimated future demands [14]. Some other characteristics of these metals may be found in Table 4.1.

*Niobium* (pure) is a soft, ductile, metal. It can be extracted from such ores like:

Niobite (Fe, Mn) (Nb, Ta)$_2$O$_6$

Tantalite (Fe, Mn) (Ta, Nb)$_2$O$_6$ (58 — 82% Ta)

Pyrochlor (Ca, Na)$_2$Nb$_2$O$_6$(F, OH, O)

Euxenite (Y, Er, Ce, U, Pb, Ca) (Ti, Nb, Ta)$_2$ (O, OH)$_6$

It oxidizes in contact with air, starting at ~ 200 °C. For use at higher temperatures it must be secured in a protective atmosphere. Niobium is a very good electricity conductor, which makes it useful in the equipment meant to generate intense magnetic fields (see also section 12.6.1).

TABLE 12.18 Physical Properties of Refractory Metals and World Reserves

| *Metal* | *Atomic mass* | *Density* ($kg/m^3$) | *Melting temperature* (°C) | *Boiling temperature* (°C) | *Amount per 10 m thick of crust* (kg) | *World reserves* (1970) (kg) | *Demand, for a capacity of $10^6$ MWe* (kg) |
|---|---|---|---|---|---|---|---|
| Nb | 92.906 | 8570 | 2468 ± 10 | 4927 | $9.4 \times 10^{13}$ | $1.0 \times 10^{10}$ | $3.3 \times 10^{9}$ |
| V | 50.942 | 5960 | 1890 ± 10 | 3380 | $6.6 \times 10^{14}$ | $1.0 \times 10^{10}$ | $2.4 \times 10^{9}$ |
| Mo | 95.94 | 10,220 | 2610 | 5560 | $6.2 \times 10^{12}$ | $5.4 \times 10^{9}$ | $2.8 \times 10^{9}$ |
| Ti | 47.90 | 4540 | 1675 | 3260 | $2.8 \times 10^{16}$ | $1.5 \times 10^{11}$ | $0.5 \times 10^{9}$ |

*Vanadium* is a white, soft and ductile metal. In air, oxidation starts at ~ 660 °C. It exhibits very good tensile properties. Both vanadium and its compounds are toxic. As powder or dust in air, a maximal concentration of 0.5 $mg/m^3$ is allowed. Small amounts may be encountered in meteorites; on Earth it may be found in:

Carnotite $K[UO_2/VO_4] \cdot 1\frac{1}{2} H_2O$

Rossite $Ca(V_2O_6).4 H_2O$

Vanadinite $Pb_5[Cl/(VO_4)_3]$ (19% of weight V)

Patronite $VS_4$.

Crude oil seems to contain quite large amounts of vanadium, so that extraction of vanadium from crude is contemplated.

*Molybdenum* is a silvery-white, hard metal. It features a high elasticity modulus. It oxidizes only at high temperatures. Molybdenum can be found in:

Molybdenite $MoS_2$

Wulfenite $PbMoO_4$

Powelite $Ca(MoW)O_4$

Nuclear properties of these materials are given in Table 12.19 [28]. The following parameters are listed: natural isotopic abundance, absorption cross-sections ($\sigma$), threshold energy values $E_{thresh}$, reaction heat $Q$, and $(d,n)$ reaction efficiency. Structural materials considered for TNRs are: Nb, V, Mo. The blanket can include such materials in proportion of 1 — 10% in weight.

Most of all, niobium is considered for the combustion chamber first wall. First wall cooling is performed by liquid Li which "washes" its surface. Niobium seems to be very compatible with liquid lithium. Still, oxygen

TABLE 12.19 Nuclear Properties of some Refractory Metals

| $Z$ | $A$ | Abundance (%) | Thermal neutron absorption cross-section (b) | $\sigma(p,n)$ at 6.7 MeV (mb) | $E_{thresh.}$ (p,n) (MeV) | Efficiency (d,n) at 15 MeV (number of n for 1 d) | $\sigma(n,p)$ at 14.5 MeV (mb) | $Q(n,p)$ (MeV) | $\sigma(n,\alpha)$ at 14.5 MeV (mb) | $Q(n,\alpha)$ (MeV) | $\sigma(n,2n)$ at 14.5 MeV (mb) | $Q(n,2n)$ (MeV) | $\sigma(n,\gamma)$ at 1 MeV (mb) |
|---|---|---|---|---|---|---|---|---|---|---|---|---|---|
| 22 | 46 | 7.93 | 0.6 ± 0.2 | | | | | | | | 53 ± 42 | | |
| Ti | 47 | 7.28 | 1.7 ± 0.3 | | | | | | | | | | |
| | 48 | 73.94 | 8.3 ± 0.6 | | | | 93 ± 33 | -2.51 | | | | | |
| | 49 | 5.51 | 1.9 ± 0.5 | | | | 6.6 ± 0.7 | - | | | | | |
| | 50 | 5.34 | 0.14 ± 0.03 | | | | | | | | | | 1.9 |
| Ti | | Natural | 6.1 ± 0.2 | | | $4.1 \times 10^{-4}$ | | | | | | | |
| 23 | 50 | 0.24 | 100.0 ± 60 | | | | | | | | | | |
| V | 51 | 99.76 | 4.8 ± 0.2 | | 1.562 | | 27 ± 4 | -0.8 | 28.6 ± 12 | -0.95 | | | 1.8 |
| V | | Natural | 5.06 ± 0.06 | | | | | | | | | | |
| 41 | 93 | 100 | 0.15 ± 0.1[1] | 1.8 | 3.7 | $3.0 \times 10^{-4}$ | | | | | | | 41 |
| Nb | | | 1.15 ± 0.05[2] | | | | | | | | | | |
| 42 | 92 | 15.84 | < 0.006[3] | | | | | | | | 190 ± 28 | -13.3 | |
| Mo | | | < 0.3[4] | | | | | | | | | | |
| | 94 | 9.04 | | | | | | | | | | | |
| | 95 | 15.72 | 14.4 ± 0.5 | 150 | 5.1 | | | | | | | | |
| | 96 | 16.53 | 1.2 ± 0.6 | { 110 | 3.6 | | | | | | | | |
| | 97 | 9.46 | 2.2 ± 0.7 | { 60 | - | | | | | | | | |
| | 98 | 23.78 | 0.14 ± 0.02 | 230 | 3.8 | | 108 ± 54 | -1.0 | | | | | 10.4 |
| | 100 | 9.63 | 0.20 ± 0.05 | | | | | | | | 3.79 ± 1.89b | -9.0 | 12.3 |
| Mo | | Natural | 2.65 ± 0.05 | | | $2.5 \times 10^{-4}$ | | | | | | | |

[1] $^{94m}Nb$; [2] $^{94}Nb$; [3] $^{93m}Mo$; [4] $^{93}Mo$.

contamination throughout a reactor's service life may enhance the adverse effects of Li. So, oxygen concentrations in excess of some hundred ppm in weight may lead to a complete penetration of niobium by the lithium, within a couple of hours.

Unlike niobium, vanadium proves able to withstand lithium penetration, even under high oxygen concentrations.

Hydrogen makes both niobium and vanadium brittle.

For thermal insulation, metallic niobium has been suggested, if coated with a film of $Al_2O_3$ (~ 0.3 mm).

Another reason for using niobium is its relatively high natural availability; molybdenum is less abundant.

One drawback with using niobium as structural material comes with its activation, accompanied by production of radioisotopes with long half-lives, which requires long-term storage, much like the radioactive wastes from fission reactors.

Refractory metals can also be used as structural materials either as alloys, or as alloying elements in steels [29]. Refractory alloys most often used, or of prospective interest, are:

Molybdenum alloys. Mo alloys are characterized by a long strain endurance, of 100 hours, at 1100 °C and ~ 350 MN/m$^2$. Lower strain values may be faced even at 1300 °C. They have a high formability and may be argon-arc welded. To protect them against corrosion, the surface may be coated with silicon or chromesilicon.

The *TZM alloy*, with 99.4% Mo, 0.5% Ti, 0.08% Zr; 0.01% C in its composition. It is a high fracture strength material, even at high temperatures ($\sigma_r$ = 380 MN/m$^2$ at 1300 °C). It oxidizes at high temperatures, which calls for protection by surface coating with sprayed Si — Al — Co particles. Another method suggested to prevent oxidation is to reinforce the outer surface with stainless steel while inserting an insulating ceramics layer between the two materials.

At high temperatures this material is compatible with both helium and liquid lithium. Moreover, the alloy is good in withstanding neutron irradiation.

Niobium alloys. Niobium alloys are able to withstand a long time stress (100 hours), at high temperatures (1000 °C) and ~ 250 MN/m$^2$. These alloys have a ductility superior to those based on molybdenum and may be both cold- and hot-processed. The alloys may be argon-arc welded. To prevent strong oxidation at high temperatures, surfaces should be silicon-coated. The best tensile properties come with the following alloys:

15% W; 3% Ti; 1% Zr; the balance — Nb
10% Mo; 3% Ti; 1% Zr; the balance — Nb.

The *Nb — 2% Ti alloy*. Addition of titanium leads to a remarkable increase in niobium's hardness, for service temperatures over 1100 °C. By oxidation, a very adherent oxide layer is formed, with superior refractory qualities, up to 1300 °C.

The *Nb — Zr — alloys*. A Nb — 1% Zr alloy that can withstand temperatures around 1000 °C is suggested for use in the blanket.

Nb — Zr alloys with higher Zr concentrations (also called ozenites) give good results in the fission reactors (Table 4.7). Niobium is used to counterbalance the effects of the C, or N impurities. When in zirconium, it has the effect of increasing tensile strength, by over 50% as compared to zircaloy, in the temperature range 300 — 400 °C. The Zr — 3% Nb — 1% Sn alloy exhibits superior resistance to corrosion and creep, in the range 350 — 400 °C, as compared with the stanium-free variety.

Niobium-poor alloys cannot be used at high temperatures. Tensile properties (such as ductility, brittleness) of Zr — Nb alloys are strongly affected in the presence of hydrogen.

Vanadium alloys. Due to brittleness and corrosion of vanadium by cooling agent impurities (Li, He), surface protection or addition of other metals have been tried. Such an alloy is V – 20% Ti.

High-temperature (600 — 1000 °C) working conditions lead to a strong deoxidation of vanadium alloys when in contact with lithium, which may significantly diminish the tensile strength.

Titanium alloys. The chief merits of Ti alloys are a low density and a high resistance to corrosion. They have small elasticity moduli (~ $10^5$ MPa); in exchange, they exhibit a quite good resistance to the effect of high temperatures (slightly inferior to that of steels).

These alloys are remarkably resistant to corrosion, even when subject to the effect of seawater or other corrosive agents, for a long time. By addition of small quantities of palladium, their resistance to corrosion may increase even more. Ti alloys with a certain admixture of chromium (3% Al; 1.5% Cr, Fe, Si) are more tensile-resistant, and can work well at 550 — 600 °C. Their hardness is inferior to that of steels.

Fracture strength ranges between 500 and 600 MPa for alloys with Al, Cr, Mo, V, Sn, and between 800 and 1400 MPa for alloys with Al, Mo, Cr, V. For instance, the $VT_8$ alloy at 600 °C features a $\sigma_r$ of 700 MPa, $\sigma_c$ = 570 MPa, and $\delta$ = 8%.

### 12.5.2. Steels

Alloyed steels and high-alloyed steels are basic materials for structures. Their main characteristics have been dealt with in Chapter 4.4. In consideration of some peculiarities with their application, we shall briefly review their characteristics of special concern when it comes to their use in TNRs. The maximal temperature at which high-alloyed steels may still be used is ~ 1200 °C. The relative shortcoming of their being affected by radioactivity, implies storage problems for relatively long periods of time (~ 50 years).

Austenitic steels with refractory metals (especially Cr) exhibit superior tensile properties at high temperatures. These steels are also corrosion-resistant.

When at extremely high temperatures, even the high-alloyed steels lose their strength and manifest a low formability. This is why, at temperatures corresponding to the upper limits, they can only be used as protection, refractory, materials.

Steels designed for service at high temperatures are [30]:

1. *In the range 400 – 550 °C* : pearlitic – ferritic steels. For use at 400 450 °C, steel must have a concentration in Cr of 1 – 3% in weight, so as to prevent cementite coagulation and graphitization. To extend the application range to 450 – 550 °C, the Cr content must increase to

3 – 3.5%. In addition, elements yielding carbides such as Mo, W, V, must be added. Within this temperature range these steels behave satisfactorily under the corrosive effect of hydrogen.

2. *In the range 500 – 600 °C.* In order to push further up the recrystallization temperature and, consequently, to prevent plastic deformation, steel should include more Cr, namely 5 — 15% in weight. To provide proper service in the upper temperature limit, it should be alloyed with Mo, W, V or Nb. For instance a steel grade containing 12% Cr behaves well under a stress of 200 MPa at 500 °C for 10,000 hours.

3. *In the range 600 – 650 °C*, the use of austenitic steels is in order. The enhanced resistance to plastic deformation of these is conferred by the austenitic phase, that features a higher recrystallization temperature. Austenitic steels are usually alloyed with 9 – 12% nickel.

Thermal strength may be improved in the presence of phases such as carbides or intermetallic compounds of the type $Ni_3(Ti,Al)$. For carbide formation, steel should be alloyed with W, and for intermetallic compounds formation with Ti and Al.

For an austenitic steel with carbides (18% Cr; 9% Ni; ~ 1% Ti), a stress of 200 MPa is acceptable for 10,000 hours of service, at 630 °C.

For structural materials, in the temperature range over 650 °C the choice comprise Ni alloys (≤ 700 °C), nickel — cobalt alloys (≤ 800 °C), and chromium alloys (≤ 1100 °C).

The so-called refractory steels are used in the temperature range 900 — 1200 °C, but they cannot serve reinforcement purposes. Their strength depends upon temperature and increases with the rise in Cr concentration. Resistance to carbide formation is obtained by Ni-additions. Examples of such steel grades are given in Table 12.20 [30].

TABLE 12.20 Refractory Steels

| *Temperature corresponding to sizeable oxidation* | *Content (%)* | | | | *Steel grade* (GOST) |
|---|---|---|---|---|---|
| | *C* | *Si* | *Cr* | *Ni* | |
| 850 | 0.35 - 0.45 | 2 - 3 | 8 - 10 | - | 40 X 9C 2 |
| 900 | ≤ 0.12 | ≤ 0.8 | 16 - 18 | - | 12 X 17 |
| 950 | 0.25 - 0.34 | 2 - 3 | 12 - 14 | 6 - 7.5 | 30 X 13H7C2 |
| 1000 - 1050 | ≤ 0.2 | 2 - 3 | 19 - 22 | 12 - 15 | 20 X 20H14C2 |
| 1100 - 1150 | ≤ 0.15 | ≤ 1 | 27 - 30 | - | 15 X 28 |

One may conclude that in fusion reactors steels cannot be employed as structural materials, or for reinforcement purposes, in extremely high-temperature operating conditions. They are bound to be used at lower temperatures (≤ 600 °C). The chemical composition of some steels of intereset for TNRs is given in Table 12.21 [30].

Figure 12.14 displays the stress curves indicating the maximal operation period under load, at a given temperature, at a plastic deformation of 0.2%, for stainless steel type 316.

TABLE 12.21 Stainless Steels

| *Stainless steel grade* | *Concentration (%)* | | | | | | | | | *Remarks* |
|---|---|---|---|---|---|---|---|---|---|---|
| | *C max* | *Mn max* | *P max* | *S max* | *Si max* | *Cr* | *Ni* | *Mo* | *Ti* | |
| 310S [1] | 0.08 | 2.00 | 0.045 | 0.030 | 1.50 | 24-26 | 19-22 | - | - | Austenitic steels with Cr, Ni not suitable for hot-treatment |
| 314 [1] | 0.25 | 2.00 | 0.045 | 0.030 | 1.5-3 | 23-26 | 19-22 | - | - | |
| 316 [1] | 0.08 | 2.00 | 0.045 | 0.030 | 1.00 | 16-18 | 10-14 | 2-3 | - | |
| 316L [1] | 0.03 | 2.00 | 0.045 | 0.030 | 1.00 | 16-18 | 10-14 | 2-3 | - | |
| 321 [1] | 0.08 | 2.00 | 0.045 | 0.030 | 1.00 | 17-19 | 9-12 | - | - | Thermally unstable |
| 403 [1] | 0.15 | 1.00 | 0.040 | 0.030 | 0.50 | 11.5-13 | - | - | - | |
| OX18 [2] | 0.012 | 1-2 | 0.040 | 0.030 | 0.8 | 17-19 | 8-9.5 | - | 0.06 | Weak resistance to corrossion, under stress |
| HST DIN 1.4970 | 0.095 | 1.81 | - | - | 0.31 | 5.1 | 15.0 | 1.3 | 0.3 | Co, Cu, V 0.02 |

[1] USA denominations
[2] GOST (USSR) denominations

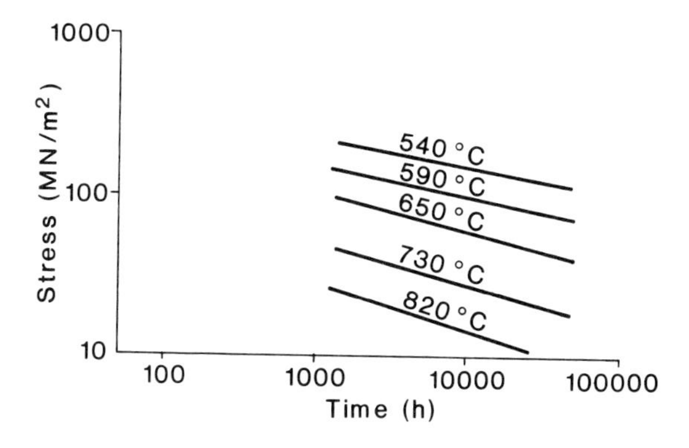


Fig. 12.14. Stress at 0.2% strain, as function of temperature and test duration, for the stainless steel 316.

### 12.5.3. Nickel-based Alloys

Unlike with fission reactors, where such materials are ordinarily avoided, nickel alloys are considered for many fusion reactor projects. The chemistry of various nickel alloys is displayed in Table 12.22. The last three formulae in this table are specified only for comparative reasons, such alloys being in fact associated with other applications [30].

Table 12.23 specifies the tensile properties of two outstanding nickel alloys.

Nickel alloys may be used — depending on their chemical composition — up to temperatures of 650 – 800 °C.

TABLE 12.22 Nickel Alloys

| *Alloy* | *Percentage content, in weight* | | | | | | | | | |
|---|---|---|---|---|---|---|---|---|---|---|
| | *Ni* | *Mo* | *Cr* | *Fe* | *Cu* | *Mn* | *Si* | *Ti* | *C* | *Co* |
| Hastelloy A | 58 | 20 | - | 22 | - | - | - | - | - | - |
| Hastelloy N | 71 | 17 | 7 | 5 | - | - | - | - | - | - |
| Hastelloy X | ~ 45 | 23-25 | 22 | 20 | 1.5-3 | 0.5 | - | - | <0.15 | - |
| NIMONIC (PE 16) | the balance | - | 18-21 | < 5 | <0.5 | < 1 | < 1 | 0.2-0.6 | <0.15 | - |
| INCONEL | 72 | - | 15.5 | 8 | - | 1 | 0.5 | - | 0.15 | - |
| CONEL | 71 | - | - | 8 | - | - | - | 3 | - | 18 |
| MONEL | 67 | - | - | 2 | 29 | 2 | - | - | - | - |
| Constantan | 39 | - | - | - | 60 | 1 | - | - | - | - |

TABLE 12.23 Tensile Properties of INCONEL and Hastelloy X

| | $\sigma_{frct.}$ (MPa) | $\sigma_{0.2}$ (MPa) | $\varepsilon_{frct.}$ (%) | *Remarks* |
|---|---|---|---|---|
| INCONEL | 700 | 310<br>230 | 44<br>42 | hot-laminated<br>annealed |
| Hastelloy X | 640 | 250 | 50 | |

Figure 12.15 shows the strain curves for a plastic deformation of 0.2% related to temperature and service life, for the alloy PE 16 (NIMONIC).

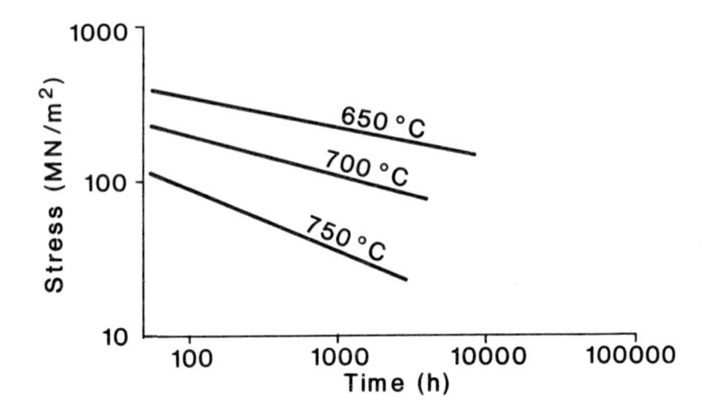


Fig. 12.15. Stress at 0.2% strain for the PE 16 (NIMONIC) alloy.

Nickel alloys do not present phase transitions, since the solid solution in phase $\gamma$ is characterized by a high recrystallization temperature. To increase alloy's strength, formation of dispersed particles in the alloy is sought, consisting either of intermetallic compounds $Ni_3(Ti,Al)$, or of nitrides (TiN) or even of carbides (TiC). To this purpose Cr, Al and Ti are added. Such alloys may be used up to 700 °C, sustaining a service stress of 250 — 280 MPa. Alloying with cobalt raises the maximal service temperature another ~ 50 °C.

### 12.5.4. Aluminium-based Materials

Aluminium-based materials cannot be used at temperatures as high as those tolerated by the other materials discussed above; yet they have the outstanding advantage of their low density. They are also considered in view of the special service conditions on the fusion reactors, featuring intense and rapidly fluctuating magnetic fields. In such contexts, Al is very convenient indeed. This metal is also used in the electric conductors of the magnetic field-generating coils.

As a structural material the metal — ceramic compound "SAP" (sinterized aluminium product) is ordinarily employed. It is obtained by sintering together $Al_2O_3$ (5 — 25%) and Al (powder) [31]. To this effect, cold-pressing or hot-pressing (~ 500 °C) is followed by sintering at 570 — 590 °C. Density of sintered material is ~ 2700 kg/m$^3$. Embedded $Al_2O_3$ particles push up recrystallization temperature, and reduce fracture strength ($\delta$ < 0.75% at 500 °C) and creep. $Al_2O_3$ improves tensile strength. The decrease in yield makes the material less elastic, thus exposing it to more easy fracture under stress. As far as compatibility with other materials, SAP compares well to the aluminium. Table 12.24 displays various compositions of some SAP grades [30].

TABLE 12.24 SAP Metal — Ceramic Compounds

| | *$Al_2O_3$ concentration (%)* | *Concentration in other elements* | *Remarks* |
|---|---|---|---|
| SAP 960 [1] | 4 | 0.1 — 0.3% Fe | |
| SAP 930 [1] | 7 | 0.05 — 0.15% Si | |
| SAP 895 [1] | 10 — 11 | 0.01 — 0.04% Zn | |
| SAP 865 [1] | 13 — 14 | 0.2 — 0.3% C | $\sigma$ = 300 MPa (100 °C) |
| CA П [2] | 20 — 22 | | $\sigma$ = 100 MPa (480 °C) |

[1] USA denominations.
[2] GOST (USSR) denominations.

Aluminium materials are praised for their low activation level, and also owing to their considerable availability throughout the world. Recent research [32] has confirmed that Al and its alloys (for instance Al — Mg — Si) meet the requirements of TNRs.

### 12.5.5. Carbon-based Materials

Carbon plays an important part in nuclear technology. As graphite or carbide it is used mainly in fission nuclear reactors. Owing to this, its characteristics are well-known.

Carbon is eligible for employment in fusion reactors, primarily owing to its high melting point. Of the structural materials considered, one could name graphite, silicon carbide, and fibrous carbon. Since graphite has been introduced in Chapter 5, we shall deal here only with some specific aspects regarding its use at high temperatures.

Carbon fibres are latecomers in the family of carbon-based materials. It is suggested that it may serve as a carbon "curtain" [33], lining the inner wall of the combustion chamber, with the role of limiting the effect of the plasma particles upon the wall. Such a curtain may well capture ions and neutral atoms escaping the plasma, and may also collect all atoms sputtered out from the wall by neutrons. Unlike heavy atoms, carbon atoms that may reach the plasma as impurities would alter, but negligibly, the fusion conditions.

The carbon curtain would reduce the effect of charged particles on the first wall. Finally, the induced radioactivity (as $^{14}C$) should be negligible. The curtain is likely to serve under the highest temperature in the reactor, namely ~ 1800 $^{o}C$.

To use graphite in contact with liquid lithium is out of the question. Use of silicon carbide brings about problems, due to brittleness and difficulties encountered in manufacturing large-size carbide structural components.

Graphite also raises questions as to its behaviour under high irradiation. About 77% of the total helium as generated in the TNR comes from the reaction $^{12}C(n,n')3^{4}He$ in the graphite structure. The consequences require further research.

### 12.5.6. Other Materials

Data in Fig. 12.13 are relevant as to the great diversity of chemical compounds (carbides, nitrides, borides, oxides, silicon compounds, sulphides) that can be considered for potential use in various high-temperature zones of the TNRs.

Thus, in hybrid reactors one envisages the employment of uranium carbide ($UC_2$), with a melting point at 2350 $^{o}C$, encapsuled in niobium (rods or spheres).

Another class of materials in use with today's fusion facilities, and also considered for power reactor projects, is that of ceramic materials [34]. These are mainly meant as electric insulators (for instance — in the plasma heating systems that use fast neutral atom injection or radiofrequency waves). Ceramics may be considered also in the manufacture of combustion chambers of certain fusion reactors and facilities.

This decade has witnessed the advent of composite materials. So far these have never been used in nuclear technology. A material made of 85% graphite and 15% boron epoxy resins has been studied. It is probable that, in the future, such heterogeneous materials would have a bearing on the proper solution of a great many problems [35].

## 12.6. MATERIALS FOR MAGNETIC DEVICES

This class covers the materials meant for the normal or superconducting coils, their support structures, thermal protection of coils and their protection against radiations, magnetic cores, or cryogenic materials. Here we shall deal only with the high-tech materials — the ones not yet commonly employed in commercial electrical engineering — namely the *superconducting and cryogenic materials*.

TNRs' magnetic systems are used to confine and heat plasma, to accumulate energy in the intervals between service pulses, or to guide charged particles. This is why they play a vital part in reactor operation.

### 12.6.1. Superconducting Materials

The most used superconducting materials nowadays are NbTi, NbZr and $Nb_3Sn$. Obviously, new superconducting materials with higher transition temperatures may lead to simpler and, therefore, more cost-effective technical solutions [36,37].

Several characteristics of the chief superconducting materials are given in Table 12.25.

TABLE 12.25. Superconducting Materials

| *Material* | *Critical temperature* $T_c$ (K) | *Critical magnetic field* $H_{c2}$ (T) |
|---|---|---|
| Al | 1.75 ± 0.002 | 0.099 |
| Ga | 1.083 ± 0.001 | 0.05 |
| Nb | 9.25 ± 0.02 | 0.274 — 0.305 |
| Sn | 3.722 ± 0.001 | 0.315 — 0.343 |
| Mo | 0.915 ± 0.005 | 0.090 — 0.098 |
| Mo* | 8.0 | 8 |
| V | 5.40 ± 0.05 | 1.10 — 1.40 |
| $Nb_3Al$ | 18.5 — 18.8 | ~ 19 |
| $V_3Ga$ | 15.1 | 22.0 |
| $Nb_3Sn$ | 18.0 | 21.5 |
| $Nb_{0.4}Ti_{0.6}$ | 9.5 | 11.0 |
| $Nb_3Ge$ | 22.0 | 37.0 |

* thin films

Currently, superconducting cables are being commercially manufactured and in service. Such cables are made of $Nb_3Sn$, $V_3Ga$, NbTi and NbTiZr. They may be tape-, or multifilament-wired. At the liquid He temperature and in ~ 10 T magnetic fields, superconducting cables may withstand current densities of $10^8$ — $10^9$ A/m$^2$, with almost no resistive losses.

The high magnetic fields (8 — 16 T) required for actual TNR operation raise difficult technical problems. These relate to: (a) stresses upon the coils; (b) superconducting phase stability; and (c) protection against sudden (accidental) variations in the magnetic field [14]. The first among these is probably the most difficult challenge and acts as a limiting factor in TNR design. Indeed, calculations prove that for a coil generating a magnetic induction field of ~ 16 T, with a diameter of the order of metres, the stress at the level of coil's circumference may reach approximately $10^9$ N/(m of coil length). Such strains must be taken over both by the structural materials and by the winding itself. Both aspects are difficult to address, the more so that the system is also subject to the effects of its own weight.

Problem (b) can actually be solved, in two ways:

1. through *cryostatic stabilization*, that consists in providing an electric circuit parallel to the superconducting one, of low electric resistance, that would take over the load in case the superconductor turns to normal state;

2. through *adiabatic stabilization* , that consists in providing for an efficient heat transfer to the cryogenic agent, so that normal (non-superconducting) spots that may accidentally occur in defective portions of the cable remain confined.

Both solutions can be implemented by devising an appropriate structure that would host superconductors in a normal conductor matrix (copper, copper-nickel), thermally permeable to both the superconductor and the cryogenic agent (liquid He).

Question (c) is in order with the magnetic system of any TNR, that can accumulate an energy up to $10^{11}$ J. At a transition to normal condition, one should appropriately provide for the dissipation of this energy without mechanical damage. This can be achieved by the provision of an external resistive load, so sized as to dissipate more than 99% of the stored energy.

There are also other challenges, such as how to thermally insulate the combustion chamber — where extremely high temperatures reign — from the cryogenic zone, which finds itself at the other temperature extreme. The transition from the combustion chamber's high temperature to the low temperature of the superconducting coil section is achieved in the wall, through the cooling agent flow, and then in the following sections made of stainless steel, boron carbide, lead and evacuated compartments (for a minimum of thermal conduction).

Boron carbide ($B_4C$) layers, as well as the lead ones, diminish the neutron and $\gamma$-radiation flux, down to the limit where there is no danger any more for the integrity of the superconductive system, since the heat released by radiation and transferred to the cryogenic system is tolerable.

Other projects conceive this insulation as made of (LiH+Pb), or ($B+H_2O+Pb$).

### 12.6.2. Cryogenic Materials

As a cryogenic agent, at least so far, *liquid helium* has been exclusively used, allowing operations below 4.2 K.

Structural materials employed in cryogenic systems (to maintain liquid helium in evacuated double-walled dewar vessels) are generally the stainless steels (see Table 12.26 [30]). Ductility and strength of these steels shall not vary significantly at such temperatures.

Ferritic (martensitic) steels may be used down to 77 K. Their low content in C and high content in Ni depress the embrittlement temperature limit and improve strength. Austenitic steels of the type Cr - Ni, Cr - Mn - Ni, etc., may also be used, down to 20 K. Also applicable at low temperatures are the copper alloys, particularly aluminium- or beryllium-bronze or even pure Al, which, on the other hand, shows less tensile strength than austenitic steels. Stainless steels also have the advantage of their low thermal conductivity, which stands as an important argument for the economy of cryogenic agent.

TABLE 12.26 Steels for Cryogenic Facilities

| *Minimum service temperature* | *Remarks* | *Composition (%)* | | | | *Type* |
|---|---|---|---|---|---|---|
| | | *C* | *Cr* | *Ni* | *Mn* | |
| 77 | For cryostats | 0.06 | - | 6-9.5 | 0.45-0.60 | Ferritic (martensitic) steel |
| | For stress-free structure | 0.10 | 13-15 | 2.8-4.5 | 13-15 | Cr - Ni - Mn steel |
| 20 | For cryostats | 0.08 | 17-19 | 9-11 | 1-2 | Austenitic steel |
| 4.2 | More ductile and resistant | 0.08 | 17-19 | 9-10 | 1-2 | Austenitic steel |
| | | 0.09-0.4 | 12-14 | 2.8-3.5 | 13-15 | Cr - Ni - Mn steel |
| | | ≤ 0.07 | 19-21 | 5-6 | 6-7.5 | Cr - Ni - Mn steel |

## 12.7. SPECIFIC PROBLEMS OF MATERIAL IRRADIATION

TNRs present problems with irradiation that are completely different from those encountered in the realm of fission nuclear reactors. Among them are erosion by sputtering, thermal evaporation and blistering of combustion chamber walls, voids and cavity formation in the material (especially due to tritium and helium production).

Fast neutron irradiation conditions are similar to the ones in fission reactors, except for the fact that the irradiated materials differ.

### 12.7.1. Erosion of the Combustion Chamber Wall Surface

Generally speaking, this issue concerns the materials of the combustion chamber first wall, these of the optical components of laser-TNRs and these of the electric components (insulators, electrodes) of TNRs devised for direct conversion of fusion energy into electric energy. Erosion affects tensile strength, thermal strength and other properties of the materials, as well as plasma purity. Consequently, the erosion must be limited. For instance, in a TOKAMAK reactor, an erosion rate of several μm per year brings about problems of plasma contamination, while a rate of 0.1 — 1 mm/year would have straightforward mechanical effects.

Erosion is due to sputtering caused by plasma particles; sputtering due to neutrons; evaporation due to walls' high temperature.

*Sputtering* is a result of high-energy particles knocking on the combustion chamber first wall. As a result, atoms are shattered from their equilibrium positions and some of them may eventually leave the wall [38]. Sputtering efficiency is a parameter of this process, and depends on factors such as the mass, energy of the incoming ion, characteristics of the knocked-on material and impact angle. Figure 12.16 gives the pulverization (sputtering) efficiency for neutrons and ions of Nb, He, T, D and H, at a normal impact on a polycrystalline niobium wall.

The relation of sputtering yield S(θ) with the atom's impact angle is described in Fig. 12.17 [39]. Yield increases considerably with the impact angle, which proves that sputtering is chiefly a matter of atomic binary- or multiple

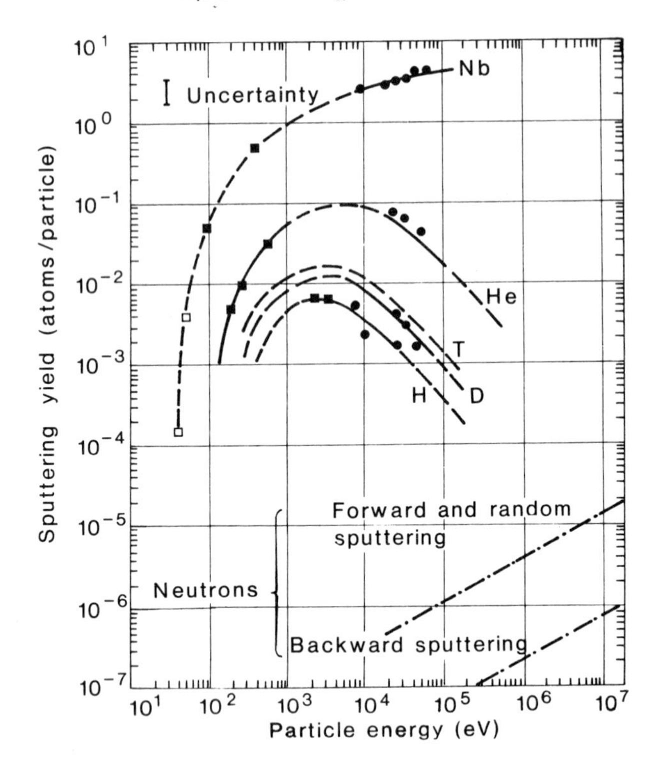


Fig. 12.16. Sputtering yield, for a polycrystalline niobium wall subject to bombardment with H, D, T, He, Nb and neutrons.

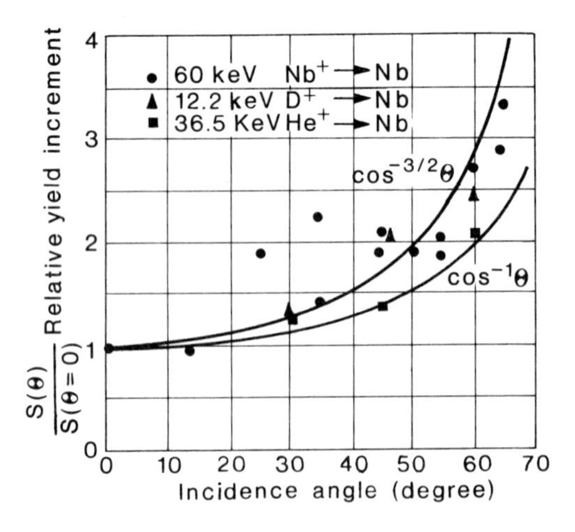


Fig. 12.17. Incremental sputtering efficiency as the incidence angle increases (starting from normal incidence), for a Nb wall.

collisions. Therefore, a decrease in the global sputtering rate may be obtained at the cost of a more complex shape of the wall, but one that would lead to an incidence angle as low as possible.

The theory on sputtering under the effect of plasma atoms and ions in a TNR suggests that molybdenum is comparable, as far as sputtering efficiency to niobium, while for vanadium and stainless steel this efficiency is twice as high. A carbon-fibre curtain practically alleviates the problem with the sputtering of the wall by the plasma ions, carbon ionization being less significant from the point of view of both plasma contamination and the wall's mechanical strength [33].

Neutron-induced sputtering is distinct from that induced by ions, both due to neutrons' higher energy ( >10 MeV) and to the fact that neutrons are neutral particles. Theory [38] indicates a sputtering efficiency of ~ $10^{-4}$ — $10^{-5}$ atoms ejected per fission neutron. Experiments evidenced higher values — of $5 \times 10^{-3}$ — $10^{-4}$, or even more.

*Erosion by wall heating* differs from pulverization, as it strongly depends on wall temperature. It may be expressed by the material's vapour pressure at a given temperature, or by the evaporation rate (atoms evaporated/$cm^2$ . s).

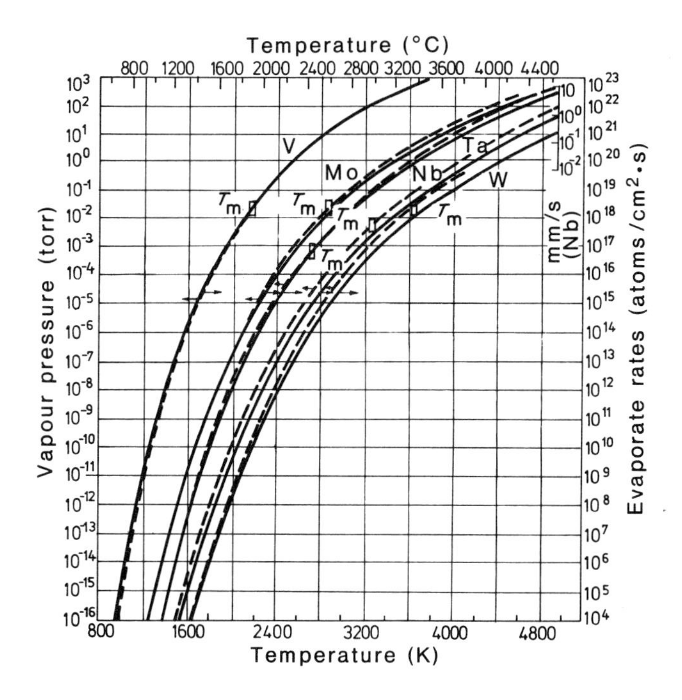


Fig. 12.18. Evaporation rate and vapour pressure for certain materials, vs.temperature; $T_m$ - melting temperature.

Figure 12.18 [2] shows the evaporation rate and vapour pressure for several materials of interest for the TNRs. To this, chemical erosion may add in certain cases, which is determined by the detachment of some molecules from the wall as a result of the formation of some compounds of wall atoms plus plasma atoms. It is the case with the erosion by chemical pulverization of graphite or tantalum carbide, under the impact of hydrogen atoms.

Table 12.27 [39] lists the characteristics of some refractory materials used in TNRs, to allow an assessment of the thermal processes at the first wall level.

TABLE 12.27 Characteristics of some Fusion Reactor Materials

| *Material* | | *V* | *Nb* | *Ta* | *Mo* | *W* | *Graphite* | *B* | *Stainless steel 403* | *TaC* |
|---|---|---|---|---|---|---|---|---|---|---|
| Atomic number *(Z)* | | 23 | 41 | 73 | 42 | 74 | 6 | 5 | | |
| Mass number *(A)* | | 50.94 | 92.9 | 180.95 | 95.94 | 183.85 | 12 | 10.82 | | |
| Density ($kg/m^3$) | | 5870 | 8570 | 16600 | 9010 | 19300 | 2250 | 2340 | 7900 | 13900<br>14500 |
| ($atoms.m^{-3}$ . $10^{-28}$) | | 6.93 | 5.56 | 5.52 | 5.56 | 6.31 | 11.3 | 13 | | |
| Melting point (K) | | 2192 | 2688 | 3269 | 2883 | 3653 | 3925 subl. | 2573 | 1400 | 4148 |
| Sublimation heat ΔH (eV/atom) at the temperature | 300 K | 5.33 | 7.50 | 8.10 | 6.83 | 8.80 | 8.2-10.4 | 5.6 | | |
| | 500 K | 5.27 | 7.43 | 8.05 | 6.78 | 8.78 | | | | |
| | 1000 K | 5.16 | 7.88 | 7.95 | 6.63 | 8.64 | | | | |
| | 1500 K | 4.96 | 7.14 | 7.77 | 6.48 | 8.50 | | | | |
| | 2000 K | 4.75 | 6.97 | 7.62 | 6.30 | 8.34 | | | | |
| Thermal conductivity *K* (W/m . K) at the temperature | 500 K | 33.1 | 56.7 | 58.2 | 130 | 149 | 80-100 | 141 | 19 | 40 |
| | 1000 K | 38.6 | 64.4 | 60.2 | 112 | 121 | 49-64 | 6.3 | | 40 |
| | 1500 K | 44.7 | 72.1 | 62.2 | 97 | 109 | 40-50 | | | 40 |
| | 2000 K | 50.9 | 79.1 | 64.0 | 88 | 100 | 30-40 | | | 40 |
| Specific heat $c_p$ (J/kg . K) at the temperature | 500 K | 500 | 280 | 145 | 254 | 135 | 1200 | 1690 | 500 | 250 |
| | 1000 K | 640 | 300 | 155 | 290 | 150 | 1670 | 2400 | | 290 |
| | 1500 K | 700 | 330 | 165 | 330 | 160 | 1800 | | | 320 |
| | 2000 K | 850 | 370 | 170 | 380 | 380 | 2000 | | | 330 |

The phenomenon of *surface swelling* of the area affected by the particle impact occurs after a critical fluency has been reached. This value depends on the radiation temperature and on the incoming particle energy, and ranges between $10^{17}$ and $10^{18}$ particles/$cm^2$ for helium impact, and $10^{18}$ — $10^{19}$ particles/$cm^2$ for hydrogen impact. To give a general idea, the swelling is a result of defects caused by radiation, which, by accumulation, may create gas-filled bubbles in the material (see also section 3.1.3). Gas bubbles may gradually expand with further accumulation of gas and, finally, blow up. This occurs mainly if the solubility of gas atoms in the material is low a case when they can diffuse through the structure and may accumulate in bubbles.

Helium is one such insoluble gas. When the particle energy exceeds ~ 250 eV (He in W) vacancies may be generated, that would accumulate helium. For instance, one vacancy in W is able to retain up to three He atoms. Under a fluency of $10^{15}$ — $10^{16}$ ions/cm$^2$, gaseous helium may form bubbles that remain stable until the temperature approaches the melting point. A rough calculation indicates that ~14% of the ions knocking-on a polycrystalline niobium wall are trapped in such bubbles. In some cases the pulverization rate may be much higher than the rate of bubble formation and, consequently, no swelling would occur.

Table 12.28 offers a few data indicative of the amplitude of the processes discussed [2].

TABLE 12.28 Sputtering and Evaporation Efficiency on a Nb — Zr Wall

| *Process* | *Ejected atoms* (atoms/cm$^2$ . s) |
|---|---|
| Neutron-induced sputtering, assuming a total power flux of 14 MeV neutrons, of 67.7 W/cm$^2$ (corresponding to a neutron flux of ~ $3\times10^{13}$ n/cm$^2$ . s) | $10^9$ |
| Other particles-induced sputtering, assuming a total power flux of particles of 10 W/cm$^2$, at: | |
| $E$ = 100 eV | $1 \times 10^{14}$ |
| $E$ = 1000 eV | $4 \times 10^{14}$ |
| $E$ = 5000 eV | $1.5 \times 10^{14}$ |
| Evaporation at | |
| 1285 K | 1 |
| 1400 K | $1 \times 10^5$ |
| 2580 K | $2 \times 10^{16}$ |

### 12.7.2. Influence of Irradiation on the Properties of Superconducting Materials

Actually, irradiation affects all decisive parameters of the superconducting materials employed in TNRs: critical temperature, $T_c$, critical field, $H_c$, and current's critical density, $J_c$.

This is to be expected as a result of the structural disorder induced by radiation into the material. There seems to exist a critical level up to which radiation does not affect the critical temperature $T_c$; also, it appears that $H_c$ depends mainly on changes in $T_c$, and that the value of $J_c$ is not altered, if originally $J_c$ is higher than $1.3\times10^6$ A/cm$^2$ (at 5 T).

As an example, Fig. 12.19 indicates the dependence on the fast neutron fluency ($E$ > 0.1 MeV) of the $T_c$ values, for $Nb_3Sn$ irradiated at a temperature of ~10 K [39].

Thus, radiation protection may become a problem worth consideration, which does not, however, constrain in any way the use of $Nb_3Sn$.

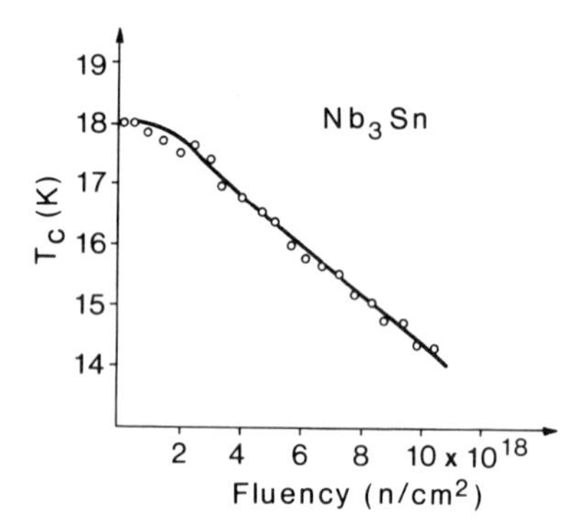


Fig. 12.19. Critical temperature of $Nb_3Sn$, vs. neutron fluency.

### 12.7.3. Radioactive Materials

The question of radioactivity and radioactive material "inventory" involved in a TNR is of an obvious interest. Current estimations indicate the superiority of the fusion systems over the fission ones. The only radioactive material required in a TNR is *tritium*. Let us discuss briefly the problems posed by this element.

12.7.3.1. Tritium activity. As said before, tritium is an essential component of a TNR. Its radioactivity poses problems concerning contamination and leakage in the environment. This is why all TNR concepts tend to limit tritium participation, and also to provide solutions to keep to a minimum normal or accidental leaks. The latter is by no means a simple matter, since most materials are tritium-permeable at higher temperatures.

For the time being there is not enough information on tritium escapes in the environment. A limit may, however, be inferred, based on current experience with the tolerated levels in the normal effluents of water-cooled reactors in operation: a dose rate of 5 mrem/year should not be exceeded, and tritium concentration in discharged water should not exceed $5 \times 10^{-3}$ Ci/m$^3$.

Such remarks lead to a maximal allowance of ~ 10 Ci/day of tritium in the environment. Yet, further research is in order in this particular case too, since it has been noticed that $T_2O$, for instance, is much more noxious than plain tritium dissipated in air.

12.7.3.2. Activity induced into structural materials. It is estimated that, due to the neutron flux generated during the operation of a TNR, the induced activity would reach a level of ~ 1 Ci/W, which, for a 1000 MWe reactor would result in ~ $10^9$ — $10^{10}$ Ci [14]. Though extremely high, this level is not dangerous in itself. To truly assess the potential hazards implied, one should consider the manner in which this activity decays in time, the kinds of radioisotopes generated, and the thermal expression of such a radioactivity. This has been deemed a major problem with fission reactors.

For an examination of these aspects, calculations have been made of the total activity that may affect the structural materials close to the combustion chamber. The results are given in Table 12.29 [14].

Potential biological hazard, expressed in km$^3$ of air per 1 W of reactor thermal power, indicates the air volume that is required to obtain a dilution

TABLE 12.29 Radioactive Materials in TNRs and Advanced Fission Reactors

| *Material* | *Major radio-nuclides* | *Half-life* | *Activity* (Ci/W(t)) | *Maximally admissible concentration in air* ($\mu$Ci/m$^3$) | *Potential biological hazard* (km$^3$ of air/W (t)) |
|---|---|---|---|---|---|
| SAP | $^{24}$Na | 15 h | 0.5 | 5x10$^{-3}$ | 0.1 |
| | $^{26}$Al | 7.5x10$^5$ yr | 2.8x10$^{-6}$ | 8x10$^{-5}$ | 3.5x10$^{-5}$ |
| | | | | | Total = 0.1 |
| V-20% Ti | $^{46}$Sc | 83.8 d | 5.1x10$^{-3}$ | 8x10$^{-4}$ | 6.4x10$^{-3}$ |
| | $^{45}$Ca | 163 d | 4.7x10$^{-3}$ | 1x10$^{-3}$ | 4.7x10$^{-3}$ |
| | $^{48}$Sc | 43.7 h | 2.3x10$^{-2}$ | 5x10$^{-3}$ | 4.5x10$^{-3}$ |
| | | | | | Total = 0.016 |
| Nb-1% Zr | $^{95}$Nb | 35.1 d | 1.55 | 3x10$^{-3}$ | 0.52 |
| | $^{93m}$Nb | 13.6 yr | 0.06 | 4x10$^{-3}$ | 0.015 |
| | $^{94}$Nb | 2x10$^4$ yr | 4.5x10$^{-4}$ | 6x10$^{-5}$ | 7.5x10$^{-3}$ |
| | | | | | Total = 0.54 |
| Stainless steel | $^{54}$Mn | 303 d | 0.13 | 1x10$^{-3}$ | 0.13 |
| | $^{60}$Co | 5.26 yr | 0.015 | 3x10$^{-4}$ | 0.05 |
| | $^{55}$Fe | 2.6 yr | 1.0 | 3x10$^{-2}$ | 0.03 |
| | $^{59}$Ni | 8x10$^4$ yr | 2x10$^{-6}$ | 2x10$^{-2}$ | 1x10$^{-7}$ |
| | | | | | Total = 0.21 |
| Advanced fission reactor | $^{239}$Pu | 2.4x10$^4$ yr | 6x10$^{-5}$ | 6.4x10$^{-8}$ | 1.0 |
| | | | | | Total isotopes Pu = 8.3 |

TABLE 12.30 Remanent Radioactive Materials, after a TNR Shut-down*

| *Material* | *Radionuclides (half-life)* | *Power density* (10$^6$ W/m$^3$) |
|---|---|---|
| SAP | $^{24}$Na (15 h) | 0.5 |
| SiC | $^{28}$Al (2.3 min) | 0.7 |
| V | $^{52}$V (3.8 min); $^{51}$Ti (5.8 min); $^{48}$Sc (1.83 d) | 0.5 |
| Stainless steel | $^{56}$Mn (2.58 h); $^{58}$Co (71.3 d); $^{55}$Fe (2.6 yr); $^{54}$Mn (303 d); $^{60}$Co (5.26 yr) | 0.6 |
| PE-16 | $^{58}$Co (71.3 d); $^{60}$Co (5.26 yr) | 0.5 |
| Nb-1% Zr | $^{92m}$Nb (10.2 d); $^{95}$Nb (35 d); $^{94m}$Nb (6.3 min); $^{95m}$Nb (90 h); $^{93m}$Nb (13.6 yr); $^{94}$Nb (2x10$^4$ yr) | 0.9 |

*Normalized to a neutron wall loading of 1 MW/m$^2$

of the total activity that would correspond to the maximal concentration allowed with the varieties of radioisotopes effectively produced. One works on the assumption of a complete discharge and evenly distributed radioactive substances.

Radioactive pollution risk is believed to be much lower with the TNRs than with the current fission reactors.

Following any reactor shut-down, the inventory of radioisotopes in the reactor would continue to generate heat as a result of radioactive emissions, thus determining a thermal pollution of the environment. For the fission reactors such a thermal pollution is intense and lasting. TNRs rank better in this respect too, as can be seen from Table 12.30 [14]. By all measures, it seems that reaching an acceptable degree of nuclear safety is, in principle, easier with a TNR than with a fission reactor.

Among the radioisotopes generated, $^{94}Nb$ is the only one that is really long-lived, thus requiring long-term storage. Use of structural materials such as SAP, SIC or V will allow a relatively rapid recycling (in less than 10 years). Recycling of stainless steel structural parts and of the nickel-based ones will, however, require a relatively long "cooling" intermission — of at least 50 years.

REFERENCES


1. Ursu, I. (1973). *Energia atomica*. Edit. Stiintifica, Bucharest.
2. Kammasch, T. (1975). *Fusion Reactor Physics*. Ann Arbor Sci.Publ., Ann Arbor, Mich.
3. Gormezano, C. (1977). *J.Phys.* 38, C3, 187.
4. Zoita, V., and others (1981). In *Plasma Physics and Controlled Nuclear Fusion Research*, IAEA, Vienna, vol. II, p. 197.
5. Ursu, I., and others (1981). *Producerea cimpurilor magnetice ultrainalte. Procese fizice si rezultate experimentale*. A 4-a Conferinta Nationala de Fizica si Tehnica Plasmei, Bucharest, June 1981.
6. Ursu, I., and others (1974). *Studiul emisiei de ioni si neutroni pe instalatia de plasma focalizata IPF-2/20*. Progrese in Fizica, Timisoara, October, 1981.
7. Gribble, R. F., and others (1974). In *Plasma Physics and Controlled Nuclear Fusion Research*, IAEA, Vienna, vol. III, p. 381.
8. Samain, A. (1979). *Annales des Phys.*, 4, 395.
9. Shafranov, V. D. (1980). *Nuclear Fusion*, 20, 1075.
10. Artsimovich, L. A. (1972). *Nuclear Fusion*, 12, 215.
11. Cohn, R. W., and others (1978). Report UWFDM-249, University of Wisconsin; Kearney, K, and others (1978). Report Ga-A 14974, General Atomic Company.
12. Ashby, D. E. T. F. (1975). *J.British Nucl.Energy.Soc.*, 14, 311.
13. Prokhorov, A. M. (1975). Fusion and Lasers. In *Proc.Third Gen.Conf.of the European Phys.Soc.*, Bucharest 1975.
14. Steiner, D. (1975). *Proc.IEEE*, 63, 1968; Steiner, D. (1971). *Nuclear Fusion*, 11, 305.
15. Leonard, Jr., R. B. (1973). *Nucl.Technology*, 20, 161.
16. Golovin, I. N., Kolbasov, B. N., Orlov, V. V. Pustunovich, V. I., and Shalatov, C. E. (1977). In *Technical Committee Meeting and Workshop on Fusion Reactor Design*, IAEA, Madison Wisconsin, USA, p. 419; Oka, Y., Hattori, H., Kondo, S., and An, S. (1981). In *Fusion Reactor Design and Technology, Proc.of the Third Technical Committee Meeting and Workshop, Tokyo 1981,* p. 581.
17. Galloway, T. R. (1978). Report UCRL-80479, University of California, Lawrence Livermore Laboratory.
18. Roth, A. (1955). *Tehnica vidului si a aparatelor electrice cu vid*. Edit. Energetica de Stat, Bucharest.

19. Sako, K., and others (1974). Design Study of a Tokamak Reactor. In *Proc. Fifth Int.Conf.Plasma Physics and Controlled Nuclear Fusion Research*, vol. 3, Tokio, pp. 535-546.
20. Cohn, R. W., and others (1974). Major Design Features of the Conceptual D-T Tokamak Power Reactor UWMAK II. In *Proc.Fifth Int.Conf.on Plasma Physics and Controlled Nuclear Fusion Research*, vol. 3, Tokio.
21. Cowles, J. O., and Pasternak, A. D. (1969). *Lithium Properties related to Use as Nuclear Reactor Coolant*, UCRL-50647, USAEC, Washington.
22. IAEA (1980). *International Tokamak Reactor. Zero Phase*, Report of the International Tokamak Reactor Workshop, IAEA, Vienna, 1979.
23. Schultz, K. R. (1979). Materials Implications of Fusion-Fission Reactor Designs. In *Proc.First Topical Meeting on Fusion Reactor Materials, Bal Harbour 1979*.
24. Kushneriuk, S. A., and Wong, P. Y. (1981). *Fissile Fuel Breeding in D – T Fusion Reactor Blankets*. AECL-7421.
25. Guggi, D., Ihle, H. R., and Neubert, A. (1976). In *Proc.Nineth Symp.on Fusion Technology, Garmish-Partenkirchen 1976*, p. 635.
26. Basu, T. K., and others (1979). *Nucl.Sci.Eng.*, 70, 309.
27. US Energy Research and Development Administration (1978). First Wall Structural Goals for Economic Fusion Power, In *Fusion Materials Bulletin*, ERDA.
28. Vlasov, N. A. (1955). *Neitrony*, G.I.T.T.L., Moskow.
29. Gold, R. E., and Harrod, D. L. (1978). Report 78-7 D2-DGRFD P1, Westinghouse R and D Center.
30. Smith, C. O. (1967). *Nuclear Reactor Materials*. Addison-Wesley Publ.Comp. Reading, Mass.; Geller, Y. A., and Rathshtadt, A. G. (1977). *Science of Materials*, MIR Publ., Moskow.
31. Lakhtine, I. (1978). *Métallographie et traitements thermiques des métaux*. Ed. MIR, Moskow.
32. Orlov, V. V., and Altovski, I. V. (1981). Report IAE-3380/8, Moskow.
33. Kulcinski, G. R., Cohn, R. W., and Land, G. (1975). *Nuclear Fusion*, 15, 327
34. Clinard, F. W. (1977). Report LA-UR 79-146, Los Alamos Scientific Laboratory.
35. Riley, R. E., and others (1979). Composite Materials for Tokamak Wall Armor, Limiters and Beam Dump Applications. In *Proc.First Topical Meeting on Fusion Reactor Materials, Bal Harbour 1979*.
36. Haubenreich, P. N. (1978). Superconducting Magnets for Fusion Reactors, In *Proc.Third Topical Meeting on the Technology of Controlled Nuclear Fusion, Santa Fe 1978*.
37. Cohn, R. W. (1981). Magnetic Fusion Reactors. In *Fusion*, Teller, E. (Ed.), Vol. 1, part B, Academic Press, p. 194.
38. Sigmund, P. (1979). *Phys.Rev.*, 184, 383.
39. Behrisch, R. (1972). *Nuclear Fusion*, 12, 695.
40. Gruen, D. M. (1979). Materials for Thermonuclear Reactors. In *Materials Science in Energy Technology*, Libowitz, G. G., and Whittingham, M. S. (Eds), Academic Press, New York.
41. Gold, R. E., and others (1981). Materials Technology for Fusion: Current Status and Future Requirements, *Nuclear Technology/Fusion*, 1, 169.
42. Bloom, E. E. (1979). *J.Nucl.Materials*, 85 - 86, 795.
43. Cairns, E. J., and others (1972). A Review of the Chemical, Physical, and Thermal Properties of Lithium that are Related to its Use in Fusion Reactors. In *The Chemistry of Fusion Technology*, Gruen, D. M. (Ed.), Chapter 3, Plenum Press, New York.

APPENDIX I

# Equivalence of some usual measurements units to those in the International System (IS)

| *Unit* | *Symbol* | *Unit IS* | *Symbol IS* | *Conversion rate* |
|---|---|---|---|---|
| *Length* | | | | |
| Fermi | F | metre | m | 1 F = $1.5 \times 10^{-15}$ m |
| Angstrom | Å | metre | m | 1 Å = $10^{-10}$ m |
| inch | in | metre | m | 1 in = $2.54 \times 10^{-2}$ m |
| foot | ft | metre | m | 1 ft = $3.048 \times 10^{-1}$ m |
| yard | yd | metre | m | 1 yd = 0.9144 m |
| *Area* | | | | |
| barn | b | square metre | $m^2$ | 1 b = $10^{-28}$ $m^2$ |
| square inch | $in^2$ | square metre | $m^2$ | 1 $in^2$ = $6.4516 \times 10^{-4}$ $m^2$ |
| square foot | $ft^2$ | square metre | $m^2$ | 1 $ft^2$ = $9.290304 \times 10^{-2}$ $m^2$ |
| square yard | $yd^2$ | square metre | $m^2$ | 1 $yd^2$ = $8.361274 \times 10^{-1}$ $m^2$ |
| *Volume* | | | | |
| barrel | - | cubic metre | $m^3$ | 1 barrel = $1.589873 \times 10^{-1}$ $m^3$ |
| cubic inch | $in^3$ | cubic metre | $m^3$ | 1 $in^3$ = $1.638706 \times 10^{-5}$ $m^3$ |
| cubic foot | $ft^3$ | cubic metre | $m^3$ | 1 $ft^3$ = $2.831685 \times 10^{-2}$ $m^3$ |
| cubic yard | $yd^3$ | cubic metre | $m^3$ | 1 $yd^3$ = $7.645549 \times 10^{-1}$ $m^3$ |
| *Mass* | | | | |
| atomic mass unit | u | kilogram | kg | 1 u = $1.660566 \times 10^{-27}$ kg |
| pound | lb | kilogram | kg | 1 lb = $2.814952 \times 10^{-2}$ kg |
| ton (short) | - | kilogram | kg | 1 ton(s) = $9.071847 \times 10^{-2}$ kg |
| ton (long) | - | kilogram | kg | 1 ton(l) = $1.016047 \times 10^{3}$ kg |

Appendix 1 (continued)

| *Unit* | *Symbol* | *Unit IS* | *Symbol IS* | *Conversion rate* |
|---|---|---|---|---|
| *Temperature* | | | | |
| degree Celsius | °C | kelvin | K | $T_K = T_C + 273.15$ |
| degree Fahrenheit | °F | kelvin | K | $T_K = (T_F + 459.67)/1.8$ |
| degree Rankine | °R | kelvin | K | $T_K = T_R/1.8$ |
| *Energy* | | | | |
| British thermal unit | btu | joule | J | 1 btu = $1.055056 \times 10^3$ J |
| calorie | cal | joule | J | 1 cal = 4.186800 J |
| electron volt | eV | joule | J | 1 eV = $1.602190 \times 10^{-19}$ J |
| erg | erg | joule | J | 1 erg = $1.0 \times 10^{-7}$ J |
| kilowatthour | kWh | joule | J | 1 kWh = $3.6 \times 10^6$ J |
| *Power* | | | | |
| British thermal unit/hour | btu/h | watt | W | 1 btu/h = $2.930711 \times 10^{-1}$ W |
| calorie/second | cal/s | watt | W | 1 cal/s = 4.186800 W |
| horse power | hp | watt | W | 1 hp = $7.456999 \times 10^2$ W |
| *Pressure* | | | | |
| standard atmosphere | atm | pascal | Pa | 1 atm = $1.01325 \times 10^5$ Pa |
| bar | bar | pascal | Pa | 1 bar = $1.00 \times 10^5$ Pa |
| pound per square inch | psi | pascal | Pa | 1 psi = $6.894757 \times 10^3$ Pa |
| torr | torr | pascal | Pa | 1 torr = $1.33322 \times 10^2$ Pa |
| *Thermodynamics, heat transfer* | | | | |
| energy/unit area | btu/ft$^2$ | joule/m$^2$ | J/m$^2$ | 1 btu/ft$^2$ = $1.135653 \times 10^4$ J/m$^2$ |

Appendix 1 (continued)

| *Unit* | *Symbol* | *Unit IS* | *Symbol IS* | *Conversion rate* |
|---|---|---|---|---|
| thermal conductivity | $\frac{btu}{h \cdot ft \cdot {}^{o}F}$ | - | W/(m . K) | 1 btu/(h . ft . $^{o}$F) = 1.7307 W/(m . K) |
| thermal diffusivity | $ft^2/h$ | - | $m^2/s$ | 1 $ft^2/h$ = 2.580640 x $10^{-5}$ $m^2/s$ |
| thermal resistivity | $\frac{{}^{o}F \cdot h \cdot ft^2}{btu}$ | - | K . $m^2/W$ | 1 $^{o}$F . h . $ft^2$/btu = 1.762280 x $10^{-1}$ K . $m^2/W$ |
| specific heat | cal/(g . $^{o}$C) | - | J/(kg . K) | 1 cal/(g . $^{o}$C) = 4.186800 x $10^3$ J/(kg . K) |
| heat flow | $\frac{btu}{h \cdot ft^2}$ | - | $W/m^2$ | 1 btu/(h . $ft^2$) = 3.155 x $10^{-8}$ $W/m^2$ |
| dynamic viscosity | cP | - | Pa . s | 1 centipoise = 1 x $10^{-3}$ Pa . s |
| *Radioactivity* | | | | |
| curie | Ci | becquerel | Bq | 1 Ci = 3.7 x $10^{10}$ Bq |
| roentgen | R | coulomb/kg | C/kg | 1 R = 2.579760 x $10^{-4}$ C/kg |
| rad | rad | grey | Gy | 1 rad = $10^{-2}$ Gy |
| rem | rem | sievert | Sv | 1 rem = $10^{-2}$ Sv |

APPENDIX 2

# Characteristics of major reactor types

| *Type* | *Acronym* | *Specific power* | | *Fuel* | | | |
|---|---|---|---|---|---|---|---|
| | | *MW/m³ of core* | *MW/tU* | *% $^{235}U$* | *Composition* | *Clad* | *Burn-up* (MW · d/t) |
| Graphite gas "classic" | GCR | 0.5-1.2 | 1-3.2 | Natural | U | Mg | 3500-4000 |
| Graphite gas "advanced" | AGR | 1.5-2 | 12-13 | 2.5 | $UO_2$ | Stainless steel | 10,000-12,000 |
| High temperature | HTGR | 8-15 | 82-123 | 12-93 | $UC_2$-$ThC_2$ | Graphite | 50,000-100,000 |
| Graphite water | - | - | 100 | 1.5-5 | $UO_2$ | Stainless steel | 4000-40,000 |
| Heavy water | HWR | 2-10 | 4-30 | Natural or < 2 | $UO_2$ | Zircaloy | 7000-12,000 |
| Pressurized water | PWR | 75-100 | 36 | 3-5 | $UO_2$ | Zircaloy | 10,000-20,000 |
| Boiling water | BWR | 25-50 | 25 | 1.5-3 | $UO_2$ | Zircaloy | 10,000-20,000 |
| Fast neutrons | FBR | 500-1000 | - | 15-50 or Pu | Various | Stainless steel or other refractory materials | 30,000-100,000 |
| Research reactors | - | - | 100-10,000 | 90 | UAl | Al | 25-30% $^{235}U$ |

REFERENCE

Sauteron, J. (1965). *Les combustibles nucléaires*. Ed. Hermann, Paris.

APPENDIX 3

# Fuel material enrichment for various types of reactors

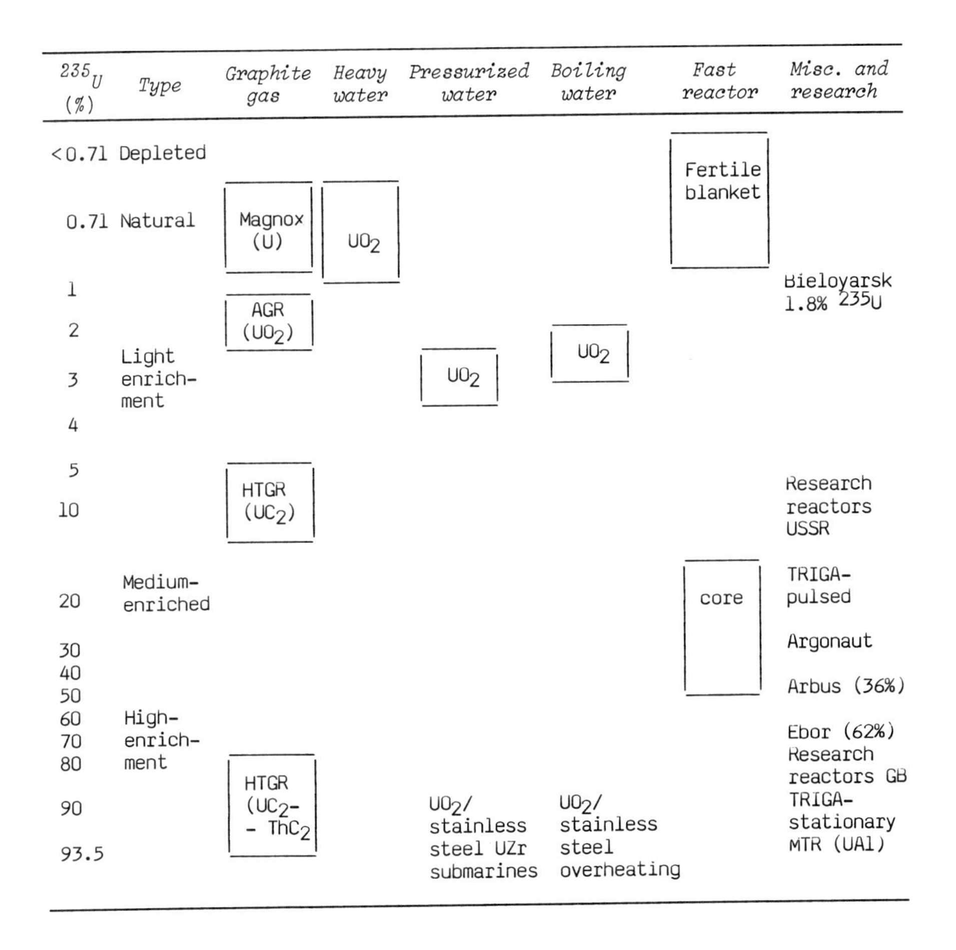

| $^{235}U$ (%) | Type | *Graphite gas* | *Heavy water* | *Pressurized water* | *Boiling water* | *Fast reactor* | *Misc. and research* |
|---|---|---|---|---|---|---|---|
| < 0.71 | Depleted | | | | | Fertile blanket | |
| 0.71 | Natural | Magnox (U) | $UO_2$ | | | | |
| 1 | | | | | | | Bieloyarsk 1.8% $^{235}U$ |
| 2 | | AGR ($UO_2$) | | | $UO_2$ | | |
| 3 | Light enrich-ment | | | $UO_2$ | | | |
| 4 | | | | | | | |
| 5 | | | | | | | |
| 10 | | HTGR ($UC_2$) | | | | | Research reactors USSR |
| 20 | Medium-enriched | | | | | core | TRIGA-pulsed |
| 30 | | | | | | | Argonaut |
| 40 | | | | | | | |
| 50 | | | | | | | Arbus (36%) |
| 60 | High-enrich-ment | | | | | | |
| 70 | | | | | | | Ebor (62%) |
| 80 | | | | | | | Research reactors GB |
| 90 | | HTGR ($UC_2$-$ThC_2$) | | $UO_2$/stainless steel UZr submarines | $UO_2$/stainless steel overheating | | TRIGA-stationary |
| 93.5 | | | | | | | MTR (UAl) |

REFERENCE


Sauteron, J. (1965). *Les combustibles nucléaires.* Ed. Hermann, Paris.

APPENDIX 4

# Electronic configuration of elements

| Z | X | Row | Configuration | | | |
|---|---|---|---|---|---|---|
| 1 | H | 1 | 1s | | | |
| 2 | He | 1 | $\vert 1s^2 \vert$ | | | |
| 3 | Li | 2 | $\vert K \vert$ | 2s | | |
| 4 | Be | 2 | $\vert K \vert$ | $2s^2$ | | |
| 5 | B | 2 | $\vert K \vert$ | $2s^2$ | 2p | |
| 6 | C | 2 | $\vert K \vert$ | $2s^2$ | $2p^2$ | |
| 7 | N | 2 | $\vert K \vert$ | $2s^2$ | $2p^3$ | |
| 8 | O | 2 | $\vert K \vert$ | $2s^2$ | $2p^4$ | |
| 9 | F | 2 | $\vert K \vert$ | $2s^2$ | $2p^5$ | |
| 10 | Ne | 2 | $\vert K \vert$ | $2s^2$ | $2p^6 \vert$ | |
| 11 | Na | 3 | $\vert K \vert L \vert$ | 3s | | |
| 12 | Mg | 3 | $\vert K \vert L \vert$ | $3s^2$ | | |
| 13 | Al | 3 | $\vert K \vert L \vert$ | $3s^2$ | 3p | |

Appendix 4 (continued)

| | Z | X | Row | Configuration | | | | | |
|---|---|---|---|---|---|---|---|---|---|
| | 14 | Si | 3 | \|K\|L\| | $3s^2$ | $3p^2$ | | | |
| | 15 | P | 3 | \|K\|L\| | $3s^2$ | $3p^3$ | | | |
| | 16 | S | 3 | \|K\|L\| | $3s^2$ | $3p^4$ | | | |
| | 17 | Cl | 3 | \|K\|L\| | $3s^2$ | $3p^5$ | | | |
| | 18 | Ar | 3 | \|K\|L\| | $3s^2$ | $3p^6$ | | | |
| | 19 | K | 4 | \|K\|L\| | $3s^2$ | $3p^6$ | | | $4s$ |
| | 20 | Ca | 4 | \|K\|L\| | $3s^2$ | $3p^6$ | | | $4s^2$ |
| T | 21 | Sc | 4 | \|K\|L\| | $3s^2$ | $3p^6$ | $3d$ | | $4s^2$ |
| T | 22 | Ti | 4 | \|K\|L\| | $3s^2$ | $3p^6$ | $3d^2$ | | $4s^2$ |
| T | 23 | V | 4 | \|K\|L\| | $3s^2$ | $3p^6$ | $3d^3$ | | $4s^2$ |
| T | 24 | Cr | 4 | \|K\|L\| | $3s^2$ | $3p^6$ | $3d^5$ | | $4s$ |
| T | 25 | Mn | 4 | \|K\|L\| | $3s^2$ | $3p^6$ | $3d^5$ | | $4s^2$ |
| T | 26 | Fe | 4 | \|K\|L\| | $3s^2$ | $3p^6$ | $3d^6$ | | $4s^2$ |
| T | 27 | Co | 4 | \|K\|L\| | $3s^2$ | $3p^6$ | $3d^7$ | | $4s^2$ |
| T | 28 | Ni | 4 | \|K\|L\| | $3s^2$ | $3p^6$ | $3d^8$ | | $4s^2$ |
| T | 29 | Cu | 4 | \|K\|L\| | $3s^2$ | $3p^6$ | $3d^{10}$ | \| | $4s$ |
| T | 30 | Zn | 4 | \|K\|L\|M\| | | $4s^2$ | | | |
| | 31 | Ga | 4 | \|K\|L\|M\| | | $4s^2$ | $4p$ | | |
| | 32 | Ge | 4 | \|K\|L\|M\| | | $4s^2$ | $4p^2$ | | |
| | 33 | As | 4 | \|K\|L\|M\| | | $4s^2$ | $4p^3$ | | |
| | 34 | Se | 4 | \|K\|L\|M\| | | $4s^2$ | $4p^4$ | | |
| | 35 | Br | 4 | \|K\|L\|M\| | | $4s^2$ | $4p^5$ | | |
| | 36 | Kr | 4 | \|K\|L\|M\| | | $4s^2$ | $4p^6$ | | |

Appendix 4 (continued)

| | *Z* | X | *Row* | *Configuration* | | | | | | | | | |
|---|---|---|---|---|---|---|---|---|---|---|---|---|---|
| | 37 | Rb | 5 | K L M | $4s^2$ | $4p^6$ | | 5s | | | | | |
| | 38 | Sr | 5 | K L M | $4s^2$ | $4p^6$ | | $5s^2$ | | | | | |
| T | 39 | Y | 5 | K L M | $4s^2$ | $4p^6$ | 4d | $5s^2$ | | | | | |
| T | 40 | Zr | 5 | K L M | $4s^2$ | $4p^6$ | $4d^2$ | $5s^2$ | | | | | |
| T | 41 | Nb | 5 | K L M | $4s^2$ | $4p^6$ | $4d^4$ | 5s | | | | | |
| T | 42 | Mo | 5 | K L M | $4s^2$ | $4p^6$ | $4d^5$ | 5s | | | | | |
| T | 43 | Tc | 5 | K L M | $4s^2$ | $4p^6$ | $4d^5$ | $5s^2$ | | | | | |
| T | 44 | Ru | 5 | K L M | $4s^2$ | $4p^6$ | $4d^7$ | 5s | | | | | |
| T | 45 | Rh | 5 | K L M | $4s^2$ | $4p^6$ | $4d^8$ | 5s | | | | | |
| T | 46 | Pd | 5 | K L M | $4s^2$ | $4p^6$ | $4d^{10}$ | | | | | | |
| T | 47 | Ag | 5 | K L M | $4s^2$ | $4p^6$ | $4d^{10}$ | 5s | | | | | |
| T | 48 | Cd | 5 | K L M | $4s^2$ | $4p^6$ | $4d^{10}$ | $5s^2$ | | | | | |
| | 49 | In | 5 | K L M | $4s^2$ | $4p^6$ | $4d^{10}$ | $5s^2$ | 5p | | | | |
| | 50 | Sn | 5 | K L M | $4s^2$ | $4p^6$ | $4d^{10}$ | $5s^2$ | $5p^2$ | | | | |
| | 51 | Sb | 5 | K L M | $4s^2$ | $4p^6$ | $4d^{10}$ | $5s^2$ | $5p^3$ | | | | |
| | 52 | Te | 5 | K L M | $4s^2$ | $4p^6$ | $4d^{10}$ | $5s^2$ | $5p^4$ | | | | |
| | 53 | I | 5 | K L M | $4s^2$ | $4p^6$ | $4d^{10}$ | $5s^2$ | $5p^5$ | | | | |
| | 54 | Xe | 5 | K L M | $4s^2$ | $4p^6$ | $4d^{10}$ | $5s^2$ | $5p^6$ | | | | |
| | 55 | Cs | 6 | K L M | $4s^2$ | $4p^6$ | $4d^{10}$ | | $5s^2$ | $5p^6$ | | 6s | |
| | 56 | Ba | 6 | K L M | $4s^2$ | $4p^6$ | $4d^{10}$ | | $5s^2$ | $5p^6$ | | $6s^2$ | |
| T | 57 | La | 6 | K L M | $4s^2$ | $4p^6$ | $4d^{10}$ | | $5s^2$ | $5p^6$ | 5d | $6s^2$ | |
| L | 58 | Ce | 6 | K L M | $4s^2$ | $4p^6$ | $4d^{10}$ | $4f^2$ | $5s^2$ | $5p^6$ | | $6s^2$ | |
| L | 59 | Pr | 6 | K L M | $4s^2$ | $4p^6$ | $4d^{10}$ | $4f^3$ | $5s^2$ | $5p^6$ | | $6s^2$ | |
| L | 60 | Nd | 6 | K L M | $4s^2$ | $4p^6$ | $4d^{10}$ | $4f^4$ | $5s^2$ | $5p^6$ | | $6s^2$ | |

Appendix 4 (continued)

| Z | X | Row | Configuration | | | | | | | | | |
|---|---|---|---|---|---|---|---|---|---|---|---|---|
| L 61 | Pm | 6 | K L M | $4s^2$ | $4p^6$ | $4d^{10}$ | $4f^5$ | $5s^2$ | $5p^6$ | | $6s^2$ | |
| L 62 | Sm | 6 | K L M | $4s^2$ | $4p^6$ | $4d^{10}$ | $4f^6$ | $5s^2$ | $5p^6$ | | $6s^2$ | |
| L 63 | Eu | 6 | K L M | $4s^2$ | $4p^6$ | $4d^{10}$ | $4f^7$ | $5s^2$ | $5p^6$ | | $6s^2$ | |
| L 64 | Gd | 6 | K L M | $4s^2$ | $4p^6$ | $4d^{10}$ | $4f^7$ | $5s^2$ | $5p^6$ | 5d | $6s^2$ | |
| L 65 | Tb | 6 | K L M | $4s^2$ | $4p^6$ | $4d^{10}$ | $4f^9$ | $5s^2$ | $5p^6$ | | $6s^2$ | |
| L 66 | Dy | 6 | K L M | $4s^2$ | $4p^6$ | $4d^{10}$ | $4f^{10}$ | $5s^2$ | $5p^6$ | | $6s^2$ | |
| L 67 | Ho | 6 | K L M | $4s^2$ | $4p^6$ | $4d^{10}$ | $4f^{11}$ | $5s^2$ | $5p^6$ | | $6s^2$ | |
| L 68 | Er | 6 | K L M | $4s^2$ | $4p^6$ | $4d^{10}$ | $4f^{12}$ | $5s^2$ | $5p^6$ | | $6s^2$ | |
| L 69 | Tm | 6 | K L M | $4s^2$ | $4p^6$ | $4d^{10}$ | $4f^{13}$ | $5s^2$ | $5p^6$ | | $6s^2$ | |
| L 70 | Yb | 6 | K L M | $4s^2$ | $4p^6$ | $4d^{10}$ | $4f^{14}$ | $5s^2$ | $5p^6$ | | $6s^2$ | |
| L 71 | Lu | 6 | K L M | $4s^2$ | $4p^6$ | $4d^{10}$ | $4f^{14}$ | $5s^2$ | $5p^6$ | 5d | $6s^2$ | |
| T 72 | Hf | 6 | K L M N | | | | | $5s^2$ | $5p^6$ | $5d^2$ | $6s^2$ | |
| T 73 | Ta | 6 | K L M N | | | | | $5s^2$ | $5p^6$ | $5d^3$ | $6s^2$ | |
| T 74 | W | 6 | K L M N | | | | | $5s^2$ | $5p^6$ | $5d^4$ | $6s^2$ | |
| T 75 | Re | 6 | K L M N | | | | | $5s^2$ | $5p^6$ | $5d^5$ | $6s^2$ | |
| T 76 | Os | 6 | K L M N | | | | | $5s^2$ | $5p^6$ | $5d^6$ | $6s^2$ | |
| T 77 | Ir | 6 | K L M N | | | | | $5s^2$ | $5p^6$ | $5d^7$ | $6s^2$ | |
| T 78 | Pt | 6 | K L M N | | | | | $5s^2$ | $5p^6$ | $5d^9$ | 6s | |
| T 79 | Au | 6 | K L M N | | | | | $5s^2$ | $5p^6$ | $5d^{10}$ | 6s | |
| T 80 | Hg | 6 | K L M N | | | | | $5s^2$ | $5p^6$ | $5d^{10}$ | $6s^2$ | |
| 81 | Tl | 6 | K L M N | | | | | $5s^2$ | $5p^6$ | $5d^{10}$ | $6s^2$ | 6p |
| 82 | Pb | 6 | K L M N | | | | | $5s^2$ | $5p^6$ | $5d^{10}$ | $6s^2$ | $6p^2$ |
| 83 | Bi | 6 | K L M N | | | | | $5s^2$ | $5p^6$ | $5d^{10}$ | $6s^2$ | $6p^3$ |
| 84 | Po | 6 | K L M N | | | | | $5s^2$ | $5p^6$ | $5d^{10}$ | $6s^2$ | $6p^4$ |

Appendix 4 (continued)

| | *Z* | **X** | *Row* | *Configuration* | | | | | | | |
|---|---|---|---|---|---|---|---|---|---|---|---|
| | 85 | At | 6 K L M N | $5s^2$ | $5p^6$ | $5d^{10}$ | | $6s^2$ | $6p^5$ | | |
| | 86 | Rn | 6 K L M N | $5s^2$ | $5p^6$ | $5d^{10}$ | | $6s^2$ | $6p^6$ | | |
| | 87 | Fr | 7 K L M N | $5s^2$ | $5p^6$ | $5d^{10}$ | | $6s^2$ | $6p^6$ | | 7s |
| | 88 | Ra | 7 K L M N | $5s^2$ | $5p^6$ | $5d^{10}$ | | $6s^2$ | $6p^6$ | | $7s^2$ |
| T | 89 | Ac | 7 K L M N | $5s^2$ | $5p^6$ | $5d^{10}$ | | $6s^2$ | $6p^6$ | 6d | $7s^2$ |
| A | 90 | Th | 7 K L M N | $5s^2$ | $5p^6$ | $5d^{10}$ | | $6s^2$ | $6p^6$ | $6d^2$ | $7s^2$ |
| A | 91 | Pa | 7 K L M N | $5s^2$ | $5p^6$ | $5d^{10}$ | $5f^2$ | $6s^2$ | $6p^6$ | 6d | $7s^2$ |
| A | 92 | U | 7 K L M N | $5s^2$ | $5p^6$ | $5d^{10}$ | $5f^3$ | $6s^2$ | $6p^6$ | 6d | $7s^2$ |
| A | 93 | Np | 7 K L M N | $5s^2$ | $5p^6$ | $5d^{10}$ | $5f^5$ | $6s^2$ | $6p^6$ | | $7s^2$ |
| A | 94 | Pu | 7 K L M N | $5s^2$ | $5p^6$ | $5d^{10}$ | $5f^6$ | $6s^2$ | $6p^6$ | | $7s^2$ |
| A | 95 | Am | 7 K L M N | $5s^2$ | $5p^6$ | $5d^{10}$ | $5f^7$ | $6s^2$ | $6p^6$ | | $7s^2$ |
| A | 96 | Cm | 7 K L M N | $5s^2$ | $5p^6$ | $5d^{10}$ | $5f^7$ | $6s^2$ | $6p^6$ | 6d | $7s^2$ |
| A | 97 | Bk | 7 K L M N | $5s^2$ | $5p^6$ | $5d^{10}$ | $5f^9$ | $6s^2$ | $6p^6$ | | $7s^2$ |
| A | 98 | Cf | 7 K L M N | $5s^2$ | $5p^6$ | $5d^{10}$ | $5f^{10}$ | $6s^2$ | $6p^6$ | | $7s^2$ |
| A | 99 | Es | 7 K L M N | $5s^2$ | $5p^6$ | $5d^{10}$ | $5f^{11}$ | $6s^2$ | $6p^6$ | | $7s^2$ |
| A | 100 | Fm | 7 K L M N | $5s^2$ | $5p^6$ | $5d^{10}$ | $5f^{12}$ | $6s^2$ | $6p^6$ | | $7s^2$ |
| A | 101 | Md | 7 K L M N | $5s^2$ | $5p^6$ | $5d^{10}$ | $5f^{13}$ | $6s^2$ | $6p^6$ | | $7s^2$ |
| A | 102 | No | 7 K L M N | $5s^2$ | $5p^6$ | $5d^{10}$ | $5f^{14}$ | $6s^2$ | $6p^6$ | | $7s^2$ |
| A | 103 | Lw | 7 K L M N | $5s^2$ | $5p^6$ | $5d^{10}$ | $5f^{14}$ | $6s^2$ | $6p^6$ | 6d | $7s^2$ |
| T | 104 | Ku | 7 K L M N | $5s^2$ | $5p^6$ | $5d^{10}$ | $5f^{14}$ | $6s^2$ | $6p^6$ | $6d^2$ | $7s^2$ |
| T | 105 | | | | | | | | | | |

| | |
|---|---|
| (boxed) | Filled electronic shell |
| T | Transition elements |
| L | Lanthanides |
| A | Actinides |

APPENDIX 5

# Some properties of elements

| *Element* | *Symbol* | *Z* | *Atomic mass* | *Density, at 20 °C* ($10^3$ kg/m$^3$) | *Melting point* (°C) | *Crystal-lographic structure* | *Atomic radius* (Å) |
|---|---|---|---|---|---|---|---|
| Actinium | Ac | 89 | 227.00 | 10.05 | 1050 + 50 | fcc | 1.88 |
| Aluminium | Al | 13 | 26.97 | 2.699 | 600.2 | fcc | 1.43 |
| Americium | Am | 95 | 243.00 | 11.7 | 990 | h | 1.82 |
| Antimony | Sb | 51 | 121.76 | 6.62 | 630.5 | r | 1.61 |
| Argon | Ar | 18 | 39.94 | 1.6604x10-3 | -189.4 | fcc | 1.92 |
| Arsenic | As | 33 | 74.91 | 5.73 | 814 | r | 1.25 |
| Astatinium | At | 85 | 210.00 | - | - | - | - |
| Barium | Ba | 56 | 137.36 | 3.5 | 704 | bcc | 2.24 |
| Berkelium | Bk | 97 | 247.00 | 13.25 | 986 | - | - |
| Beryllium | Be | 4 | 9.02 | 1.85 | 1315 | hcp | (1.13) |
| Bismuth | Bi | 83 | 209.00 | 9.80 | 271 | r | 1.82 |
| Boron | B | 5 | 10.82 | 2.3 | 2000-2300 | r | 0.97 |
| Bromine | Br | 35 | 79.91 | 3.12 | -7.2 | o | 1.19 |
| Cadmium | Cd | 48 | 112.41 | 8.85 | 321 | hcp | 1.52 |
| Calcium | Ca | 20 | 40.08 | 1.55 | 850 | fcc | 1.97 |
| Carbon | C | 6 | 12.01 | 2.22 | 3700 | h | 0.77 |
| Cerium | Ce | 58 | 140.13 | 6.78 | 804 | fcc | 1.82 |
| Cesium | Cs | 55 | 132.91 | 1.9 | 28 | bcc | 2.70 |
| Chlorine | Cl | 17 | 35.45 | 2.9476x10-3 | -101 | t | (1.07) |
| Chromium | Cr | 24 | 52.01 | 7.19 | 1890 | bcc | 1.28 |
| Cobalt | Co | 27 | 58.94 | 8.8 | 1495 | hcp | 1.25 |
| Columbium (Niobium) | Nb | 41 | 92.91 | 8.57 | 2415 | bcc | 1.47 |
| Copper | Cu | 29 | 63.57 | 8.96 | 1083 | fcc | 1.28 |
| Curium | Cm | 96 | 247.00 | 7 | 1340 | - | - |
| Dysprosium | Dy | 66 | 162.46 | 8.56 | 1500 | hcp | 1.77 |
| Erbium | Er | 68 | 167.20 | 9.15 | 1440 | hcp | 1.75 |
| Europium | Eu | 63 | 152.00 | 5.22 | 1100-1200 | bcc | 2.04 |
| Fluorine | F | 9 | 9.00 | 0.3742x10-3 | -223 | | 0.71* |
| Francium | Fr | 87 | 223.00 | 2.48 | - | | 2.77 |
| Gadolynium | Gd | 64 | 156.90 | 7.95 | 1312 | hcp | 1.78 |
| Galium | Ga | 31 | 69.72 | 5.91 | 29.78 | fco | 2.7 |

P & T-P

Appendix 5 (continued)

| *Element* | *Symbol* | *Z* | *Atomic mass* | *Density, at 20 °C* ($10^3$ kg/m$^3$) | *Melting point* (°C) | *Crystal-lographic structure* | *Atomic radius* (Å) |
|---|---|---|---|---|---|---|---|
| Germanium | Ge | 32 | 72.60 | 5.36 | 958 | dc | 1.39 |
| Gold | Au | 79 | 197.20 | 19.32 | 1063 | fcc | 1.44 |
| Hafnium | Hf | 72 | 178.60 | 13.36 | 2130 | hcp | 1.59 |
| Helium | He | 2 | 4.00 | $0.1663\times10^{-3}$ | -271.4 | hcp | 1.22 |
| Holmium | Ho | 67 | 164.94 | 8.76 | 1500 | hcp | 1.76 |
| Hydrogen | H | 1 | 1.008 | $0.08381\times10^{-3}$ | -259.4 | h | 0.46 |
| Indium | In | 49 | 114.76 | 7.31 | 156.4 | fct | 1.57 |
| Iodine | I | 53 | 126.92 | 4.93 | 114 | o | (1.36) |
| Iridium | Ir | 77 | 193.10 | 22.5 | 1454 | fcc | 1.35 |
| Iron | Fe | 26 | 55.85 | 7.87 | 1539 | bcc | 1.28 |
| Kripton | Kr | 36 | 83.70 | $3.4797\times10^{-3}$ | -157 | fcc | 1.97 |
| Lanthanum | La | 57 | 138.92 | 6.19 | 920 | hcp | 1.87 |
| Lead | Pb | 82 | 207.21 | 11.34 | 327.4 | fcc | 2.38 |
| Lithium | Li | 3 | 6.94 | 0.53 | 186 | bcc | 1.57 |
| Lutetium | Lu | 71 | 174.99 | 9.74 | 1650-1750 | hcp | 1.73 |
| Magnesium | Mg | 12 | 24.32 | 1.74 | 650 | hcp | 1.60 |
| Manganese | Mn | 25 | 54.93 | 7.43 | 1245 | c | (1.60) |
| Mercury | Hg | 80 | 200.61 | 13.55 | -38.87 | r | 1.55 |
| Molybdenum | Mo | 42 | 95.95 | 10.2 | 2622 | bcc | 1.40 |
| Natrium | Na | 11 | 22.997 | 0.97 | 97.7 | bcc | 1.92 |
| Neodymium | Nd | 60 | 144.27 | 6.89 | 1024 | hcp | 1.82 |
| Neon | Ne | 10 | 20.18 | $0.8389\times10^{-3}$ | -248.6 | fcc | 1.60 |
| Neptunium | Np | 93 | 237.00 | 20.49 | 640 | o | 1.50 |
| Nickel | Ni | 28 | 58.69 | 8.90 | 1455 | fcc | 1.25 |
| Nitrogen | N | 7 | 14.00 | $1.1641\times10^{-3}$ | -210.0 | h | 0.71 |
| Osmium | Os | 76 | 190.20 | 22.5 | 2700 | hcp | 1.35 |
| Oxygen | O | 8 | 16.00 | $1.3303\times10^{-3}$ | -218.8 | c | 0.60 |
| Paladium | Pd | 46 | 106.70 | 12.0 | 1551 | fcc | 1.37 |
| Phosphorus | P | 15 | 30.98 | 1.82 | 44.81 | sc | 1.28 |
| Platinum | Pt | 78 | 195.23 | 21.45 | 1773.5 | fcc | 1.38 |
| Plutonium | Pu | 94 | 239.00 | 19.8 | 639 | m | 1.62 |
| Polonium | Po | 84 | 210.00 | 9.51 | 252 | m | (1.40) |
| Potasium | K | 19 | 39.09 | 0.86 | 63 | bcc | 2.38 |
| Praseodymium | Pr | 59 | 140.92 | 6.78 | 950 | hcp | 1.83 |
| Promethyum | Pm | 61 | 145.00 | - | 1027 | - | 1.81 |
| Protactinium | Pa | 91 | 231.00 | 15.37 | 3000 | | 1.63 |
| Radium | Ra | 88 | 226.05 | 5.0 | 700 | | 2.35 |
| Radon | Rn | 86 | 222.00 | $9.229\times10^{-3}$ | -71 | fcc | 1.82 |
| Rhenium | Re | 75 | 186.31 | 21.02 | 3170 | hcp | 1.37 |
| Rhodium | Rh | 45 | 102.91 | 12.41 | 1966 | bcc | 1.34 |
| Rubidium | Rb | 37 | 85.48 | 1.53 | 39 | bcc | 2.51 |
| Ruthenium | Ru | 44 | 101.70 | 12.20 | 2500 | hcp | 1.34 |
| Samarium | Sm | 62 | 150.43 | 7.53 | 1300 | r | 1.81 |
| Scandium | Sc | 21 | 45.10 | 2.5 | 1530 | hcp | 1.60 |
| Selenium | Se | 34 | 78.96 | 4.81 | 220 | h | 1.16 |
| Silicon | Si | 14 | 28.06 | 2.33 | 1430 | dc | 1.17 |
| Silver | Ag | 47 | 107.88 | 10.5 | 960.5 | fcc | 1.44 |
| Stanium | Sn | 50 | 118.70 | 7.298 | 231.9 | tetr | 1.58 |
| Strontium | Sr | 38 | 87.63 | 2.50 | 770 | fcc | 2.15 |
| Sulphur | S | 16 | 32.06 | 2.07 | 119.0 | fcc | (1.04) |
| Tantalum | Ta | 73 | 180.88 | 16.60 | 2996 | bcc | 1.47 |
| Technetium | Tc | 43 | 99.00 | 11.5 | 2150 | hcp | 1.34 |
| Tellurium | Te | 52 | 127.61 | 6.24 | 450 | h | (1.43) |
| Terbium | Tb | 65 | 159.20 | 8.33 | 1364 | hcp | 1.77 |
| Thallium | Tl | 81 | 204.39 | 11.85 | 300 | hcp | 1.71 |
| Thorium | Th | 90 | 232.12 | 11.71 | 1690 | fcc | 1.80 |

Appendix 5 (continued)

| *Element* | *Symbol* | *Z* | *Atomic mass* | *Density, at 20 °C* ($10^3$ kg/m$^3$) | *Melting point* (°C) | *Crystal-lographic structure* | *Atomic radius* (Å) |
|---|---|---|---|---|---|---|---|
| Titanium | Ti | 22 | 47.90 | 4.51 | 1690 | hcp | 1.47 |
| Tulium | Tm | 69 | 169.40 | 9.35 | 1550-1650 | hcp | 1.74 |
| Uranium | U | 92 | 238.07 | 19.10 | 1133 | o | 1.53 |
| Vanadium | V | 23 | 50.95 | 6.10 | 1710 | bcc | 1.36 |
| Wolfram | W | 74 | 183.92 | 19.2 | 3395 | bcc | 1.41 |
| Xenon | Xe | 54 | 131.30 | $5.486 \times 10^{-3}$ | -112 | fcc | 2.18 |
| Ytterbium | Yb | 70 | 173.04 | 7.01 | 1800 | fcc | 1.93 |
| Yttrium | Y | 39 | 88.92 | 5.51 | 1509 | hcp | 1.81 |
| Zinc | Zn | 30 | 65.38 | 7.133 | 419.46 | hcp | 1.37 |
| Zirconium | Zr | 40 | 91.22 | 6.5 | 1824 | hcp | 1.60 |

sc : simple cubic
fcc : face-centred cubic
fco : face-centred orthorhombic
bcc : body-centred cubic
hcp : hexagonal close-packed
tetr: tetragonal
fct : face-centred tetragonal

dc: diamond cubic
r : rhomboedric
o : orthorhombic
h : hexagonal
t : trigonal
m : monoclinic
c : complex

*ionic radius

REFERENCES

Tipton, C. R. Jr. (1960). *Reactor Handbook*, Vol. 1, *Materials*. Interscience Publ. New York.
Pop, I. (1968). *Magnetismul paminturilor rare*. Edit. Academiei, Bucharest.
Nenitescu, C. D. (1979). *Chimie generala*. Edit. Didactica si Pedagogica, Bucharest.

APPENDIX 6

# Irradiation testing of structural materials for CANDU — 600 nuclear power station

## A. STAINLESS STEELS AND ZIRCONIUM ALLOYS FOR STRUCTURAL COMPONENTS

### 1. Tensile Test

*Purpose*: determination of irradiation-induced brittleness.

*Irradiation device*: capsule for materials testing that reproduces neutronic and thermal conditions featuring the reactor operation.

*Testing method*: tensile test on standard-size irradiated samples.

*Information and data provided* : stress – strain curve, elongation, tensile strength.

### 2. Impact Strength (Resilience) Test

*Purpose*: determination of changes in material's brittleness and ductile-to-brittle transition temperature, as an effect of irradiation.

*Irradiation device*: capsule for materials testing that reproduces neutronic and thermal conditions featuring the reactor operation.

*Testing method*: Charpy-impact test on irradiated samples.

*Information and data provided* : brittle-to-ductile transition curves, post-irradiation transition temperature, fracture pattern (ductile fracture proportion), energy absorbed by shear in the fracture region.

### 3. Other Tests

Hardness determination (microhardness and global hardness).

Corrosion resistance determination.

Fatigue tests.

Creep strength determination.

Evaluation of cracking tendency.

### B. ZIRCONIUM ALLOYS USED FOR FUEL CLADS

#### 1. Tensile Test

*Purpose*: determination of irradiation effect on flow, and rupture, strength, and elongation.

*Irradiation device*: capsule for materials testing.

*Testing method*: tensile tests on standard-size irradiated samples.

*Information and data provided*: flow limit, rupture stress, uniform elongation, total elongation.

#### 2. Fatigue Test

*Purpose*: determination of clad behaviour under cyclic mechanical loading.

*Irradiation device*: capsule for materials testing.

*Testing method*: repeated bending (number of cycles, frequency of cyclic loading).

*Information and data provided*: number of cycles till rupture, total post-cycling elongation, total sample bending, plastic strain, microstructural characteristics (electron scanning microscopy - ESM, electron transmission microscopy - ETM).

#### 3. Relaxation Test

*Purpose*: determination of stress relaxation under stationary load.

*Irradiation device*: capsule for materials testing.

*Testing method*: loading as in tensile test, followed, after a pre-established time, by recording of stress relaxation.

*Information and data provided*: stress – strain curves, initial stress values, stress values at certain time intervals, microstructral characteristics (ETM).

#### 4. Creep Test under Internal Pressure

*Purpose*: resistance to creep determination, in correlation with temperature, internal pressure, fast neutrons flux and fluency.

*Irradiation device*: materials testing capsule allowing pressurization of clad samples during irradiation, or irradiation on originally pressurized clads.

*Information and data provided*: outer diameter and clad thickness vs. time, microstructural characteristics (ETM).

5. Creep Test under External Pressure

*Purpose*: collapse by creep and instantaneous collapsing pressure determination.

*Irradiation device*: capsule for materials testing.

*Testing method*: recording of circular strains dynamics under the effect of external pressure.

*Information and data provided*: outer diameter (maximum and minimum) variation vs. time, wall thickness variation.

6. In-pile Relaxation Test

*Purpose*: determination of stress relaxation under the combined effect of temperature and neutron flux.

*Irradiation device*: capsule for materials testing.

*Testing method*: application of a normal constant load on samples with different curvature radii (that generates different stresses) and post-irradiation measurement of stress relaxation by determination of modifications in the curvature radii.

*Information and data provided*: sample profile (relaxation) evolution, microstructural characteristics (ETM).

7. Primary Creep and Relaxation Tests on Conic Samples

*Purpose*: clad behaviour simulation in regions neighbouring fuel fissures ($UO_2$).

*Irradiation device*: capsule for materials testing.

*Testing method*: tensile test on conically-shaped samples (of variable section) that ensures a linear stress distribution vs. sample length.

*Information and data provided* : stress – strain curves, stress relaxation factors, microstructural characteristics (ESM and ETM).

REFERENCE

ICEFIZ (1979). *Program de testare la iradiere a materialelor structurale pentru CNE-CANDU*, a technical report.

APPENDIX 7

# Neutron cross-sections

A

| | *Element (compound)* | *Data in Fig.* | *Neutron energy ($E_n$)* |
|---|---|---|---|
| 1 | H | A.7.1 | 0.001 – 1000 eV |
| | | A.7.2 | 0.01 – 300 MeV |
| 2 | $H_2O$ | A.7.1 | 0.001 – 1000 eV |
| 3 | D | A.7.3 | 0.1 – 300 MeV |
| 4 | $D_2O$ | A.7.4 | 0.001 – 10 eV |
| 5 | Li | A.7.5 | 0.01 – 10 eV |
| 6 | Be | A.7.6 | 0.001 – 1000 eV |
| 7 | B | A.7.7 | 0.01 – 1000 eV |
| 8 | C | A.7.8 | 0.01 – 300 MeV |
| 9 | N | A.7.9 | 0.01 – 4000 eV |
| 10 | O | A.7.10 | 0.01 – 400 eV |
| 11 | F | A.7.11 | 0.01 – 400 eV |
| 12 | Na | A.7.12 | 1 – 10,000 eV |
| 13 | Mg | A.7.13 | 2 – 100 MeV |
| 14 | Al | A.7.14 | 0.01 – 4000 eV |
| 15 | K | A.7.15 | 0 – 0.6 MeV |
| 16 | Ca | A.7.16 | 0 – 0.6 MeV |
| 17 | Ti | A.7.17 | 0 – 1.5 MeV |
| 18 | V | A.7.18 | 0 – 1 MeV |
| 19 | Cr | A.7.19 | 0.01 – 4000 eV |
| 20 | Fe | A.7.20 | 0.01 – 4000 eV |
| 21 | Co | A.7.21 | 0 – 1 eV |
| 22 | Ni | A.7.22 | 0.01 – 4000 eV |
| 23 | Li | A.7.23 | 0 – 1.4 MeV |
| 24 | Zr | A.7.24 | 0.01 – 400 eV |
| 25 | Nb | A.7.25 | 0.01 – 1000 eV |
| 26 | Ru | A.7.26 | 0.01 – 1000 eV |
| 27 | Ag | A.7.27 | 0.01 – 1000 eV |
| 28 | Cd | A.7.28 | 0.001 – 1000 eV |
| 29 | In | A.7.29 | 0.01 – 1000 eV |
| 30 | Sn | A.7.30 | 0.01 – 1000 eV |
| 31 | I | A.7.31 | 0.01 – 1000 eV |
| 32 | La | A.7.32 | 0.01 – 1000 eV |
| 33 | Nd | A.7.33 | 0.04 – 10 eV |
| 34 | Sm | A.7.34 | 0.01 – 100 eV |
| 35 | Eu | A.7.35 | 0.01 – 100 eV |
| 36 | Gd | A.7.36 | 0.001 – 1 eV |
| 37 | Dy | A.7.37 | 0.001 – 30 eV |
| 38 | Pb | A.7.38 | 0.01 – 1000 eV |
| 39 | Bi | A.7.39 | 2 – 50 MeV |
| 40 | U | A.7.40 | 20 – 300 MeV |
| 41 | $^{237}Np$ | A.7.41 | 0 – 3 MeV |

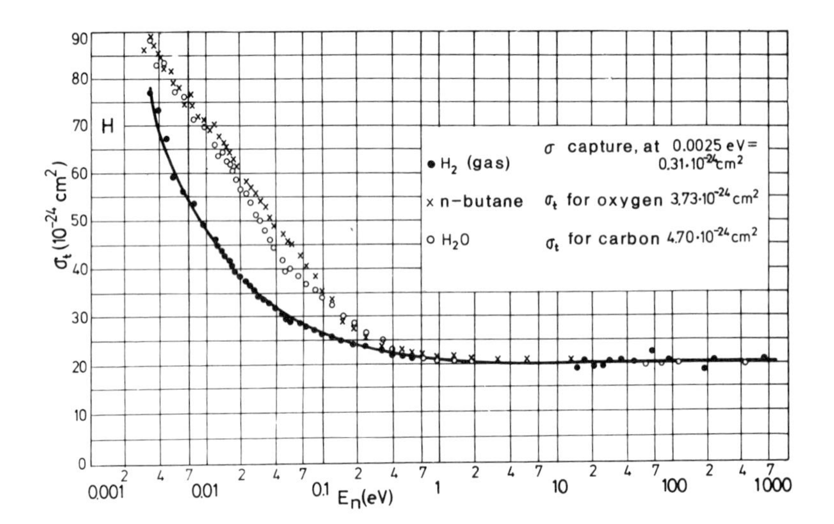


Fig. A.7.1

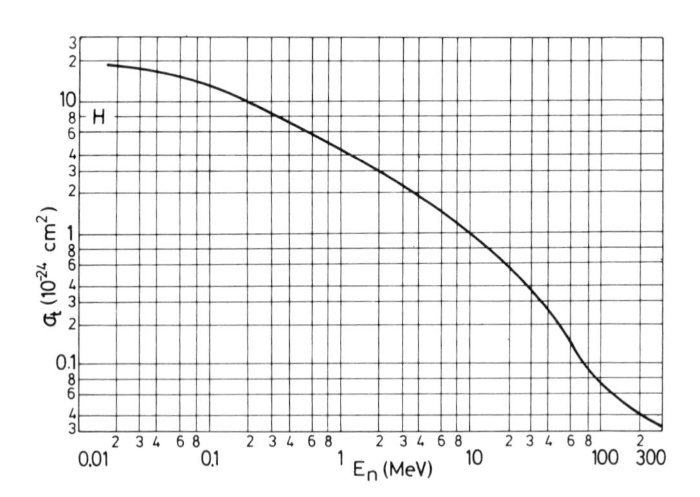


Fig. A.7.2

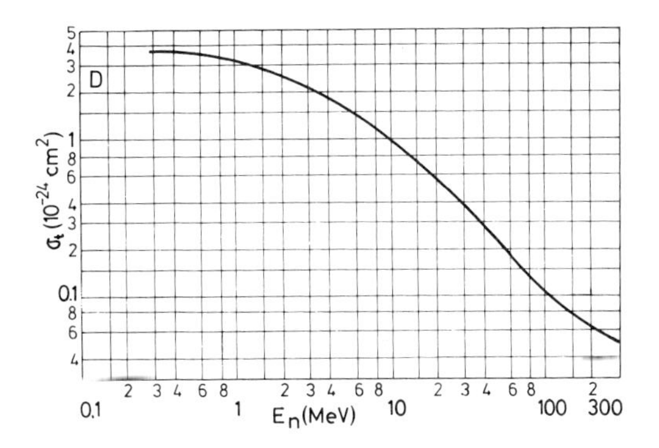


Fig. A.7.3

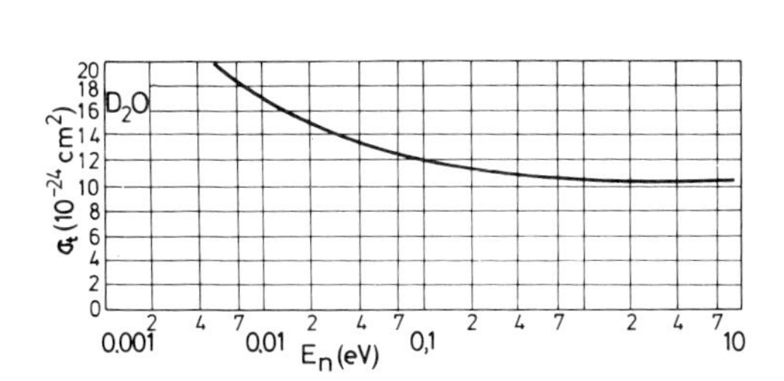


Fig. A.7.4

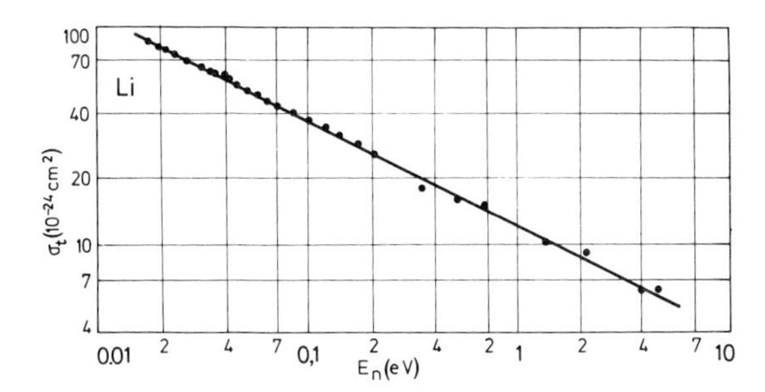


Fig. A.7.5

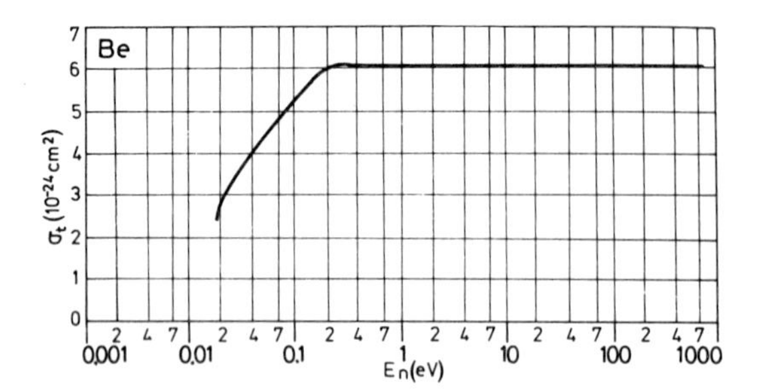


Fig. A.7.6

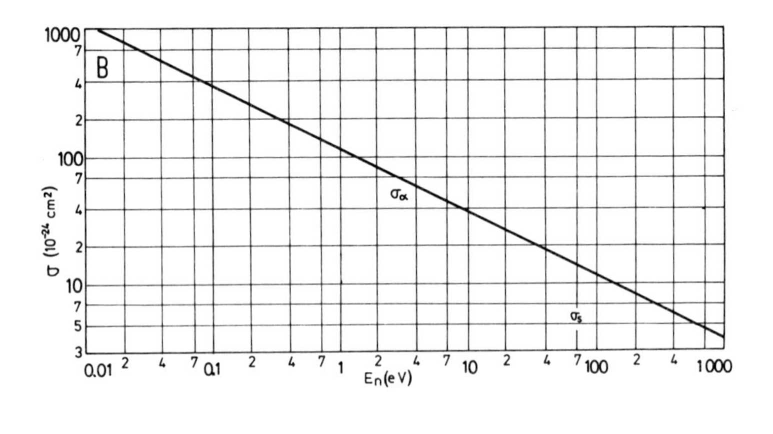


Fig. A.7.7

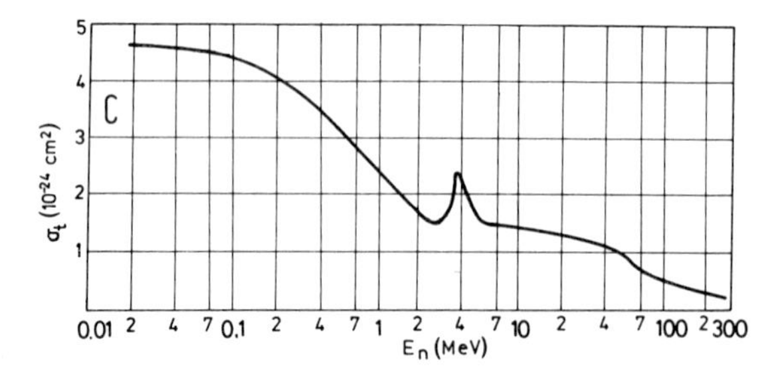


Fig. A.7.8

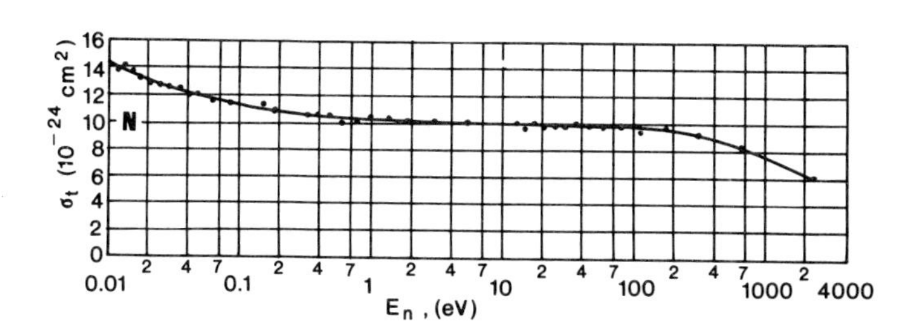


Fig. A.7.9

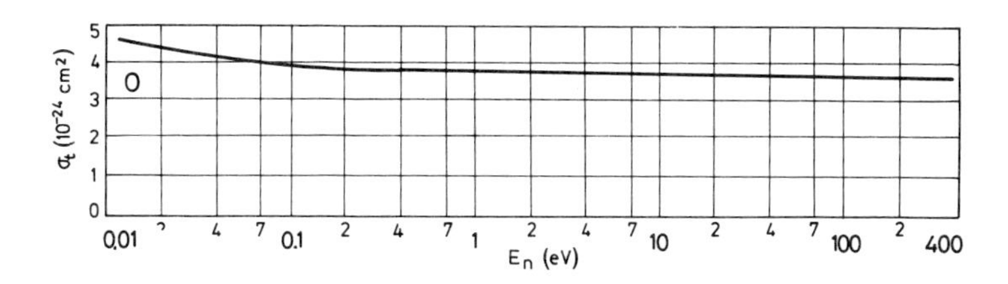


Fig. A.7.10

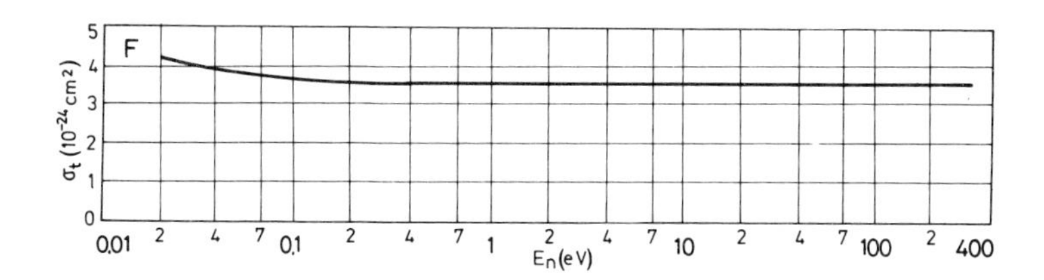


Fig. A.7.11

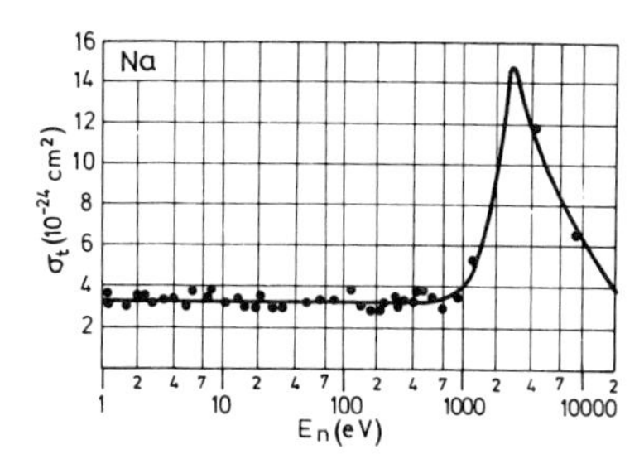


Fig. A.7.12

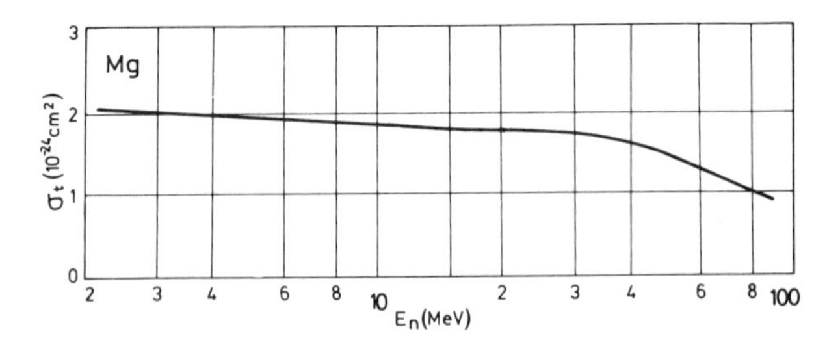


Fig. A.7.13

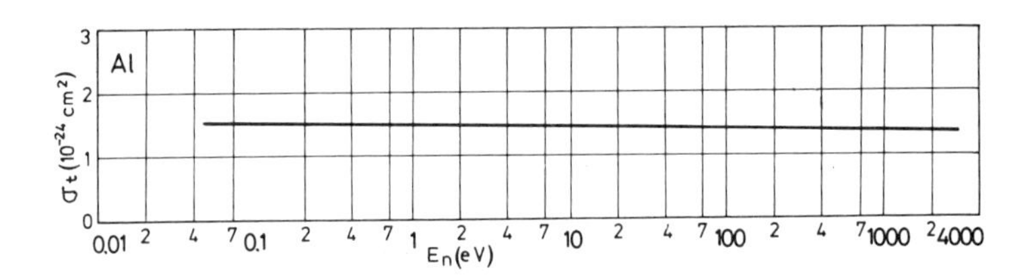


Fig. A.7.14

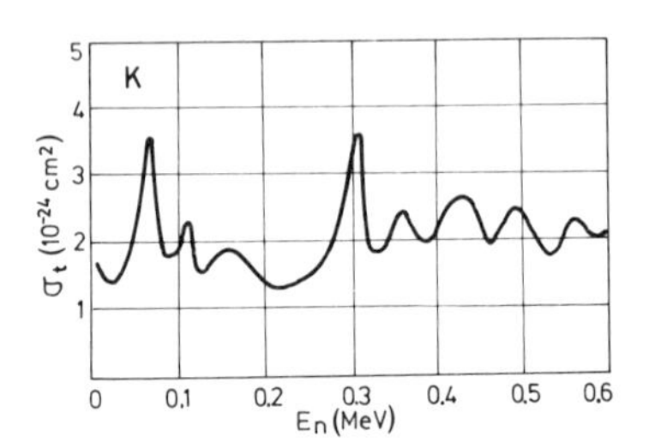


Fig. A.7.15

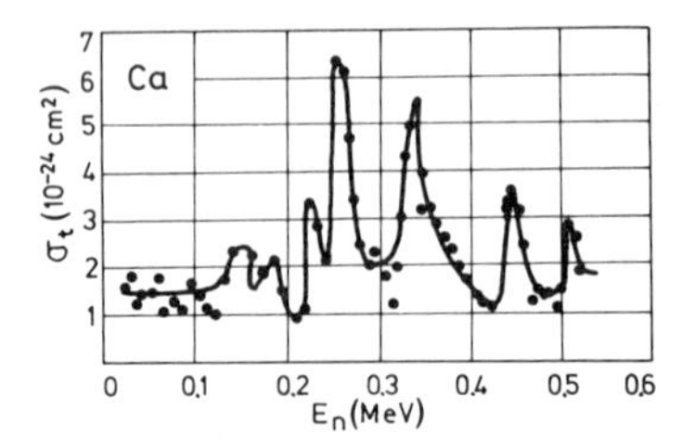


Fig. A.7.16

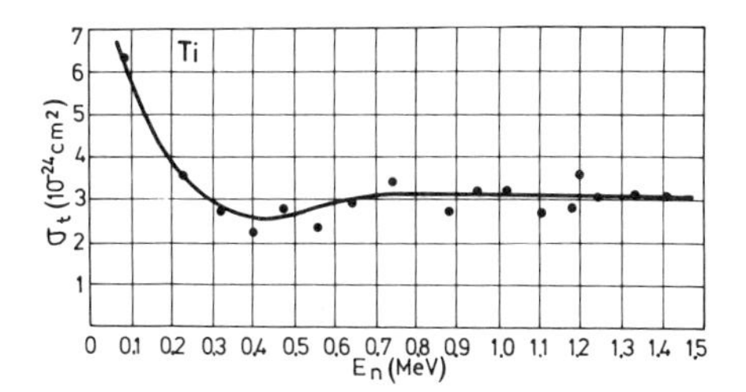


Fig. A.7.17

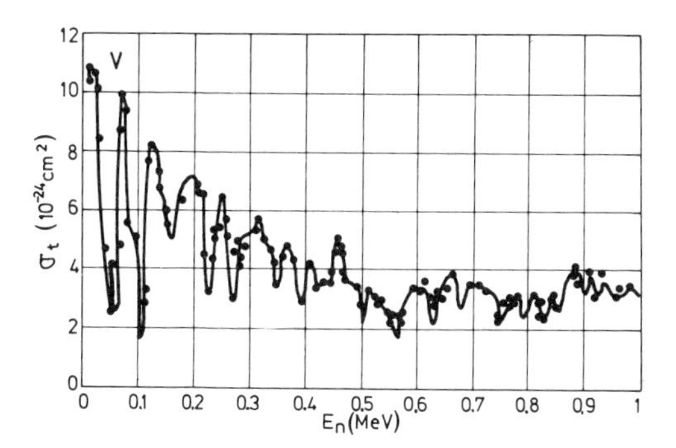


Fig. A.7.18

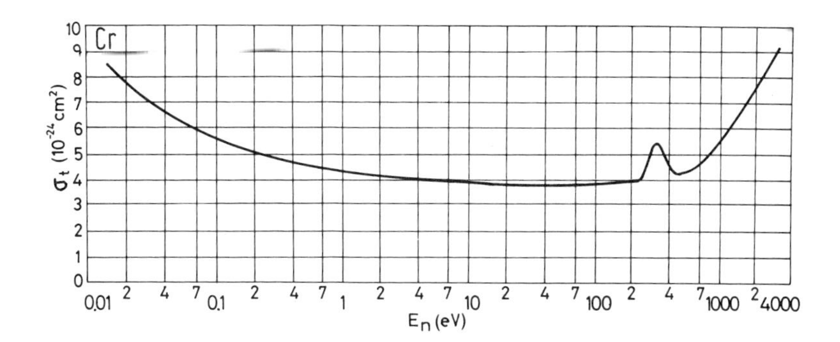


Fig. A.7.19

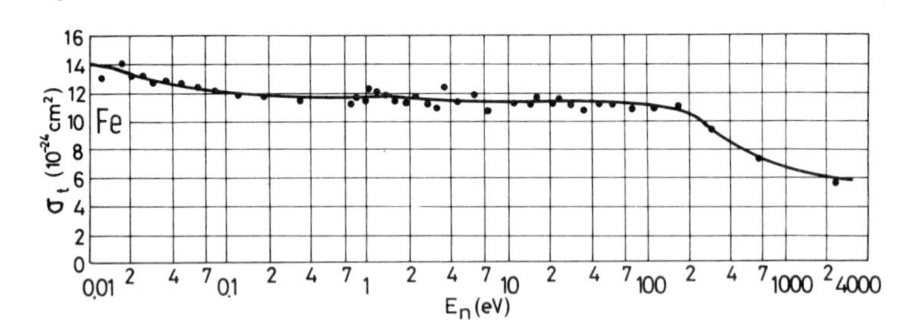


Fig. A.7.20

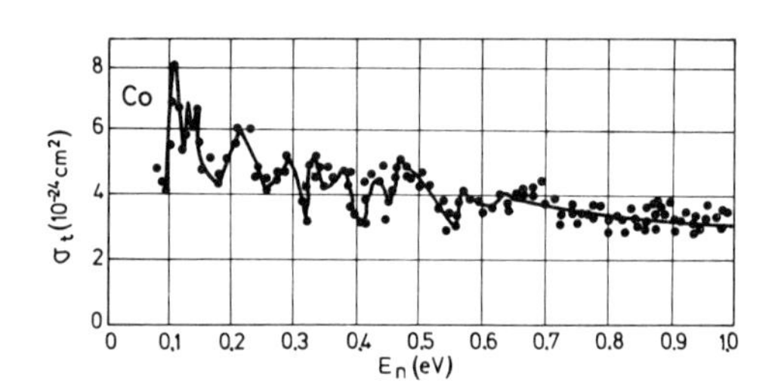


Fig. A.7.21

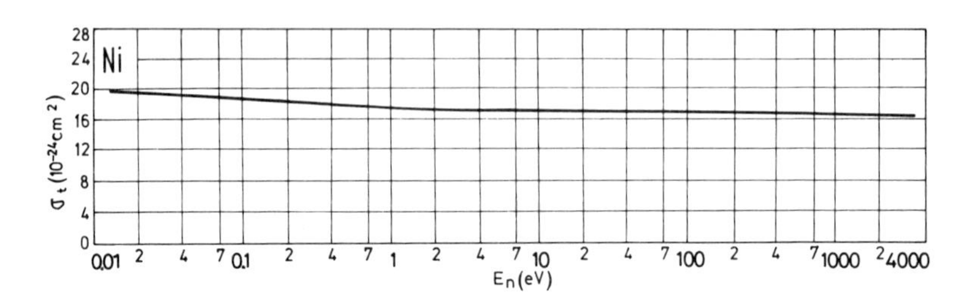


Fig. A.7.22

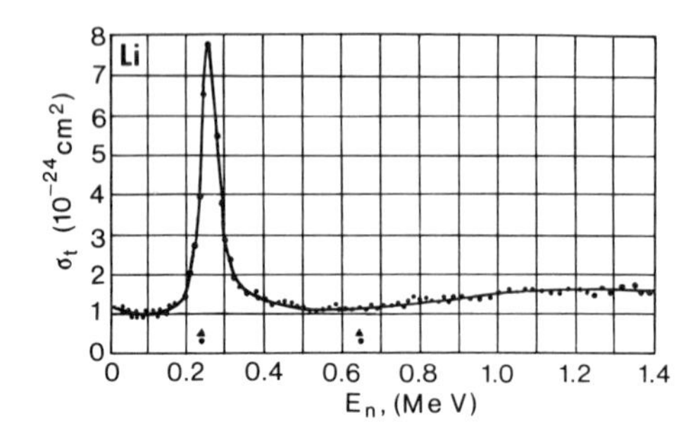


Fig. A.7.23

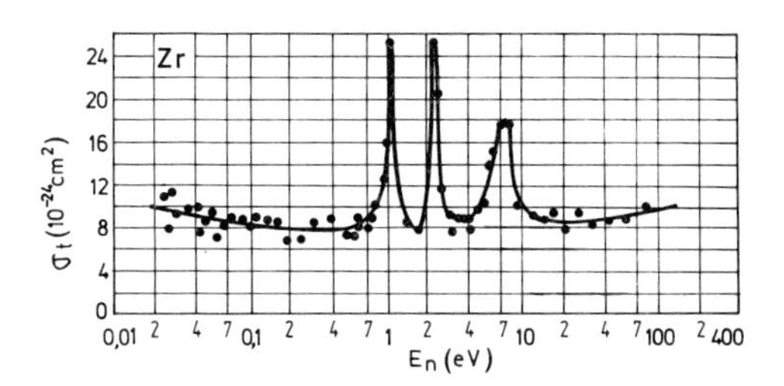


Fig. A.7.24

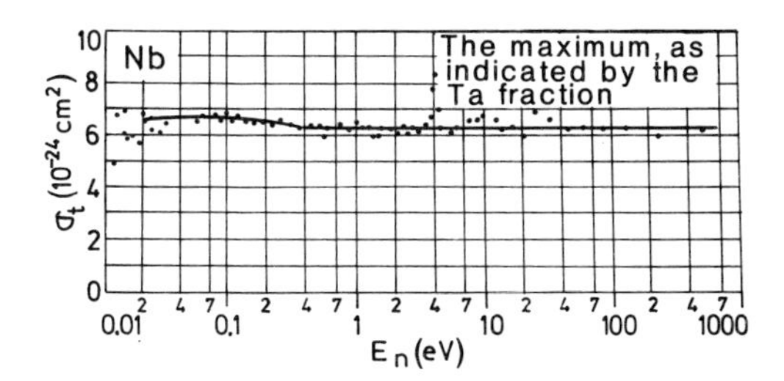


Fig. A.7.25

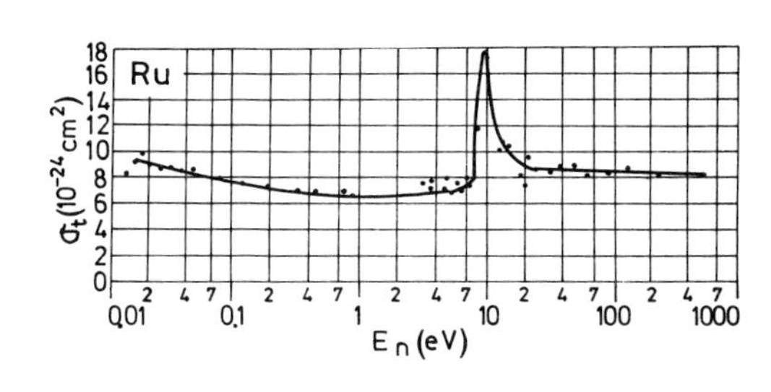


Fig. A.7.26

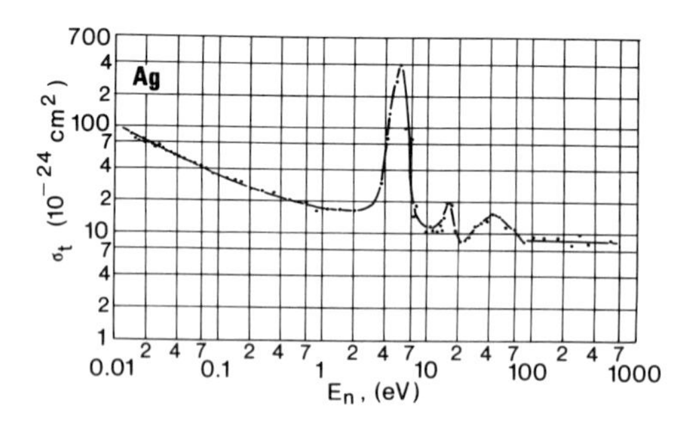


Fig. A.7.27

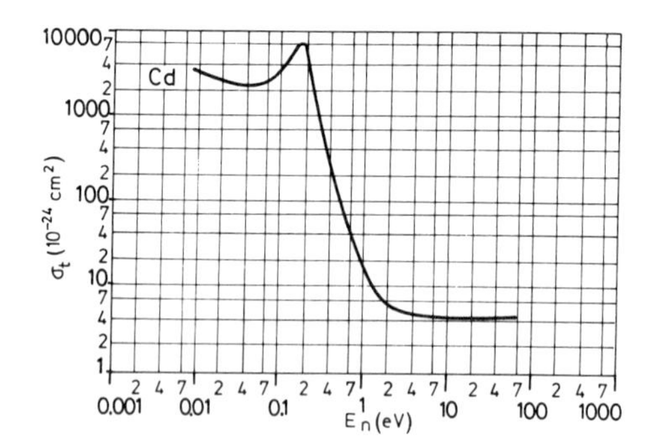


Fig. A.7.28

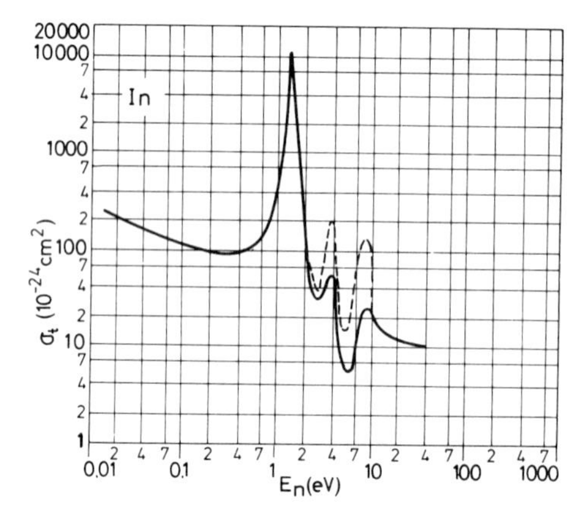


Fig. A.7.29

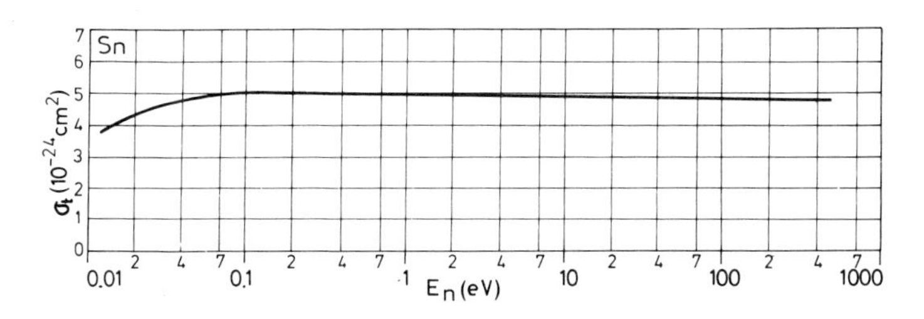


Fig. A.7.30

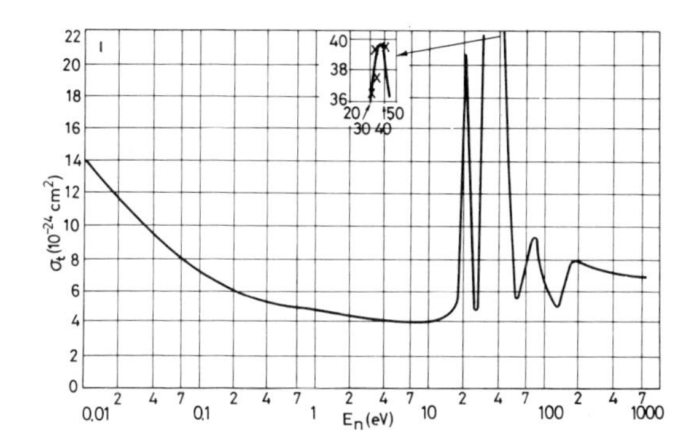


Fig. A.7.31

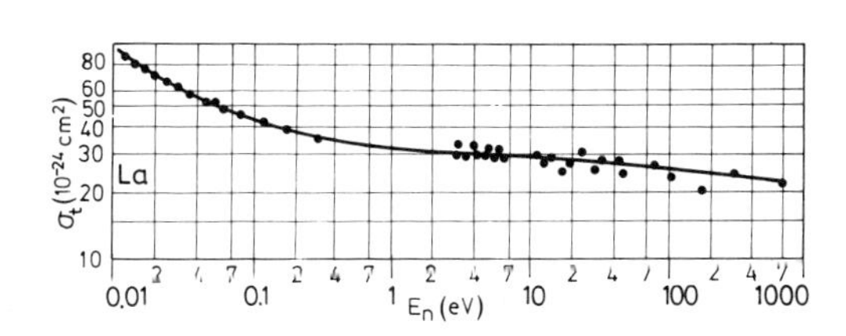


Fig. A.7.32

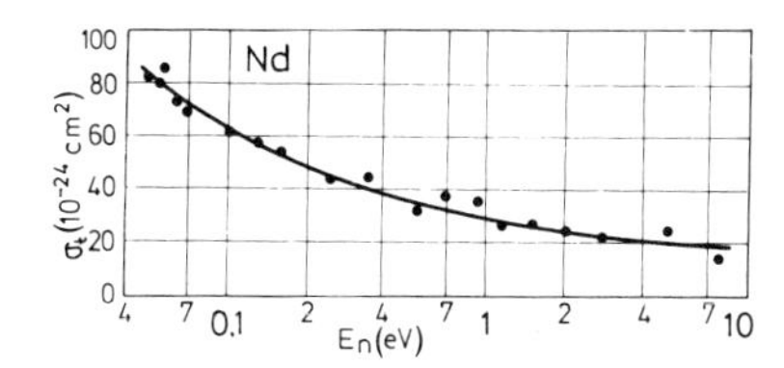


Fig. A.7.33

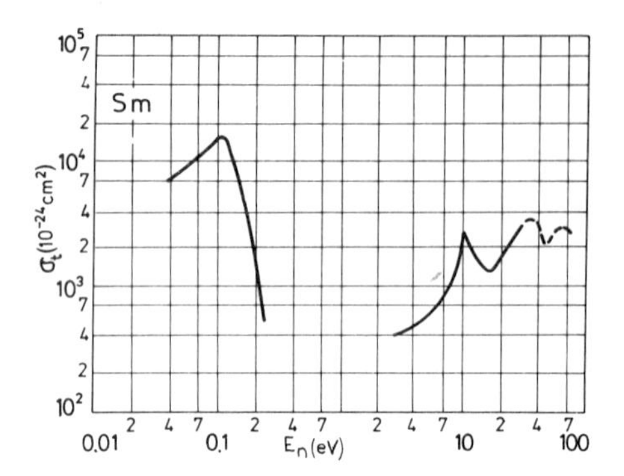


Fig. A.7.34

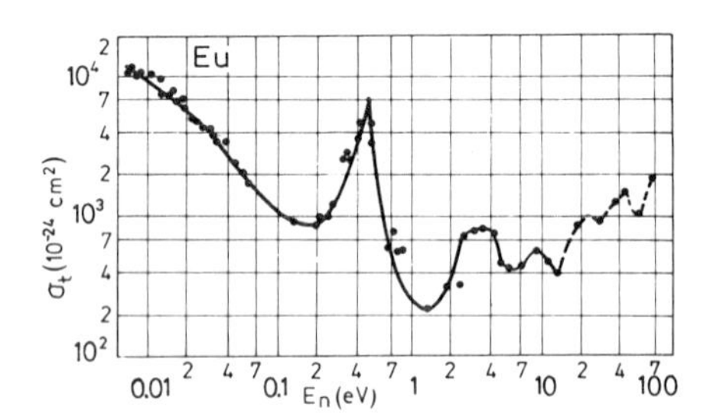


Fig. A.7.35

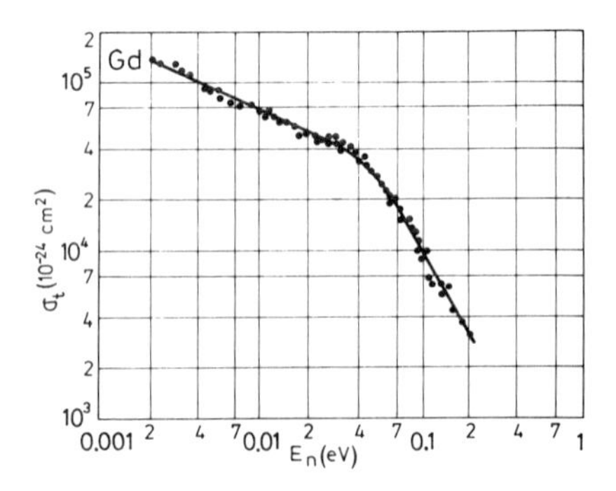


Fig. A.7.36

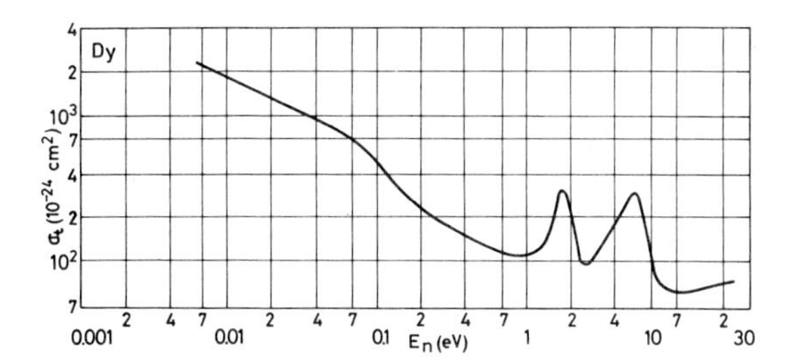


Fig. A.7.37

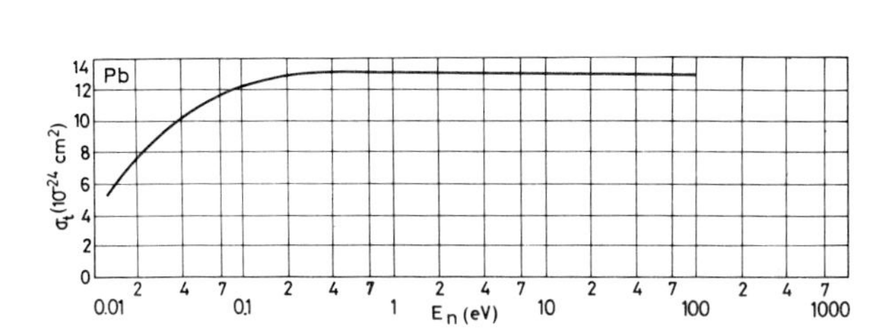


Fig. A.7.38

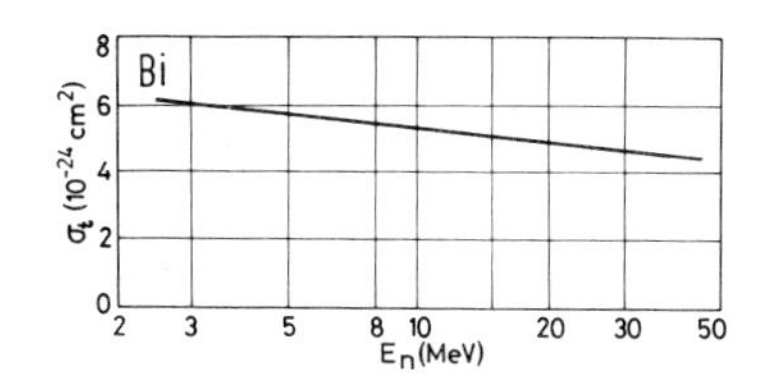


Fig. A.7.39

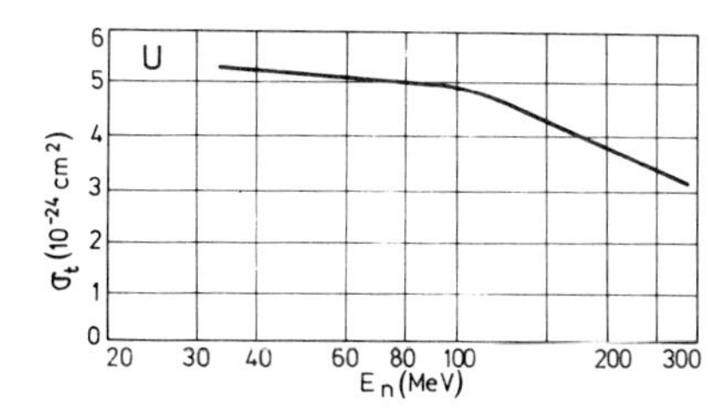


Fig. A.7.40

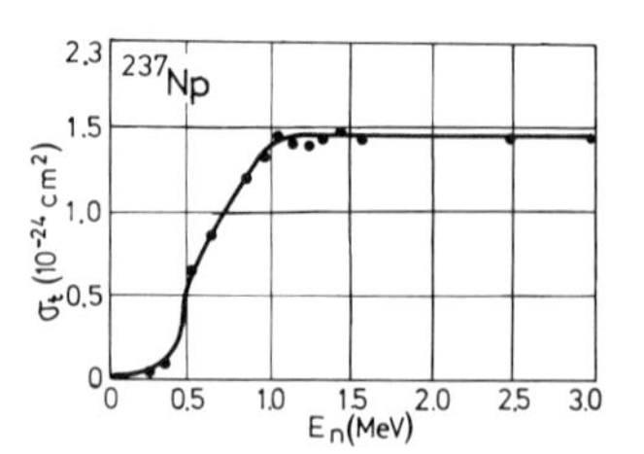


Fig. A.7.41

B

| | Capture cross-sections for thermal neutrons $\sigma_\gamma$ (b) | Scattering cross-sections $\sigma_s$ (b) | Capture integral, at resonance $I_\gamma$ (b) |
|---|---|---|---|
| H | 0.332 ± 0.002 | 20.436 ± 0.023 | — |
| D | 0.00053 ± 0.00002 | 3.390 ± 0.012 | — |
| He | < 0.05 | 0.76 ± 0.01 | — |
| Li | 70.7 ± 0.7 | — | — |
| Be | 0.0092 ± 0.001 | 6.14 ± 0.008 | 0.004 ± 0.001 |
| B | 759 ± 2 | 3.6 ± 0.2 | 341 ± 2 |
| C | 0.0034 ± 0.0002 | 4.75 ± 0.02 | 0.0015 ± 0.0002 |
| N | 1.85 ± 0.05 | 10.6 ± 0.5 | 0.90 ± 0.05 |
| O | 0.00027 ± 0.000025 | 3.76 ± 0.02 | 0.31 ± 0.04 |
| F | 0.0095 ± 0.0007 | 4.0 ± 0.1 | 0.0176 ± 0.003 |
| Ne | 0.038 ± 0.006 | 2.42 ± 0.01 | |
| Na | 0.400 ± 0.030 | 3.2 ± 0.2 | 0.311 ± 0.010 |
| | 0.530 ± 0.005 | | |
| Mg | 0.063 ± 0.003 | 3.416 ± 0.006 | 0.038 ± 0.004 |
| Al | 0.230 ± 0.003 | 1.49 ± 0.03 | 0.17 ± 0.01 |
| Si | 0.160 ± 0.02 | 2.2 ± 0.1 | — |
| P | 0.180 ± 0.007 | — | 0.08 ± 0.02 |
| S | 0.520 ± 0.030 | 0.975 ± 0.006 | |
| Cl | 33.2 ± 0.5 | — | 12 ± 2 |
| Ar | 0.678 ± 0.009 | 0.644 ± 0.004 | 0.42 ± 0.05 |
| K | 2.10 ± 0.10 | 1.5 ± 0.2 | 1.0 ± 0.1 |
| Ca | 0.43 ± 0.02 | — | 0.20 ± 0.02 |
| Sc | 26.5 ± 1.0 | 24 ± 2 | 11.3 ± 1.0 |
| Ti | 6.1 ± 0.2 | 4.0 ± 0.3 | — |
| V | 5.04 ± 0.04 | 4.93 ± 0.03 | 2.7 ± 0.1 |
| Cr | 3.1 ± 0.2 | 3.8 ± 0.3 | 1.7 ± 0.2 |
| Mn | 13.3 ± 0.2 | 2.1 ± 0.2 | 14.0 ± 0.4 |
| Fe | 2.55 ± 0.03 | 10.9 ± 0.2 | 1.4 ± 0.2 |
| Co | 37.2 ± 0.2 | 6.7 ± 0.3 | 75.5 ± 1.5 |
| Ni | 4.43 ± 0.16 | 17.3 ± 0.5 | 2.2 ± 0.2 |
| Cu | 3.79 ± 0.03 | 7.9 ± 0.2 | 3.2 ± 0.3 |
| Zn | 1.10 ± 0.04 | 4.2 ± 0.2 | 2.3 ± 0.3 |
| Ga | 2.9 ± 0.1 | 6.5 ± 0.2 | 18.7 ± 1.5 |
| Ge | 2.3 ± 0.2 | 7.5 ± 0.2 | 6.1 ± 1.0 |
| As | 4.3 ± 0.1 | 7.0 ± 1.0 | 60 ± 4 |
| Se | 11.7 ± 0.2 | 9.7 ± 1.0 | 13 ± 2 |
| Br | 6.8 ± 0.1 | 6.1 ± 0.2 | 90 ± 6 |
| Kr | 25 ± 1.0 | 7.5 ± 0.13 | 53 ± 7 |
| Rb | 0.37 ± 0.03 | 6.20 ± 0.3 | 6.0 ± 0.5 |
| Sr | 1.21 ± 0.06 | 10 ± 1 | 11 ± 2 |

Appendix 7 B (continued)

| | Capture cross-sections for thermal neutrons $\sigma_{\gamma}$ (b) | Scattering cross-sections $\sigma_s$ (b) | Capture integral, at resonance $I_{\gamma}$ (b) |
|---|---|---|---|
| Y | 1.28 ± 0.02 | 7.60 ± 0.06 | 1.0 ± 0.2 |
| Zr | 0.185 ± 0.003 | 6.40 ± 0.07 | 1.10 ± 0.15 |
| Nb | 1.15 ± 0.05 | | 8.5 ± 0.5 |
| Mo | 2.65 ± 0.08 | 5.8 ± 0.2 | 22 ± 2 |
| Ru | 2.56 ± 0.13 | — | 42 ± 4 |
| Rh | 150 ± 5 | | 1100 ± 50 |
| Pd | 6.9 ± 0.4 | 5.0 ± 0.3 | 90 ± 5 |
| Ag | 63.6 ± 0.6 | — | 747 ± 20 |
| Cd | 2450 ± 20 | 5.6 ± 0.3 | |
| In | 193.5 ± 1.5 | | 3200 ± 50 |
| Sn | 0.63 ± 0.01 | | 8.5 ± 2.0 |
| Sb | 5.4 ± 0.6 | 4.2 ± 0.2 | 175 ± 10 |
| Te | 4.7 ± 0.1 | | 54 ± 3 |
| I | 6.2 ± 0.2 | | 147 ± 6 |
| Xe | 24.5 ± 1.0 | 4.30 ± 0.02 | |
| Cs | 29.0 ± 1.5 | | 415 ± 15 |
| Ba | 1.2 ± 0.1 | | 7.5 ± 1.0 |
| La | 9.0 ± 0.3 | 9.3 ± 0.7 | 12.2 ± 0.6 |
| Ce | 0.63 ± 0.15 | 4.7 ± 0.3 | 3.0 ± 0.8 |
| Pr | 11.50 ± 0.3 | 3.3 ± 0.5 | 14.1 ± 0.2 |
| Nd | 50.5 ± 2.0 | 16 ± 1 | 45 ± 5 |
| Sm | 5800 ± 100 | | 1400 ± 200 |
| Eu | 4600 ± 100 | 8.0 ± 1.0 | 2430 ± 200 |
| Gd | 49,000 ± 1000 | | 390 ± 10 |
| Tb | 25.5 ± 1.1 | 20 ± 2 | 430 ± 40 |
| Dy | 930 ± 20 | 100 ± 10 | 1600 ± 100 |
| Ho | 66.5 ± 3.3 | 9.4 ± 0.2 | 700 ± 20 |
| Er | 162 ± 8 | 11.0 ± 0.8 | 740 ± 10 |
| Tm | 103 ± 3 | 12 ± 2 | 1720 ± 40 |
| Yb | 36.6 ± 2.0 | 25.0 ± 0.8 | 182 ± 10 |
| Lu | 77 ± 3 | 8 ± 2 | 900 ± 50 |
| Hf | 102 ± 2 | 8 ± 2 | 2000 ± 100 |
| Ta | 21.0 ± 0.7 | 6.2 ± 0.6 | 710 ± 20 |
| W | 18.5 ± 0.5 | | 352 ± 30 |
| Re | 88 ± 4 | 11.3 ± 0.5 | 850 ± 50 |
| Os | 15.3 ± 0.7 | | 210 ± 10 |
| Ir | 426 ± 4 | 14 ± 2 | 2250 ± 200 |
| Pt | 10.0 ± 0.2 | 11.2 ± 1.0 | 140 ± 6 |
| Au | 98.8 ± 0.3 | | 1560 ± 40 |
| Hg | 375 ± 5 | | 73 ± 10 |
| Tl | 3.4 ± 0.5 | 9.7 ± 0.4 | 12 ± 1 |
| Pb | 0.170 ± 0.002 | 11.4 ± 0.2 | 0.16 ± 0.04 |
| Bi | 0.033 ± 0.004 | | 0.19 ± 0.03 |
| Th | 7.40 ± 0.08 | 12.67 ± 0.08 | 85 ± 3 |
| $^{235}U$ | 98.6 ± 1.5 | 13.8 ± 0.5 | 114 ± 6 |
| $^{238}U$ | 2.70 ± 0.02 | 8.90 ± 0.16 | 275 ± 5 |

REFERENCES

Mugbabghab, S. F., and Garben, D. I. (1973). BNL, 3rd edition, Vol. II.
Adair, R. (1950). Rev.Mod.Phys., 22, 249.
Maikova, E. I. (Ed.) (1951). *Neitronnye effectivnye sechenya elementov.* Izd.Inostrannoy Literatury, Moskow.

# APPENDIX 8

## I. QUALITY REQUIREMENTS FOR HEAVY WATER USED AS MODERATOR IN NUCLEAR REACTORS

The minimal requirements on heavy water quality enabling its use as moderator in nuclear reactors are specified, *inter alia*, in the US Standard ASTM D.2032-68 (revised in 1975); in comparison to those currently practised in Romania, these are:

| *Characteristic* | *Accepted levels as per* | |
|---|---|---|
| | *ASTM D.2032-68 (rev.1975)* | *Romanian norms* |
| 1. Deuterium isotopic content, percentage in atoms[1] | ≥ 99.75 | ≥ 99.75 |
| 2. Electric conductivity at 25°C (µS/cm) | < 15 | < 15 |
| 3. $KMnO_4$ (g/ml) | $< 1 \times 10^{-5}$ | $< 1 \times 10^{-5}$ |
| 4. Slit charge, ppm $SiO_2$ | < 5 | < 5 |
| 5. Total solids, ppm | < 12 | < 5 |
| 6. Chlorides, ppm | < 0.1 | < 0.1 |
| 7. pD | - | 6.5 – 7.5 |
| 8. $D_2O$ to $H_2O$ specific weight, at 25 °C | - | ≥ 1.1075 |

[1] Deuterium isotopic content expressed as 100 . D/(H+D).

## A. The Isotopic Content in Deuterium

*Purpose*: isotopic title determination.

*Measuring devices* : IR spectrophotometer; digital densimeter, pycnometer; nuclear magnetic resonance spectrometer; mass spectrometer; optical interferometer; refractometer.

*Analysis methods*: regardless of analysis method, in handling and storing heavy water samples special precautions are in order to avoid the degradation of the isotopic title by absorption of ordinary water from the atmosphere and the vessel, or of other properties, such as: turbidity, electric conductivity, content in organic substances, etc.

1. Infra-red absorption spectrophotometry. HDO absorption line intensity at 2.9 μm ( ~ 3400 $cm^{-1}$) is measured and compared to the intensity of the line that is not affected by this absorption, at 2.6 μm ( ~ 3800 $cm^{-1}$). To establish the instrument's sensitivity heavy water standards of known concentration are used. The accuracy achieved is 0.01% D/(D+H). The amount required in sample is 0.5 ml; the duration of analysis is 5 min.

2. Densimetry. The frequency of the proper oscillation mode of a glass tube containing a sample of 1 ml of water is measured. The frequency — concentration relation is established, observing corrections for ambient humidity and pressure with and without the sample in the tube, during analysis. For a digital densimeter, the achieved accuracy is 0.02% D/(D+H) and the duration of analysis is 30 min.

3. Picnometry. Density is determined with pycnometers and a $10^{-6}$ g-precision balance. Controlled environment and special handling procedures are in order. Accuracy as obtained is 0.02% D/(D+H); the amount in the sample is 30 ml; and the analysis duration is 60 min. The result is an absolute value.

4. Nuclear magnetic resonance spectrometry. The sample is placed inside a tube that is co-axial with a second tube containing the reference specimen, both being placed in the magnetic field of a high-resolution NMR facility. The minimum volume required in the sample is 1 ml. The NMR spectrometer should be equipped with a data storage and processing system. Accuracy is 0.02% D/(D+H), and the duration of analysis is 15 min.

5. Mass spectrometry. The $D_2O$ — $H_2O$ — HDO liquid phase is reduced on a hot metal, usually uranium (600 °C) or zinc (450 °C). With a double focusing magnetic analyser the composition of the resulted $H_2$, HD, $D_2$

gaseous mixture is determined, in comparison with that of reference samples. Accuracy is 0.01% D/(D+H); necessary amount in the sample is 1 ml; and the analysis duration is 30 min.

6. Optic interferometry. One measures the difference in the optical path of a light beam that passes through the sample in comparison with a reference one. This is related to the $D_2O$ sample concentration. With ordinary equipment the accuracy is 0.02% D/(D+H). The sample should contain 20 ml; the analysis duration is 20 min. Informations and data provided: heavy water concentration $c \geqq 99.75\%$.

B. Electric Conductivity

*Purpose* : heavy water impurification level determination.

*Gauge* : conductivemeter, of a precision of ± 0.1 μS/cm in the 0-20 μS/cm range.

*Measurement method* : as per ASTM D.1125; in Romania equally valid is STAS* 7722-68, observing the norms in effect concerning handling of high-purity heavy water.

*Information and data provided* : the electric conductivity of heavy water, typically < 15 μS/cm at 25 °C.

C. Determination of Content in Organic Substances

*Purpose* : evidencing the content in organic carbon of heavy water, either directly or expressed via the consumption of $KMnO_4$.

*Gauge* : TOC (total organic carbon) analyser and other appropriate chemical instrumentation.

*Measurement method* : the chemical method is applied as per ASTM D.2033 or STAS 3002/61, observing the norms regulating the handling of high-purity heavy water. In the TOC analyser, heavy water is passed through a combustion reactor, then through a methanation reactor, data being gathered with an infra-red detector. The accuracy is 0.25 mg/l, and the duration of analysis is 5 min.

*Information and data provided* : the equivalent $KMnO_4$ consumption, typically $< 1 \times 10^{-5}$ g/ml.

---

*STAS — Government-enacted standard.

## D. Turbidity

*Purpose*: heavy water turbidity determination.

*Gauge*: turbidimeter with a $\pm 0.5$ ppm accuracy in the 0-5 ppm range. A spectrophotometer equipped with a nephelometric device may also be used.

*Measurement method*: as per ASTM D.1889 or STAS 6323/76, observing the norms regulating handling of high-purity heavy water. The duration of analysis is 5 min.

*Information and data provided*: heavy water turbidity, typically $< 5$ ppm $SiO_2$.

## E. Total Content in Solids

*Purpose*: determination of the total content in solid substances of the heavy water.

*Gauge*: analytic balance, $\pm 10^{-6}$ accuracy, in a thermostabilized chamber, at $105 \pm 2$ °C.

*Measurement method*: the content in solid substances is measured, as per ASTM D.1888 or STAS 6953-64 — observing the norms regulating handling of high-purity heavy water. Duration of analysis is 30 min.

*Information and data provided*: the content in solid substances is obtained; for the heavy water made in Romania, typically $< 5$ ppm.

## F. Content in Chlorides

*Purpose*: measurement of the total content in chlorides of the heavy water.

*Working device*: a customized ionometer.

*Working method*: as per ASTM D.512 or STAS 6363-75. Duration of analysis is 10 min.

*Information and data provided*: the content in chlorides, typically $< 0.1$ ppm.

## G. pD

*Purpose*: determination of deuterium ions concentration.

*Gauge*: pD meter, $\pm 0.1$ pD.

*Measurement method*: as per STAS 6325-75.

*Information and data provided*: for the heavy water made in Romania the pD ranges between 6.5 and 7.5 pD units.

II. MEASUREMENT METHODS AND PERFORMANCES IN DETERMINING

| | *Measurement method* | *Substance* | *Aggregation state* | *Concentration range* | *Type of measurement* |
|---|---|---|---|---|---|
| *0* | *1* | *2* | *3* | *4* | *5* |
| 1 | Mass spectrometry | $H_2$ | Gas | 50 – 150 ppm | With on-line computer data processing |
| | | | | | Without computer |
| | | | | 150 – 1000 ppm | Discontinuous |
| | | | | 1000 ppm – 1% | Discontinuous |
| | | $H_2O$ | Liquid | 50 – 150 ppm | With on-line computer data processing |
| | | | | | Without computer |
| | | | | 150 – 1000 ppm | Discontinuous |
| | | | | 1000 ppm – 1% | Discontinuous |
| | | $H_2S$ | Gas | 50 ppm – 1% | With on-line computer data processing |
| | | | | | Without computer |
| | | $NH_3$ | Liquid | 50 ppm – 1% | With on-line computer data processing |
| | | | | | Without computer |
| 2 | Infrared spectrophotometry | $H_2O$ | Liquid | 145 – 155 ppm | Continuous |
| | | | | 0.1 – 0.3% | Continuous |
| | | | | 0.5 – 2% | Continuous |
| | | | | 99 – 99.6% | Continuous |
| | | | | 99.6 – 99.9% | Continuous |
| | | | | 99.6 – 99.9% | Discontinuous |
| | | $H_2S$, $H_2$<br>$NH_3$ | Gas<br>Liquid | 99.6 – 99.9% | Discontinuous |
| 3 | Vibratory densimeter | $H_2O$ | Liquid | 1 – 99% | Discontinuous |
| | | $H_2$, $H_2S$<br>$NH_3$ | Gas<br>Liquid | 1 – 99% | Discontinuous |

DEUTERIUM CONCENTRATION IN HYDROGEN, $H_2O$, $NH_3$ AND $H_2S$

| *Accuracy* | *Sample volume* | *Duration of analysis* | *Sample preparation* | *Source of perturbation* |
|---|---|---|---|---|
| *6* | *7* | *8* | *9* | *10* |
| ± 0.05 ppm | 0.5 l NTP | 10 min | Drying + solid impurities retention | Traces of water |
| ± 0.1 ppm | 0.5 l NTP | 20 min | Drying + solid impurities retention | |
| ± 0.5%* | 0.5 l NTP | 20 min | Drying + solid impurities retention | |
| ± 1%* | 0.5 l NTP | 30 min | Drying + solid impurities retention | |
| ± 0.05 ppm | 1 µl | 20-40 min | Elimination of dissolved gases, vacuum purification and total conversion to $H_2$ | "Isotopic memory" affecting analysis duration when the sample is changed |
| ± 0.1 ppm | | | | |
| ± 0.5%* | | | | |
| ± 1%* | | | | |
| ± 1%* | 0.5 l NTP | 15 min | Drying, total convertion to $H_2$, or combustion to $H_2O$ and neutralization | Corrosion of electric contacts, toxicity |
| | | 30 min | | |
| ± 1%* | 1 ml | 20 min | Phase transition and conversion to $H_2$, or combustion to $H_2O$ and neutralization | Catalyst transport throughout circuits |
| | | 40 min | | |
| ± 10 ppm | 10 ml/min | 30 s | Retention of impurities in suspension and thermostabilization at ± 0.01 °C | Interference, presence of some salts |
| ± 0.01% | 10 ml/min | | | |
| ± 0.05% | 10 ml/min | | | |
| ± 0.02% | 10 ml/min | | | |
| ± 0.005% | 10 ml/min | | | |
| ± 0.01% | 0.5 ml | 5 min | | |
| ± 0.01% | 2 l NTP | 30 min | Combustion to $H_2O$, neutralization, vacuum purification | |
| ± 0.02% | 2 ml | 30 min | Purification, thermostabilization at ± 0.01 °C, normalization over 90% | Any kind of impurity, $H_2O^{18}$ traces |
| ± 0.02% | 2 l | 50 min | Combustion, neutralization, vacuum purification | Vacuum quality (in vacuum purification) |

*Percentage of measured value

Appendix 8.II (continued)

| | *Measurement method* | *Substance* | *Aggregation state* | *Concentration range* | *Type of measurement* |
|---|---|---|---|---|---|
| *0* | *1* | *2* | *3* | *4* | *5* |
| 4 | Gravimetry | $H_2O$ | Liquid | 1 – 99.5%<br>99.5 – 99.9% | Discontinuous<br>Discontinuous |
| | | $H_2,H_2S$<br>$NH_3$ | Gas<br>Liquid | 1 – 99.5%<br>99.5 – 99.9% | Discontinuous<br>Discontinuous |
| 5 | Refractometry | $H_2O$<br>$H_2,H_2S$ | Liquid<br>Gas | 1 – 99%<br>95 – 99.5% | Discontinuous<br>Continuous |
| | | $NH_3$ | Liquid | 1 – 99% | Discontinuous |
| 6 | Nuclear magnetic resonance | $H_2O$ | Liquid | 99.6 – 99.9% | Discontinuous |
| 7 | Gas chromatography | $H_2$<br>$H_2O$ | Gas<br>Liquid | 1 – 20%<br>1 – 20% | Discontinuous<br>Discontinuous |
| 8 | Floating gauge densimetry | $H_2O$ | Liquid | 99.6 – 99.9% | Continuous |
| 9 | Optic interferometry | $H_2O$ | Liquid | 1 – 99.5%<br>99.5 – 99.9% | Discontinuous<br>Discontinuous |
| 10 | Emission spectroscopy | $H_2$ | Gas | 2 – 90% | Discontinuous |

Among the gauges employed to determine deuterium concentration in $H_2O$, $H_2$, $NH_3$ and $H_2S$, as presented in Table II, accuracy recommends the following ones:

1. *In the 50 – 150 ppm range,* for liquid $H_2O$ – the mass-spectrometer, customized for deuterium determinations (covering 2 – 3 masses), equipped with facilities for conversion to $H_2$ and computer data processing.

2. *In the 1 – 5% range,* for gaseous $H_2$ and liquid $NH_3$ – the mass spectrometer, customized for deuterium determinations, equipped with special ion-collecting facilities at concentrations over 1% D and – for $NH_3$ – with a unit for conversion to $H_2$; a desk-top computer can take care of data acquisition and processing, and of the instrument's control.

3. *In the 1 – 99% range,* for $H_2O$, $H_2$, $NH_3$, $H_2S$ the vibratory densimeter is recommended, that provides a good accuracy; the sample must be in liquid

| *Accuracy* | *Sample volume* | *Duration of analysis* | *Sample preparation* | *Source of perturbation* |
|---|---|---|---|---|
| *6* | *7* | *8* | *9* | *10* |
| ± 0.1%<br>± 0.02% | 10 ml<br>30 ml | 60 min<br>60 min | Purification, thermostabilization at ± 0.01 °C, normalization over 90% | Any kind of impurity, $^{18}OH_2$ traces |
| ± 0.1%<br>± 0.02% | 2 l | 100 min | Combustion, neutralization, vacuum purification | Any kind of impurity $^{18}OH_2$ traces |
| ± 0.3%<br>± 0.06% | 1 ml<br>10 ml/min | 5 min<br>30 s | Thermal prism thermostabilization at ± 0.01 °C | |
| ± 0.3% | 2 l | 40 min | Combustion, neutralization, vacuum purification | |
| ± 0.02% | 1 ml | 15 min | Thermostabilization | Impurities, interference |
| ± 1%<br>± 1.5% | 500 µl<br>2 ml | 10 min<br>40 min | Conversion to $H_2$, on $CaH_2$ | |
| ± 0.01% | 10 ml/min | 30 s | Thermostabilization at ± 0.01 °C; securing a high purity | Presence of $^{18}OH_2$ requires separate corrections |
| ± 0.1%<br>± 0.02% | 10 ml<br>10 ml | 20 min<br>20 min | Vacuum purification, thermostabilization, normalization over 90% | |
| ± 0.1% | 2 l | 40 min | Vacuum purification | |

state (water) and hence the $H_2$, $NH_3$, $H_2S$ analysis requires prior combustion; neutralization (for $NH_3$ and $H_2S$); then vacuum purification by distillation; if the primary object is $H_2O$, the densimeter can be coupled on-line, at key check-points in the isotopic separation installation.

4. *In the 99 – 100% range* , convenient for $NH_3$, $H_2$ and $H_2O$ is the infrared spectrophotometer, tuned on the wavelength of 2.6 – 2.9 µm, that can operate either when required or on-line (for water); sample must be in liquid state (water) and, therefore, prior combustion of $NH_3$ and $H_2$ in oxygen is in order.

This instrumentation is rather sophisticated and requires laboratory conditions. The simplest apparatus for isotopic determinations on a wide range, and with an acceptable accuracy when it comes to fast investigations, is the immersion refractometer; it has also the advantage of being portable.

III. MEASUREMENT METHODS AND CURRENT PERFORMANCES, FOR THE DETERMINATION OF THE $H_2/N_2$ RATIO IN GASEOUS PHASE, IN THE 3:1 RANGE.

| *Measurement method* | *Measurement type* | *$N_2$ determination accuracy* | *Sample volume* | *Duration of analysis* | *Restrictive conditions* |
|---|---|---|---|---|---|
| 1. Chromatography in gaseous phase | Discontinuous | ±0.1% $N_2$ | 500 μl NTP | 15 min | |
| 2. Industrial analyser | Continuous | ±0.2% $N_2$ | 25 l/h | - | |
| 3. Optic interferometry | Discontinuous | ±0.05% $N_2$ | 2 l NTP | 40 min | Presence of two gases only: $N_2$ and $H_2$ |

Chromatography in gaseous phase and the use of industrial analysers equipped with thermal conductivity cells are recommended. The method based on optic interferometry is sophisticated and demanding.

## IV. TURBINE FLOWMETERS

Flow measurement with turbine flowmeters has gained prominence, especially since 1945. Presently, the method is extensively used both in grass-root and top industries (aircraft, space, nuclear, etc.), since it offers several advantages, such as:

1. Superior performances: 0.25% accuracy; 0.025% reproducibility; ±0.5% linearity, over a wide range.

2. Fast response.

3. Easy conversion of the information to digital form.

4. Good stability of calibration.

Turbine flowmeter manufacturers must provide individual calibration to each item delivered, which comes with its own characteristic flow coefficient, including its variation with fluid viscosity (the universal viscosity curve). In this way accuracy is guaranteed.

Also, by improvements with bearings, rotors, materials and lubrication, as well as by the advent of bearing-free rotors, the variety of fluids that can be inspected with these devices widened (including corrosive and cryogenic fluids), together with the measurable flow ranges, and the maximum pressure accessible increased (above 20 MPa).

The following table hints at the best-suited methods and devices, for various flow ranges.

| *Fluid* | *Flow range* | *Measurement method* |
|---|---|---|
| 1. Water | 200 – 400 $m^3/h$ | Differential pressure flowmeters (D.P.) with: diaphragms, nozzles, Venturi tubes, turbine flowmeters |
| 2. Liquid ammonia | 5 – 1000 kg/h | Turbine flowmeters<br>Encased metallic rotameters with magnetic transmission |
| | 1000 – 10,000 kg/h | Turbine flowmeters<br>D.P. flowmeters with: diaphragms, nozzles, Venturi tubes |
| 3. Liquid hydrogen | 500 – 1000 kg/h | Turbine flowmeters |
| Gas, high pressure (100-300 bar) | 500 – 1000 kg/h | Turbine flowmeters<br>Encased metallic rotameters with magnetic transmission |
| Gas, low pressure (10 bar) | 500 – 1000 kg/h | Rotameters<br>D.P. flowmeters with: diaphragms, nozzles, Venturi tubes<br>Annubar flowmeters |

REFERENCES

Steflea, D., and Pavelescu, M. (1977). Report ICEFIZ.
Kirillin, V. A. and others (1963). *Tyazhelaya Voda, Teplofizitcheskye svoystva*. Gosenergoizdat, Moskow.
Pavelescu, M. (1977). Report ICEFIZ.
Ursu, I., Peculea, M., Simplaceanu, V., and Demco, E. (1976). Metoda de determinare a imbogatirii izotopice a apei grele. Romanian Patent no 67839/1978.
ASTM (1976). *Annual Book of ASTM Standards*, Part 31, *Water*, American Society for Testing and Materials, Philadelphia.
Stanciu, V. (1977). Report ICEFIZ.
Stefanescu, D. (1977). Report ICEFIZ.
Ursu, I., Peculea, M., Pavelescu, M., Steflea, D., Stanciu, V., Hirean, I., and Demco, D. (1982). Report AIEA no 3036/RB.

## APPENDIX 9

# The main fission products of thermal neutrons-irradiated uranium, which determine the fuel radioactivity

| *A* | *Z* | *Element* | *Fission yield* (%) | *Half-life* | *Radiation energy* (MeV) | | *Post-irradiation activity** (Ci/tU) | |
|---|---|---|---|---|---|---|---|---|
| | | | | | β | γ | β | γ |
| 85 | 36 | Krypton | 0.3 | 10.3 yr | 0.7 | - | 500 | - |
| 89 | 38 | Strontium | 4.7 | 53 d | 1.5 | - | | - |
| 90 | 38 | Strontium | 5.8 | 28 yr | 0.6 | - | 45,000 | - |
| 90 | 39 | Yttrium[1] | - | 64 h | 2.2 | - | | - |
| 91 | 39 | Yttrium | 5.8 | 60 d | 1.5 | - | 60,000 | - |
| 95 | 40 | Zirconium | 6.3 | 63 d | 0.4 | 0.7 | 70,000 | 65,000 |
| 95 | 41 | Niobium[1] | - | 35 d | 0.16 | 0.75 | 110,000 | 105,000 |
| 99 | 42 | Molybdenum | 6.1 | 67 h | 1.2 | 0.04-0.8 | - | - |
| 99 | 43 | Technetium[1] | - | $2x10^5$ yr | 0.3 | - | - | - |
| 103 | 44 | Ruthenium | 2.9 | 40 d | 0.22 | 0.5 | 55,000 | 20,000 |
| 106 | 44 | Ruthenium | 0.4 | 1 yr | 0.04 | - | | - |
| 106 | 45 | Rhodium[1] | - | 30 s | 3.5 | 0.5 | - | - |
| 129 | 52 | Tellurium[2] | 0.3 | 33 d | - | 0.11 | - | - |
| 131 | 53 | Iodine | 3.0 | 8 d | 0.6 | 0.36 | - | - |
| 133 | 54 | Xenon | 6.5 | 5.3 d | 0.34 | 0.08 | 3000 | - |
| 137 | 55 | Cesium | 6.2 | 30 yr | 0.5 | - | 2300 | - |
| 137 | 56 | Barium[1,2] | - | 2.6 min | - | 0.66 | - | - |
| 140 | 56 | Barium | 6.3 | 12.8 d | 1.0 | 0.16-0.5 | - | - |
| 140 | 57 | Lanthanum[1] | - | 40 h | 0.8-2.2 | 0.1-2.5 | - | - |
| 141 | 58 | Cerium | 5.7 | 33 d | 0.4 and 0.6 | 0.14 | 170,000 | 12,000 |
| 144 | 58 | Cerium | 6.0 | 284 d | 0.17-0.3 | 0.03-0.13 | | - |
| 144 | 59 | Praseodymium[1] | - | 17.5 min | 3.0 | 0.7-2.2 | | - |
| 143 | 59 | Praseodymium | 6.2 | 13.7 d | 0.9 | - | 12,000 | |
| 147 | 60 | Neodymium | 2.6 | 11.3 d | 0.4-0.8 | 0.1-0.5 | | - |
| 147 | 61 | Prometium[1] | - | 2.6 yr | 0.22 | | | |

* After an irradiation at 1000 MWd and 100 days of decay.

[1] Direct offsprings, from the β-decay of the isotopes immediately above in the table.

[2] $^{129}$Te and $^{137}$Ba do not undergo β-decay; following an internal rearrangement of nucleons a γ-de-excitation takes place.

REFERENCE

Sauteron, J. (1965). *Les combustibles nucléaires*. Ed. Hermann, Paris.

APPENDIX 10

# Crystalline structures and magnetic properties of some uranium compounds

| *Compound* | *Lattice type* | *Lattice constant* ($10^{-7}$ m) | *Curie temperature* $T_C$ (K) | *Paramagnetic Curie temperature* θ(K) | *Magnetic moment* ($\mu_B$) | |
|---|---|---|---|---|---|---|
| | | | | | *Effective* | 4.2 K |
| Ferromagnetic | | | | | | |
| α-$UH_3$ | $BiF_3$ | 41.60 | 182 | 174 | 2.8 | 0.9 |
| β-$UH_3$ | β-W | 66.31 | 170 | 173 | 2.44 | 0.9-1.2 |
| β-$UD_3$ | β-W | 66.20 | 169 | 177 | 2.44 | 0.98 |
| US | NaCl | 54.86 | 176 | 173-180 | 2.22-2.31 | 1.20-1.76 |
| UTe | NaCl | 61.51 | 105 | 104 | 2.84 | 1.10-2.20 |
| β-$U_2N_3$ | $La_2O_3$ | $a$ = 37.00<br>$c$ = 58.26 | 186 | | 1.94 | |
| $UGa_2$ | $AlB_2$ | $a$ = 42.38<br>$c$ = 46.64 | 130 | 125 | 3.25 | 1.50 |
| $UGe_2$ | $ZrSi_2$ | $a$ = 40.60<br>$b$ = 158.00<br>$c$ = 39.70 | 52 | -89 | 3.03 | 0.8 |
| $UFe_2$ | $MgCu_2$ | 70.65 | 160 | 170 | 2.0 | U = 0.06<br>Fe = 0.59 |
| $UNi_2$ | $MgZn_2$ | $a$ = 49.59<br>$c$ = 82.45 | 30 | 45 | 2.5 | 0.12 |
| Antiferromagnetic | | | | | | |
| UN | NaCl | 48.895 | 53 | -320 | 0.75 | 3.1 |
| USb | NaCl | 61.805 | 230 | 95 | 2.8 | 3.8 |
| $UO_2$ | $CaF_2$ | 54.691 | 30.8 | -220 | 1.8 | 3.2 |
| $UPb_3$ | $AuCu_3$ | 47.91 | 32 | -170 | - | 3.2 |

Appendix 10 (continued)

| *Compound* | *Lattice type* | *Lattice constant* ($10^{-7}$ m) | *Curie temperature* $T_C$ (K) | *Paramagnetic Curie temperature* θ(K) | *Magnetic moment* ($\mu_B$) *Effective* | 4.2 K |
|---|---|---|---|---|---|---|
| Paramagnetic | | | | | | |
| UC | NaCl | 49.56 | - | - | - | - |
| $UAl_3$ | $AuCu_3$ | 42.87 | - | - | - | - |
| $USi_3$ | $AuCu_3$ | 40.40 | - | - | - | - |
| $UGe_3$ | $AuCu_3$ | 41.96 | - | - | - | - |
| $UF_6$ | Pnma | *a* = 99<br>*b* = 89.62<br>*c* = 52.07 | | | | |
| Diamagnetic | | | | | | |
| $UO_2(NO_3)_2 \cdot 6H_2O$ | Cmc $2_1$ | *a* = 131.97<br>*b* = 80.35<br>*c* = 114.67 | | | | |
| $UO_2SO_4 \cdot 3H_2O$ | Orthorhombic | *a* = 125.8<br>*b* = 170.0<br>*c* = 67.3 | | | | |

## REFERENCES


Ursu, I. (1979). *Rezonanta magnetica in compusi cu uraniu*, Edit. Academiei, Bucharest.

Brodsky, M. B. (1978). *Rep.Progr.Phys.*, 41, 1457.

Vainstein, B. K., Chernov, A. A., and Shuvalov, L. A. (1979*, 1980**). *Sovremennaya krystallographya*. Vol. 1 and 2*, Vol. 3**. Izd.Nauka, Moskow.

APPENDIX 11

# Properties of materials, of chief interest in fuel element design

| *Property* | *Material* | *Parametric expressions, symbols and units* |
|---|---|---|
| 1 Specific heat | $UO_2$ (solid) | $$c_p = 15.496 \left[ \frac{k_1 \theta^2 e^{\theta/T}}{T^2 [e^{\theta/T} - 1]^2} + 2k_2 T + \frac{k_3 E_D}{RT^2} e^{-\frac{E_D}{RT}} \right]$$ $298 < T < 3100$ K<br>where $c_p$ is the specific heat (J/kg.K); $k_1$ - a constant (19.145 cal/mol.K); $k_2$ - a constant ($7.8473 \times 10^{-4}$ cal/mol.K$^2$); $k_3$ = $5.6437 \times 10^6$ cal/mol; $\theta$ = 535.285 K; $E_D$ = 37694.6 cal/mol; $R$ - the ideal gas constant (cal/mol.K). |
| | $(U,Pu)O_2$ (solid) | i.q. $UO_2$, where the constants should read:<br>$k_1$ = 19.53 cal/mol.K; $k_2$ = $9.25 \times 10^{-4}$ cal/mol.K$^2$; $k_3$ = $6.02 \times 10^6$ cal/mol; $\theta$ = 539 K; $E_D$ = 40,100 cal/mol; |
| | Zircaloy (hydrided) | $c_p$ = 281 — 375, 300 < $T$ < 1090 K<br>$c_p$ = 502 — 816, 1093 < $T$ < 1173 K<br>$c_p$ = 770 — 356, 1193 < $T$ < 1248 K<br>$$\Delta c_p = \frac{ABC}{T^2} e^{-B/T} \left[ e^{\frac{T-T_s}{0.02T_s}} + 1 \right]^{-1}$$ $300 < T < 2200$ K<br>where $c_p$ is the specific heat (J/kg.K); $T$ - the temperature (K); $\Delta c_p$ is the addition to true specific heat due to hydride solution (J/kg.K); $T_s$ is the minimum temperature for complete solution of $H_2$ in the material, $T_s = B/\ln(A/H)$; $A$ = $1.332 \times 10^5$ ppm $H_2$; $B$ = $4.40 \times 10^3$ K; $C$ = 45.7 J/kg.ppm $H_2$; $H$ is the $H_2$ concentration (ppm). |

Appendix 11 (continued)

| *Property* | *Material* | *Parametric expressions, symbols and units* |
|---|---|---|
| 2 Thermal conductivity | $UO_2$ | $K = \frac{1-\beta(1-D)}{1-\beta(1-0.95)}\left[\frac{40.4}{464+T} + 1.216\times10^{-4}e^{1.867\times10^{-3}T}\right]$ $0 < T < 1650\ ^{o}C$ |
| | | $K = \frac{1-\beta(1-D)}{1-\beta(1-0.95)}\left[0.0191 + 1.216\times10^{-4}e^{1.867\times10^{-3}T}\right]$ $1650 < T < 2840\ ^{o}C$ |
| | | where $K$ is the thermal conductivity (W/cm $^{o}C$); $D$ - the fraction from fuel theoretical density; $T$ - the temperature ($^{o}C$); $\beta$ - the porosity coefficient, $\beta = 2.58 - 0.58\times10^{-3}\,T$. |
| | $(U,Pu)O_2$ | $K = \frac{D}{1+\beta(1-D)}\,\frac{1+0.04\beta}{0.96}\left[\frac{33}{375+T} + 1.54\times10^{-4}e^{1.71\times10^{-3}T}\right]$ $0 < T < 1550\ ^{o}C$ |
| | | $K = \frac{D}{1+\beta(1-D)}\,\frac{1+0.04\beta}{0.96}\left[0.0171 + 1.54\times10^{-4}e^{1.71\times10^{-3}T}\right]$ $1550 < T < 2840\ ^{o}C$ |
| | | where $\beta = 1.43$. |
| | Zircaloy-4 | $K = 7.848 + 2.2\times10^{-2}T - 1.676\times10^{-5}T^2 + 8.712\times10^{-9}T^3$ $570 < T < 1770$ K |
| | | where $K$ is the thermal conductivity (W/m.K). |
| | $ZrO_2$ | $K = 1.9599 - 2.41\times10^{-4}T + 6.43\times10^{-7}T^2 - 1.94\times10^{-10}T^3$ $500 < T < 1700$ K |
| | | where $K$ is the thermal conductivity (W/m.K). |
| 3 Spectral emissivity | $UO_2$ | $E = 19.324 - 6.68\times10^{-2}T + 9.396\times10^{-5}T^2 - 6.397\times10^{-3}T^3 + 2.10\times10^{-11}T^4 - 2.667\times10^{-15}T^5$ $1000 < T < 2370$ K |
| | | where $E$ is the spectral emissivity. |

Appendix 11 (continued)

| Property | Material | Parametric expressions, symbols and units |
|---|---|---|
| 4 Thermal expansion | $UO_2$ | $\frac{\Delta l}{l} = -4.972\text{x}10^{-4} + 7.107\text{x}10^{-6}T + 2.581\ T^2 + 1.140\text{x}10^{-13}T^3$ <br> $500 < T < 2700\ ^oC$ <br> where $\Delta l/l$ is the relative thermal expansion. |
| | $(U,Pu)O_2$ | $\frac{\Delta l}{l} = -3.9735\text{x}10^{-4} + 8.4955\text{x}10^{-6}T + 2.1513\text{x}10^{-9}T^2 + 3.7143\text{x}10^{-16}T^3$ <br> $500 < T < 2700\ ^oC$ <br> where $T$ is the temperature ($^oC$). |
| | Zircaloy-4 (in axial direction) | $\frac{\Delta l}{l} = -1.708257\text{x}10^{-3} + T[5.223606\text{x}10^{-6} + T(1.204285\text{x}10^{-9})]$ <br> $T < 1073$ K <br> $\frac{\Delta l}{l} = -8.208037\text{x}10^{-3} + 9.683099\text{x}10^{-6}T$ <br> $1073\ K < T < 2125$ K |
| | Zircaloy-4 (in radial direction) | $\frac{\Delta D}{D} = 8.2076\text{x}10^{-4} + T\{-7.856\text{x}10^{-6} + T[1.9236\text{x}10^{-8} - T(6.1409\text{x}10^{-12})]\}$ <br> $T < 1050$ K <br> $\frac{\Delta D}{D} = -6.6513772\text{x}10^{-3} + 9.700599\text{x}10^{-6}T$ <br> $1050\ K < T < 2125$ K |
| 5 Elastic modulus | $UO_2$ | $E = 2.262\text{x}10^{11}(1-1.131\text{x}10^{-4}T)[1-2.62(1-D)]$ <br> $0 < T < 1600\ ^oC$ <br> where $E$ is the Young modulus (Pa); $T$ - temperature ($^oC$); $D$ - fraction from theoretical density. |
| | $(U,Pu)O_2$ | $E = 2.520\text{x}10^{11}(1-7.843\text{x}10^{-4}T)[1-2.03(1-D)]$ <br> $0 < T < 1700\ ^oC$ |

Appendix 11 (continued)

| Property | Material | Parametric expressions, symbols and units |
|---|---|---|
| 6 Creep deformation rate | $UO_2$ | $\dot{\varepsilon} = \frac{(9.728\times10^{6} + 3.24\times10^{-12}\dot{F})\,\sigma\, e^{-\frac{90{,}000}{RT}}}{(-87.7 + D)G^2} + \frac{1.376\times10^{-4}\sigma^{4.5} e^{-\frac{132{,}000}{RT}}}{-90.5 + D} + 9.24\times10^{-28}\sigma\dot{F}e^{-\frac{5200}{RT}}$ <br> $710 < T < 2100$ K <br> where $\dot{F}$ is the fission rate (fissions/$m^3$.s); $\sigma$ - the local stress (psi); $G$ - the grain dimension ($\mu$m); $R$ - the ideal gas constant (cal/mol.K); $D$ - the fraction from theoretical density. |
| | $(U,Pu)O_2$ | $\dot{\varepsilon} = \frac{2.5\times10^{6} + 1.88\times10^{-12}\dot{F}}{G^2} e^{-\frac{110{,}000}{RT}} \sigma\, e^{-33.3(1-D)} \cdot e^{3.55\,C} + (4.37\times10^{-4} + 4.5\,\dot{F})\,\sigma^{4.5} e^{-\frac{140{,}000}{RT}} \cdot e^{10.3(1-D)}\, e^{3.56\,C}$ <br> $1100 < T < 1700$ °C <br> where $C$ is the $PuO_2$ content (the other symbols have the same meaning as in the $UO_2$ case). |
| | Zircaloy | $\dot{\varepsilon} = 5.129\times10^{-29}\,\Phi(\sigma + 7.252\times10^{2}\, e^{4.967\times10^{-8}\sigma}) \cdot e^{-\frac{10{,}000}{RT}}\, t^{-1/2}$ <br> $370 < T < 600$ K <br> where $\dot{\varepsilon}$ is the deformation rate (m/m.s); $\Phi$ - the fast neutron flux, $E > 1$ MeV (n/$m^2$.s); $\sigma$ - the applied stress (Pa); $R$ - the ideal gas constant, 1.987 cal/mol.K; $t$ - the duration of stress application (s). |
| 7 Tensile strength | $UO_2$ | $\sigma_f = 1.56\times10^{8}[1-2.62(1-D)]^{1/2}\, e^{-380/(1.987\,T)}$ $\quad 273 < T < 773$ K <br> $\sigma_f = 1.43\times10^{8}$ $\quad 773\text{ K} < T$ <br> where $\sigma_f$ is the cracking stress (Pa); $D$ - the fraction from theoretical density; $T$ - the temperature (K). |

Appendix 11 (continued)

| Property | Material | Parametric expressions, symbols and units |
|---|---|---|
| 8 Swelling | $UO_2$ | $\Delta V/V = 0.28$    $T < 1673$ K<br>$\Delta V/V = 0.28[1+0.00575(T-1673)]$    $1673 < T < 2073$ K<br>$\Delta V/V = 0.28[3.3-0.004(T-2073)]$    $2073 < T < 2473$ K<br>$\Delta V/V = 0.476$    $2473 \text{ K} < T$<br><br>where $\Delta V/V$ is the swelling rate per $10^{20}$ fissions/$cm^3$. |
| 9 Fractional fission gas release | $UO_2$ | $F = 1 - (1-k')\frac{1 - e^{-kt}}{kt}$<br>$1000 < T < 2000$ K<br><br>where $F$ is the release fraction; $k$ and $k'$ depend on local temperature and fraction $D$ from theoretical density, respectively:<br>$k = 0.25\ e^{-14,800/T}$<br>$k' = e^{-6920/T} + 33.95 - 0.338\ D$ |
| 10 Clad elongation (irradiation effect) | Zircaloy-4 | $\frac{\Delta l}{l} = 1.407\text{x}10^{-16}\ e^{240.8/T}(\Phi t)^{1/2}(1-3f_z)(1+2a)$<br>$40 < T < 360$ °C<br><br>where $\Delta l/l$ is the relative length increment; $\Phi$ - the fast neutron flux (n/$m^2$.s), $E > 1$ MeV; $t$ - the time (s); $f_z$ - the texture factor; $a$ - the degree of cold-working (fraction of cross-sectional area reduction, in $m^2/m^2$). |
| 11 Clad oxidation | Zirconium alloys | $X = [19.9\text{x}10^{12}\ At\ e^{-15,660/T} + X_o^3]^{1/3}$<br>pre-transition stage<br><br>$X = 23.0\text{x}10^{8}\ At\ e^{-14,400/T} - 215\ e^{-1110/T} +$<br>$+ 1.16\text{x}10^{-3}\ X_o^3\ e^{1260/T} + X_T$<br>post-transition stage: $520 < T < 670$ K |

Appendix 11 (continued)

| *Property* | *Material* | *Parametric expressions, symbols and units* |
| --- | --- | --- |
| | | where $X$ is the amount of $O_2$ absorbed in the material ($mg/dm^2$); $X_o$ - the amount of $O_2$ in the initial $ZrO_2$ layer ($mg/dm^2$); $t$ - the time (days, at temperature $T$(K)); $X_T$ - the $O_2$ amount absorbed during transition, $X_T = 123.1\ e^{-790/T}$. Constant $A$ is: $A = 4.211\text{x}10^7\ e^{-2.608\text{x}10^{-2}T}$, for a BWR environment; $A = 3.110\text{x}10^3\ e^{-1.195\text{x}10^{-2}T}$, for a PHWR environment; $520 < T < 670$ °C |
| 12 Hydrogen absorption | Zirconium alloys tubes | $H = \frac{0.0154}{\tau} \frac{B}{8A} (X-X_o)$ ; pre-transition stage |
| | | $H = H_T + \frac{0.0154}{\tau} \frac{2.27\ B}{8A} (X-X_T)$ ; post-transition stage |
| | | where $H$ is the concentration of hydrogen (ppm) absorbed from coolant; $H_T$ is the concentration of hydrogen (ppm) absorbed when $X = X_T$; and $B$ is the hydrogen fraction generated by corrosion during pre-transition stage; $X$, $X_o$, $X_T$, $A$ - as defined above. |

REFERENCE

McDonald, P., and Thompson, L. B. (1976). *MATPRO - Versions 06 and 09 - A Handbook of Nuclear Material Properties for Use in Analysis of Light Water Reactor Fuel Rod Behaviour*, Report TREE - NUREG - 1005, EC & G Idaho, Inc.

APPENDIX 12

# Nuclear reactors types and net electric power, as registered at the International Atomic Energy Agency

*(a) Reactors in operation*

Units<br>MWe

| *Country* | *Reactor types* | | | | | | | | | |
|---|---|---|---|---|---|---|---|---|---|---|
| | PWR | BWR | GCR | AGR | PHWR | LWGR | HTGR | FBR | Others | Total |
| Argentina | | | | | 1<br>335 | | | | | 1<br>335 |
| Belgium | 4<br>2561 | | | | | | | | | 4<br>2561 |
| Brazil | 1<br>626 | | | | | | | | | 1<br>626 |
| Bulgaria | 4<br>1632 | | | | | | | | | 4<br>1632 |
| Canada | | | | | 10<br>5168 | | | | 1<br>250 | 11<br>5418 |
| Czechoslovakia | 2<br>762 | | | | | | | | | 2<br>762 |
| Finland | 2<br>840 | 2<br>1320 | | | | | | | | 4<br>2160 |
| France | 22<br>19,205 | | 6<br>2050 | | | | | 1<br>250 | 1<br>70 | 30<br>21,575 |
| German DR | 5<br>1694 | | | | | | | | | 5<br>1694 |

Appendix 12 (a) (continued)

Units/MWe

| Country | PWR | BWR | GCR | AGR | PHWR | LWGR | HTGR | FBR | Others | Total |
|---|---|---|---|---|---|---|---|---|---|---|
| FR of Germany | 7/6590 | 5/3159 | | | 1/52 | | 1/13 | 1/17 | | 15/9831 |
| India | | 2/396 | | | 2/413 | | | | | 4/809 |
| Italy | 1/242 | 1/875 | 1/150 | | | | | | | 3/1267 |
| Japan | 11/7504 | 13/8787 | 1/158 | | | | | | 1/150 | 26/16,599 |
| R of Korea | 1/564 | | | | | | | | | 1/564 |
| Netherlands | 1/450 | 1/51 | | | | | | | | 2/501 |
| Pakistan | | | | | 1/125 | | | | | 1/125 |
| Spain | 2/1053 | 1/440 | 1/480 | | | | | | | 4/1973 |
| Sweden | 2/1715 | 7/4700 | | | | | | | | 9/6415 |
| Switzerland | 3/1620 | 1/320 | | | | | | | | 4/1940 |
| Taiwan | | 4/3110 | | | | | | | | 4/3110 |
| USA | 48/38,899 | 27/18,842 | | | | | 1/330 | | | 76/58,071 |
| UK | | | 27/4446 | 5/3105 | | | | | 1/92 | 32/7643 |
| USSR | 11/4734 | 1/50 | | | | 22/9926 | | 3/746 | | 37/15,456 |
| Yugoslavia | 1/632 | | | | | | | | | 1/632 |
| Total | 128/91,323 | 65/42,050 | 35/7284 | 5/3105 | 15/6093 | 22/9926 | 2/343 | 5/1013 | 4/562 | 281/161,699 |

*(b) Reactors under construction*

Units / MWe

| Country | PWR | BWR | GCR | AGR | PHWR | LWGR | HTGR | FBR | Others | Total |
|---|---|---|---|---|---|---|---|---|---|---|
| Argentina | | | | | 2 / 1291 | | | | | 2 / 1291 |
| Belgium | 3 / 2914 | | | | | | | | | 3 / 2914 |
| Brazil | 2 / 2490 | | | | | | | | | 2 / 2490 |
| Bulgaria | 1 / 1000 | | | | | | | | | 1 / 1000 |
| Canada | | | | | 12 / 8040 | | | | | 12 / 8040 |
| Cuba | 1 / 408 | | | | | | | | | 1 / 408 |
| Czechoslovakia | 6 / 2520 | | | | | | | | | 6 / 2520 |
| France | 25 / 27,375 | | | | | | | 1 / 1200 | | 26 / 28,575 |
| German DR | 8 / 3276 | | | | | | | | | 8 / 3276 |
| FR of Germany | 4 / 5087 | 3 / 3748 | | | | | 1 / 296 | 1 / 280 | | 9 / 9411 |
| Hungary | 4 / 1632 | | | | | | | | | 4 / 1632 |
| India | | | | | 6 / 1320 | | | | | 6 / 1320 |
| Italy | | 2 / 1964 | | | | | | | 1 / 35 | 3 / 1999 |
| Japan | 5 / 4467 | 5 / 4766 | | | | | | 1 / 250 | | 11 / 9483 |
| R of Korea | 7 / 6227 | | | | 1 / 629 | | | | | 8 / 6856 |
| Mexico | | 2 / 1308 | | | | | | | | 2 / 1308 |

Appendix 12 (b) (continued)

Units/MWe

| *Country* | *Reactor types* PWR | BWR | GCR | AGR | PHWR | LWGR | HTGR | FBR | Others | Total |
|---|---|---|---|---|---|---|---|---|---|---|
| Philippines | 1/620 | | | | | | | | | 1/620 |
| Poland | 1/440 | | | | | | | | | 1/440 |
| Romania | | | | | 2/1320 | | | | | 2/1320 |
| S Africa | 2/1842 | | | | | | | | | 2/1842 |
| Spain | 8/7350 | 3/2806 | | | | | | | | 11/10,156 |
| Sweden | 1/915 | 2/2110 | | | | | | | | 3/3025 |
| Switzerland | | 1/942 | | | | | | | | 1/942 |
| Taiwan | 2/1814 | | | | | | | | | 2/1814 |
| USA | 45/49,952 | 23/24,934 | | | | | | | | 68/74,886 |
| UK | | | | 9/5533 | | | | | | 9/5533 |
| USSR | 16/14,840 | | | | | 7/8000 | | | | 23/22,840 |
| Total | 142/135,169 | 41/42,578 | | 9/5533 | 23/12,600 | 7/8000 | 1/296 | 3/1730 | 1/35 | 227/205,941 |

REFERENCE

International Atomic Energy Agency (1982). *Nuclear Power Reactors in the World*, Reference Data Series No 2. IAEA, Vienna.

APPENDIX 13

# Hydrometallurgy of uranium

To obtain uranium from ores, two hydrometallurgical treatments are in use: the acid process and the alkaline process [1,2]. The basics in the technological flow-sheets of these methods are introduced in Figs A.13.1 and A.13.2 [3], respectively.

*The acid method* (Fig. A.13.1) suits ores with a low content in carbonates ( < 5-8%). The process starts with preparation operations: crushing, wet grinding, sieving — sorting, followed by ore solution in sulphuric acid (1 - 1.5 M). Thereby uranium is transferred in solution as $UO_2SO_4$, $UO_2(SO_4)_2^{2-}$ and $UO_2(SO_4)_3^{4-}$. Solution efficiency is 95 - 99%. Maximum sulphuric acid consumption is 50 - 60 kg/t. Residual ore is neutralized with calcium oxide, then disposed of. Uranium is extracted from solution in columns, by means of ion-exchange resins. Then it is diluted in a hydrogen nitrate solution and precipitated with ammonia, as ADU. ADU is then filtered by pressure filters, to obtain ultimately the so-called "yellow cakes".

*The alkaline method* applies mainly to ores with a high carbonate content and a low sulphide content, respectively; for ores presenting also a high sulphide content, prior removal of sulphides is in order (by flotation) to diminish sulphide concentration down to less than 1%. After preliminary preparation, as described for the acid method, the ore is dissolved in sodium carbonate ($Na_2CO_3$) in the presence of air (which oxidizes $U^{4+}$ to $U^{6+}$), forming sodium-uranil three-carbonate ($Na_4UO_2(CO_3)_3$). Extraction efficiency is 80 - 90%.

The solution is performed as in the acid process, in special reaction vessels known in the professional idiom as "pachucas". These are 3 - 4 m in diameter and 10 - 15 m in height. Uranium extraction from solution is done with ion-exchange resins. Then the resin is washed, and uranium is eluated as sodium-uranil three-carbonate. To split this compound the solution is neutralized and uranium is precipitated with ammonia, as ammonium diuranate. Like in the acid process, by filtration "yellow cakes" are obtained. The "yellow cakes" are the raw material for the $UO_2$ powder manufacture — by refining, or for $UF_6$ preparation — by fluorination, in view of enrichment.

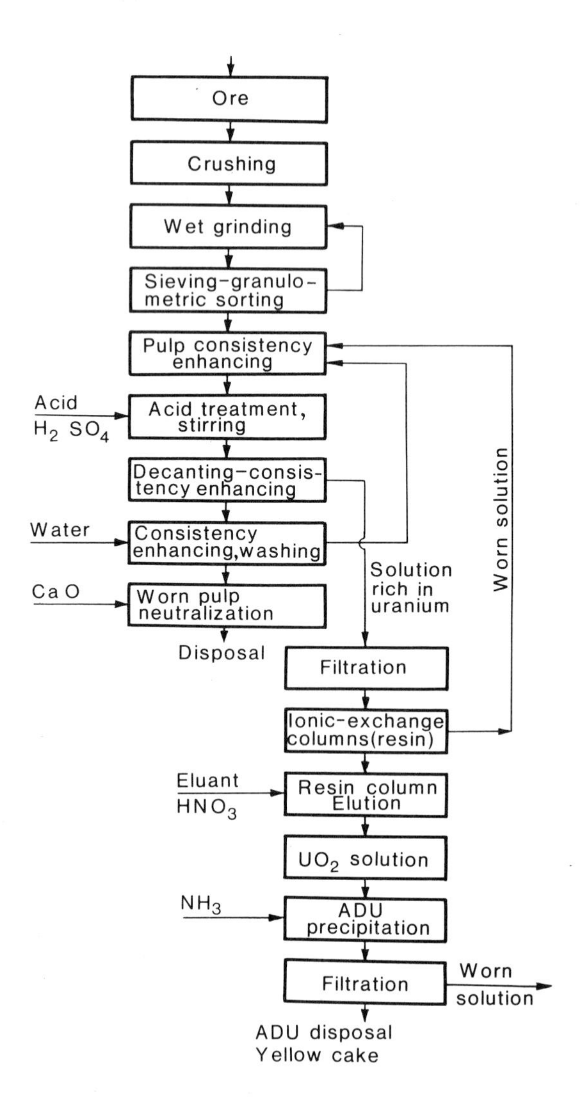
Ore
Crushing
Wet grinding
Sieving-granulo-metric sorting
Pulp consistency enhancing
Acid
$H_2 SO_4$
Acid treatment, stirring
Decanting-consistency enhancing
Water
Consistency enhancing,washing
Ca O
Worn pulp neutralization
Disposal
Solution rich in uranium
Worn solution
Filtration
Ionic-exchange columns(resin)
Eluant
$HNO_3$
Resin column Elution
$UO_2$ solution
$NH_3$
ADU precipitation
Filtration
Worn solution
ADU disposal
Yellow cake

Fig. A.13.1

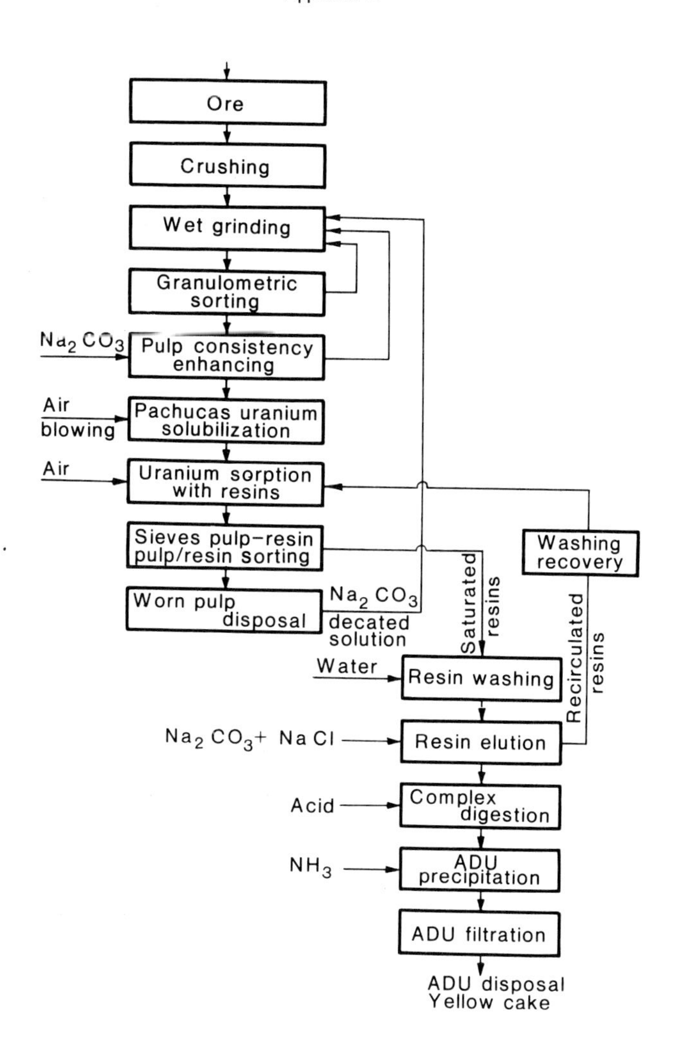


Fig. A.13.2

REFERENCES

Sauteron, J. (1965). *Les combustibles nucléaires*. Ed.Hermann, Paris.

Bunus, F. (1981). *Actinidele si aplicatiile lor*. Edit. Stiintifica si Enciclopedica, Bucharest.

OECD, Nuclear Energy Agency and International Atomic Energy Agency. *Uranium Extraction Technology*, a report.

## APPENDIX 14

# Nuclear fuel cycles

*The fuel cycle* comprises all successive transformations which the fissionable material intended for a nuclear fuel has to undergo. The material is used as the name of the cycle. Currently, four fuel cycles are outstanding:

1. the natural uranium cycle,
2. the enriched uranium cycle,
3. the plutonium cycle,
4. the thorium cycle.

In the variants — either with natural or with enriched uranium — uranium cycles are mainly based on the fission of $^{235}U$ — the only natural nuclide fissionable. Irradiation of $^{238}U$ leads to plutonium generation (see section 10.1). As the fuel burns a fraction of the plutonium as generated in the reactor (which depends on the $^{238}U/^{235}U$ ratio, reactor type and burn-up) undergoes fission; so plutonium contributes, besides $^{235}U$ and $^{238}U$ (the latter by fission with fast neutrons) to the total energy generation in the fuel.

The plutonium cycle starts from $^{238}U$, whereas the thorium cycle starts from $^{232}Th$ (see chains presented in section 10.1). These cycles, also known as "advanced cycles", imply re-processing of irradiated fuel and require an initial inventory of fissionable material, namely $^{235}U$ of natural origin or Pu resulting from the spent fuel discharged from reactors operating on uranium cycles (natural or enriched).

The fissionable material required in order to constitute the initial inventory may also be obtained from fertile material irradiated whith neutrons in fission facilities, or — as a prospective avenue — by means of particle accelerators (by spalation).

In the sequel, the four fuel cycles introduced above are briefly presented.

### THE NATURAL URANIUM CYCLE

The main types of reactors using this cycle are:

1. natural uranium, graphite-moderated, $CO_2$-cooled reactors (for instance Magnox, EDF);
2. natural uranium, heavy water moderated-and-cooled reactors (for instance PHWR-CANDU).

The main phases in the natural uranium cycle are listed in Fig. A.14.1. This is an "open" ("once-through") cycle. The burnt fuel elements still contain 0.40 - 0.45% $^{235}U$ and 0.15 - 0.20% fissionable Pu — in the graphite-moderated reactors, and 0.20 - 0.25% $^{235}U$ and 0.25 - 0.28% fissionable Pu — in the heavy water-moderated reactors. The fissionable Pu production per energy generated is considerable. For the graphite-moderated reactors, at a burn-up of 3300 MWd/t one obtains 0.78 kg/MW(e).year, and for heavy water-moderated reactors, at a burn-up of 7300 MWd/t one obtains 0.47 kg/MW(e).year. Although the specific production of plutonium is high, plutonium concentration in the burnt elements is low. As the cost of re-processing is proportional to the amount of material re-processed, the obtained plutonium may credit only partially the re-processing. It is for this reason that the residual value of the burnt elements is not usually taken into account in evaluating fuel cycle costs.

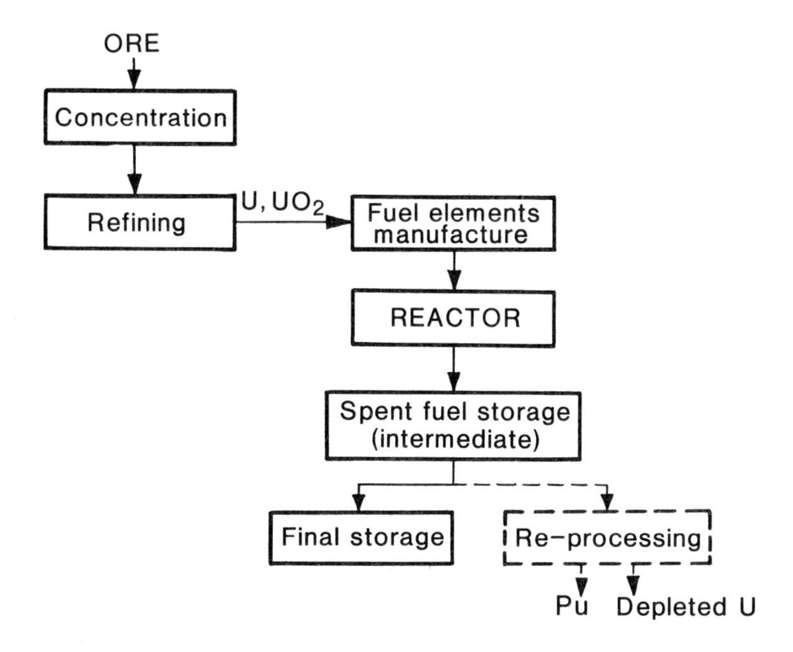


Fig. A.14.1. Natural uranium cycle.

## THE ENRICHED URANIUM CYCLE

The main types of reactors using this cycle are:

1. power thermal reactors LWR, HTR and HWR;
2. research and materials testing reactors.

As compared to the natural uranium cycle, the $^{235}U$-enriched cycle implies, in addition, an enrichment phase — Fig. A.14.2. This phase is preceded by the conversion of refined $UF_4$ to $UF_6$. The enrichment phase is followed by re-conversion of $UF_6$, which now contains enriched uranium, to $UO_2$ and, usually, by conversion to uranium of the uranium hexafluoride that contains depleted uranium (0.2-0.3% $^{235}U$).

In most variants of the enriched-uranium cycle, in the burnt elements uranium has an isotopic enrichment in $^{235}U$ higher than that of natural uranium. This enhances the interest for uranium recovery by re-processing. So, for the current PWRs and BWRs, uranium in burnt elements has a content in $^{235}U$ of 0.82 - 0.85% and 0.96% $^{235}U$, respectively [2]. In HTRs, the cycles using uranium at an enrichment similar to the above one would lead to a content in $^{235}U$ of the burnt elements higher than 1%. The fuel cycle version that implies $^{235}U$ recycling is showed in Fig. A.14.2 by a dashed line.

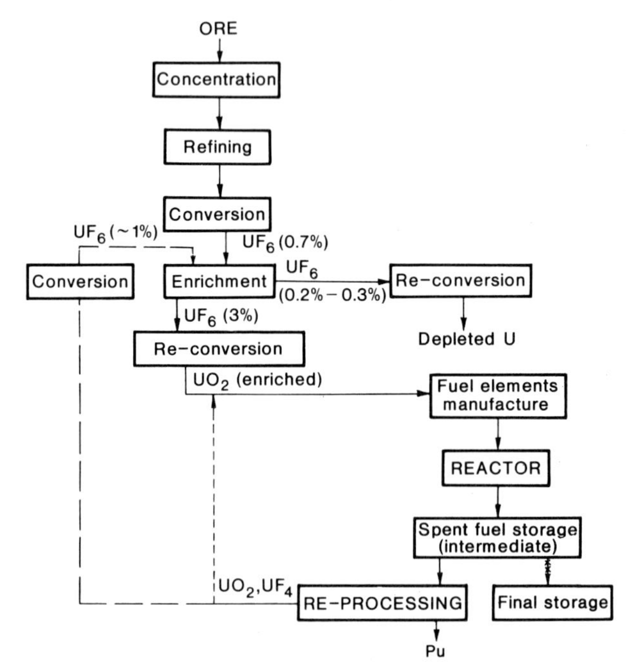


Fig. A.14.2. Enriched uranium cycle: (— — — —), with natural uranium re-enrichment; (-----), with enrichment adjustment by enriched uranium addition.

The situation is different with the HWRs. The use of lightly-enriched uranium (~ 1.2%) determines that the content in $^{235}U$ of the burnt elements is only 0.074 - 0.10% — lower than that of the depleted uranium resulting from enrichment 0.2 - 0.3%. In this case, uranium recycling is no longer of interest.

Uranium recovered by re-processing may be used in the fabrication of fuel elements, on the following alternative:

(a) uranium is converted to $UF_6$, in view of enrichment;

(b) an enrichment adjustment is done, by admixture of fresh $UO_2$ powder having a higher enrichment.

The first variant requires access to an enrichment facility that can be conveniently adapted to accepting a raw material of an enrichment different from that of the natural uranium.

## THE PLUTONIUM CYCLE

The plutonium cycle, featuring the nuclide chain displayed in section 10.1 permits, at equilibrium, obtaining of a surplus of neutrons per fission act, sufficiently high to generate from the fertile material ($^{238}U$) a quantity of fissionable material ($^{239}Pu$, $^{241}Pu$) larger than the one burned by fission. This surplus can be obtained only with fast neutrons.

Figure A.14.3 shows the plutonium cycle in FBRs. The initial plutonium inventory required in order to start this cycle is 3 - 5 t/GW(e), including immobile stocks, and it is obtained in the uranium cycles described above. The initial supply in depleted or natural uranium as well as the plutonium supply, are indicated in Fig. A.14.3 by dashed lines. During burning, the content in plutonium of the fuel elements in the central zone of the core diminishes. However, the newly generated Pu compensates this loss, so that a net production of 0.1 - 0.3 t Pu/GW(e).yr is obtained.

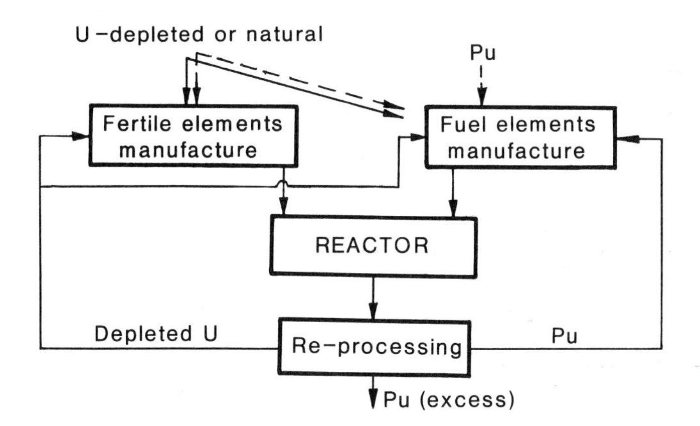


Fig. A.14.3. Plutonium cycle in FBRs: (——), equilibrium cycle; (— — — —), initial inventory.

To reproduce the fissionable burnt material and to produce the excess plutonium, depleted or natural uranium are used. The limited requirements of fed-in uranium 0.7 - 0.9 t/GW(e) makes this cycle practically independent of the price of natural uranium. Even in the present state of the art with fuel fabrication and re-processing, the plutonium cycle is competitive with natural or enriched uranium cycles, provided the burn-up is appropriately high. Current projects require, for fuel elements in the central zone, a burn-up of 60 - 100 MW(e)/kg.

In this cycle, applied to fast reactors, plutonium content at equilibrium is 75% $^{239}Pu$, 22% $^{240}Pu$, 2% $^{241}Pu$ and approximately 1% $^{242}Pu$ [3].

For the plutonium cycle in thermal reactors the neutron balance associated to cycle's nuclide chain, at equilibrium, would permit only a partial reproduction of burnt fissionable material, which still can eventually make the cycle self-sustaining, but not over-productive. As a consequence, it is necessary to supply the reactor with fissionable material. The breeding effect remains, however, important. The breeding fraction of fissionable nuclides may, in some cases, reach about 0.9 of consumption (in HWRs). The yearly uranium savings that can be obtained by plutonium recycling in PWRs and PHWRs are indicated in Table 10.3. In PHWRs uranium consumption may be reduced by as much as 55%.

The plutonium cycle in thermal reactors implies initial inventories of fissionable materials much smaller than in the case of fast reactors.

Looking forward to the advent of commercial FBRs, questions of their optimal co-existence with thermal reactors are currently investigated. Various scenarios involving mutually supportive PHWRs, PWRs and FBRs are assessed in connection with the respective fuel cycles, including considerations of uncertainty and risk.

## THE THORIUM CYCLE

The nuclide chain that features the thorium cycle (given in section 10.1) provides, at equilibrium, a neutron surplus per fission event, sufficiently high to allow generation from fertile material ($^{232}Th$), of a quantity of fissile material ($^{233}U$, $^{235}U$) greater than that consumed in the fast neutron case and, in certain conditions, in the thermal neutron case too.

In fast reactors the thorium cycle is similar to that described in Fig. A.14.3, provided one takes thorium as the fertile material and uranium (especially $^{233}U$) as the fissile one.

The thorium cycle shows outstanding performances in thermal MSBRs. In this case one obtains a doubling time of the core inventory of about 20 years (as with the plutonium cycle in FBRs) by continued re-processing of the fuel that is dissolved in the primary coolant (see Chapter 7).

In thermal HWRs, LWRs and HTRs, the thorium cycle provides for the so-called "self-sustaining" cycles, in which fissionable material production equals fissionable material consumption. Such a cycle is shown in Fig. A.14.4.

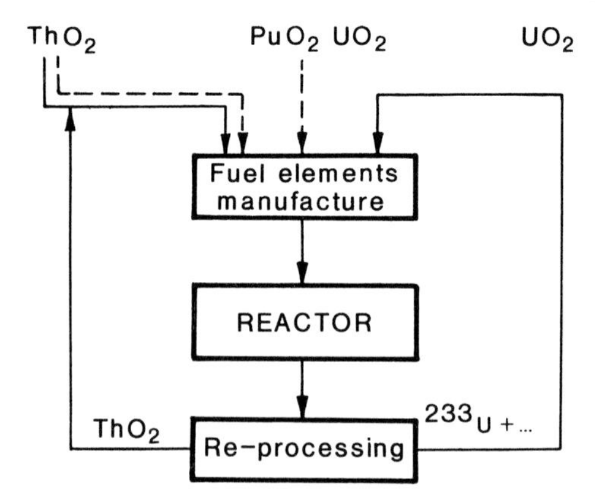


Fig. A.14.4. Thorium cycle (the self-sustaining variant) in PHWRs and PWRs: (——), equilibrium cycle; (— — — —), initial inventory.

Self-sustaining cycles feature lower burn-ups. This makes them competitive only as long as the uranium price is kept high.

An example of self-sustaining cycle that is, in principle, suitable for use in PHWR-CANDU reactors is described in Table 10.6. Its uranium demand concerns, in fact, only the initial inventory.

For the current PHWR-CANDU reactors a thorium cycle with a higher burn-up, that is 30 MWd/kg, was proposed; it is believed that it may become competitive in the near future. This particular cycle is to be sustained by a certain supply with fissionable material. Overall, the expected savings (including initial inventory) from the implementation of various thorium cycles in PHWR-CANDU reactors follow from Table 10.6.

## REFERENCES

Sauteron, I. (1965). *Les combustibles nucléaires*, Ed. Hermann, Paris.
IAEA (1980). *INFCE, International Nuclear Fuel Cycle Evaluation*, Final Report, IAEA, Vienna.
Naruki, K., Kono, K., Satoh, H., and Uematsu, K. (1982). Experience of Plutonium and Uranium Mixed-Oxide Fuel Fabrication in Japan, In *Proc.Int.Conf.on Nucl.Power Exp.*, IAEA-CN-42/293, Vienna.
Ursu, I., and Pavelescu, M. (1981). The Analysis of Some Nuclear Power Systems. *Energy Research*, 5.
Pavelescu, M., and Ursu, I. (1982). A Decision Problem under Uncertainty for Nuclear Power System Development. *Energy Research*, 6.
Bailly, H., Chalony, A., Guillet, H., Leclere, J., Mikailoff, H., and Weinberg, G. (1982). In *Proc.Int.Conference on Nucl.Power Exp.*, IAEA-CN-42/441, Vienna.

# Index